Railway Track and Maintenance

A Manual of Maintenance of Way and Structures

By E. E. Russell Tratman
with
Introduction by Tony D. Steele

A.C. Kalmbach Memorial Library
National Model Railroad Association, Inc.®

This is a reprint of the 1926, fourth edition of *Railway Track and Maintenance* by E. E. Russell Tratman, published by McGraw-Hill Publishing Company. The work was originally published in 1897 under the title *Railway Track and Trackwork*. This edition contains a new introduction by Tony D. Steele. The original book, and other fine volumes on railroad history and modeling, can be found at the A.C. Kalmbach Memorial Library.

A special note of appreciation is extended to Tony Steele, Steve Rosnick of Destination Printing Services, NMRA Intellectual Property Counsel James McClain, NMRA Vice President John Roberts, Library Operations Associate Susan Miscio, and Library Director Brent Lambert, all of whom provided valuable assistance in the production of this reprint.

ISBN: 0-9647050-6-0

A publication of the A. C. Kalmbach Memorial Library
National Model Railroad Association, Inc.®
4121 Cromwell Road
Chattanooga, TN 37421
http://www.nmra.org

Cover Design by Brent Lambert
Typesetting by Susan Miscio

Printed in the United States by Phillips Brothers Printers

Introduction

Long ago there was a transportation mode that was at once industrial, com mercial, and institutional in nature. It affected the development of nations, the business of the world, and the culture of everyone with whom it came into contact. It was as pervasive as a second government, as modern as tomorrow and as reliable as a best friend. This system of transportation was the railroad, and, simply stated, it was indispensable to our way of life.

For a time, nearly one in seven working men labored for a rail-oriented enterprise, meaning that the railroad industry employed more people than all government agencies combined. And about one in six of those men labored to maintain or expand the physical plant of the railways. From bootstrap beginnings, the art of applying the steel wheel to a steel rail evolved into an engineering science. Simple tracklayers were joined by structural builders, signal mechanics, and a host of additional skilled laborers.

A crew works the track with this ATSF #360 crane. Photo from the Anton Krapek collection.

Although rivalries between railroad companies were often fierce, these same companies paused their competitive struggles to share vital technical secrets with one another, for the common (and public) good. Books, such as *Railway Track and Maintenance*, were published to disseminate that knowledge, and fortunately, to preserve it for consideration by future generations. The wisdom contained herein was intended for the immediate use by practitioners in that bygone age, but it still holds an attraction for readers today. It was presented in a fairly pure and unvarnished form,and the explanations remain as fresh as if the ideas and concepts were brand new. There was no air of "We've always done it this way".

Further study and reflection by interested modelers will separate the arcane from the useful. The underlying reasoning and causes for their development should aid the reader in the crafting of his or her own creations even more than the included plans and diagrams. The precise information required for a modeling effort may not be contained within, but by referencing this book, one may be able to think like a designer from another century. A sense of place and time will be comfortably wrapped around the finished project.

Content

The thrust of this book was to shed light on fairly involved technical matters, so that individuals, ranging from higher management to field gangs, could appreciate their part, and their stake, in the development of rational maintenance standards and practices. The content focus was aimed at the practical worker, not the design engineer. There was a thorough presentation of background information on all subject matter. Applicable alternative theories, concepts, and design approaches were reviewed as simply as possible. Understanding the author's reasons for choosing a particular solution is easily absorbed, not shrouded in layers of numbing text. The solutions were generic with no brand-name advertising involved. Many case-history examples were given and typical dimensions and physical geometries were displayed, with the origin railroad and its particular approach usually cited.

There is a good presentation of practical economics in every section. The best *first-cost* considerations are defined. Frequently the costs of *continuing operation and maintenance* for alternative measures are explored. Finally, the attractive *long-term paybacks* for the higher-quality practices are presented as the "best management practices" of that time. It is not all academic exercises; practical intuition on such matters was promoted and encouraged.

The author supplied many comparisons of standard designs and practices for validation and orientation purposes. His recommendations were contrasted with those of the American Railway Engineering Association (now the American Railway Engineering and Maintenance-of-Way Association) and various lines in North America. There also was input from some international (mostly European) railways and selected transit, industrial, and military railways. The format is very similar to that of the ageless manual of the AREA, in style of both content and format. There are brief, effective, and encompassing text descriptions, accompanied by comparative charts, lists, and tables.

Timing

This book was originally published in 1897 as *Railway Track and Trackwork*. It was revised and reproduced in 1900 and also 1908. It was completely revamped and renamed when the fourth edition was published in 1926.

This particular edition was issued well after the great North American Railroad Rebuild era (1900-1915). The great steel trestles, soaring concrete viaducts, monumental tunnels, and gargantuan grade-reduction, second-track construction projects had upgraded the rail network from ocean to ocean. The book also comes just after World War 1 (1916-1920), when burgeoning war traffic resulted in the temporary (and misguided) federal takeover of all national rail operations Bristling with corporate indignation, the rail lines had recently taken back their properties and launched into badly needed programs of equipment and physical plant upgrades.

There are some fundamental differences between current railroad designs and practices and those articulated by the author. Strength factors for track and structures, therefore, are based on the wheel-load distribution and the impacts from counterbalancing designs of the heaviest expected steam and electric locomotives. Railcar wheel loadings would not rival them for many decades. The superelevation and clearance factors, including the banking of tracks on curves and the distances from railtops to overhead and lateral structures, were referenced primarily to those locomotives and the long passenger cars they drew.

D & H rails stacked at a welding plant, Carbondale, PA, 1974. Photo by Tony Steele.

Train-Control Techniques

Communication and control systems were fairly primitive then. Automatic train-stop systems had been mandated on a number of railroads by an order of the Interstate Commerce Commission (ICC) in 1922. This was one of the first governmental safety responses to the challenges of high-speed railroading (in excess of 79 mph). Neither Centralized Traffic Control (CTC) nor sophisticated grade crossing automatic signaling was described since they would be a while in coming. Radio was not available yet.

Semaphore signals were being upgraded from 2-position to 3-position, even while some lines were replacing them with light signals (color, position, color-position, etc.). While the electric operation of remotely-located switches and signals was being recognized as superior, mechanical operation of such was still common in new design with pipe connections being considered more reliable than wire ones.

Telegraph orders, delivered via reasonably adjacent stations, towers, or offices, were the implied system of coordination aside from telephone usage, which had grown to rival the use of the telegraph for railroad business communications. Train-order, block, and interlocking signals were the control systems most used in this era.

Roadbed

The standards for roadbed cross-sections were many, about one for every chief engineer who ever lived. But the origins of the current 12 inches of ballast on 12 inches of sub-ballast with wide shoulders outside the tie ends are evident in the 12-inch thick sub-ballast layer of the Pennsylvania Railroad (PRR) and the 12-inch ballast shoulder of the Union Pacific (UP).

Crushed stone is still the best ballast, but many other materials have been used. Cinders, chats (mine tailings), and blast-furnace slag all were utilized for economic reasons at different times. The regional use of burned clay, sand, seashells, chert (disintegrating limestone), or even dirt as very light-duty ballast has been documented as well.

Track

During this era, railcar wheel loads of 36,000 to 39,000 pounds and (average) coupled car lengths of 65 feet were not even a gleam in a shipper's eye. The author reported loads of 15,000 to 20,000 pounds and lengths of 42 feet at that time.

Narrow-gauge crossing diamond under construction, Allaire State Park, NJ, 1989. Photo by Tony Steele.

To support these loads, rail sections had grown remarkably. The 75 to 85-pound per yard rail sections then in main track were being supplanted by the "new" 100-pound per yard rail. But already, heavier sections, with much of the extra material being placed primarily in the vertical "web" to increase the strength against bending, were being rolled for select customers. This range included the 136 and 140-pound per yard rail sections that are currently favored for main line use.

Joint bars varied in size from 24-inch (4-hole) to 38-inch (6-hole). Choices made for various rail sections may have had more to do with the quality of roadbed section and railroad maintenance philosophies than anything else. Compromise joints (cast offset bars) were in common usage, between different rail sections, to keep the gage corners of the different-height rails properly aligned. However, special taper rails (forged with the different sections at each end) were also in use on a few lines.

Materials for new main track construction were a little lighter and easier to handle than exist now. Rails were commonly cut to 33-foot lengths, versus the 39-foot just becoming recommended then. Used rails might be trimmed to as short as 15 feet, and still be used in yards and sidetracks, turnouts, and other special trackwork. Tie numbers of 17 to 20 per 33-foot rail (23.3 inch to 19.8 inch spacings) were quoted. Compare that to the contemporary 80-foot-long rails (before welding into quarter-mile strings) and 24 wooden ties per 39-foot rail (19.5-inch spacing) or the much heavier concrete ties.

Main track wooden ties (or crossties) were already being specified at their current dimensions: 7 inches thick by 9 inches wide by 8 ½ feet long. Smaller ties were still used though: 6 inches by 8 inches by 8 feet. Somewhat surprisingly, some enterprising individuals had been engaged in experimental importation of tropical hardwood ties (precursors to the current Azobe materials). Longer lives had to be balanced against the vagaries of supply, cost, and necessity to bore for spiking. Concrete and steel ties had been tried as substitutes for hardwood and softwood ties, particularly for industrial and street railway purposes.

The use of tie plates and preservative treatment was still a topic for persuasive reasoning. Depending upon weather and traffic-wear effects, a doubling of tie life was expected then to an average of 15 to 20 years. If used, the tie plates of the era measured 7 inches by 9 inches, a far cry from the 7 ¾ inches by 14 inches to 15 ½ inches specified today. The need for all sorts of spiking patterns was recognized in response to varying curvatures, gradients, and levels of traffic experienced. Even then, there were concerns over the holding power of cut spikes in wood ties, but their relative economy, ease of installation, and maintainability have continued to render them preferable to the screw spike.

Organization of Forces

Regular inspection, small-scale spot maintenance, and large-scale production efforts are mainstays of a railroad's annual maintenance budget. A number of natural and other emergency tasks are also timeless: train wreck clean-up operations, clearing floating ice and debris jams, and dealing with washouts, riverbank protection needs, and temporary ("shoofly") trestles and tracks. The winter specialties of snow plowing, clearing, and flanging are still planned yearly also, along with the readying of associated snow shed and fence structures.

It was reported that some rail renewal programs were being scheduled during the winter months, a time when there was less traffic. Since jointed rail work had little susceptibility to ground freeze-up conditions, there would be few downsides to such a strategy. There were recommendations in the book for year-round gangs with balanced duties consisting of cold-weather maintenance versus warm weather mass-production renewal projects. This was done in order to avoid the seasonal hiring and laying-off of many experienced workers. The book also provides a recommendation for the rather labor- intensive seasonal removal of snow fencing, cattleguard wings, and the planks of minor grade crossings as regular winter-related duties.

On the labor front, while the major crafts had long been organized, many other rail workers (particularly in maintenance tasks) had just been unionized. This was done during the federal stewardship of the industry as a bureaucratic measure to eliminate dealings with unrepresented individuals. But work rules and rights for the workers were in their infancy, especially for those in the field. Long work days and weeks away from home, usually in company "camp cars" - and little personal mobility except that presented by passing trains - made for great strains on the family lives of maintenance workers. Disciplinary systems and motivational methods, fairly extreme relative to those in factory environments, would be slow to evolve into higher forms of corporate/union enlightenment.

Railway Track and Maintenance

It was a common motivational practice for annual prizes of $25 to $100 to be presented for the best track maintenance sections in a division or supervisor's territory. The money probably represented at least 1 to 4 months pay for the foreman or supervisor affected. Track surface smoothness comprised at least half of the judging basis. Undoubtedly a certain amount of politics and a good deal of luck was required to win. For instance, did the section just receive a full rehabilitation by production gangs?

D & H track gang correcting a "sun kink" track buckling, Richmondville, NY, 1977. Photo by Tony Steele.

The use of outside contractors for many track, bridge, or signal tasks was still in its infancy and usually was applied to the most specialized services or complex construction projects. The railroad force-account system (use of railroad crews only) was viewed as the most reliable method of accomplishing the work while protecting the property and operations of the railroad. Thus, a very wide latitude of crafts and equipment was still required by railroad forces.

A careful overview of typical engineering and maintenance-of-way organizations was presented, covering the positions and duties at all levels. For instance, the position of Section Trackwalker was described as on-foot inspections of the same limited segments of railroad on a daily (except Sunday) basis. In modern times, only a few railroads are believed to have retained this position; all others found the costs unsupportable.

Many of this era's manual tools have survived, remarkably unchanged. Off-track construction machinery was still outfitted with horses and wagons. Complex mechanization for construction and maintenance work amounted to rail-mounted cranes, derricks, pile drivers, spreaders, and snowplows. On-track maintenance machinery was either very large (tracklaying machines) or very compact (air compressors and their pneumatic tools), with no automation or fast cycle rates. High production rates were the result of huge gangs.

The mass-production manpower need of these organizations was tremendous, compared to the lean organizations of today. Project daily schedules reflect short work "windows" (activity time between frequent trains), requiring large gangs with manual tools and only the largest of on-track equipment for lifting, plowing, spreading, etc.

During the production season, immense tracklaying and ballast gangs (60 to 126 men) would swing into action with specially equipped trains and much human activity. Rail renewal gangs ranged from 36 to 260 men, equipped with manual tools, cranes, and air compressors, mostly. Two turnout gangs of 40 and 52 men (plus a pair of wreck cranes) were cited for yard ladder and slip switch replacement where they worked 4 hours and 1 hour 18 minutes, respectively.

Other specialties required sizeable forces, too. Bridge gangs consisted of 6 to 20 men, centered on self-propelled derrick cars capable of setting girders of up to 95' in length without the use of falsework. Tunnel repair gangs (12 to 24 men) operated from dedicated cars with special rooftop scaffolds to work on the tunnel roofs. Pole line gangs had 8 to 18 men, but might be supplemented with horses for wire-pulling. Extensive use was made of work trains (little availability of off-track equipment), camp trains (for the large project gangs used everywhere), and contractors' construction trains (for specialized or major improvement projects).

Track inspection cars were well used for annual, or more frequent, inspections of lines and particularly for track quality. Some had theater-style viewing out the back, and others, called "dynagraph" cars, were equipped with more sophisticated sensors, paint-markers, and chart-recorders to document the geometry of the rails. These were the antecedents of the modern track geometry cars.

Modeling Thoughts

There was quite a bit of other model-relevant information put into this book. A good deal of the items covered may have survived to current decades in untouched railroading "backwaters". The less local expansion or pressure for upgrading, the longer it may have lasted. In my estimates of duration, track geometry, cuts and fills, culverts, and bridges could realize a 100-year life. Depots, shops, and other industrial buildings can reach a 50 to 100-year life. For many track and signal components, a 10 to 30-year life is possible (in first position), although some have lasted much longer through refurbishing and recycling.

Various right-of-way appliances, typical structures, and other details had insightful descriptions and instructions included; many were accompanied by detailed plans. Your attention is kindly directed to the sections on snow, station, and right-of-way fencing; cattle guards; signs; mail cranes, bridge warning "tell-tales"; bumping posts; bridge and station decks; building materials and painting; section tool houses (single-gang versus double-gang); section and bunk houses; cabins shanties, and toilet buildings; trainsheds, station platforms, and station shelters; water tanks, standpipes, columns, track tanks (pans), pumping plants, and water treatment facilities; coaling trestles, inclined conveyors, and elevated coaling stations; fuel oil stations, sanding facilities, and ash pits; engine pits, turntables, transfer tables, and wye-tracks; track scales, truck scales, and wagon scales.

Some caution is advised as to the literal use of dimensions given, unless constructing finescale, era-specific models. NMRA standards should be referenced, also, as a practical reality check. Flangeway widths and depths are usually necessarily larger in models, to complement the manufactured tolerances of many items purchased "off-the-shelf". The same can be said about guardrail gage-to-gage dimensions at frogs, crossings, and bridges - as well as clearances (lateral and overhead) from railtops. Many formulas printed in the book are field-expedient approximations for use by gang foremen and supervisors - but they are still "good enough" for most modeling purposes.

D & H Signal Department camp car, Colonie, NY, 1971.
Photo by Tony Steele

Many railroads established elaborate standards for designs and practices that may have differed noticeably from the items in this book. If desired, sometimes copies of such standards may be obtained through historical and technical societies, the commercial model or railroad press, and/or the NMRA library. Very specific subjects for modeling may adhere to no particular standard, so one's research should be conducted with cautious good humor.

Tony D. Steele, P.E.
2003

RAILWAY TRACK AND MAINTENANCE

RAILWAY TRACK
AND
MAINTENANCE

A Manual of Maintenance-of-Way and Structures

FORMERLY PUBLISHED UNDER THE TITLE OF
RAILWAY TRACK AND TRACK WORK

BY
E. E. RUSSELL TRATMAN, A. M. Am. Soc. C. E.
Member American Railway Engineering Association,
Western Editor of Engineering News-Record

Fourth Edition
Second Impression

McGRAW-HILL BOOK COMPANY, Inc.
NEW YORK: 370 SEVENTH AVENUE
LONDON: 6 & 8 BOUVERIE ST. E. C. 4
1926

PRINTED IN THE UNITED STATES OF AMERICA

THE MAPLE PRESS COMPANY, YORK, PA.

PREFACE

Railway engineering was considered for many years as being limited mainly to location and construction, while the maintenance and repair to keep the lines in condition were considered as matters for foremen and laborers. It came to be recognized, however, that both the track construction and the state in which it is maintained have an important bearing upon the efficient and economic handling of traffic, and as a result there has been a marked development in the position and work of the engineer in relation to maintenance of way. A small book on "Track," written in 1886 by Wm. B. Parsons, was practically the first book dealing with the subject from the engineering point of view, and written for the use of the engineer rather than for that of the section foreman or the "practical" roadmaster. On this account the book had at once an extensive sale among engineers.

The evident demand among engineers for a technical book on track construction and track work (or maintenance-of-way), led the writer to prepare a book on "Railway Track and Track Work," with the purpose of making a comprehensive work specially adapted for the use of engineers in the maintenance-of-way department and of young engineers intending to enter this department. The first edition was published in the autumn of 1897. Within a few months the greater part of the edition had been sold, and it was practically exhausted by the summer of 1900. As the recognition of the track and its maintenance as an important feature in railway work continued to develop, there was a continued and increasing demand for the book. This came from railway engineers in all departments and also from the engineering schools, many of the schools adopting it as a textbook in railway engineering.

The demand was met by the preparation of a second edition, which was published in 1901. This was almost entirely rewritten, in view of many changes and developments since the preparation of the first edition. New chapters on signals, interlocking plants, and electric railways were added, as the increased use of signals was leading to the development of the signal department in railway service, while interurban electric railways were opening a new field for engineers. By 1907 the second edition had become nearly exhausted and the writer undertook the revision for a new edition which appeared in 1908. This also was practically rewritten, owing to the many changes and improvements that had been introduced. The new material was influenced also by the fact that on many railways maintenance-of-way was gradually being put in the hands of a separate engineering organization, more or less independent of the general engineering department, the work of the latter relating mainly to construction and improvements.

For 18 years this book continued in steady demand, but in some respects it was getting out of date as to practice, although the principles presented were still unaltered. In 1923, therefore, the writer undertook to rewrite the book entirely, bringing it up to date, including new developments in material and methods, and producing a book not only for engineers of maintenance-of-way but also for railway engineers in general, for railway officials and for engineering students. It includes the bridge, signal, telegraph, and other special departments, shows the relation of the track and its maintenance to railway operation, and presents their financial and economic aspects.

In addition to the general principles and purposes underlying track design and maintenance, and the systems of practice which are everywhere applicable, the book gives numerous details as to equipment, material, appliances, and methods as used by individual railways in different parts of the country and under various conditions of traffic and climate. This comprehensive and representative treatment has been adopted as a characteristic of the book, and has met with such commendation as to prove that it meets a definite demand. That the scope of the book is wide, may be seen from the table of contents, and the various subjects are treated in both a descriptive and critical manner. For this reason the original title has been changed slightly to indicate that the book covers much more than track work.

Throughout the book, frequent reference is made to the work of the American Railway Engineering Association, which has carried on extended investigations as to track and maintenance-of-way, and which has presented numerous standard designs and recommendations for practice. To the numerous engineers, railway officers, and manufacturers of railway material, who have furnished information for use in this new book, the writer extends his thanks for their assistance. He also expresses his appreciation of the several criticisms, corrections, and commendations which have been sent to him at various times.

E. E. Russell Tratman.

Wheaton, Ill.,
January, 1926.

CONTENTS

PAGE

PREFACE v

PART I

TRACK MATERIALS, APPLIANCES AND STRUCTURES

CHAPTER I

INTRODUCTION 1

Railways and Track—Maintenance-of-way—Rolling Stock and Traffic—Track Studies—Ordinary and Special Designs—Cost of Track—The Rate Problem—Railway Statistics.

CHAPTER II

ROADBED AND BALLAST 17

Grading and Sodding—Track Ditches—Concrete Roadbed—Ballast Materials and Cross Sections—Typical Practice—Bridges, Tunnels and Subways.

CHAPTER III

TIES AND TIE-PLATES. 35

Timber Resources—Wood for Ties—Life of Ties—Purpose and Economics of Preservative Treatment—Preservative Processes—Metal Tie-plates—Tie Renewals and Records—Metal and Concrete Ties.

CHAPTER IV

RAILS 60

History and Development—Chemical Composition—Specifications and Manufacture—Special Steels—Expansion and Creeping—Canted Rails—Stresses in Rails—Wheels and Rails—Life and Wear—Reconditioning Old Rails.

CHAPTER V

RAIL JOINTS AND RAIL FASTENINGS 79

Requirements of Rail Joints—Design and Wear of Splice Bars—Bolts and Bolt Spacing—Special Joints—Bridge Expansion Joints—Drive and Screw Spikes—Bolts and Wedges—Tie Bars, Rail Braces and Rail Anchors.

CHAPTER VI

SWITCHES AND FROGS. 99

Turnouts—Split, Stub, Spring and Slip Switches—Types of Frogs—Guard

PAGE

Rails—Track Crossings—Switch Ties and Timbers—Continuous-rail Switches and Frogs—Switchstands and Lamps—Derails.

CHAPTER VII

FENCES AND GRADE CROSSINGS 119
Right-of-way Fences—Posts and Gates—Snow Fences—Cattleguards—Highway Crossings—Paving and Protection—Track Crossings—Tracks in Paved Streets.

CHAPTER VIII

TRACK ON BRIDGES AND ELEVATED STRUCTURES 134
Open, Solid and Ballasted Bridge Floors—Guard Rails and Rerailers—Elevated Railways—Station Floors—Track-elevation Bridges.

CHAPTER IX

TRACK SIGNS AND ACCESSORIES 148
Typical Railroad Signs—Bridge Warnings—Mail Cranes—Bumping Posts—Sand Track.

CHAPTER X

RAILWAY BUILDINGS 162
Tool Houses—Section-men's Houses—Trainsheds—Platforms and Shelters—Clearance Limits—Miscellaneous Buildings—Painting.

CHAPTER XI

WATER AND FUEL STATIONS; TURNTABLES; SCALES 169
Railway Water Supply—Tanks and Water Columns—Track Tanks—Water Softening Plants—Coal and Fuel-oil Stations—Ashpits—Turntables and Transfer Tables—Track and Wagon Scales.

CHAPTER XII

TERMINALS, YARDS AND SIDINGS 187
Freight Yard Design and Operation—Gravity and Flat Switching—Hump Grades—Track Brakes—Engine Terminals—Passenger Terminals—Passing Tracks and Industry Tracks.

CHAPTER XIII

TRACK TOOLS, STORES AND LABOR-SAVING APPLIANCES 205
Tool Equipment for Section Gangs and Bridge Gangs—Economy of Labor-saving Machinery—Pneumatic and Electric Tools—Distributing Stores and Supplies.

CHAPTER XIV

ELECTRIC RAILWAY TRACK AND TRACK WORK 223
Interurban and Street Railways—Overhead-wire and Third-rail Systems—Street Railway Tracks—Rail Corrugation—Maintenance and Reconstruction—Organization of Forces.

PART II

MAINTENANCE WORK AND ECONOMICS

CHAPTER XV

PAGE

ENGINEERING AND MAINTENANCE-OF-WAY ORGANIZATION 235
Railway Organization Systems—Maintenance Engineers—Roadmasters and Foremen—Force and Labor—Organization for Track Work—Permanent Forces and Winter Work—Extra Gangs—Bridge, Water and Signal Departments.

CHAPTER XVI

TRACKLAYING AND BALLASTING. 258
Force and Methods—Tracklayer Machines—Ballast Cars—Construction and Maintenance Work—Distributing Ballast—Cleaning Old Ballast.

CHAPTER XVII

RAIL RENEWAL. 269
Laying Rails under Traffic and with Track Closed—Single-track Operation —Care of Rail Joints—Machinery, Force and Methods—Typical Practice—Rail Handling Devices.

CHAPTER XVIII

DRAINAGE AND DITCHING: HOLDING LOOSE SAND 280
Culverts and Waterways—Track Ditches and Drains—Water Pockets and Wet Cuts—Sliding Ground—Machine Ditching—Control of Drifting Sand.

CHAPTER XIX

TRACK WORK FOR MAINTENANCE 289
Main and Branch Line Practice—Tunnels and Bridges—Protection of Gangs —Train Signals—Seasonal Track Work—Routine and Special Work—Oiling and Weed-killing Machines—Policing and Scrap Disposal.

CHAPTER XX

GAGE, GRADES AND CURVES. 316
Various Gages—Grades and Maintenance—Grade Compensation—Vertical Curves—Curve Economics—Wheel Action—Transition Curves—Superelevation—Sharp Curves—Switchbacks.

CHAPTER XXI

SWITCH WORK AND TRACK LAYOUTS 342
Turnouts, Crossovers and Track Crossings—Speed at Turnouts—Terminal and Yard Plans—Field Work—Turnout Tables.

CHAPTER XXII

BRIDGE WORK AND TELEGRAPH WORK 363
Bridge Inspection, Painting and Repair—Clearance Diagrams—Bridge Classification—Record Systems—Tunnel Repairs—Maintenance of Telegraph Lines—Power Wires Crossing Tracks.

CHAPTER XXIII

PAGE

RAILWAY IMPROVEMENTS AND BETTERMENTS 380
Traffic Developments—Double Tracking—Grade and Curve Reduction—Filling Trestles—Grade Separation—Resurveys—Summit Tunnels—Handling Construction Trains—Work Train Service—Camp Trains—Contractors' Trains.

CHAPTER XXIV

SIGNALING AND INTERLOCKING 398
Train Dispatching and the Block System—Types of Signals—Automatic Train Stops—Interlocking Plants—Electric Railway Signals.

CHAPTER XXV

RAILWAY EMERGENCY WORK 408
Washouts and Ice Jams—Temporary Trestles—Shoofly Tracks—Piledrivers and Derrick Cars—Wrecking Trains and Operations—Snow Fences, Snowsheds and Snow Plows—Clearing Terminals—Electric Railway Methods.

CHAPTER XXVI

TRACK INSPECTION AND PRIZES 426
Routine and Special Inspections—Dynagraph Cars—Marking for Track Conditions—Awards and Prizes.

CHAPTER XXVII

MAINTENANCE RECORDS, REPORTS AND ACCOUNTS 432
Track Charts—Bridge and Building Records—Valuation and Improvements—Routine Track Reports and Time Books—Labor Distribution and Cost Records—Requisitions and A.F.E. Forms.

CHAPTER XXVIII

ECONOMIC ASPECTS OF TRACK MAINTENANCE 458
Budget Systems—Work Schedules—Equating Track Conditions—Time Studies of Track Work—Labor Economics—Maintenance by Contract—Standardization.

INDEX 469

PART I

TRACK CONSTRUCTION; MATERIALS, APPLIANCES AND STRUCTURES

RAILWAY TRACK AND MAINTENANCE

CHAPTER I

INTRODUCTION

Railways and Track.—The track constitutes the finishing surface of a railway and forms the path or "rail-way" to carry and guide the wheels of locomotives and cars, these wheels having inside flanges which keep them in place upon the rails. With the railway system of the United States aggregating 250,220 miles of line and 385,855 miles of track (including sidings and yards), it is evident that the construction and maintenance of track is an important item in railway work. A glance at the table of contents will indicate the variety of materials and appliances included in track construction and the varied character of work involved in track maintenance. Furthermore, it is necessary to recognize the care required in devising both the character of construction and the methods of maintenance best adapted to the varying conditions of service and traffic on different railways. Railway officers should realize clearly these points: (1) the importance of the track and its maintenance in relation to the efficient and economical operation of the railway; (2) the large proportion of the operating expenses represented by this maintenance work; and (3) the economic importance of good track for railways carrying fast and heavy trains and dense traffic.

There is general recognition of the fact that the track and track forces should be largely under the management of men of engineering training, who understand not only the practical work but also the principles upon which this work is based, as well as the economics of railway operation. Without knowledge of these principles and economics, little improvement could be made in track materials or track work in order to meet developments and changing conditions in railway service. And unless the track is in the hands of engineers, the economic aspects of such matters as heavier rails, better ballast, treated ties, drainage work and labor-saving appliances are not likely to be presented adequately or convincingly to the executives or higher officials whose authority is needed for the work and expenditures. This statement applies even more forcibly, of course, in questions of general improvement work, such as double tracking and line or grade revision.

Maintenance-of-way.—Maintenance is the continual work of keeping in proper condition for service a structure subject to wear, corrosion, damage, deterioration, and the disturbance or distortion resulting from its use or service as well as from wind, weather, and age. Thus "maintenance-of-way," or maintaining railway track in condition, includes tightening bolts and fastenings, replacing worn rails and ties, raising low joints, adjusting the track to line and surface, keeping curves true and with their proper superelevation, repairing or replacing switches and frogs, cleaning and dressing ballast, inspecting and policing the line, and performing a multitude of other detail duties. Since labor represents from 50 to 70 per cent of maintenance expenses, as against 30 to 50 per cent for material, the regulation of work on a systematic basis and the introduction of labor-saving appliances are evident and important means towards economy.

Bridges must be painted and perhaps repaired or strengthened at intervals. Buildings also must be repaired and painted. As these structures are related more or less directly to the track, they are included under the general classification of "maintenance-of-way and structures." Maintenance, however, does not include reconstruction, such as the laying of heavier rails, the replacement of an old bridge with a new structure adapted for heavier loading, or the erection of a new water station. Such work is classed under betterments or general improvements.

Although maintenance is composed largely of a great variety of details, it is a continual work requiring a large force, so that the aggregate expense for a railway or a division is a considerable item. This expense will be at a minimum where the track construction is in proper relation to the loads and traffic which it carries, but in any case it represents a material percentage of the operating expenses. Since about 50 per cent of the expense is absorbed in the regular everyday work of roadway repair, not including rail and tie renewals, propositions have been made to reduce this expense by the wider use of labor-saving machinery and by introducing new types of track construction. The first method is a very practicable economy, but the few experiments as to the second method have not proved very favorable, as noted later.

That maintenance is essential to safety, as well as to efficiency and economy of railway transportation is very evident. A loose bolt, bent switch point, decayed ties, rails not properly secured during renewals, a rail defect which is overlooked, or a choked culvert which is left for later attention, may be the cause of a derailment or washout with more or less disastrous and costly results.

Deferred and Inadequate Maintenance.—Deferred maintenance means that work which is really needed, or the replacement of material which has reached the economic (or perhaps the safe) limit of life, is postponed through inability to secure the necessary authority or funds. This situation may be due to a real financial crisis, but is more likely to result from failure of the executives to recognize the economics of expenditures. In either case, the ultimate result is excess cost. Thus the withholding of $100 at the proper or critical time will probably mean the ultimate expenditure of $200 or $500 to restore the track or structure to its normal condition. Beyond this critical time, wear and deterioration will proceed more and more rapidly, until a condition is reached necessitating delays and slow movements by traffic, and perhaps resulting eventually in serious and costly accidents. With a structure such as railway track, subjected to severe stresses and various disturbing influences, it is of economic importance that renewals, repairs, and maintenance should proceed regularly and systematically. It may be said emphatically that financial difficulties are no excuse for dangerous conditions. If a railway company actually has not the money to put its track in reasonably good shape, it must reduce speeds and loads to the requirements of safety. This situation is rare, however, and where a dangerous condition of track exists, it is the result usually of carelessness or of failure to provide the necessary funds.

The results obtained from expenditures on maintenance-of-way are in some cases unsatisfactory, as where money is spent over and over again on a stretch of poor track without producing any improvement. This may be due to such defects as light track or soft spots in the roadbed, or it may be due to a large increase in loads and traffic. It would be economy to investigate the conditions, determine the cause and remedy, and then carry out some permanent improvement, such as laying heavier rails or draining the wet ground. But such action would require a definite authority and expenditure, which the executives may consider cannot be allowed. The consequent result, therefore, is a continual but unnoticed expense in maintenance and temporary work. Expenditures on permanent improvements may effect important economies in both maintenance and operation, but the economics of track construction and maintenance and their relation to the economics of railway operation are not usually understood or appreciated at their full value outside of the engineering and maintenance departments.

Track and Rolling Stock.—From 60 to 75 per cent of wear and tear of track is due to engines or engine mileage, owing to the great wheel loads, the concentration of such loads on a short length of track, and the destructive effects of bad counterbalancing, slipping of wheels, and excessive use of sand on the rails, besides the occasional damage to rails, frogs, and switches by badly worn tires. The car is usually a passive rolling weight, although wheels with flat spots and trucks that do not swivel readily may impose destructive effects. Freight trains are usually more severe upon the track, as their engines are likely to be pulling hard and the cars ride more heavily, owing to their loads and to their spring rigging being more rigid than that of passenger cars.

Day cars weigh 40 to 70 tons, and sleeping cars 60 to 85 tons. The lighter cars on four-wheel trucks and the heavier cars on six-wheel trucks give 10,000 and 14,800 lb. per wheel, respectively. Freight cars weigh

20 to 30 or even 40 tons empty, and 60 to 100 or even 160 tons loaded, with wheel loads of 15,000 to 20,000 lb., and with a wheel base of 6 to 8 ft. for the truck and 20 to 35 ft. for the car. Some of the largest freight cars have six-wheel trucks in order to distribute the load. Locomotives have from 4 to 10 driving-wheel loads of 12 to 15 or even 20 tons, and an aggregate of 120 to 200 tons on a driving-wheel base of 14 to 22 ft. These are merely static loads, but their dynamic effect with trains in motion as well as their relations to track conditions must be taken into account, for loads that could be carried safely and without excessive wear on substantial track might be excessive for light rails and inferior track. The transportation department should have rigid rules limiting the speed at which "dead" engines, not under steam, may be hauled, as the hammer blows of their counterweights may be very destructive. All cases of damage to rails and track by locomotives in bad condition should be reported promptly.

Traffic and Maintenance.—Increase in traffic causes increased wear and tear of track and increased maintenance expense, but the relation is indefinite, being influenced by many variable factors both as to weight and intensity of traffic and character or condition of track. Furthermore, different items of track work are affected in different degrees. Ballast, line and surface, and wear and renewals of rails and fastenings are affected directly by traffic variations. Tie renewals are affected to some extent, since decay may limit the life of ties under light traffic, while with heavy traffic it may be limited by mechanical wear. Items affected little or not at all include such work as repair of fences and signs, mowing and clearing right-of-way, cleaning ditches and paving road crossings.

Track maintenance is affected to some extent by increased train speeds, since a higher standard of maintenance is required with high speeds, while the faster trains also cause more damage and disturbance to the track. According to a report of the American Railway Engineering Association, the increase in wear is greater with freight train speeds than with passenger train speeds, and it is naturally greater on curves than on tangents. Assuming an increase of freight train speed from 20 to 25 m. p. h. on a division having 75 per cent tangent and 25 per cent curves, it is estimated that the consequent increase in maintenance expense would be about 2.4 per cent, or approximately 0.5 per cent for each mile-per-hour increase in speed. If the division carries considerable passenger traffic, this would establish the standard of maintenance on account of the necessity of making an easy-riding track, so that the maintenance cost may be assumed to increase with the proportion of passenger trains.

Track and Traffic.—The relation between the track structure and the traffic which it carries is sometimes overlooked. For new lines the selection of weight of rails, kind and spacing of ties, and kind and depth of ballast may be made arbitrarily and influenced more or less by the question of cost. On existing lines, also, when heavier engines, cars, and trains, higher speeds or more numerous trains become necessary to handle the business, little consideration may be given to the character of the existing track or its ability to carry the increased loads and traffic with economy and safety. In this respect it is desirable to have much closer relations than usually exist between the engineering, maintenance, mechanical, and operating departments. A track that is light in proportion to the traffic conditions will have excessive expense in maintenance. Rails are worn more rapidly, joints become low and loose, spikes are loosened, ballast is crushed and displaced, frogs and switches are battered, and the whole track is shifted from proper line and surface. Renewals will be more frequent and more costly, and there will be a marked increase in all the items of maintenance. In addition, there is likely to be some increased wear of rolling equipment due to rough track.

It has been shown by the investigations of the Committee on Stresses in Track (noted below) that different stresses in rails are caused by different classes of locomotives and by different designs of engines of the same class, this latter being due to variations in wheel spacing, counterbalancing, and distribution of wheel loads. Furthermore, C. T. Ripley, Chief Mechanical Engineer of the Santa Fe Ry., has pointed out that if the locomotive designer will give consideration to the matter of reducing the stresses in rails he will aid in reducing both locomotive maintenance and track maintenance.

Many engineers and roadmasters have had experience of the increasing difficulty in maintaining track which has not been improved or strengthened in proportion to the heavier service imposed upon it. Lack of proper relation between track and traffic is evidence of the necessity of improvement in the track, but does not constitute any argument against operating methods. The business of the railway is to carry traffic, and it would be illogical as well as useless to complain about the effects of increased loads and speeds with a view to any modification in operating conditions. On the contrary, the aim must be to fit the track to the traffic conditions, but it will be the duty of the engineer to show the desirability of improving the track to meet the more severe conditions in order to secure higher efficiency and economy in the operation of the railway.

During recent years there has been a tendency to strengthen the track on main lines, in view of increased loads and traffic. Standards of main track construction have been adopted by many railways, but the extent to which the standards are applied is variable and they are rarely applied to all parts of the line which carry heavy equipment or fast trains. A railway having several hundred miles of track will include, as a rule, various kinds of track construction and traffic conditions, while the quality of track maintenance is likely to vary equally with the character or quality of construction. Consequently, maximum wheel loads, maximum trains, fast-freight trains, and heavy high-speed passenger trains (with their locomotives heavy in axle loads and total weight) may run for considerable parts of their journeys upon track which is not fully adapted for such service. The result is shown in wear and tear of equipment and track, involving a high maintenance cost and perhaps a low factor of safety. This is indicated by accidents occurring under such conditions, even on main-lines, as shown occasionally by the accident investigations of the Interstate Commerce Commission.

A classification of track on the basis of traffic carried, as adopted by the American Railway Engineering Association, is as follows: Class A, all districts of a railway having more than one main track; also districts having a single main track with traffic exceeding a freight-car mileage of 150,000 per mile per year, or a passenger-car mileage of 10,000 per mile per year with a maximum passenger-train speed of 50 miles per year. Class B, all districts having a single main track with a traffic less than that of class A, but exceeding a freight-car mileage of 50,000 or a passenger-car mileage of 5,000 miles, with a maximum passenger train speed of 40 m. p. h. Class C, all districts with traffic conditions below those of classes A and B.

Track and Train Resistance.—That the resistance to traction of locomotives and cars varies widely with the condition of track was pointed out by the late Julius Kruttschnitt, of the Southern Pacific Ry. Train resistance is taken commonly at 6 lb. per net ton for low speeds; but this assumes track in good condition and the resistance may be twice or three times this amount on track having light rails, bad surface, low joints, and a poor degree of maintenance. The heavier the rail and its tie and ballast supports, the easier it is to maintain track in good line and surface, the less will be the deflection under loads, and the less will be the work required to pull the wheels out of the depressions at low and yielding joints.

Studies of Track.—Railway track as a unit, or as a complete structure for carrying moving loads or trains, has received comparatively little attention in research, although much work has been devoted to its component parts, such as forms of rails, preservative treatment for ties, and the holding power of spikes.

Investigations of the track, as a whole, were carried out by P. H. Dudley in connection with his extensive studies for the improvement of rail sections. His work showed the relation of tie bearing and ballast to rail stresses and deflections under load, and to the distribution of the load through ties and ballast to the roadbed. Somewhat similar studies were made in France by Mr. Cuenot when a general increase in train speeds on the Paris, Lyons & Mediterranean Ry. made it necessary to determine the stresses in track, its behavior under these stresses, and the methods of strengthening required to withstand these stresses. He arrived at these conclusions: (1) Heavy rails should be used and secured by screw spikes. (2) The rails should be laid on steel tie-plates having outside ribs to support the back of the spike head. (3) The ties should be much more rigid (ties in Europe are generally thinner and more flexible than those in main-line tracks of American railroads). (4) The ballast should be so tamped as to give a specially firm bearing for not more than 15 in. on each side of the rail, since pressure transmitted through the tie to the ballast and subgrade diminishes rapidly with the distance from the load. (5) The rail joints should be of the three-tie supported type. (6) By supporting the joints and increasing the rigidity of the ties, the resistance to traction would be reduced materially and trains could be run at higher speeds without reduction in comfort or safety.

By far the most thorough and extensive investigations of this kind are those which have been conducted for some years by the Committee on Stresses in Track, representing the American Railway Engineering Association and the American Society of Civil Engineers. These studies have covered particularly the following features: (1) the stresses in rails under various conditions of rail weight, loading, and track condition; (2) the distribution of pressure from the rails through ties and ballast to the roadbed. They include, also, the efficiency of rail joints and the effects of canting the ties inward. Extended reports have been made, giving information of great value. Few definite conclusions have been presented, however, since the investigations are still in progress. All the above studies have been made on railway tracks in actual service. For experiments with different kinds of rail joints, ties, and ballast the German government railways at one time had an oblong test track with 820-ft. sides and end curves of 650 ft. radius. On this track a locomotive could be run continuously to intensify the loading or traffic conditions and thus shorten the test period.

Track Design.—On the railway system of the entire world the track is of a uniform type, consisting of tee or flange rails on wood cross-ties in a bed of ballast. The few and limited exceptions include: (1) double-headed rails held in cast-iron chairs on the ties; (2) steel and concrete ties; (3) pairs of metal or concrete bowls or blocks in place of cross-ties; and (4) concrete slabs as a substitute for ballast and having usually wood blocks or longitudinals to carry the rails. Although track consists essentially of the simple elements noted above, it is in practice a somewhat complicated structure, with numerous designs of rail-joint splices, attachments of rails to ties, and the switches and special construction required at turnouts, junctions, crossings, yards, and terminals.

In weight of rail, American railways have an extreme range, from 50 to 140 lb. per yard, but 75 to 85 lb. is about the minimum for main tracks, the lighter rails being on tracks of minor importance. Many railways are using rails of more than 100 lb. per yard. In foreign practice, 100 lb.

per yard is about the maximum, but the maximum loads on foreign railways do not equal those handled in this country. European engineers formerly criticized American practice for the extensive use of rails of comparatively light weight, but they failed to allow for the larger number of ties as compared with European practice. Since the whole track deflects under load by compression of the ballast and roadbed, in addition to the deflection of the rails between the ties, a light rail supported by 17 to 20 ties may form a better track than heavier rails on 10 to 14 ties.

It follows that for the same loads, modern American track with 85- to 100-lb. rails supported by 18 or 20 ties per rail length, and with ties 6 to 7 in. thick, is smoother and more substantial than track having rails of the same weight with only 10 or 12 ties of the 5-in. thickness which is common on foreign railways. This matter of rail support is sometimes overlooked, and it has even been reasoned erroneously that in introducing heavier rails the number of ties can be reduced without reducing the quality of the track.

Rail joints and the attachments of rails to the ties are weak points in the track structure. The difficulties of making and maintaining a splice or joint device that will be equal to the solid rail in strength, and will not be a large item in maintenance cost, have led to the use of longer rails in order to reduce the number of joints. But the rail joint problem still remains to be solved. As to rail fastenings, no entirely satisfactory substitute for the common drive spike has been discovered. The screw spike has advantages, but there are certain difficulties in its use, and it is used on a relatively small scale in this country, although it is a standard fastening on many foreign railroads. In spite of the general use of the drive spike it cannot be considered an adequate fastening for heavy track carrying heavy traffic. Probably few outside of the maintenance-of-way department realize the amount of work and expense involved in redriving and renewing spikes to keep a main track in reasonably good condition, even under moderate traffic. In addition to this, there is the damage thus done to the ties, with consequent increased cost for materials and labor represented by tie renewals.

Tamping the ballast under the ties so as to give a firm support to the rails is one of the most important features in track maintenance, but it is manifestly impossible to approach uniformity of support with about 3,000 ties per mile, or even with the 17 to 20 ties in a rail length. In theory, the ties are sometimes considered as rigid piers carrying the rail as a continuous girder. In practice, the ties are supports of varying and unknown stability, yielding unequally under pressure. This yielding increases the span of the girder (or rail), so that it may be subjected to stresses which it was not designed or intended to carry. The yielding of the ties tends to increase by the compression, shifting, or disintegration of the ballast, until the next period of tamping. But any increase in track deflection is a decrease in the quality or efficiency of the track as a structure for carrying rolling loads. Heavy rails are of advantage, therefore, in bridging over the weaker spots and preventing undue deflection under loads.

Improvement is desirable now in these directions: (1) extending first-class track construction over all main lines carrying fast and heavy traffic; (2) extending a high degree of maintenance uniformly over such lines, and eliminating points where track is not maintained in normal condition in relation to its service; (3) introducing labor-saving machinery and closer control of work to reduce the expense of maintenance; (4) extending the clearing and policing of track and right-of-way, in connection with maintenance, for the sake of neatness and good appearance of the railway; (5) introducing methods of determining unit costs of the different kinds of track work; (6) introducing better systems of accounting, to facilitate more effective analysis and distribution of expenditures on maintenance of way and structures.

Special Designs of Track.—To eliminate troubles due to intermittent cross-tie supports, longitudinal supports for the rails have been tried on a small scale. Several forms of concrete longitudinals have been designed and some years ago the Pennsylvania R. R. laid about 1,000 ft. of track with steel longitudinals designed by G. Lindenthal. These latter were of double channel section carrying oak blocks on which the rails were laid. A track designed by J. W. Schaub had 10-in. I-beams, $8\frac{1}{2}$ ft. apart, placed in the ballast and connected by tie rods; upon these were the track ties, secured by hook bolts. A further development has been the use of concrete slabs on the roadbed to take the place of ordinary loose ballast, the rails being carried by various arrangements of wood blocks or wood or steel longitudinals in or upon the concrete. Experiments with such construction on the Northern Pacific Ry. have not given much promise, even in reduction of maintenance expense, which is one of the main objects of a more permanent track structure. A still later proposition is a design by F. H. Alfred, president of the Pere Marquette R. R., for a reinforced-concrete slab having embedded in it steel longitudinal members to carry the rails. In most of such designs, however, the first cost would far outweigh any saving in maintenance cost. The above statement refers to ordinary railway track construction. Special types of track have been used extensively for tunnels and underground railways, but there the conditions are very different from those of open roadbed. Nevertheless, there is a field for the development of new types of track construction, as well as for improvement in the various detail parts of the type of track now in universal use.

Except perhaps in regard to the amount of maintenance required, the typical track construction already outlined appears to meet all requirements, providing its several parts are proportionate in strength and stability to the speeds and loads of the traffic carried. With such a proper relation of track to traffic, the maintenance work is at a minimum, but even so, this work is a continual task and expense.

Almost the only means proposed for more substantial construction to insure reduced maintenance is that of providing a rigid concrete or concrete and metal platform to carry the rails, dispensing with cross-ties and ballast and giving a better distribution of load upon the subgrade, as noted above. Even with such construction, variations in surface would no doubt develop due to settlement and disturbance of the roadbed, which necessarily is more or less unstable and subject to influences of weather and climate. It remains to be determined whether the expense of such construction, giving a smooth and theoretically permanent surface, would be compensated by reduced cost of maintenance of track, engines, and rolling equipment. Results of the very limited experiments thus far made have not been promising in this respect.

Portable Track.—Of a very special type is the portable, narrow-gage track which is used extensively on construction work and in mines, plantations, and industrial plants, and which also did such good service for field railways during the World War. Such track is usually of 24-in. gage with rails of 16 to 25 lb. per yard, in lengths of 15 to 16 ft., bolted or riveted to steel plate or shallow trough ties 36 to 42 in. long, and having six to eight ties per section. Slotted holes in the ties enable T-head bolts to be inserted from the top of the tie. For locomotives, or for wheel loads of more than 1½ tons, 20-lb. rails should be the minimum. Rail joints may be spliced by fish plates attached to one end of each rail, the adjacent rail slipping between the projecting plates and being secured in the field by bolts or special forms of attachment. A sleeve joint to eliminate all loose parts has short angle splice bars riveted to a joint tie, so that the ends of the rails can be inserted and held securely. This joint tie may be riveted to the ends of the rails of one track section, or it may be a separate piece, the ends of both track sections being slipped into the sleeves as the track is laid.

Complete track sections 15 ft. long weigh about 230 to 250 lb. with 16-lb. rails, 300 to 325 lb. with 20-lb. rails, and 350 or 375 lb. with 25-lb. rails. Short sections, 7½ ft. long, provide for adjustments at turnouts. Curves of 30 to 60 ft. radius are supplied in 15- and 7½-ft. lengths. Switches and track crossings are also provided in special sections.

Both riveted and bolted track are used for industrial lines, the former being the simpler and dispensing with loose parts. For military purposes, the separate rails and ties may be transported more conveniently to field headquarters, where they are bolted up into track sections ready for laying. Ballast is frequently omitted on industrial lines but is desirable for military lines, while additional bearing on soft ground may be given sometimes by wood ties, about 4 by 8 by 48 in., placed between the steel ties. Grades are limited to $2\frac{1}{2}$ to 4 per cent, as a rule, but are sometimes as high as 5 or even 6 per cent. On military lines where rails are laid on wood ties, as in ordinary track, curves are as short as 100-ft. radius, and grades may be compensated at the rate of 0.02 per cent per degree of curvature. Steam locomotives of 7 to 20 tons and gasoline locomotives of 3 to 15 tons are used on portable railways, operating trains at 5 to 10 m. p. h. with 15 m. p. h. as the maximum.

Railway Improvements.—In addition to improved track construction and a higher standard of track maintenance, many general improvements are needed to facilitate the movement of traffic with the least delay and the least liability of accident. Such improvements include additional double-track and passing-siding facilities, yard and terminal facilities, grade and line revision, water and fuel stations, block signals along the open line, and interlocking plants at junctions, crossings, passing sidings, and yard entrances. The important relation which such facilities bear to the prompt and efficient handling of freight traffic has been forced upon the attention of railway authorities by the enormous increase in traffic and by several periods of traffic congestion during recent years.

The double tracking of main lines to relieve such congestion, and the construction of new cut-off lines to avoid busy points, may be of greater importance than the building of new branches and extensions which bring in more traffic and thus increase the business of the main lines. As to yards, the delay of a car in any one division yard may be from 10 to 20 hours. This may be increased greatly under adverse conditions, especially where the car has to be transferred between the yards of different railways at a connecting or terminal point. In a long trip, the cumulative detention may thus amount to several days. These detentions may be reduced materially by improvements in yard design and yard operation, and also by improvements in operating methods, such as running through trains between terminal points, without breaking up and reassembling them at intermediate division yards.

Cost of Track and Maintenance.—Actual cost of track will depend upon the character of construction and the market prices, but the relation of cost of track to total cost of railway is indicated by Table A. A word of caution may be given as to the prices in this table, as they do not afford any basis for comparisons of cost of construction. The costs of the Baltimore & Ohio R. R. represent a substantially built, heavy-traffic main line as it actually exists, including both main tracks and side tracks, and with prices as of 1925. The figures for the other roads, however, are those tentatively allowed in the Interstate Commerce Commission's valuations, and cover lighter construction and structures at assumed low prices of earlier years, besides which they usually provide for some second-hand material and less grading, so that they would not represent actual costs for a main line at this time. In fact, the railways

TABLE A.—COST OF TRACK AND RAILWAY CONSTRUCTION
(Per Mile of Single Track)

	Baltimore & Ohio R. R.		Interstate Commerce Commission Figures					
			Chicago, Burlington & Quincy R. R.		Great Northern Ry.		Seaboard Air Line Ry.	
	Cost per mile, 1925	Per cent of total cost, 1914	Cost per mile, 1917	Per cent of total cost	Cost per mile, 1914	Per cent of total	Cost per mile, 1914	Per cent of total cost
Rails	$ 6,041	8.88	$ 4,741	12.20	$ 5,000	12.3	$ 4,129	11.70
Ties and switch ties	5,450	5.45	4,102	10.50	3,000	7.4	1,851	5.24
Frogs and other track material	3,035	3.02	1,501	3.85	1,560	3.8	961	2.72
Ballast	9,947	8.49	2,885	7.40	1,700	4.2	1,259	3.57
Tracklaying and surfacing	7,719	6.66	2,610	6.70	2,560	6.3	1,782	5.05
Fencing	345	0.42	385	1.00	235	0.6		
Road crossing, cattleguards, signs	1,123	1.50	477	1.20	476	1.2	276	0.78
Signals and interlocking	1,208	1.28	303	0.80	136	0.4	41	0.12
Stations	4,905	5.04	1,892	4.90	1,400	3.4	1,390	3.94
Water stations	812	1.08	649	1.70	670	1.7	192	0.54
Coaling stations	280	0.29	....	.....	....	....	75	0.21
Bridges, trestles, culverts	10,133	13.10	4,627	11.90	3,500	8.7	3,264	9.25
Total	50,998	55.21	24,172	62.15	20,237	50.0	15,219	43.12

have in many cases taken exception to these valuation figures as being much too low. It must be borne in mind also that these costs do not include engineering, general expenditures, contingencies or overhead items of any sort. Of greater value are the percentages of total cost, as these do not fluctuate with prices.

Maintenance-of-way and structures, which is a specific classification in railway accounting, represents a considerable proportion of the total operating expenses and operating revenue, as shown by typical examples in Table B. Its cost varies according to the character of track and traffic and the relations between these two factors. On the Santa Fe Ry., this cost increased from $1,428 per mile in 1914 to $3,463 in 1920, and dropped only to $3,103 in 1924. This sudden and sustained rise was due mainly to the great increase in cost of labor and materials resulting from the World War. As an indication of the various items involved in "maintenance-of-way and structures," Table C is given, showing the distribution on a road where the total amount was about $5,500,000, or 12 per cent of the operating expenses. This table gives also the distribution of the main items on the principal railways of the United States for the 3-year period ending with June, 1917. In dividing this item of expenses, maintenance-of-way represents from 65 to 85 per cent of the total, with 35 to 15 per cent for structures. Furthermore, about 50 per cent of the expense for maintenance-of-way is for the routine work of roadway repairs, exclusive of rail and tie renewals.

TABLE B.—RELATIONS OF MAINTENANCE AND OPERATING EXPENSES; 1924

Railways	Percentage of operating expenses			Percentage of operating revenue			Ratio of operating expenses to operating revenue
	Maintenance-of-way and structures	Maintenance of equipment	Transportation	Maintenance-of-way and structures	Maintenance of equipment	Transportation	
Bessemer & Lake Erie	13.49	46.15	36.09	.....	.....	.....	.....
Boston & Maine	.....	.....	.....	12.80	20.70	43.00	81.20
Chicago & Eastern Illinois	12.95	35.81	45.09	.....	.....	.....	.....
Chicago & Northwestern	18.72	25.37	50.19	.....	.....	.....	80.65
Cleveland, Cincinnati, Chicago & St. Louis	.....	.....	.....	12.31	21.87	37.06	76.09
Delaware & Hudson	.....	.....	.....	11.84	27.10	37.07	81.59
Erie	.....	.....	.....	11.53	24.82	38.69	70.81
Illinois Central	.....	.....	.....	13.76	22.64	36.47	77.10
Lehigh Valley	.....	.....	.....	11.00	24.00	40.00	80.00
Maine Central	.....	.....	.....	16.11	19.52	42.07	81.92
Michigan Central	.....	.....	.....	12:76	18.78	34.81	70.95
Norfolk & Western	21.18	32.62	41.82	15.15	23.33	29.90	73.88
St. Louis—San Francisco	18.34	28.07	47.99	12.98	19.88	33.98	.....
Southern	19.05	24.85	48.75	13.73	17.91	35.13	.....
Wheeling & Lake Erie	17.39	33.27	43.60	13.67	26.16	34.27	.....

TABLE C.—DISTRIBUTION OF EXPENDITURES FOR MAINTENANCE-OF-WAY AND STRUCTURES

Items	One railway, 1924, per cent	United States railways, 1914–1917, per cent	Items	One railway, 1924, per cent	United States railways, 1914–1917, per cent
Superintendence	7.16	5.69	Wharves and docks	0.02	.....
Roadway maintenance	9.00	10.96	Coal and ore wharves	0.09	.....
Tunnels; subways	0.06	.....	Telegraph and telephone lines	1.25	.....
Bridges; trestles; culverts	4.60	.....	Signals and interlocking	4.94	.....
Ties	15.24	15.41	Power-plant buildings	0.38	.....
Rails	8.02	2.91	Miscellaneous structures	0.92	.....
Other track material	6.98	4.52	Roadway machines	0.82	0.45
Ballast	0.18	1.88	Small tools and supplies	1.08	1.13
Applying track material	11.70	25.91 (Applying track material and Track maintenance together)	Removing snow, ice, and sand	1.68	1.76
Track maintenance	8.25		Assessments for public improve-	0.01	
Fences; snow and sand			ments	0.01	.....
fences	0.67	0.98	Injuries to persons	1.42	.....
Crossings and signs	2.23	1.30	Insurance	1.20	.....
Stations and office buildings	2.80	.....	Stationery and printing	0.25	.....
Roadway buildings	0.34	.....	Other expenses	0.87	.....
Water stations	0.71	.....	Maintaining joint tracks, yards		.....
Fuel stations	0.81	.....	and other facilities	1.03	.....
Shops and engine houses	5.35		Total	100.00	72.98

Expenses, Revenue, and Rates.—Railways are built for the purpose of furnishing transportation (or carrying passengers and goods from place to place), and bringing some return to the people who invest their money in the enterprise. To provide the service requires: (1) the construction and maintenance of a roadbed with rails, or a "rail-way"; (2) the provision and maintenance of rolling equipment (cars and locomotives); (3) the organization of an operating department to plan and control the train service. For the three main divisions of operating expenses the proportions in 1923 were as follows: maintenance-of-way and structures, 16.62 per cent; maintenance of equipment, 29.93 per cent; conducting transportation, 47.18 per cent, or a total of 93.73 per cent. Of the amount for maintenance of-way and structures, 62 per cent was for labor, 30 per cent for material, and 8 per cent for depreciation and miscellaneous. Operating expenses absorb a large proportion of the revenue from operation. In 1916, this operating ratio was 65.64 per cent; under government operation during the World War it increased to 94.32 per cent in 1920, but in 1924 (4 years after the railways had been returned to private ownership) the ratio had declined to 76.14 per cent. For the railways listed in Table B, the operating ratio ranged from 70.81 to 81.92 per cent.

These figures indicate how little money is left to the railways to pay out for the numerous taxes imposed, for dividends to investors and for the improvements necessary to keep pace with increasing traffic and to provide increased facilities to the public. This undesirable condition is due largely to the exceptional restrictions and high taxes imposed upon railway companies by the federal government and various state governments and local authorities. No other large industry has its rates or charges regulated or rather reduced so that its net returns approach the vanishing point.

Under the Transportation Act passed by Congress in 1920 it is the duty of the Interstate Commerce Commission so to regulate the rates as to enable the railways to earn a fair return. The commission has determined on $5\frac{3}{4}$ per cent as a "fair return." But it has failed to establish rates adequate to provide this amount, so that few railways have approached this figure, and receiverships and financial loss have resulted. This situation is the effect of a hostile attitude assumed unfortunately by legislative bodies and also by shippers and the general public who largely fail to realize the true relation of railway service to their welfare and interest. Other transportation companies, public utility concerns, banks, manufacturers, and commercial concerns are free, in the main, to fix their rates and prices to produce a much higher return than $5\frac{3}{4}$ per cent, while the railways are restricted and harassed in various ways, although they constitute one of the greatest industries and probably the one which has done most for the growth and welfare of the country. It must be remembered also that railway securities are not held exclusively by great financial concerns, but are held largely as investments by the general public and by banks and other concerns in which the public invests its money, so that any depreciation in values of the securities has a widespread and depressing effect upon the public welfare and prosperity.

Railway Statistics.—In conclusion of this introductory chapter and in order to illustrate the extent and importance of the railway system, the author gives the following summarized statistics for the railways of the United States, compiled from the report of the Interstate Commerce Commission for the year 1923.

RAILWAYS OF THE UNITED STATES, 1923

	Total	Electrified
Miles of line, all railways[1]	250,222	.....
Miles of line, class I railways[1]	235,574	1,535
Miles of second track	33,127	513
Miles of third track	3,030	194
Miles of fourth track	2,088	169
Miles of other main tracks	328	33
Miles of yards and sidings	111,708	812
Miles of total tracks	385,855	3,256

Block system, automatic, miles (1925)	43,840
Block system, non-automatic, miles (1925)	64,250
Locomotives, steam, 69,005; electric	409
Cars, passenger, 57,159; company service	112,426
Cars, freight, 2,401,077; private freight	230,220
Floating equipment; steamers, tugs, barges	2,602
Passengers (average travel, 38.46 miles)	986,913,075
Freight, revenue tonnage	2,333,600,764
Freight; non-revenue tonnage	276,455,373
Freight, average haul per ton, miles	176.86
Receipts per passenger, $1.16; per pass-mile	3.018 ct.
Receipts per ton, $1.974; per ton-mile	1.116 ct.
Train mileage (class I roads), passenger	560,984,411
Train mileage, freight	620,329,534
Train mileage, total, including mixed and special	1,207,714,249
Train mileage, work-train service	37,384,812
Capital; stock, $9,092,933, 214; funded debt	$11,964,580,105
Capital; total	21,057,513,319

[1] The length of "all railways" includes switching and terminal railways, but is exclusive of 798 miles in Alaska and 254 miles in Hawaii. The maximum railway mileage was 254,251 miles in 1916, since which time the total has been reduced by the abandonment of numerous branches and small lines owing to financial conditions and loss of traffic. All other figures are for class I railways only, excluding the switching and terminal lines and the 22,510 miles of minor railways of classes II and III (except as noted). The total mileage of track includes 1,365 miles of narrow gage. Class I railways are those having an annual revenue of more than $1,000,000.

RAILWAYS OF THE UNITED STATES, 1923—*Continued*

	Class I railways		Switching and terminal lines
	Total	Per mile	Total
Operating revenue	$6,289,580,027	$26,743	$66,737,117
Operating expenses	4,895,166,819	20,814	47,698,024
Net revenue	1,394,413,208	5,929	19,039,093
Taxes, etc	333,857,117	1,419	4,489,011
Operating income	1,060,556,091	4,510	14,550,082
Total income	1,060,914,996	4,511	16,382,766

Operating expenses:
- Maintenance-of-way and structures .... $ 813,688,760; 16.62 per cent
- Maintenance of equipment .... 1,465,156,595; 29.93 per cent
- Conducting transportation .... 2,309,608,766; 48.18 per cent

Ratio of operating expenses to operating revenue ... 77.83 per cent

Additions and betterments (excluding equipment) .. $ 336,606,000

New lines and extensions 29,952,809

Railway employees:
- Executive, professional, clerical & general ..... 302,794; 16.11 per cent
- Maintenance of way and structures .... 401,723; 21.37 per cent
- Maintenance of equipment and stores .... 589,879; 31.38 per cent
- Transportation (engine, train and yard men) .. 588,374; 31.14 per cent
- Total 1,879,770

Employees: Maintenance of Way and Structures:
- Roadmasters and general foremen 3,294
- Asst. general foremen and inspectors 1,281
- Section foremen 40,080
- Section men (over 50 per cent of total) 219,314
- Bridge-and-building foremen 5,440
- Bridge-and-building labor force 55,481
- Extra-gang and work-train laborers 58,310
- Signal foremen, inspectors, maintainers and helpers 18,523
- Total 401,723

Rails, main track, weight per yard, class I lines:

	Miles		Miles
Under 50 lb	1,228.67	100 to 109 lb	38,799.91
50 to 69 lb	45,500.52	110 to 119 lb	393.23
70 to 79 lb	41,116.78	120 to 129 lb	644.64
80 to 89 lb	66,159.76	130 to 139 lb	3,296.36
90 to 99 lb	60,168.23	140 lb. and over	30.10

Rails, average weight; eastern lines, 93.37 lb.; southern lines .... 82.30 lb.

Rails, average weight; western lines, 79.99 lb.; all lines .... 84.32 lb.

Rails laid in replacement and betterment:
- New, 1,729,696 tons; second hand 1,409,276

Ties laid for renewals:
- Untreated; hardwood, 23,067,273; softwood 19,004,867
- Treated, 41,655,616; second hand 260,227
- Other than wood, 447,002; total 84,434,985

Bridge and switch ties, 277,615,107 ft. b. m.

CHAPTER II

ROADBED AND BALLAST

A continuous surface or roadbed must be provided for the track, but it is assumed here that the general work of railway construction has been completed, with cuts and fills brought up to subgrade level and with bridges, tunnels, and other works ready for the track.

Preparation of Roadbed.—In completing the earthwork or grading, and in making ready for the tracks, it is desirable for practical and economic reasons to provide ample width of roadbed on cuts and in fills. With narrow embankments or fills, the edges soon become washed and broken down, so that ballast begins to slip away, and additional maintenance is then required continuously to keep the track in condition. In narrow cuts, it is difficult to maintain ditches and drainage, while in cold regions the snow settles readily and is difficult to remove. Such cuts and fills are the cause of much additional maintenance and of subsequent improvement work for widening them. It is very desirable that the roadbed should be compact enough to provide a firm foundation for the track, but in most new construction this condition is not obtained, and much ballast is lost by being driven into the comparatively soft and loose surface. In exceptional cases, the roadbed has been compacted by 10-ton steam rollers, with very satisfactory results. A firm roadbed is specially desirable where heavy track and traffic are concerned, since the laying of heavier rails does not compensate for a yielding foundation of track.

To provide for drainage, the new roadbed is usually finished with a crown of 3 to 4 in. for single track, or 4 to 6 in. for double track, the surface sloping up from each side to the center line or to a central flat portion. This crowned surface does not last long, even on well-compacted roadbed, for the ballast works down into the soil and clay, or earth works up into the ballast, while the traffic load tends to make the surface concave rather than crowned. Present practice favors the use of a heavy bed of sub-ballast between the ballat proper and the roadbed.

Track Spacing.—Main tracks are spaced usually 13 ft. center to center; sometimes 13½ or 14 ft., while in Illinois a law specifies 13½ ft. In stations it is sometimes reduced to 12½ ft. The distance between centers of a main track and a parallel sidetrack, a passing track, or a yard lead should be not less than 16 ft. ($18\frac{1}{2}$ ft. in Illinois), or 13 ft. on branch lines, but may be as much as 20 ft., especially where water columns, signal posts, signal bridge supports, or high switchstands are placed between the main and side tracks. Ladder tracks may be 17 ft. c. to c. from similar track or main tracks. For sidings and yard tracks the spacing is usually 13 ft. (in Illinois, $13\frac{1}{2}$ ft.), but some roads space yard tracks alternately 13 and 15 ft. On curves, the spacing between centers of tracks must be increased to maintain the clearance between cars. Taking cars 85 ft. long over all, with trucks $59\frac{1}{2}$ ft. c. to c., the Pennsylvania R. R. gives this increase as 1 in. for each half degree, beginning with 1 in. for a 30-min. curve and ending with 15 in. for an 8-deg. curve.

Roadbed and Ditches.—On fills, the width of roadbed is from 28 to 38 ft. for double track and 16 to 25 ft. for single track. In cuts, the width between ditches is 29 to 34 ft. for double track and 16 to 20 ft. for single track, but where there is no defined ditch and the roadbed slopes to the toe of the cut, the width may be as much as 40 ft. for double track and 25 or 27 ft. for single track. Some roads have a uniform roadbed width in both cuts and fills. For class *A* roads the minimum width should be 20 ft., with at least 18 in. of roadbed shoulder or berm beyond the ballast toe. These shoulders need to be maintained by occasional filling; cinders are suitable for this purpose, but if earth is used, drainage should be provided for the enclosed space along the middle of the roadbed.

Drainage of the roadbed is an important factor in economical maintenance of track, its importance increasing in proportion to rainfall and traffic. The ballast provides for draining water away from the track, and side ditches in cuts carry off the water from ballast, roadbed, and slopes. Track ditches are of three types: (1) a flat-bottomed trench; (2) a V-ditch with steep slope from roadbed to toe of cut; (3) a broad, shallow ditch formed by extending the roadbed on a flat slope to the toe of cut. This last is adapted for dry regions and light soil. In any case the ditch section depends upon local conditions, but it should provide ample capacity for disposing of occasional heavy rains. In wet cuts, a tile drain may be laid in the ditch and covered with stone (see "Drainage and Ditching").

The ditches must have such depth and grade as to carry off water quickly and keep the subgrade well drained. At the end of the cut the ditch should be diverted away from the track and increased in width, so as to prevent the flow of water from gullying the slopes of the adjacent fill. In wet and soft cuts, especially if of narrow width, it may be necessary to form the ditch of concrete or to protect its sides by planks and old ties held in place by cross-struts. At road crossings, the ditch may be continued by cast-iron or wrought-iron pipe of suitable size, but not less than 6 in. in diameter. Clay pipe may be used if there is ample depth of cover to protect it from breakage.

Sodding Ditches and Slopes.—Where ditches are lined with sod, as practiced on some roads in order to prevent erosion, the grass and weeds should be kept well trimmed. Another practice is to maintain a 12-in. strip of sod along the edge of the roadbed to prevent breaking down and gullying, but in some cases this is claimed to interfere with drainage. A width of 8 to 12 in. should be left between the sod and the ballast toe, and for good appearance this space may be covered with cinders. More attention ought to be given to sodding or seeding the slopes of cuts and fills in order to prevent washing and erosion, and thus maintain the roadbed width on fills and avoid continual clogging and cleaning of ditches in cuts. The cost of sodding is so high that such work is rarely seen except on suburban lines and near stations; but sowing seeds and planting vines is less expensive. Where springs break through the slopes of cuts, a broad drain of stone or concrete may be built on the slope to carry the water to the track ditch.

Concrete Roadbed.—In view of the inevitable degree of instability in an earth roadbed, numerous plans and some experiments have been made with concrete roadbed to provide a more permanent foundation for

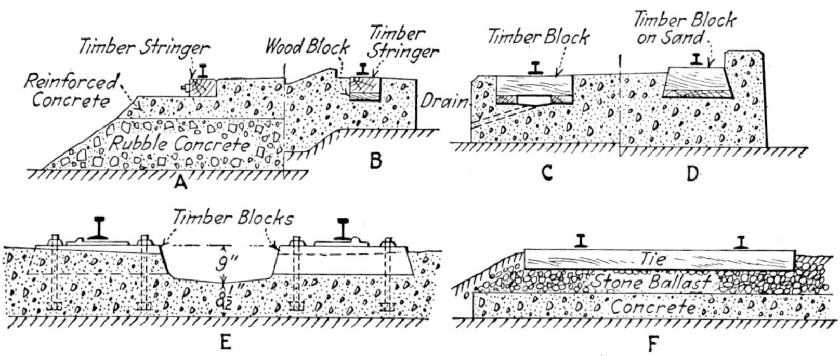

Fig. 1.—Concrete roadbed construction.

the track. This roadbed construction, of which some variations are shown in Fig. 1, is of two general types: (1) a concrete base on which are laid wood blocks or stringers to carry the rails, thus dispensing with ballast; (2) a concrete slab under the ordinary bed of ballast, thus separating it from the earth roadbed. An early design is shown at *A*, Fig. 1. A similar design, but with blocks of wood in pockets in the shallower side portions of the slab, has been tried on the Chicago Junction Ry. A combination of reinforced-concrete slab with steel longitudinals embedded in it has been proposed by F. H. Alfred, president of the Pere Marquette R. R.

The Northern Pacific Ry. has three forms of concrete unballasted track construction, as shown, covering about 2,000 ft. of track: *B* (Fig. 1), 8 by $1\frac{1}{2}$ ft., in $16\frac{1}{2}$-ft. lengths, with troughs for wood stringers 6 by 10 in., resting on 4- by 6-in. blocks embedded in the concrete, and with inside curbs or guard rails formed on the top of the slab; *C*, similar slabs 23 in. thick, with wood blocks 6 by 8 by 24 in. spaced 16¾ in. c. to c., or 11 in. at joints, the blocks resting on small sills, forming a central drain which has outlets at intervals; *D*, slabs 33 ft. long, 8½ ft. wide, and 18½ in. thick, with pockets for wood blocks resting on 3 in. of sand, these slabs having outside curbs 8 in. high. In 9 years' service the maintenance expense was greater than for adjacent ballasted track, while the track on concrete was much more rigid. The slabs settled slightly, but the maintenance work was in adjusting and renewing the rail supports. A design used in several large passenger terminals, shown at *E*, provides a rigid base and a floor that can be flushed and cleaned readily.

Concrete slabs as foundations for the ballast have been used to a limited extent where lines of heavy traffic are laid on soft roadbed. These installations include suburban lines of the Long Island R. R., main lines on the New York Central R. R. (*F*, Fig. 1), and the approaches to the Chicago Union Station. In this last case, a 10-in. slab on a 12-in. bed of cinders has its upper surface formed in drainage planes and is covered with stone ballast 8 to 10 in. deep under the ties. On a smaller scale, this plan has been used at railroad and highway grade crossings. Such foundation construction is too expensive for general use, even allowing for the consequent reduction in maintenance work and renewals, while the same advantages may be obtained to some extent and at relatively little expense by means of a good bed of sub-ballast. For very heavy traffic such methods would be worth considering.

BALLAST

The softness of an earth roadbed and the rigidity of base in rock cuts make it necessary to interpose a bed of loose material or ballast between the roadbed and the load-bearing elements of the track. This ballast is an important factor in securing good track, with economy of maintenance. Its principal purposes are: (1) to support the ties and distribute the traffic load over the roadbed; (2) to provide drainage under and around the ties; and (3) to provide for surfacing and adjusting the track. Pressure under the tie is both vertical and lateral and at a certain depth the pressure begins to distribute itself over an increasing area. One main function of the ballast bed, therefore, is to carry the load, without displacement of the material, to the depth at which the lateral distribution begins. The coarser kinds of ballast have greater ultimate load capacity, which is of particular advantage in the upper part of the ballast, carrying vertical load.

Any kind of ballast will gradually work down into the roadbed and eventually form a good foundation for the additional material which must be applied at intervals to keep the track at grade. A better way to form such a foundation is to consolidate the roadbed and then cover it with a bed of sub-ballast of gravel, slag, or cinders, which will assist drainage, distribute the load, and prevent loss or waste of the more expensive ballast. Shrinkage of ballast is the difference between the quantity measured loose in the cars and that measured in the track after the material has been compacted by tamping and by the passage of trains. Tests recorded by H. E. Hale indicate the following percentages of shrinkage: stone, 16.6; sand, 14.4; chats, 8.5; gravel, 21.8; slag, 24.1; cinders, 27. For a given quantity and section of ballast in track, a correspondingly greater quantity must be ordered. Payment is based on the quantity measured loose in the cars.

If the relation of traffic conditions to the resistance of ballast and roadbed is overlooked, and the depth of ballast is insufficient to give adequate support for the weight and density of traffic, then it would be real economy to apply a greater quantity and perhaps a better quality of ballast. A depth of 8 to 12 in. of ballast under the ties is usually specified for main track, but 12 in. should be the minimum where traffic is heavy. A greater depth is exceptional, although the Pennsylvania R. R. standard is 18 in. of ballast on 12 in. of cinders as sub-ballast. Other roads, using sub-ballast, put 6 or 8 in. of stone on 6 in. of gravel. On curves, the standard depth should be maintained under the tie at the low rail.

Broken stone, bank gravel, washed gravel, slag, and cinders are most generally used. Burned clay, sand, shells, and other materials are used locally; even earth is sometimes used, but this is suitable only for very light and infrequent traffic. The best materials are those which are durable, retain their solidity and position under the disturbing effects of wind and weather, give good drainage, are free from dust, and make an easy-riding track. Stone and washed gravel will drain readily while clean; but bank gravel, burned clay, sand, and earth will retain water, more or less, and are liable to heave in frosty weather. Where water is held under the ties, the ballast will deteriorate, and the movements of the ties will churn it into mud. On the other hand, dusty material is hard on axle bearings and machinery, besides causing discomfort to passengers. The spraying of oil from perforated pipes on tank cars has been tried with success on dusty ballast, but the practice is now little used. Grass is sometimes allowed to grow on track with sand ballast to prevent displacement by wind and rain.

Broken Stone.—This is, in general, the best material for ballast, as it meets the requirements noted above and can be handled or tamped in any weather. The most suitable stone is that which is hard and tough, breaks up into cubical or angular fragments instead of into thin, flat pieces, and does not disintegrate or crush under traffic or weather conditions or by tamping in the track. Granite and trap rank first, with limestone second, but limestone varies and is sometimes too soft for good ballast. Suitable stone would weigh about 175 lb. per cubic foot for trap and 168 lb. for limestone. It would be tested for weight, wear, toughness, soundness, and cementing quality.

As to size, practice varies considerably, but, in general, it ranges from that which will pass a $2\frac{1}{2}$-in. screen to that which will not pass a $\frac{3}{4}$-in. screen; this is the recommendation of the American Railway Engineering Association. A few roads use 3-in. maximum (with $\frac{1}{2}$-, $\frac{3}{4}$-, or 1-in. minimum), but others prefer 2-in. and a few put it as low as $1\frac{1}{2}$-in. The smaller maximum sizes are preferable with such hard rock as trap and granite. On the other hand, there is some preference for the larger maximum where freight traffic is heavy and passenger traffic light. An increasing tendency to reduce the maximum limit below $2\frac{1}{2}$ in. is due largely to the wide introduction of machine tamping, which works more effectively with the smaller sizes. Further, these sizes make less noisy track, can be tamped more easily, and require less work to obtain a good surface. For bridge decks, a range of $\frac{3}{4}$ to 1 in. is sometimes specified.

Gravel and Washed Gravel.—Bank or pit-run gravel is usually poor material for main tracks, as it has too much sand and dirt and too great a variation in size of the larger aggregate. Such ballast does not drain well and makes dusty or muddy track. Cementing gravel, with a large percentage of clay, tends to pack solid. As to the proportions of fine material, bank gravel should contain not more than 2 per cent dust or 40 per cent sand, except that for less important lines the proportions may be 3 and 60 per cent.

Cleaned and graded gravel, however, ranks next to crushed stone in quality and is usually much cheaper, but under heavy traffic it requires more work to maintain track in surface. The gravel may be simply screened, but it is better both washed and screened, and then graded by mixing the several sizes as required, a certain amount of fine material being used to form a compact mass. Screened gravel should contain not less than 25 nor more than 40 per cent of sand, while in washed gravel these proportions should be 15 to 25 per cent, sand being classed as material which will pass a No. 10 and be retained on a No. 50 screen. The maximum size is 2 to $2\frac{1}{2}$ in. In screening, the fine sand is first removed, and then the stones above maximum size, the latter rejections going to a crusher. Rotary screens grade the remainder (including the crusher product); a general range of sizes is $1\frac{1}{4}$, $\frac{3}{4}$, and $\frac{5}{16}$ in., the sizes being delivered separately or mixed to storage bins. Before and during the screening the material is washed by water jets. Freshly washed material containing 20 per cent of sand weighs about 3,000 lb. per cubic yard.

Slag.—Blast-furnace slag, used extensively in the vicinity of iron and steel plants, is almost as durable as crushed stone, but its weight averages only about 2,000 lb. per cubic yard. In the sharp and angular form of the pieces it has an advantage over ordinary gravel. Air-cooled slag from dump piles or slag pits is excavated by steam shovels and loaded on cars for crushing and screening, the ballast sizes ranging generally from $\frac{3}{4}$ to $2\frac{1}{2}$ in. Granulated or water-cooled slag, which is about the size of coarse sand, weighs only about 1,500 lb. per cubic yard and is too light for main line ballast; it also has a tendency to compact or "set" in spots, becoming hard like concrete. This material, however, is used for sidetracks and yards and makes a good sub-ballast under some heavier and more serviceable top ballast.

Burned Clay.—This ballast, used mainly in the Southwest, may be made from almost any clay that has a low proportion of sand, but the best material is gumbo or black-wax soil which is entirely free from sand. Since it weighs only 1,800 to 2,300 lb. per cubic yard, it should be used liberally, with a depth of 8 to 10 in. under the ties, or 6 to 8 in. if laid on sub-ballast. This latter arrangement is preferable, since the burned clay holds moisture and thus tends to keep the roadbed damp and soft. The sub-ballast may be a 4- to 6-in. bed of gravel.

Burned clay ballast is usually leveled just below top of tie, shouldered 6 to 12 in. from the ends of ties, and then sloped 1 on $1\frac{1}{2}$ or 1 on 2 to the roadbed. Its quality varies with the clay and the burning, but in general it drains readily (if the roadbed is well drained or covered with sub-ballast), does not crush or heave, is fairly free from weeds for a few years, is not dusty if burned with good coal (lignite and poor coal leave considerable ash), and will not require renewal for 8 to 20 years, according to quality and traffic. The size is usually $\frac{1}{2}$ to $2\frac{1}{2}$ or 3 in., but if it burns in large lumps, it may have to be run through a crusher, both to reduce the maximum size and to give a more uniform product. It is usual to specify that absorption shall not exceed 15 per cent by weight, but actually it is more nearly 25 or sometimes 30 per cent.

For burning, a site 500 to 4,500 ft. long on the clay deposit is cleared, and a row of cordwood and old ties about 6 ft. wide and 3 ft. high is covered with a small amount of nut or slack coal; upon this is a layer of clay 9 to 12 in. thick, which may be placed by an excavator. The wood is then lighted. After a few days, a $1\frac{1}{2}$- to 2-in. layer of coal is placed and covered with 9 to 12 in. of clay. Other layers of coal and clay are added at intervals of 4 to 6 days, until the heap is about 10 ft. high and 20 ft. wide. A ton of coal will burn about 4 cu. yd. of clay, and as the clay swells in burning, a cubic yard of it will make $1\frac{1}{4}$ to $1\frac{1}{2}$ cu. yd. of ballast. The burning is done usually by contract, the railroad furnishing the coal, land, and sidetrack, and loading the ballast on cars. Partial estimates are given on kiln measurement, and the final estimate is from car measurement. Unless the material is thoroughly burned (as shown by its red color) and by a clean fire, it is likely to disintegrate and turn to mud or dust.

Cinders.—Engine cinders make a serviceable ballast for light traffic and in yards and sidings, and are cheap if no long haul is required. But they have not the qualities desirable for heavy main-line service. Being porous, the ballast drains well. It does not heave with frost, and it prevents weeds from growing. The best quality comes from water-filled ashpits or where the cinders are washed out by jets, so that the fine material is removed, as, otherwise, the cinders will make a dusty track. Cinders may be used to advantage for leveling fills to subgrade, for filling pockets where mud has been dug out, and for sub-ballast.

Miscellaneous Materials.—Sand makes a fairly good ballast under light traffic, but is dusty and requires considerable maintenance work, since it is displaced by wind and rain and tends to flow away from the ties as they move under passing loads. It may be held by oiling or blanketing with some heavier material, or grass may be allowed to grow on it, as noted above, but for heavy traffic lines in sandy country it may pay to haul in some better material. Shells are used to some extent near the coasts. Chert is a disintegrating limestone found mainly in the Southwest and handled as bank-run material; its maximum size after excavating and handling is about 3 in. It frequently contains clay and thus resembles cementing gravel. Owing to its continual disintegration and the difficulty of working it in the track, it is not a satisfactory material. **Chats,** or the rock tailings from lead and zinc mills, resemble clean, coarse sand up to $\frac{3}{8}$-in. particles. Disintegrated granite, found in the Rocky Mountains, resembles clean, fine gravel, except that the particles are angular instead of rounded. It is used as bank-run material. Both materials differ from chert in that they are stable and not disintegrating.

Ballast Cross-sections.—The form of cross-section given to the ballast bed varies with the material, climatic conditions, and the opinions of engineers. Several forms have been noted and are shown in Figs. 2 to 13. Shouldering out the ballast 6 to 9 in. beyond the ends of the ties is designed to hold the track in line, especially on curves. In Europe, it has been general to use a much wider shoulder than in this country, but there is now a tendency to reduce this width, as its effect appears to be doubtful. It is desirable to keep the ballast clear of the base of rail to allow free drainage, especially where a track circuit is used for signal equipment, as the ballast, particularly if wet, may cause leakage of current from the rails.

Stone is generally kept a little below the top of the tie, and then shouldered out and sloped 1 on 2; but with ballast only 6 to 8 in. deep under the ties, it may be filled level with the tops of the ties, and for 6 to 9 in. beyond their ends. On double track, the ballast is usually level for the entire top width, sometimes with a row of coarse stones between the tracks to serve as a drain; occasionally the ballast is sloped between the tracks in the same way as on the outside. Stone and other coarse ballast should be handled with forks instead of shovels in order to prevent putting dirt into the track.

A similar arrangement is employed with the same depth of chats, cinders, and gravel, but with slopes of 1 on $2\frac{1}{2}$. Where these materials are 12 in. deep, the top may have a flat slope of about 1 on 24 from the top of the tie at center of track to end of tie, clearing the rail, then rounded out, and sloped 1 on $2\frac{1}{2}$ or 1 on 3. Sand and chats may be shaped the same as gravel, from center of track to about 1 in. below top of tie at end, but with a wider shoulder; and may hold in position better if rounded instead of sloped (usually 1 on 2) to the roadbed. Light sand is sometimes filled level with top of ties and for 6 in. beyond their ends, to check displacement by wind or vibration. Such a shoulder prevents material slipping from under the ends of the ties. Shell and chert ballast may be level with top of tie for 3 ft., then sloped to bottom of tie and 1 on 5 to roadbed, avoiding a thin toe by rounding it off on a 4-ft. radius to give a total width of 14 or 15 ft. with ballast 8 or 12 in. deep under the ties. The quantities then will be 1,735 and 2,910 cu. yd. per mile. A combination section used by the Kansas City Southern Ry. has 8 in. of shell for the length of tie, covered with gravel finished to the normal gravel section. This requires 1,735 cu. yd. of shell and 1,375 cu. yd. of gravel per mile.

Ballast should not be filled in over the ties, as this tends to keep them damp; but in a very hot and dry climate it may be desirable to protect the ties in this way. Earth ballast, however, is usually filled 3 or 4 in. above the ties for a width of 12 to 24 in., in order to give a steep drainage slope to bottom of end of tie, beyond which the roadbed is sloped 1 on 8 to the ditch or toe of cut. Cementing gravel is sometimes given a similar top surface and then sloped 1 on 3 from the end of the tie to the roadbed.

On curves, ballast that drains readily may be level with the top of the tie on the high side, and sloped to 3 in. below the top on the low side. With inferior ballast there may be a 9-in. shoulder on the high side, then a slope to bottom of tie on the low side, and then a continuation of the slope of the tie, ending with a rounded toe at the roadbed.

To line up the ballast toe, a board is used as a guide, against which is placed a row of the larger pieces. The Southern Pacific Ry. uses a row of small boulders, cobbles, or blocks of stone about the size of small, paving blocks, these being tamped into the roadbed so as to be flush with the surface. Such toe lines give a neat appearance; they serve as a guide for the section men in dressing the ballast and roadbed shoulder; they indicate by displacement any settlement or deformation of the roadbed, and their condition indicates to officials on trains the general quality of maintenance work.

EXAMPLES OF BALLAST CROSS-SECTIONS

American Railway Engineering Association.—The standard design for class *A* or heavy-traffic railroads with stone ballast provides a flat roadbed 36 ft. 3 in. wide on fills and between ditches in cuts, the ditches to be V-shaped, with steep slope from edge of roadbed to toe of cut and lined with sod. A 12-in. bed of sub-ballast, sloped off at the ends, as in Fig. 2, is covered by stone ballast 12 in. deep under ties $8\frac{1}{2}$ ft. long. This stone ballast is level with the top of the tie at the middle, slopes to a clearance of $\frac{1}{2}$ to 1 in. under the rails, and to 2 in. below the top of the tie, whence a curve of 4 ft. radius is followed by a slope of 1 on 2. The width over the ballast toes is 19 ft. 3 in. on single track, and 33 ft. 3 in. on double track, the toes being 3 in. beyond the sub-ballast and

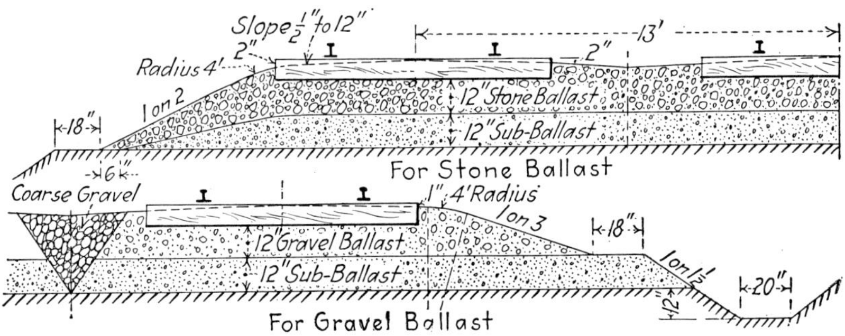

FIG. 2.—Ballast sections; A. R. E. A.

leaving 18-in. roadbed shoulders. On curves the top ballast is 12 in. deep under the tie at the low rail, and the width of roadbed will vary, the 1 on 2 slopes and 18-in. shoulders being constant.

With gravel ballast the sub-ballast extends the full width of the roadbed, which is 38 ft. 8 in. on fills and between ditches, or 48 ft. over ditches. These ditches are of trench shape, 12 in. deep and 20 in. wide, with slopes of 1 on $1\frac{1}{2}$. Where extra drainage is necessary, a tile drain of 6-in. minimum diameter is laid with its top not less than 12 in. below the bottom of the ditch and covered with cinders. The surface of the gravel ballast slopes to 1 in. below the top of the tie, has an 8-in. shoulder, a curve of 4 ft. radius for a distance of 14 in., and then a slope of 1 on 3, bringing the ballast toe 18 in. inside the edge of the sub-ballast. Between the tracks a V-shaped space extending down to the roadbed is filled with coarse gravel, with the middle 2 in. below the top of the tie.

For class *B* track, the roadbed width is 16 ft., with 9-in. ballast under 8-ft. ties. Stone and slag are shaped as above. Gravel, chats, and cinders have slopes of 1 on 3 from 2 in. below top of tie or from bottom of tie; in this last case the 4-ft. curve is inward from end of tie. For class *C* track, the roadbed width is 14 ft., with gravel or chert ballast 6 in. deep under the ties and sloped 1 on 3. For track of classes *B* and

C, the roadbed ditch is not a trench but is formed by a slope of about $1\frac{1}{2}$ on 1 to the toe of the cut. Ballast quantities per mile for class *A* track are given below. If the material is measured at the point of loading, an allowance for shrinkage is made of 12 per cent for stone and 4 to 15 per cent (according to quality of material) for gravel and sub-ballast.

Track	Stone, cubic yards		Gravel, cubic yards	
	Top	Sub-ballast	Top	Sub-ballast
Double	7,320	5,206	7,626	7,268
Single	3,995	2,664	4,144	4,726

Illinois Central R. R. (Figs. 3 and 4).—For double track of class *A* with stone or slag ballast, the roadbed width is 34 ft. on fills and between ditches in cuts, or 41 ft. between slopes in cuts. From the ends of the

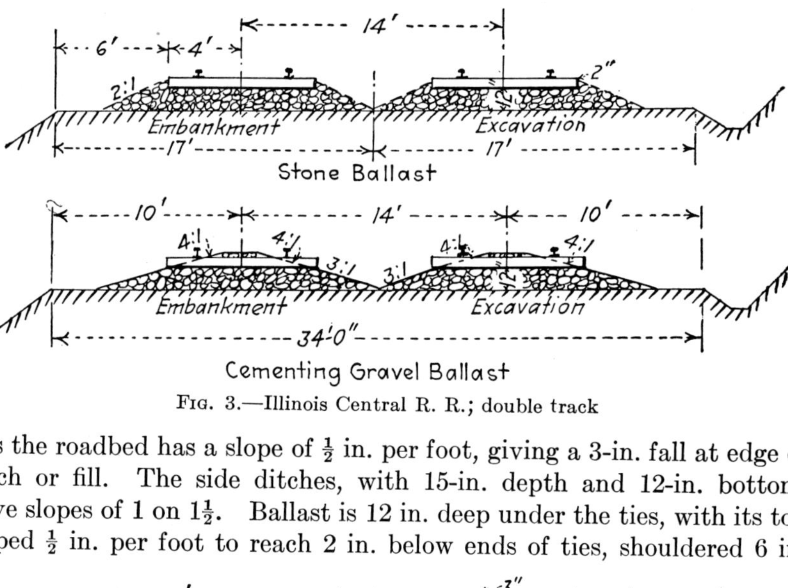

FIG. 3.—Illinois Central R. R.; double track

ties the roadbed has a slope of $\frac{1}{2}$ in. per foot, giving a 3-in. fall at edge of ditch or fill. The side ditches, with 15-in. depth and 12-in. bottom, have slopes of 1 on $1\frac{1}{2}$. Ballast is 12 in. deep under the ties, with its top sloped $\frac{1}{2}$ in. per foot to reach 2 in. below ends of ties, shouldered 6 in.

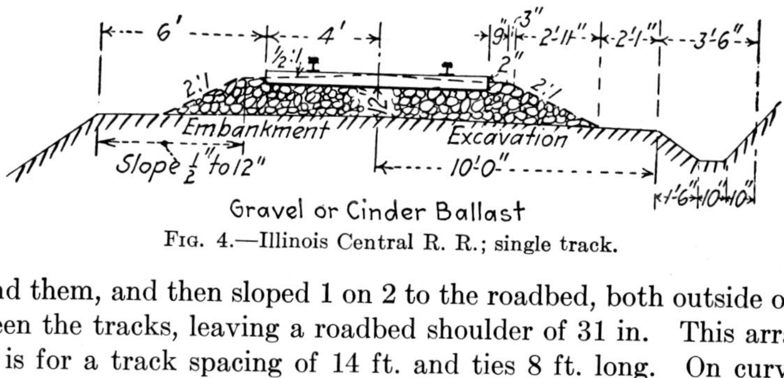

FIG. 4.—Illinois Central R. R.; single track.

beyond them, and then sloped 1 on 2 to the roadbed, both outside of and between the tracks, leaving a roadbed shoulder of 31 in. This arrangement is for a track spacing of 14 ft. and ties 8 ft. long. On curves, a 12-in. depth of ballast is maintained under the lower ends of the ties.

Gravel and cinder ballast is given a similar section but with a 12-in. ballast shoulder and 25-in. roadbed shoulder. Cementing gravel is filled 3 in. above center of tie, flat for 2 ft. in the middle and then sloped 1 on 4 to bottom of tie and 1 on 3 to roadbed, leaving a shoulder of 31 in. Grass sod on roadbed shoulder and in ditches has been discontinued, for though it gives a neat appearance and prevents erosion of ditches and edges of fills, it was found to prevent water from draining freely.

For track of class *B*, the ballast depth is 10 in. and the roadbed width 32 ft., while class *C* track has 8 in. of ballast and a 30-ft. roadbed. On single track the ballast and ditch sections are the same, but the roadbed width is 20, 18, and 16 ft. for classes *A*, *B*, and *C*, with corresponding widths of 27, 25, and 23 ft. over ditches in cuts. The ballast quantities per mile are as follows:

BALLAST QUANTITIES: ILLINOIS CENTRAL R. R.

Class of track	Ballast	Double track		Single track	
		Tangents, cubic yards	Addition for each inch of super-elevation, cubic yards	Tangents, cubic yards	Addition for each inch of super-elevation, cubic yards
A.............	Stone, slag	5,860	458	2,375	246
A.............	Gravel, cinders	6,321	488	3,235	256
A.............	Cementing gravel	5,376	474	2,729	258
B.............	Stone, slag	4,951	425	2,510	233
B.............	Gravel, cinders	5,367	451	2,738	251
B.............	Cementing gravel	4,410	450	2,262	239
C.............	Gravel, cinders	3,943	418	2,002	213

Union Pacific R. R. (Figs. 5 and 6).—For double track with stone or slag ballast, the roadbed width is 29 ft. on fills and 28 ft. between ditches in cuts, crowned to 1½ in. under the center of each track. In cuts, the width between slopes at roadbed level is 34 ft. in earth and 31 ft. in rock. The ballast is 15 in. deep below the top of the tie at the center, its surface sloped from the middle of the tie to give 1-in. clearance under the rail, and 6-in. shoulders beyond the ends of the ties. Between tracks it is depressed to 4 in. below the base of the rail. In rock cuts, a flat-bottomed ditch 12 in. wide and 12 in. deep has a slope of 4 on 1 from the edge of the roadbed. In earth, a V-shaped ditch is used, 12 in. deep and 36 in. wide. With gravel and disintegrated granite the width is 30 ft. on fills and 29 ft. in cuts, with a total width between slopes in cuts of 35 ft. in earth and 32 ft. in rock. The ballast section is as above, but 27 ft. wide over the toes.

On single track with stone ballast, the width is 16 ft. on fills and 15 ft. between ditches in cuts, with a crown of $1\frac{1}{2}$ in.; the width between slopes at roadbed level on cuts is 21 ft. in earth and 18 ft. in rock. For gravel ballast the width is 17 ft. on fills and 16 ft. in cuts, giving a width between slopes of 19 ft. in rock and 22 ft. in earth. Ballast quantities are as

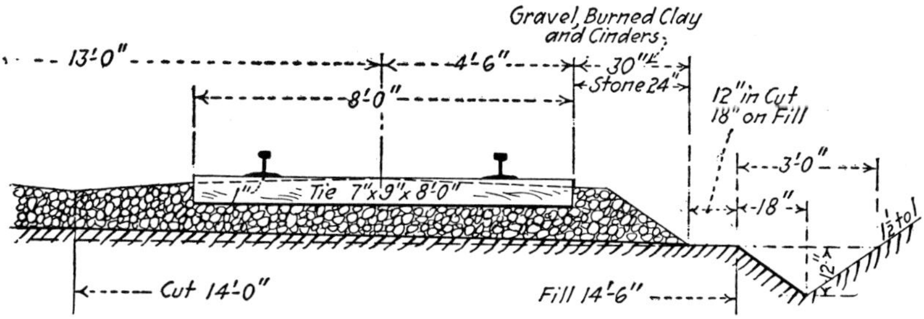

FIG. 5.—Union Pacific R. R.

follows, including 2 per cent above the theoretical to allow for loss and for depressions in roadbed; double track, 4,787 cu. ft. for stone and slag, 4,918 cu. ft. for gravel; single track, 2,356 and 2,492 cu. ft. The ties are 7 by 9 in. by 8 ft. On curves the depth of ballast is 19 in. under the low rail, with both low rails in the same plane. The ballast is shouldered out 12 in. beyond ends of ties and sloped not more than $1\frac{1}{2}$ on 1 on the high side. For heavy-traffic main line the width of roadbed on curves

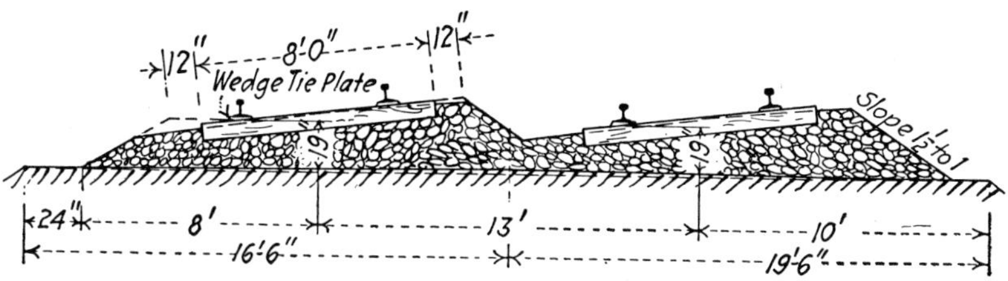

FIG. 6.—Union Pacific R. R.; track on curves.

is 33 ft. on fills and between ditches, with a total width of 39 ft. in cuts. On curves of 2 deg. or over, wedge tie-plates are placed under the low rails to keep them vertical.

Pennsylvania R. R.—Unusually heavy ballasting is a feature of the standard roadbed cross-section (Fig. 7). The stone ballast is 18 in. deep under the ties, laid on a 12-in. bed of sub-ballast extending to the edge of roadbed on fills and retained at the ditches in cuts by a toe of stone ballast. This arrangement is different from the sub-ballast section shown in Fig. 2. The top ballast is $23\frac{1}{8}$ in. deep at the center of the roadbed, with a level surface ½ in. below the top of the tie and a slope of about 1 on 1½ from the ends of the ties. The track spacing is 13 ft. c. to c., increased on curves by 2 in. for each degree of curvature.

The surface of the roadbed is crowned at ¼ in. per foot, and in cuts a steeper slope extends from the ballast to the toe of the cut, 6 in. below and 36 in. beyond the ballast toe (or 11 ft. 3 in. from the rail). This section forms a wide and shallow ditch. With four tracks, the outer ties are 3 in. below those of the inner tracks, and have 18 in. of ballast under them. Gravel ballast on single track is rounded off so as to be clear of the rail base and about an inch above the bottom of the tie, with the same toe width as the stone ballast.

Slopes of cuts are trimmed to the angle of repose of the material and then sodded or covered with vines to prevent erosion. To check washing

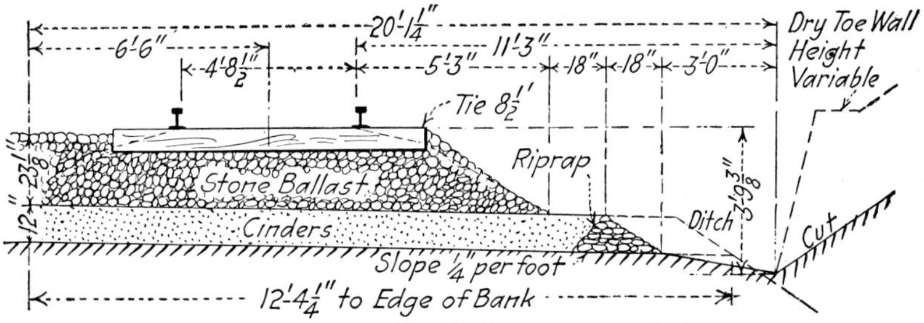

FIG. 7.—Pennsylvania R. R.; stone ballast.

of cuts, ditches are cut along the high side, at least 5 ft. from the top of the slope. Where it is desired to limit the width of the cut, dwarf retaining walls of dry masonry are built to a suitable height along the side ditch, as indicated in Fig. 7. In soft places, cinder ballast is placed until the roadbed has settled. At industrial spurs or sidings, the track spacing is not less than 16 ft., and the cinder sub-ballast is extended to carry the ties. The center of this track is 10 ft. 10½ in. from edge of fill, or 12 ft. 10¼ in. from toe of cut. Roadbed widths and ballast quantities are tabulated below.

Number of tracks	Width of roadbed				Ballast			
	Cuts, between toes		Fills		Width of top ballast over toes		Quantity per mile	
							Stone top ballast	Cinder sub-ballast
	Feet	Inches	Feet	Inches	Feet	Inches	Cubic yards	Cubic yards
1	27	2½	24	8½	15	2½	4,800	4,100
2	40	2½	37	8½	32	11	9,000	6,400
4	66	2½	63	8½	68	2	18,600	11,200

Delaware, Lackawanna & Western R. R.—The double-track roadbed (Fig. 8) is 33 ft. wide, crowned 5 in. at the middle, and in cuts the width between slopes is 41 ft. The ties are 7 by 9 in. by $8\frac{1}{2}$ ft. Ballast is 27 ft. wide for stone and $27\frac{1}{2}$ ft. for washed gravel. Both materials have a level surface 1 in. below the top of the tie and are rounded to form a

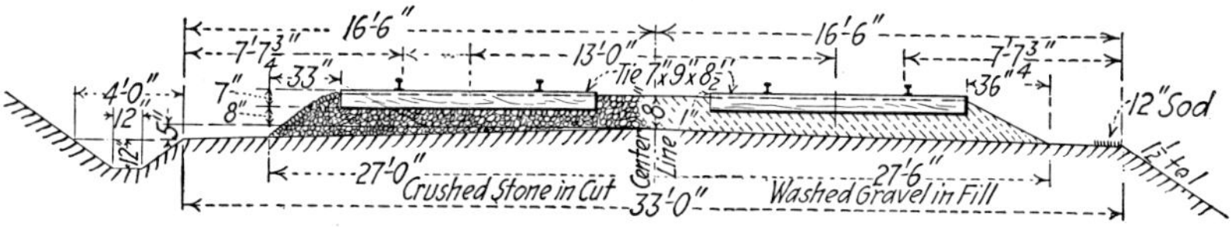

FIG. 8.—Delaware, Lackawanna & Western R. R.; double track.

ballast shoulder, the slope then leaving a wide berm or roadbed shoulder between ballast toe and edge of roadbed. On fills, a 12-in. strip of sod is laid along the outer edge of this berm. The ditch differs from that of the Pennsylvania R. R.,(Fig. 7) in being a defined trench, 12 in. deep below the edge of the roadbed and having a bottom width of 12 in. in earth and 27 in. in rock. On single track, however, the roadbed has a

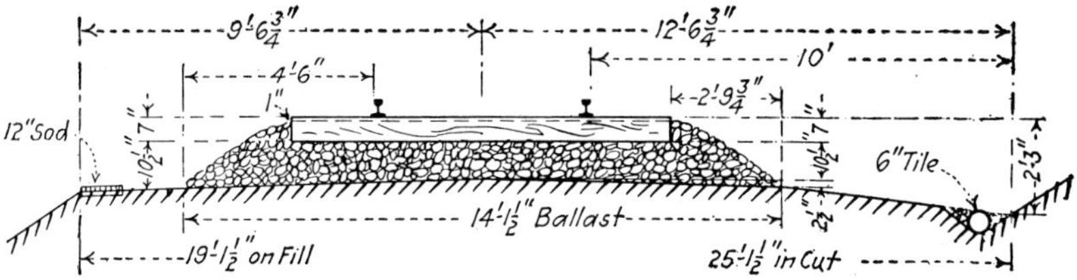

FIG. 9.—Delaware, Lackawanna & Western R. R.; single track.

flat slope to the toe of the cut, 27 in. below the base of the rail, but a 6-in. tile drain may be laid along the toe of the cut in a trench filled with broken stone, as in Fig. 9. The width of the roadbed for single track is 19 ft. $9\frac{1}{2}$ in. on fills and 25 ft. $1\frac{1}{2}$ in. between slopes in cuts, with a ballast width of 14 ft. $1\frac{1}{2}$ in., in which width the roadbed has a fall of $2\frac{1}{2}$ in. from the middle. Ballast is $10\frac{1}{2}$ in. deep under the middle of the ties.

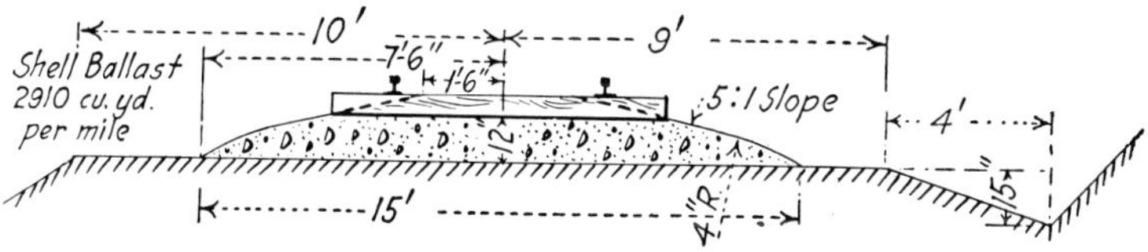

FIG. 10.—Kansas City Southern R. R.; shell ballast.

Kansas City Southern Ry.—For shell ballast this road has a form of cross section differing from those described above. As shown by Fig. 10, it is level with the tops of the ties for a width of 3 ft. and has then a slope of 1 on 5 ending in a curve of 4 ft. radius to the roadbed.

Masonry Bridges and Viaducts.—In arched structures with filling between spandrel walls, this filling is shaped to drainage planes, and tile drains lead to weeper holes at the haunches, extending either horizontally through the spandrel wall or vertically through the arch ring. The viaduct approach to the Thebes bridge over the Mississippi River has the space between the spandrel walls filled with red gravel compacted by flooding, its surface sloping down from the arch crown to the piers and from the walls to the center. In the longitudinal valley and covered by the ballast is a 6-in. tile drain connecting at each pier with a 6-in. vertical drain formed in the pier and having a lateral outlet just above the ground level. Where no filling is used, but a flat concrete deck is carried by spandrel columns, the drainage is a simple matter (see "Bridge Floors").

Tunnels and Subways.—In tunnels a good bed of clean, coarse ballast should be laid, and ample provision made to take care of water and keep it away from the ballast. In many cases a considerable flow of water has to be dealt with, requiring large drains or side ditches. In rock tunnels all drains may be laid in trenches and covered with gravel or broken stone.

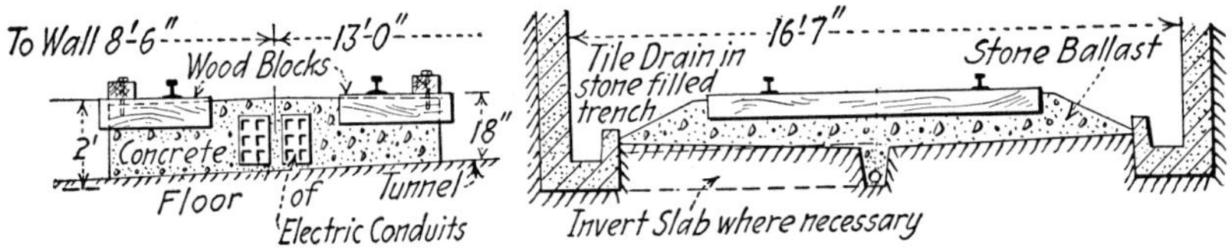

FIG. 11.—Roadbed in tunnels.
At left: concrete base; Delaware, Lackawanna & Western R. R.
At right: ballasted track; Union Pacific R. R.

For tunnels on the Union Pacific R. R. (Fig. 11), the standard practice is to use a tile drain in the center line and to form side ditches in the footings of the concrete walls by means of low curbs which hold the ballast. Where a concrete floor is used, it includes the trench and tile drain and has two low, curb walls to retain the ballast and form side ditches or gutters. In the double-track tunnel of the Baltimore Belt R. R., a box drain 24 in. wide and 16 in. deep, of dry masonry with side openings and top slabs, was built on the concrete invert, and stone packing filled to the level of the top, above which was the 24-in. bed of stone ballast.

Special systems of roadbed construction are employed for rapid-transit underground railways and for circular iron-lined tunnels. In the rectangular subways at New York, ordinary ballasted track is laid on the flat concrete floor, with about 6 in. of stone ballast under the ties and between curbs or footings 11 ft. apart. Where this width is greater than 11 ft. or where there are no footings between the tracks, the ballast is sloped 1 on 2 from top of tie. In the circular tunnels under the East

River, the concrete lining is shaped to form shelves supporting the ends of the ties, these supports being just outside the rails, as shown in Fig. 12. This construction is the direct opposite of a plan employed in some of the "tube" subways in London, where the ties rest on a central ridge of concrete and have their ends supported on stone ballast. On the New York lines, a half-tile drain is laid on the invert, and 3-in. stone ballast or round field stone is filled under and between the ties. On

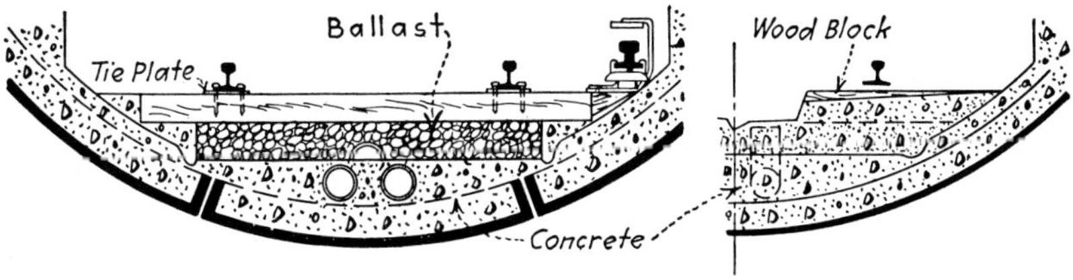

FIG. 12.—Track in New York tube subways.

curves the shelves are arranged to provide the necessary superelevation, with the ties raised at one end and lowered at the other. This construction is somewhat flexible and makes an easy-riding track. At subway stations the concrete floor is shaped to carry two rows of rail blocks with a broad drain between them, as at *E*, in Fig. 1. This permits ready cleaning and drainage, and in several cases the deep space between the rails has saved the lives of passengers who have fallen or jumped in front of trains.

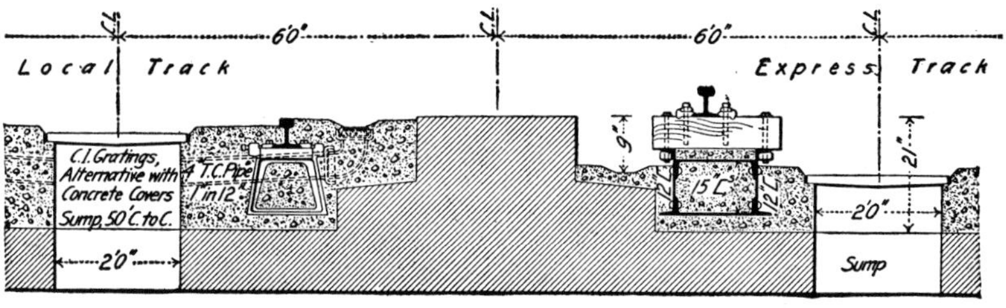

FIG. 13.—Track in Philadelphia subway.

In the flat-floor, rapid-transit subway at Philadelphia (Fig. 13), the local tracks have the rails supported on cast-iron chairs embedded in the concrete floor, and concrete is filled in to the rail heads. The express tracks have steel longitudinals embedded in the floor, each consisting of a pair of 12-in. channels back to back and 15 in. apart. To the flanges of each longitudinal are bolted wood blocks 6 by 10 in., 24 in. long, to which the rails are spiked. The entire floor can be flushed and washed.

Since tunnel dimensions often permit only a thin bed of ballast, and the maintenance of ballasted track in tunnels is difficult and expensive, plans have been proposed for unballasted construction. The plan carried out in the double-track Bergen tunnel of the Delaware, Lackawanna & Western R. R. (Fig. 11) provides two parallel concrete beds 8 ft. wide and 13 ft. c. to c., built on the crowned rock floor of the tunnel so that the thickness is about 18 in. on the inner, and 24 in. on the outer, side. In pockets along each side are fitted creosoted tie blocks about 8 by 8 in., 30 in. long, spaced 22 in. c. to c., and secured in the pockets by side wedges. The 90-lb. rails are held by screw spikes and clips. Outside guard timbers 5 by 8 in. are secured by bolts screwed into the tie blocks and by anchor bolts embedded in the concrete. In the middle of each concrete roadbed block are embedded conduits for electric wires and cables, which are accessible by manholes at intervals.

CHAPTER III

TIES AND TIE-PLATES

Ties or cross-ties laid on the ballast to carry the rails, have other functions to perform besides serving as supports and attachments for the rails. They have to maintain the gage; to act as transverse beams, and to distribute the load from the base of the rail over a large area of the ballast. This load distribution is not uniform along the length of the tie, but is concentrated in a distance of 14 to 16 in. on each side of the center line of each rail. It is under these portions of the tie that the ballast has to be firmly tamped. As the load is applied on the rails, the tie bends vertically, so that its top is convex at the center and concave in the end portions. The extent of the bending depends largely upon the character and condition of the ballast, being accentuated if the track is centerbound or has a hard ridge of compacted ballast under the middle of the ties. A general depression in the track under load occurs even in well-tamped ballast, while with loose ballast there may be play or movement between the ties and the ballast.

Ties and Timber Resources.—The importance of the question of supply of railway ties is shown by the enormous amount of timber involved. With an average of 2,700 ties per mile, the 400,000 mi. of track on 250,000 mi. of railway in the United States represent 1,100,000,000 ties in place. Electric railway track, with 47,280 mi., brings the total to about 1,234,000,000 ties. An annual addition of about 3,850 mi. of track for steam railways (including yards and sidings), and 475 mi. for electric railways, requires about 11,500,000 ties. Annual renewals at 200 ties per mile for steam railway track require 91,000,000 ties, besides 6,000,000 for electric railway track at 125 per mile. In addition there is about 271,100,000 ft. b. m. annually for bridge ties and switch timbers, or the equivalent of about 9,000,000 ties.

Thus the aggregate annual consumption is approximately 117,500,000 ties for additions and renewals. Taking 120,000,000 ties in round numbers, at 12 cu. ft. per tie (including all waste), this total represents 1,440,000,000 cu. ft. of forest-grown timber. In stands of young timber of uniform age the annual cut may be 80 ties per acre, but including the cutting in farm woodlots and among saw-timber trees, it is estimated by the U. S. Forest Service that probably 15,000,000 acres of timber are culled annually to produce the tie supply.

To grow the annual supply of tie timber under present conditions, with an estimated growth rate of 24 cu. ft. per acre, would require 60,000,000 acres or 24 per cent of the present growing area of 250,200,000 acres. Under ideal conditions of intensive forest management with a yield of 55 cu. ft. per acre, the required area for annual tie production would be 26,200,000 acres or 5.6 per cent of the future growing area of 469,500,000 acres, exclusive of Alaska. The difference between the two growing areas is represented by: (1) areas of mature standing timber in which growth is offset by disease and decay; and (2) by 81,000,000 acres of devastated lands producing no forest growth.

For many years the total forest area has been in course of reduction by reckless use, destruction, and fire waste. Some steps have been taken to protect the remaining resources, but those for reforestation are negligible. The Pennsylvania R. R. has plantations aggregating about 1,000 acres planted with black locust, chestnut, tamarack, pin oak, red oak, and Scotch pine for ties (to be cut in 40 to 60 years); also catalpa and locust for fence posts (15 to 30 years). The red oak and pine were selected because their growth is more rapid than white oak, while if protected by preservatives and metal tie-plates, these woods will give about twice the life of untreated white oak ties.

Wood for Ties.—As local sources of timber become exhausted, the railways have to secure their supplies from more distant fields, and as the haul from point of supply to point of use now averages 700 miles, the cost of transportation is a factor in cost of ties. Woods are used in about the following proportions: oak, 52 per cent; southern pine, 14; Douglas fir, 7; cedar, 5; chestnut, 5; cypress, 4; tamarack, 3 per cent; with 1.5 per cent each for western yellow pine and lodgepole pine; 1 each for western larch, beech, maple, hemlock, and redwood; 0.5 each for gum and birch, and 1 per cent for all other kinds. Most woods have various subdivisions, such as the numerous kinds of oaks and pines, black and honey locust, red and yellow fir, and black and white walnut.

In the specifications of the A. R. E. A. the following woods are listed, those in each group having similar characteristics so that they may be mixed together (other species are used locally):

1. Ties which may be used without treatment (these are all "heart" ties); catalpa, cedar, chestnut, cypress, fir (Douglas), locust (black), mulberry (red), oak (white), pine, redwood, sassafras, and walnut (black).

2. Ties which should be treated (those marked * are "sap" ties); group 1, ash, hickory, black locust,* honey locust, red oak, white oak,* black walnut;* group 2, cedar,* cypress,* Douglas fir,* hemlock, larch, pine,* redwood; group 3, beech, birch, cherry, gum, hard maple; group 4, catalpa,* chestnut,* elm, hackberry, soft maple, mulberry,* poplar, sassafras,* spruce, sycamore, white walnut.

Oak and Pine Ties.—White oak has long been regarded as the best wood for untreated ties, as it ranks high in both hardness and durability, but so much of the best oak has been cut that the quality of white oak ties in general is much lower than is usually assumed, and some other timbers have greater durability. For this reason and because of the increasing proportion of sapwood in many of the ties now delivered, there is a growing tendency to treat even white oak ties to increase their economic life. Its varieties include burr, post, and rock oak. Chestnut oak is sometimes classed with white oak, but belongs rather to the red oak class, which is of lower durability. This latter class includes black, pin, red, water, and yellow oak.

Pine is used extensively in its numerous varieties; yellow and white pine are the best, for although soft, they resist decay and ties will last 5 to 7 years under heavy traffic or up to 9 years under light traffic. Southern yellow pine includes such varieties as longleaf, shortleaf, loblolly, pond, slash, and spruce pine. Shortleaf southern yellow pine resembles the Baltic fir largely used for ties in Europe, but the latter is harder as it is grown in a colder country. Tapping pine for turpentine has little effect upon the quality of the timber. Of the pine ties produced, more than 75 per cent are of southern yellow pine, about 20 per cent western yellow pine and lodgepole pine, and 5 per cent white and Norway pine. Longleaf pine is generally preferred to oak for bridge ties, as it is stronger and stiffer and less inclined to warp and twist when exposed. Oak and pine are usually preferred for switch ties.

Various Woods.—Chestnut is equal to white oak in durability, but is not so hard and has a tendency to split. Cedar is very durable, but being soft it does not hold spikes well and is soon cut by the rails if tie plates are not used. Hemlock and tamarack also require similar protection. Redwood (used mainly in the Pacific states) and cypress (in the South) are both durable but very soft, their life ranging from 5 years without tie plates to 12 years if thus protected. Cypress is so soft that traffic will drive tie-plates down into the wood, but a life of 20 to 25 years is known under light traffic. Gum has been objected to on account of rapid decay, promoted by excessive checking, but with special attention it can be used successfully. On the Santa Fe Ry., which uses sweet, black, and cotton or tupelo gum, the ties are brought out of the woods as soon as they are cut, stacked 2 by 7 on the right-of-way for seasoning and inspection, and then sent to the tie yards promptly. Before being creosoted they are adzed and bored and have a slice cut off each end to show if the wood is sound. The treated ties are handled carefully with tongs and are protected by tie-plates.

Old and Second-growth Timber.—Second-growth timber may or may not be equal in quality to the original stand which it replaces, depending upon the conditions. First-growth timber has grown usually in a standing forest, while second-growth timber is more or less scattered over the cleared area, so that it has more moisture and sunshine, resulting in more rapid growth and consequently a larger proportion of sapwood. In hardwood trees, the best wood is put on when a definite rate of growing is maintained, and rapid second-growth material may be equal to that of the first stand in density and mechanical properties. But in conifers, such as southern yellow pine, the rapid second growth under open conditions is apt to produce wood of lower density and quality. Second-growth timber in such open conditions is also likely to have more limbs and branches and, therefore, more defects. Density is, in general, the guide to quality in timber.

Ties from old trees are sometimes considered the more liable to decay. But young trees and the ties made from them have usually a larger proportion of sapwood, which is not durable. Ties cut from dead trees killed by insects or fire are usually as good as those from live trees, if the wood has not been injured subsequently by decay or insects. In dead trees left standing too long and in logs cut from live trees but left in the woods, the sapwood (and eventually the heartwood), may decay or be infested by boring insects. But sound wood from dead trees should be as strong and durable as that from live trees.

In a living tree, the heartwood and part of the sapwood is actually dead, so that specifications should not exclude dead trees but should rather specify a limit of decay or insect attack. As to sap conditions, there is little difference in moisture content of green timber in summer and winter. But if timber cannot be hauled out of the woods promptly it is better to cut it in the fall or winter, since it is then less liable to attack by fungus and insects and less likely to check in seasoning. It will also be partly seasoned and less susceptible to attack when warmer weather comes.

Hewed and Sawed Ties.—In recent years the use of sawed ties has increased so rapidly that they will soon represent nearly 50 per cent of the output. Opinion favorable to hewed ties is due largely to the fact that hewers are likely to select the best material and leave inferior sticks which are difficult to shape by hand but may readily be shaped by sawing. The idea that the action of the axe or adze closes the pores of the wood is erroneous; on the contrary, the axe marks or score marks afford more lodgement for moisture than the woolly surface left by sawing. Hewed ties have an advantage in greater volume and wider bearing area for a given size, but a tree will produce more sawed ties than hewed ties, as the waste in hewing is considerable.

Sawed ties are handled more easily and with less labor cost; their uniform dimensions facilitate laying and renewal, and they do not require adzing in track to give a level seat for the rail. Quality of ties depends upon inspection of product rather than upon the method of production.

Foreign Timbers.—Importation of hardwood ties from Australia and South America has been suggested, but the few experiments have not given promising results. The supply from the former is limited, while as to the latter there is little definite knowledge of the woods themselves and no organization to insure a supply of ties that would be acceptable as to size, species, or quality. Furthermore, the experiments indicate that ties which are satisfactory in tropical countries are likely to fail by splitting or decaying in this country. Of Australian timbers, jarrah lasts about 15 to 20 years; ironbark, 20 to 30; and red gum and stringbark, 20 years. In Cuba and the Philippine Islands, several kinds of native hardwoods last from 5 to 25 years, without treatment or tie plates, but under light loads and traffic. Many of these timbers are so hard that they have to be bored for spikes.

Specifications for Ties.—Good specifications closely adhered to and supported by careful inspection are important in the interests of the railways. In fact, with uniform specifications, joint inspection has possibilities, each inspector acting for all the railroads purchasing ties at the point where he is stationed, and thus avoiding difficulties due to competitive purchases.

Under the specifications of the A. R. E. A., ties must be free from decay, large splits or shakes, large or numerous holes or knots, or grain with a slant of more than 1 in 15. For straightness, a line along the top between the middle points of the ends must be entirely within the tie, while a similar line along the side must be more than 2 in. from top and bottom at all points. The thickness should not vary more than $\frac{1}{2}$ in. between the two sides or the two ends. Extracts from these specifications are given below.

Ties from needle-leaved woods shall be of compact wood throughout the top fourth of the tie, where any inch of any radius from the pith shall have not less than one-third summerwood in six or more rings of annual growth, or not less than one-half summerwood in fewer rings. Ties of coarse wood having fewer rings or less summerwood will not be accepted unless specially ordered. Ties for use without preservative treatment shall not have sapwood wider than one-fourth the width of the top of the tie between 20 and 40 in. from the middle, and will be designated as "heart" ties. Those with more sapwood will be designated as "sap" ties. All ties, except those of . . . shall be made from trees which have been felled not longer than 1 month. Ties shall be straight, well hewed or sawed, cut square at the ends, have bottom and top parallel, and have bark entirely removed.

Defects in Wood.—A shake is a separation of the annual rings and should not be more than $\frac{1}{4}$ in. wide or 4 in. long. A split is a break across the rings and should not be more than 10 in. long unless an anti-splitting device is used. Checking or end splitting is likely to occur with rapid drying of ties containing excessive moisture, and may be prevented by driving S-shaped or channel-shaped pieces of strap iron into the ends of the ties. Where such checks have opened wide, the Northern Pacific Ry. closes them by means of a 10-ft. lever having a projecting shoe which fits under a log chain wrapped around the end of the tie. When the separated parts are thus drawn together, they are secured by driving the S-irons.

Names of Ties.—A heart tie has no sapwood wider than one-fourth the top width at any point between 20 and 40 in. from middle of tie. A sap tie has sapwood wider than the above. A pole tie is made from a tree whose diameter or cross-section is sufficient for only one tie. A slab tie is made from the outside of the log. A split tie is made from a cross-section sufficient to yield two or more ties. A half-round tie is sawed or hewed on top and bottom only, with the bottom of greater width. A halved tie is sawed or hewed on all sides and with the pith of the tree near the bottom at about midwidth. A quartered tie is similar to the preceding but with the pith at or near one lower corner. A "doty" tie is infected with fungus; in cypress it is called a "pecky" tie. A cull is a tie which does not meet the requirements of the specifications. Some railroads select hardwood ties for curves and place softer woods on tangents.

Protection from Decay and Wear.—Ties are subjected to two destructive influences: decay and mechanical wear. The latter includes the cutting and disintegration of the surface of the tie by the rail and the cutting of the wood fiber by the spikes. Both forms of wear favor decay by the admission of moisture. This latter effect is aggravated by continual redriving of spikes, which are gradually lifted by the vertical motion of the rail, forced outward by thrust on the rail or forced laterally when they are driven in slots of the angle bars to prevent creeping of the rails. On joint ties the excessive wear may reduce the life to 50 or 75 per cent of that of intermediate ties.

Great economy in the use of ties is being effected by the use of preservative processes to prevent decay and of metal tie-plates to prevent wear. The increased life thus obtained also gives a more permanent track, requiring less maintenance and less frequent disturbance of the ballast bed for tie renewals, which is important also in view of the increasing cost of labor. Such protective measures enable cheap and inferior woods to be used economically. Additional economy results from close adherence to good specifications; also from careful inspection of new ties for acceptance and of old ties proposed for renewal.

Life of Ties.—Classification of tie timbers by length of life is of little value, since their life in track varies so greatly with the different varieties and grades of even the same species, and with the varying conditions of roadbed, track, traffic, and climate. In addition, records of life and kind of tie are incomplete and of only slight value for direct comparison, although this condition is changing. For general purposes, however, the average life of untreated ties is given in Table I; this table also gives comparisons in which the several species are graded in relation to white oak, rated at 100.

TABLE I.—PROPERTIES OF WOOD TIES

	Average life, years	Ratio to white oak = 100		Crushing strength, pounds per square inch
		A. R. E. A.	U. S. F. S.	
Hardwoods:				
Ash	4– 5	72	40– 55	828
Aspen	3– 4	...	25– 35	
Basswood	4	...	30– 40	209
Beech	4– 6	70	40– 50	607
Birch	4– 5	68	35– 50	454
Butternut	....	21	50– 70	258
Catalpa	8–10	...	125–175	
Cherry	4– 8	90	....	444
Chestnut	6–10	101	100–120	
Cottonwood	3	...	30– 40	
Elder	....	...	25– 35	
Elm	4– 5	68	50– 75	350– 696
Gum, black, cotton (tupelo)	3– 4	...	30– 50	451
Gum, red	3– 5	62	65– 75	
Hackberry	4	32	....	525
Hickory	3– 5	...	40– 56	900–1,100
Locust, black, yellow	9–12	214	150–250	1,426
Locust, honey	7–10	32	80–100	1,684
Magnolia	....	...	40– 50	
Maple, hard	4– 5	47	40– 50	742
Maple, soft	3			
Mulberry, red	....	85	150–200	
Oak, white oak group	8–10	100	100	853
Oak, red oak group	3– 5	51	40– 55	778
Oak, burr	8–10	...	....	836
Oak, chestnut	6– 8	...	70– 90	
Osage orange	....	...	200–300	2,260
Poplar, yellow	5	...	40– 55	
Sassafras	5– 7			
Sycamore	3– 4	37	35– 45	433
Walnut, black	8–12	141	100–120	
Willow	....	...	30– 40	193
Soft woods (conifers):				
Cedar, eastern red (juniper)	6–12	...	150–200	
Cedar, southern white	....	...	80–100	
Cedar, other species	....	156	125–175	
Cypress, bald	5– 9	121	125–175	
Cypress, black	8–12	...	....	548
Fir, Douglas	6– 9	75	75–100	427
Fir, true	3– 4	90	25– 35	
Hemlock	4– 6	63	35– 55	
Larch, western	7– 8	90	75– 85	
Pine, jack	3– 4	96	35– 45	
Pine, loblolly, lodgepole	4– 5	41	35– 50	
Pine, longleaf, slash (Cuban)	6– 8	...	75–100	491
Pine, Norway	....	...	45– 60	
Pine, pitch, sugar	....	...	45– 65	358
Pine, shortleaf	....	...	60– 80	410
Pine, southern yellow	6–10	77	80–100	
Pine, white	6– 7	...	70– 90	314
Pine, western white	3– 6	...	65– 80	
Pine, western yellow	4– 5	71	35– 50	345
Redwood	5– 6	112	125–175	578
Spruce, red, 262; white	5– 7	77	35– 50	545
Tamarack	3– 6	63	75– 85	450
Yew, western (Pacific)	....	...	170	

These comparisons are for decay only and would be very different if based on hardness, since the softer ties (without tie plates) would not attain the normal life before being rendered useless by mechanical wear. Thus, as indicated in the table, oak, locust, and walnut are both durable and hard; cedar, cypress, and redwood are durable but soft; beech is hard, but not durable, while hemlock is neither hard nor durable. Further changes would be caused if allowances were made for the use of tie plates and preservative processes. In this table there is no definite relation between the columns of years and percentages, the figures of life in years representing experience reported from various individual railroads.

Size of Ties.—Since the tie acts more as a beam than as a block or support, its stiffness and, consequently, its depth are important, while reduction in strength by excessive spiking must be considered. Ties 8 and 8½ ft. long are in general use. To secure proper distribution of load outside of the rails a length of 9 ft. may be desirable, especially where canted tie-plates give an outward diagonal direction of load. Although the tie does not develop load for about 4 in. under its extreme ends, this part of the length is necessary to confine the ballast under the adjacent part which does develop load.

Until about 1910 the common size for ties was 6 in. thick, 8 in. wide, and 8 ft. long, but since then there has been increasing use of greater thickness for ties in main track to secure higher beam strength in view of the greatly increased loads to be sustained. For main track, therefore, a size of 7 by 9 in. by 8½ ft. is becoming general. Where the width is not uniform, as in hewed ties, the wider ties may be selected for rail joints. Standard sizes adopted by the A. R. E. A. include six grades, varying by cross-sectional dimensions measured in the parts between 20 and 40 in. from the middle of the tie in both directions, as shown in Table II. For each grade the size differs in ties sawed or hewed on four sides (class *A*) or on two sides (class *B*). If both sizes of class *B* in grade 3 should be ordered, the 7 by 7-in. size would be designated as class 3-*A*. Ties of No. 6 grade, 7 by 10 in., are little used, and the width should not exceed 10 or 12 in. to allow of proper tamping. Ties exceeding specified dimensions by more than 1 in. in thickness, 3 in. in width, or 2 in. in length will be regraded or rejected.

TABLE II.—STANDARD SIZES OF TIES

Grade	Class *A*		Class *B*	
	Thickness	Top width	Thickness	Top width
1.......	none accepted		6	6
2.......	6	7	6	7
3.......	{ 6	8	6	8 }
	..	..	7	7 }
4.......	7	8	7	8
5.......	7	9	7	9
6.......	7	10	7	10

Tie Spacing.—With the ordinary 33-ft. rail, it is usual to employ 18 to 20 ties per rail length on main track and 16 to 18 ties on branch lines. This gives an average spacing of 22 to 20 in. in main track. At joints and shoulders the spacing is often closer than for intermediate ties, but a minimum clear spacing of 8 in. is required to give proper room for hand tamping, or 6 in. for machine tamping. With its three-tie supported joint, the New York Central R. R. spaces the three joint ties $14\frac{1}{2}$ in. c. to c., and the intermediates $21\frac{1}{8}$ or $24\frac{1}{2}$ in. for 20 or 18 ties per 33-ft. rail. With a two-tie suspended joint, the joint spacing is 16 in., with intermediate spacing of 26 in. for 16 ties per rail. On running and storage sidetracks there are 16 and 14 ties, respectively, for 30-ft. rails. This road uses $8\frac{1}{2}$-ft. ties, 7 by 8 to 7 by 10 in. for yellow pine and 6 to 7 by 7 to 10 in. for oak and chestnut.

This relatively close spacing of ties has been a subject of criticism by railway engineers in Europe, where a spacing of 30 to 34 in. is common. Thus the 45-ft. rails commonly used in England have usually 18 ties, giving an average spacing of 30 in. c. to c., with joint ties 22 to 25 in. c. to c. As these ties are only 5 in. thick, with a 10-in. width and 9-ft. length, this arrangement obviously provides less bearing area and beam strength than the American plan, while the ties give longer spans and greater deflection for the rails and less effective bearing area on the ballast. On English railways, however, wheel loads are considerably lighter than in this country.

Both rails and ties in track shift to some extent under the influence of traffic, thus disturbing the original relations at joints. If these relations are to be maintained or restored, considerable work may be required in respacing the ties. But some roads consider that no such relations are necessary, that it makes no difference whether the joint comes on or between joints, and that creeping may be prevented by the use of rail anchors.

Seasoning and Stacking Ties.—New or green ties should be seasoned for 6 to 12 months before being used or treated, and for this purpose they are stacked at a tie yard. Treated ties are best left to dry and season for 2 to 6 months; this is specially desirable for those treated with water solutions, but creosoted ties are sometimes loaded for shipment as soon as they come from the retorts.

New ties for inspection are usually stacked on the right-of-way, 8 to 10 ft. from the rail, in dry, shady, and airy places. They may be in alternate layers of two and nine or seven (known as 2 by 9 or 2 by 7), with the bottom layer carried on sills or blocks so as to be at least 6 in. above the ground. In the "two" layers the ties may project 2 or 3 in. beyond the others, and these ties, as well as the outer ties of the "seven" layers, may be placed on edge to give more air space. The stacks should have not more than 12 layers (54 to 66 ties) and should be 5 ft. apart for convenience of inspection. A single transverse tie may be placed below the top layer so that these ties will be inclined so as to shed water. Ties stacked on end are not acceptable.

For seasoning at the yards, the ties may be stacked as above, or sometimes 1 by 7 and 1 by 9, as shown in Fig. 14, with about 100 ties in each stack, but not more than 20 layers high and with 4 ft. spaces between the stacks for fire protection. For slow seasoning of ties having a large moisture content or treated with water-solution preservatives, a stacking 7 by 7 or 9 by 9 may be used to give restricted air spaces. Creosoted ties in storage may be stacked solid, or in close layers, 10 by 10, with a layer of earth on top for fire prevention. A close top layer with earth covering can also be used with untreated ties stacked along the track. To prevent decay and fungus growth, the yard surface should be cleared of rubbish and vegetation, and covered with a good bed of cinders.

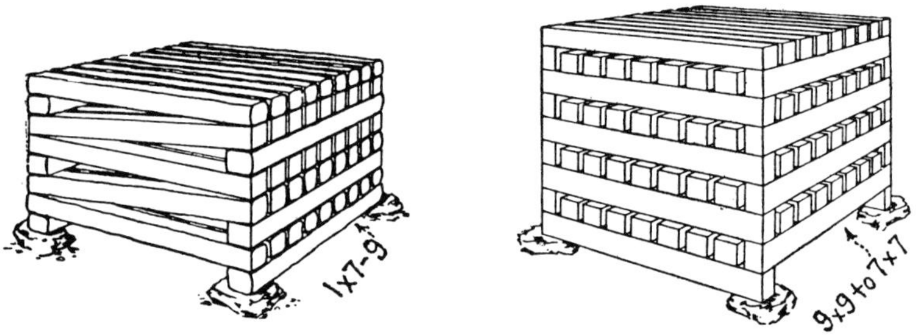

FIG. 14.—Ties piled for seasoning.

Machine-dressed Ties.—There is increasing use of mechanical appliances to dress ties level at rail seats of hewed ties and to bore them for spikes. All such dressing should be done before treatment. The tie, moved by conveyors, passes first between saws which take a thin slice off each end to facilitate penetration of the preservative and to show if the wood is sound. It then goes to an adzing machine with revolving cutters which face the surface at each rail seat; and then to a machine which bores the spike holes at required spacing. Finally, a branding machine marks the date at one end, and at the other end indicates the character of rail or tie plate for which the tie is bored. For Douglas fir ties, an additional machine with toothed rollers punches the surface with incisions about $\frac{3}{4}$ in. deep and 2 in. apart to assist penetration of the preservative in this dense wood. This mechanical treatment may be applied at a stationary plant or at a portable plant consisting of a large box car fitted with the machines, engine, and boiler.

Longitudinals.—Longitudinal timbers as supports for the rails have been used occasionally on bridges and in large passenger stations, where they have a rigid support (see "Track on Elevated Structures"). Steel longitudinals are used in one of these stations; they were used extensively on open track in Europe at one time and have been tried on a small experimental scale on the Pennsylvania R. R., as noted in Chapter I.

Tie Plugs.—When spikes are drawn from the ties during track work, the holes should be filled at once with creosoted wood plugs, whether or not a new spike is to be driven. Machine-made plugs, which are preferable, are made from elm, white ash, basswood, birch, and sometimes oak or Douglas fir. An alternative method devised in Australia is to fill the hole with a hot bituminous composition, into which the new spike is driven and which then hardens so as to hold the spike effectively.

Old Ties.—Decayed ties removed and discarded as useless (after due inspection) are usually burned on the right-of-way. To a limited extent they are used for fence posts and cribbing, or they may be cut up by saws or shears to serve as fuel at roundhouses, although it does not pay to transport them far. Section men may generally have the ties for firewood, but it is rarely advisable to sell them to farmers, as their removal involves trespassing which may lead to accidents or the taking of new ties. For burning, the ties should not be in large piles or too close to the tracks, as the heat may injure the varnish of cars. Ashes of fires and refuse at the saws or shears should be raked over to recover spikes and other iron which has a scrap value.

PRESERVATIVE TREATMENT

Purpose and Economics.—The principle of applying preservative processes to timber is to fill or coat the wood cells with a material which is toxic or poisonous to the germs or fungus growths causing decay. The economic and practical advantages of applying such treatment to wood ties to increase their life have been proved abundantly, but until recent years the general use of treated ties made but slow progress in this country. A leading cause of this situation may be seen in the higher first cost due to treatment, as expenditures and appropriations are apt to be considered largely on the basis of cost rather than economy. As to whether the excess cost for treatment should be charged to capital account or maintenance, there is difference of opinion, but in the accounting system of the Interstate Commerce Commission this cost is chargeable to maintenance. About 50 per cent of the new ties laid since 1920 are treated, but in future this proportion should be increased materially.

Life of Treated Ties.—Since this treatment protects the ties against decay and thus gives them a longer life, it has resultant effects in reduced annual purchases for ties, reduced track work for tie renewals, and improved condition of track due to its less frequent disturbance for such renewals. Annual tie renewals have been reduced 50 to 60 per cent in this way. Furthermore, treatment enables softer and cheaper kinds of wood to compare to advantage with more expensive ties of high class but untreated wood. Many woods averaging a life of only 3 to 6 years in track will last 15 to 20 years when treated. In one case where it was estimated that treatment increasing the life to 8 years would be highly economical, the ties when treated actually lasted 15 years or more. For the cost of ties in track, first cost is only one item; allowance must be made for inspection, freight, treatment, distribution, placing, and perhaps tie-plates and fastenings.

Although untreated white oak has been considered the highest quality of wood for ties, much white oak gives a life of little more than 6 years, so that there is an increasing tendency to treat even white oak. Many woods are superior to white oak in both economy and efficiency if they are treated with preservative and equipped with tie-plates to protect them against decay and wear. In fact, tie-plates are usually desirable to obtain the full advantage of treatment, since a tie giving 5 years' service, if untreated, may last 8 years if treated, and 12 to 15 years if both treated and tie-plated. Cases of unsatisfactory results are due largely to defective wood or improper treatment, and are insignificant in comparison with satisfactory experience in all parts of the country. For example, of zinc-treated ties laid on the Chicago, Milwaukee & St. Paul Ry., in 1906, 85 per cent remained in 1923 and were expected to give a life of 22 years; of zinc-treated beech ties, 42 per cent remained after 12 years; untreated white oak laid in 1903 averaged 12 years, but of zinc-treated pine laid at the same time (though not tie-plated until 1911), 57 per cent were in track in 1923 and were expected to average 22 years of life. On the Chicago, Burlington & Quincy R. R., with a score of wood species and with three different treating processes, 80 to 96 per cent of the treated ties were in use after 11 years, while only 6 per cent of the untreated ties remained.

Extensive experiments and experience on the Baltimore & Ohio R. R. indicate an average life of 16 years for treated ties, as against 9½ years for untreated ties. In test tracks on other roads, where untreated white oak ties have lasted only 5 years, treated ties have given a life of 10 years for red oak, hemlock, beech, and gum, and 10 to 15 years for hard maple. Of untreated pine ties, 93 per cent were removed within 10 years, while only 10 to 30 per cent of similar ties treated with zinc tannin had been removed. In a lot of fir ties, 60 per cent of those untreated had been removed in 6 years and all were gone in 8 years, while of zinc-treated ties, 85 per cent were in use after 15 years and 55 per cent after 20 years' service. Of 24,000 creosoted ties laid in a test track on the Santa Fe Ry. in 1906, only 6.67 per cent had been removed in 1923; the others had already a life of 17 years and were expected to average more than 20 years. Again, with red oak and water oak, giving ordinarily a life of only 3 or 4 years, ties have lasted 10 to 15 years with zinc treatment and 15 to 20 years when creosoted. Removal of treated ties does not necessarily indicate decay or failure of the treatment, since they may have been damaged by fire or derailment. Thus, in 12 years service on the Chicago, Burlington & Quincy R. R., 88 per cent of the untreated ties were removed on account of decay and 7 per cent for other causes, while of different kinds of treated ties, from $1\frac{1}{2}$ to 12 per cent were removed for decay and $4\frac{1}{2}$ to 13 per cent for other causes.

Preservatives.—Creosote oil (or dead oil of coal tar) and zinc chloride are the preservatives mainly used. The former is the more effective, but also the more expensive, and with the full-cell process the cost may be too high in proportion to the gain, since the ties may be worn out by spiking and rail cutting before the wood has reached its limit of life. In this process, all the creosote forced into the wood is left in place. To reduce the cost, the empty-cell process has been devised, which extracts some of the oil forced into the wood cells but leaves the cells thoroughly coated. This empty-cell process may be applied also with the zinc treatment.

The cost and availability of creosoting are limited by manufacturing conditions, the domestic production of creosote oil (as a by-product) of the required quality being below requirements, and the importation of foreign oil being irregular. On the other hand, there are geographical or climatic limits to the use of zinc, since this is a soluble salt which is likely to leach out of the ties in damp climate. The leaching is not so rapid as is sometimes supposed, and may be checked or prevented by auxiliary treatment with oil. In general, however, zinc-treated ties are not considered advantageous in the eastern and southern states on account of the relatively high rainfall and humidity. The zinc solution is said to cause corrosion of spikes, but investigation indicates that this may be due to various other causes, such as cinder ballast, brine dripping from refrigerator cars, or the chemical composition of the steel.

Petroleum or crude oil has been tried extensively, but mainly in combination with other materials, as it has little preservative effect. On the other hand, it tends to reduce checking. When added to creosote, it reduces the cost and makes the solution more fluid, but there must be sufficient creosote to kill the germs which cause decay. On the Santa Fe Ry. and the Southern Pacific Ry. a mixture of 70 per cent creosote and 30 per cent petroleum is applied by the Rueping empty-cell process, leaving about 7 lb. of mixture per cubic foot of timber. But the Philadelphia & Reading R. R. has used 30 per cent of creosote to 70 per cent petroleum, with a retention of 3 lb. of creosote per cubic foot. In the hot climate of India, ties treated with a mixture of equal parts of creosote and petroleum were in good condition after 10 years, and experiments are being made with mixtures having 67 and 25 per cent of petroleum. To prevent the leaching of zinc chloride, ties have been given a supplementary treatment with petroleum and other materials in order to seal the preservative in the wood, while one of the latest developments is the use of an emulsion of zinc chloride and petroleum in order to eliminate the expense of applying separate treatments.

Processes of Treatment.—For any preservative, the ties are treated or impregnated under pressure in a horizontal steel cylinder or retort, 6 to

8 ft. in diameter and 120 to 200 ft. long, having a narrow-gage track inside for a train of small steel cars or trams loaded with ties. The ties should be seasoned for 3 to 6 months before treatment, and also for 2 or 3 months after treatment to let the ties dry and allow chemical actions to take place. With zinc-treated ties, this latter seasoning increases the strength and resistance to wear, reduces leaching, and also reduces the disturbance of signal circuits by leakage of current from rails to ties.

All cutting and boring of the ties should be done in advance, as otherwise such work removes wood that has been treated at some expense and exposes less fully treated surfaces to influences which may induce decay. Such work on treated ties may reduce their life by 30 to 50 per cent. For this reason the Northern Pacific Ry. furnishes section gangs with creosote oil to be applied when it is necessary to adze or trim the rail seats of ties in the track. Boring spike holes and slicing the ends of ties, as described already, also assists the penetration of the preservative into the wood. Following is a summary of different processes, as applied to ties in the treating cylinder.

Creosote (Full-cell).—After an initial vacuum of 22 in. to free the wood of air and moisture, the creosote oil is introduced without breaking the vacuum. Pressure is then raised and is maintained at about 125 lb. for sufficient time to give the required absorption, which may vary from 10 lb. per cubic foot for ties with little sapwood to 24 lb. for piling which has wide sapwood. The temperature of the preservative is 170 to 200°F., averaging at least 180°F. After pressure has been released and the cylinder emptied, a 22-in. vacuum is created, so that the ties will not be dripping wet when removed. For piling and bridge timbers this full-cell process is generally required, giving 14 lb. of oil per cubic foot for timbers and 16 or 24 lb. for piles used on land and marine work, respectively. All such material is air seasoned before treatment.

Creosote (Empty-cell).—In the Rueping process, the ties in the empty retort are first subjected to air pressure for 30 min. to force and compress air in the wood cells, this pressure varying from 40 to 60 lb. for red oak and 70 to 90 lb. for pine with a large proportion of sapwood. Then the preservative is introduced and a pressure of about 150 lb. is maintained to give the required absorption. As the pressure is released, the expanding air forces out surplus oil and then a 22-in. vacuum is created, as noted above. This process is designed to give an average of 5 lb. of oil per cubic foot. The Lowry process differs from the above in omitting the air-pressure stage and relying upon the air compressed by the oil pressure to expand and force out surplus oil when this pressure is released. It is designed to give an average of 6 lb. per cubic foot for ties and 10 lb. for other material. As operated on the Northern Pacific Ry. for 7-lb. retention (0.8 gal.), the cylinder is filled with oil at 170 to 185°F. in 3 min.; pressure is pumped up to 100 or 150 lb. and held for 1 to 6 hr., according to the kind and condition of ties. Then the solution is drawn off in 10 to 15 min. and a vacuum of 22 to 24 in. is maintained for 1½ hr., the time of treatment ranging from 4 to 6 hr. This is for birch, red oak, hard maple, and rock elm ties, which if untreated have a life of only about 3 years, but which, when treated and equipped with tie-plates, last about 20 years.

Zinc-chloride.—In the Burnett process, or burnettising, air-seasoned material is steamed from 1 to 2 hr. at not more than 20 lb. pressure, and then a 22-in. vacuum is maintained for about 30 min. to free the wood of air and moisture. After draining the condensate and while the vacuum is held, the preservative is introduced and a pressure of 100 to 125 lb. is applied and maintained until the required absorption is obtained. When the solution has been drawn off, a vacuum is created to help dry the wood. The solution is at a temperature of 130 to 190°F., averaging at least 150°F. Its strength is not to exceed 5 per cent and the absorption should average 0.4 to 0.5 lb. of dry zinc chloride per cubic foot. Experiments on the Union Pacific Ry. indicate that with lodgepole pine and Douglas fir ties, the penetration is increased 50 to 100 per cent when the temperature of the solution is 175 to 200°F.

Zinc-creosote.—In the Card process, using a mixture of zinc chloride and oil, the solution is proportioned to give an average of 0.5 lb. of dry zinc chloride and 3 lb. of creosote oil per cubic foot. The steaming, vacuum, and pressure are the same as in the straight zinc process, but the temperature of the solution averages 180°F. The Rueping process may be applied with a zinc-chloride solution followed by the creosote oil.

Zinc-tannin.—In the Wellhouse three-stage process, the first treatment and absorption are the same as in the straight zinc process, but with a minimum temperature of 140°F. After the zinc solution has been drawn off and the ties have been left for 15 min. to drain, a 2 per cent solution of tannic acid is held at 100 lb. pressure for half an hour, being then withdrawn and replaced by a 1 per cent solution of glue for the same pressure and time, the glue and tannin serving to seal up the pores or cells and prevent leaching of the zinc. A two-stage process used at one time by the Rock Island Lines consisted of a first treatment with a solution of 1½ to 3 per cent zinc chloride and 0.5 per cent glue, applied at 125 lb. pressure for 3 to 6 hr.; after draining the cylinder, a tannic-acid solution of 0.5 per cent was applied at 125 lb. for about 2 hr. to fix the glue. This process was for hemlock and tamarack ties having a life of 5 to 6 years untreated.

Preliminary Steaming.—There is difference of opinion as to the value of steaming ties in the cylinder, preliminary to treatment. Experiments on the Union Pacific Ry. indicate that 2 hr. steaming increased the absorption and depth of penetration on lodgepole pine but not in Douglas fir; if extended for 4 hr. it caused serious checking of heartwood. The steaming can be omitted if a solution of higher temperature is used, averaging 200°F. This effects a considerable saving, but with such a temperature the pressure should not exceed 140 to 150 lb. to avoid injury to the wood. In other experiments, steaming was found to increase the absorption by 12 per cent in red oak and 50 per cent in sap pine. The specifications of the American Wood Preservers Association provide that if steaming is required by the purchaser, it is not to exceed 30 lb. pressure, 10 hr. time, or 274°F. temperature, the maximum pressure and temperature not to be reached in less than 2 hr. A 22-in. vacuum is then to be maintained for 15 min. Steaming is used frequently with green piling for marine purposes, where an impregnation of 24 lb. of creosote per cubic foot is desired.

METAL TIE-PLATES

Purposes of Tie-plates.—As unprotected wood ties soon begin to deteriorate at the rail seats, owing to the abrasion and cutting which result from the continual motion of the rails, with consequent entrance of moisture to aid decay, there has been an extensive introduction of metal tie-plates which are placed between the rail and tie in order to distribute the load over a larger area of the tie and to protect the wood from wear by the motion of the rails. In this way and at small expense, the life of ties of durable but soft timber (treated or untreated) is increased materially, with consequent reduction in maintenance work and in disturbance of track for renewals. Where unprotected ties have had to be renewed in 3 years on account of rail cutting, ties of the same kind but fitted with tie-plates have lasted 6 to 9 years.

These plates also prevent the widening of gage which occurs, particularly on curves, from the outward tilting of the rail as lateral pressure causes the outer edges of the rail base to cut into the tie. In this way they eliminate much work in gaging and spiking. Tie-plates are also a safety factor, since ties badly cut and frequently respiked are weakened and do not afford a secure grip for the track spikes. An additional advantage is that the plate ties together the inner and outer spikes, so that when the latter is subjected to lateral thrust it may be assisted to some extent by the pull of the plate against the inner spike.

With soft ties and heavy traffic, the plates may be applied on all ties. With hardwood ties, they may be necessary only on curves, to resist the cutting and overturning action of the rail. They may be placed on every tie on all curves where traffic is heavy, or only on curves of more than 3 deg. with medium traffic. In some cases they are used only at rail joints and on four to eight ties per rail; they are of particular value where the wear is severe, as at rail joints and turnouts and in yards and terminals.

Types of Tie-plates.—Tie-plates of various designs are made of steel, wrought iron, and malleable iron; the spike holes are arranged to suit individual rail sections and are sometimes notched to permit of spiking rails of two or more different widths. Flat-bottomed plates were first used, but being small and thin they buckled, shifted out of place, and made a noisy track. To avoid this difficulty, plates were made with sharp ribs or chisel-point lugs on the underside, so as to bite into the wood and make the ties self-attaching. Objections as to injuring the wood and permitting the entrance of moisture led to modifications of these details. Some modern plates are shown in Fig. 15.

In present practice, the plates are either flat on the bottom or have a few ribs (longitudinal, transverse, diagonal, or in both directions) only about $\frac{1}{8}$ in. in depth, so as simply to form depressions in the wood to hold the plate in place but without cutting the fibers. Thus a tie-plate 7 by $9\frac{1}{4}$ in. used by the Chicago & Northwestern Ry. is $\frac{1}{2}$ in. thick at the ends but $\frac{9}{16}$ in. at the middle, giving a camber of $\frac{1}{16}$ in. in the direction of the

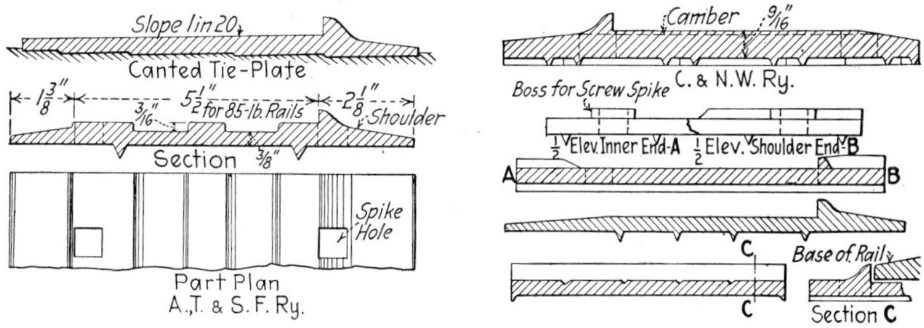

FIG. 15.—Metal tie-plates.

length of the rail. Diagonal blunt ribs $\frac{5}{32}$ in. deep extend in both directions across the bottom, while the top has a shoulder $\frac{3}{8}$ in. high to hold the outer side of the rail base. In any case the plates are of sufficient thickness to carry the load without buckling and to distribute the load on the rail over the entire area covered by the plate. With flat plates, separate spikes are sometimes used to attach the plate to the tie, independent of those which fasten the rail, but it must be remembered that too many spike holes will weaken both the plate and the tie.

Shoulder tie-plates, having a rib on the top surface to fit against the outer edge of the rail base, are used more generally than flat-topped plates, since they afford additional resistance to thrust and also relieve the spikes from both pressure and abrasion. In fact, in some cases the body of the spike stands clear of the edge of the rail. These plates are particularly adapted for maintaining gage at curves and are recommended by the A. R. E. A. as preferable to rail braces, except for frog-guard rails and for stock rails at switches. In some designs the shoulder laps over the rail base and serves as a fastening, but this arrangement is little used. To prevent rusting, especially on the under side, the New York Central R. R. is making extensive use of steel tie-plates having

0.25 per cent copper, the results more than compensating for the extra cost of about $2 per ton.

Tie-plates with inclined top surfaces to tilt or cant the rails inward have been introduced in recent years in experimenting with the European practice of canting the rails so as to be perpendicular to the coned surface of the wheel tread. In the standard M.C.B. (or A.R.A.) wheel design, this coning is 1 in 20 and the inclination of the tie-plate is usually the same. Some roads, however, use a coning of 1 in 38 or 1 in 13, and have the tie-plates made accordingly (see "Canting of Rails"). Tests on the Pennsylvania R. R. have shown little difference of wear of rails and ties fitted with level and inclined tie-plates, and the matter is still in the experimental stage.

Grooves or corrugations are sometimes formed in the rail seat of the tie-plate in order to reduce the amount of metal and to permit the escape of sand and dirt which may work between the rail and the plate. For use with screw spikes, the plates may have projections so shaped as to support the outer side of the mushroom head of the spike, thus resisting displacement and preventing the spike head from being bent over by the pressure of the rail. In certain cases these projections are formed at the backs of holes for cut spikes; they are also sometimes so made as to prevent the spike head from being driven down in close contact with the rail.

As to the best size of tie-plate, the A. R. E. A. specifies a width not less than 6 in. and as much more as the standard width of tie will permit, while the length should be not less than the safe-bearing area of the wood divided by the width of plate. The greater proportion of the length is outside the center line of the rail, to provide sufficient bearing to resist the overturning tendency already mentioned. Thickness will be proportional to the length in order to prevent buckling and to insure proper distribution of load on the tie. In general, the width is 6 to $7\frac{1}{2}$ in. and the length 9 to $10\frac{1}{2}$ in., with a thickness of $\frac{3}{8}$ to $\frac{5}{8}$ in. and a weight of 6 to 8 lb.

Special Forms of Tie-plates.—Heavy cast base plates or rail chairs with bolt fastenings have been tried, but with results not commensurate with cost, each plate weighing about 16 lb. Wood tie-plates, tried for economy, have given unsatisfactory results, partly by cutting into the tie but mainly by crushing and splitting, especially where the surface of the tie was uneven. They would also shift unless nailed to the tie. These plates were largely of beech 5 by 8 by $\frac{3}{8}$ in. and creosoted white oak 6 by 7 by $\frac{1}{4}$ in. Felt pads set in pockets on metal tie plates have been tried in terminal stations to reduce noise. They are used also on some English railways, the pads being placed between the ties and the cast-iron chairs.

Among special forms of tie-plates combined with rail fastenings is a design having two shoulders, fitting the width of rail base, the outer one hooking over the edge of the rail. On this side, the spikes serve only to hold the plate, the rail being held by a vertical setscrew which passes through the shoulder and can be set so tightly as to prevent creeping of the rail. On the inner side, the rail is held directly by the spikes or by hooked clamps which fit into dovetailed holes in the plate; behind these clamps the spikes are driven, with their heads bearing on the clamps. Another plate has somewhat similar clamps on both sides of the rail, but the spikes behind them are in reversed position, so that their heads rest on the plate and hold it in place. To prevent creeping of the rail, a thin wedge may be driven horizontally between the edge of the rail base and lugs on the tie-plate.

Application of Plates.—As a rule, the tie-plates are placed when the ties are in the track, their exact position being determined by the rails, especially with shoulder plates and with ties bored for the spikes. Attempts to apply flanged plates by machine presses at the shops have not been successful, and with the shallow flanges now employed it is doubtful if plates so applied would hold to the tie. The plates should be thoroughly bedded on the ties when laid, and with shallow flanges and soft ties this may be done by the weight of traffic before sand and dirt can work under the plates and prevent their coming to a firm seat. On hewed ties and old worn ties, a smooth seat should be trimmed, but with as little adzing as possible.

For applying tie-plates before rails are laid (as with plates secured by independent fastenings), the position of one plate may be marked at the line side of the tie, and when this is placed, the position of the second plate can be determined by measurement or a gage. To insure accurate placing, the Buffalo, Rochester & Pittsburgh Ry. uses the Ware tie-plate gage, consisting of a 1-in. gas-pipe rod having at one end a fixed flat head about the size of a tie-plate, and at the other end a similar sliding head secured by a thumbscrew, this head being adjusted according to the size of plate to be laid (see "Track Tools").

After using the flat heads to indicate if the seats have been adzed to give a level bearing, the bar is turned with these heads on edge, forming guides for the positions of the two tie-plates. The company's rules require that before ties are placed, both tie-plates are to be applied, using the Ware surfacer as a gage and driving them with a steel follower plate and sledge, or with a wooden beetle $7\frac{3}{4}$ in. diameter and 15 in. long, having a 28-in. handle; or they may be driven by striking on opposite corners with mauls or sledges, but they must be driven to a bearing and not left to be seated by traffic.

For attaching the more heavily flanged plates, a steel or wood block may be placed on the plate and struck with a sledge or wooden beetle. If plates are placed under the rails, smaller striking blocks may be placed on the projecting ends. Hydraulic tie-plate setters have also been devised. An improvised machine used on the Union Pacific Ry. consisted of a hand car having at each side a drop hammer working in leads like a small pile driver; this hammer struck a U-shaped block set astride of the rail with its legs resting on the ends of the tie- plates.

TIE RENEWALS AND RECORDS

In view of the great number of ties required annually for renewals and the cost involved in renewals, with consequent disturbance of track conditions, it is evident that the use of means to increase the life of ties is desirable and that great care should be taken to secure the full economic life of ties in the track. The annual expenses for ties have exceeded those for rails for several years. On 10 important railroads, the amount spent for ties in 1924 ranged from three to twelve times that spent for rails, with an average of three times for all the roads. Of two railroads which have ties lasting 6 and 10 years, the former has almost double the expense for ties and labor, with an additional expense for spikes to replace those lost or broken during renewals, besides the additional maintenance work due to more frequent disturbance of the ballast and track.

Tie renewals which formerly averaged 250 to 350 or even 400 ties per mile per year have been reduced to 120 to 150, or even less than 100, by more careful selection of ties, more careful inspection of ties in track, and the use of preservative processes and metal tie-plates to protect the wood against decay and wear. Renewals have been reduced on the Santa Fe Ry. from 336 per mile in 1898 to 128 in 1923, and on the Lehigh Valley R. R. from 336 in 1900 to 77 in 1922 and 91 in 1923. Annual renewals have been assumed to average 10 per cent for main-line track and 7 per cent for branch lines, but in modern practice even main-line renewals have been reduced to 3 or 5 per cent. The difference in track condition where track is disturbed at 300 points and at 100 points per mile may be appreciated readily. Records of reduced numbers of ties renewed do not always or necessarily indicate increase in the real life of the ties, since purchases of new ties may be influenced by financial conditions more than by track conditions. If sufficient new ties are not provided, then many old ties which should be removed must be continued in service, with probably increased expense for maintenance of track.

The selection of ties requiring renewal is made first by the section foreman, as a rule, and is then checked by the roadmaster or a tie inspector, who marks the ties that may be removed during the next season. A further inspection may be made by the division engineer. The roadmaster's estimate for new ties is examined by the engineer and compared with previous records before being submitted for approval. An important check is a careful inspection of all ties removed before they are disposed of, this being made by the roadmaster or by a tie inspector who reports on ties improperly removed and discusses the report with the roadmaster before sending it to the engineer or the head of the tie department. The roadmaster takes up the matter with the section foremen who are responsible, and the good ties thus reclaimed are either put back in main track or laid in sidetracks. Such inspection has a marked effect in lessening the number of ties removed prematurely.

Estimates for new ties may be required in the fall, so that requisitions can be made out for the purchasing agent and ties obtained for the next spring. But the final selection and marking or "spotting" of ties to be removed should be done early in the spring, as this is nearer the time for the work and the selection will be closer to the actual needs than if done 6 months previously.

Forecasting tie renewals, as practiced on the Baltimore & Ohio R. R., is based on a careful tabulation of ties used in renewals and new construction from 1907 to 1914, and on an average life of $9\frac{1}{2}$ and 16 years for untreated and treated ties as determined from a curve developed by the U. S. Forest Products Laboratory and checked by actual experience. These renewals declined from 309 per mile (268 estimated) in 1915 to 207 (214 estimated) in 1923. Assuming that only treated ties will now be used, the forecast table indicates a gradual reduction to 134 renewals per mile in 1934, followed by a rise until the cycle of renewals of all treated ties is covered, reaching a maximum of 180 per mile in 1942, beyond which the average is 158 per mile per year. That the total renewals on the road were about the same as in 1907, although the mileage had increased 24 per cent, is evidence of the benefits of treated ties, which now represent 50 per cent of all ties in renewals. Forecasts and records of this kind form a valuable check on individual counts and estimates.

Tie Records and Marking.—With the mounting costs of ties and track work and wider recognition of the economic value of ties, more careful methods are being practiced for keeping accurate record of the life of ties, so that comparisons may be made between different methods of treatment or protection. Attempts to keep records of all ties soon showed this practice to be unreliable, owing to the great number and variety involved, the number of records from foremen as individual observers, and the great variations in track, traffic, and climatic conditions.

To avoid this complication, a method of test sections has been introduced, by which a stretch of not less than a mile is selected on some one track section on each operating division. This stretch is then laid with new ties, using various kinds for comparison. All the ties are marked and given a careful inspection annually; this is done usually by the engineer and representatives of the tie department, but sometimes outside experts are employed. The test sections should represent average conditions of construction and maintenance, should be free from switches, and not likely to be disturbed by extensive rail renewals. For each kind of wood or treatment there should be not less than 100 ties. Each test section should have also 100 test ties of a specified grade of untreated white oak, to serve as a basis of comparison between different sections. No tie on the test section must be removed without authority from the official in charge of tests; if such removal is made in emergency, report must be made and the tie carefully preserved until inspected by this official. In each test section the record includes conditions of climate, rainfall, drainage, ballast, track construction, and traffic, while the records show the condition of each tie at each annual inspection.

Every treated tie, whether in ordinary track or on a test section, should be marked with a dating nail driven into the top of the tie in a designated position and indicating the year the tie was put in track. On test tracks, every untreated tie should be similarly marked, but with a nail having a distinctive form of head. The form of dating nail specified by the A.R.E.A. is $2\frac{1}{2}$ in. long and $\frac{1}{4}$ in. in diameter, with a $\frac{5}{8}$-in. flat head in which are recessed two figures indicating the year. Ties in test tracks are also sometimes numbered with white paint for convenience. Reliable records of life of ties cannot be kept without such permanent marking of the initial date, which also tends to check premature removal where treated ties are laid in ordinary track sections.

For test sections on the Santa Fe Ry., a monthly report made by the foreman and checked by the roadmaster shows, for main and side tracks separately, and for treated and untreated ties separately, the number of ties removed and placed. This form gives the number of ties, the year mark, kind of wood, kind of treatment, and reason for removal (rotten, burned, rail cut, shattered, or broken through accident). It shows also the number of tie-plates applied for renewal and on ties not previously plated. To insure accurate record, the foreman is required to enter the year mark in his memorandum book at the time the tie is removed. For service records of treated ties, the U. S. Forest Products Laboratory uses a form which gives the following information:

Record of Treated Ties

Species of wood....................	Where grown........	
Sawed or hewed....................	Dimensions..........	
Where placed......................	Grade..............	per cent
Curve.......................deg.	Tangent............	
Number of ties....................	Spacing............	
Seasoned or unseasoned............	Bored for spikes......	
Preservative......................	Average absorption...	lb. per tie
Process..........................	Average absorption...	lb. per cubic foot
Ballast..........................	Weight of rail........	
Spikes, cut or screw................	Tie plates, kind......	; size..........
Date inspected....................	Traffic..............	tons per year
Number remaining.................	Cause of removal.....	
Present condition..................	Average life.........	

Reasons for unsatisfactory results with treated wood.............

METAL AND CONCRETE TIES

Substitutes for Wood.—Ties of steel, cast iron, and concrete are used extensively and with satisfactory results in many foreign countries, on main lines as well as on secondary lines and pioneer railways. With a decreasing supply of good tie timber in this country, there have been innumerable designs for substitute ties, many of which have had limited experimental trial, but, taken as a whole, the results have not been promising as to cost, efficiency, or durability. On the other hand, greater care in the use of wood ties and extensive increase in the adoption of preservative processes, increasing the life at comparatively little extra expense, have made the substitute tie of little more than academic interest at present. Detailed records of the use of metal ties in various countries are given in the author's reports on "The Use of Metal Ties for Railways" (U. S. Forestry Division, 1890 and 1894). More recent records as to steel and concrete ties in this country are given in the A. R. E. A. annual reports, which in 1923 also included a statement of fundamental requisites for substitute ties.

That it is desirable to experiment more extensively with such ties, is the opinion advanced by the above association, which takes the view that while no substitute ties can compete with treated wood ties on an annual cost basis at present prices, still a practicable substitute tie should be developed with a view to possible changes in relative prices. A tentative design proposed for trial by the association utilizes a structural steel channel, for while this is not the best section for the purpose, the rolling of special sections for ties is out of the question at the present time.

A life of 30 to 40 years has sometimes been assumed for steel ties, but there are few reliable records applicable to the conditions of heavy wheel loads and traffic in this country, where experience indicates a limit of 15 years. In addition to general corrosion, the life is affected by local wear and corrosion at bolt holes and fastenings; in case of derailment, steel ties are usually damaged more severely than wood tics. A serious objection to steel, in view of the rapid increase in the use of track circuits for signal and train control systems, is the difficulty of providing effective insulation between rails and ties at reasonable cost for construction and maintenance. In track work it appears that the cost and labor for attention to rail fastenings, gaging, and lining would be somewhat less than with wood ties, but higher for surfacing and ballasting, owing to the crushing effect of the steel tie under load. As to comparative cost, it has been estimated that a 200-lb. steel tie at $4.50 complete and having a life of 13 years would be as economical as a wood tie costing $2.25 in place (with labor, spikes, and tie-plates) and lasting 7 years. But a treated wood tie costing $2.50 and lasting 12 or 15 years would be still more economical.

American Substitute Ties.—The most extensive experience in this country has been with a steel tie of I-beam section (Fig. 16), weighing

19½ to 24 lb. per foot, and having bolted clip fastenings. Although about 1,500,000 ties of this kind have been laid at different times, they are not now made or purchased, owing to high prices of steel. Of the ties removed,

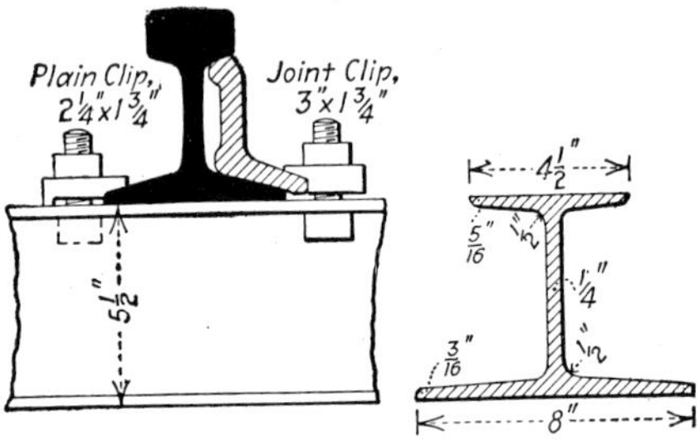

FIG. 16.—Steel I-beam tie.

about half were defective in various ways and the others were damaged by derailments. These and other steel ties are used to a limited extent

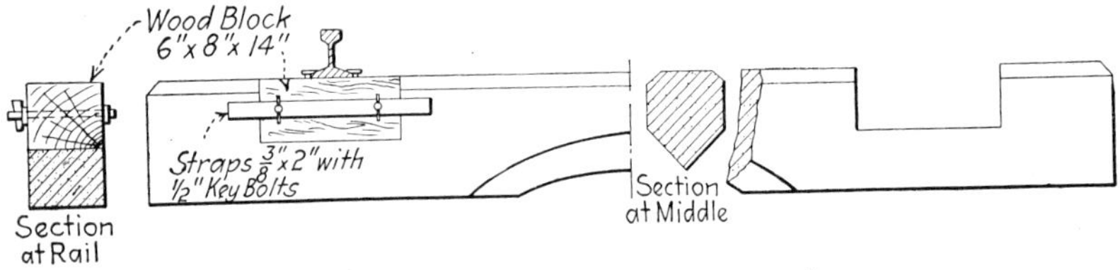

FIG. 17.—Shearer reinforced-concrete tie.

in street railway track, where they are usually embedded in concrete, while for mine and industrial or portable narrow-gage track there is con-

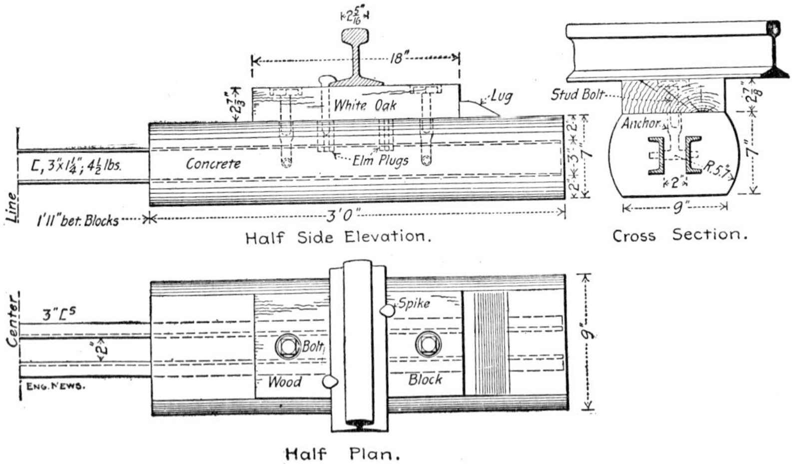

FIG. 18.—Kimball combination tie with concrete blocks.

siderable use of shallow trough or flat plate ties stiffened by corrugations along the top.

Reinforced-concrete ties of various designs have been tried, usually on a small scale, but while some have been satisfactory, the greater number have failed by shattering or disintegration of the concrete. They sometimes make a hard-riding track and are thus considered objectionable in main lines, but for sidings or yards a durable concrete tie at reasonable cost may have advantages in reducing maintenance work. Ties of this kind weigh usually 300 to 600 lb. About 20,000 concrete ties have been laid, including innumerable designs. A typical tie, tried on the Southern Pacific Ry., is shown in Fig. 17, while Fig. 18 shows a compound tie composed of a pair of concrete blocks connected by light steel channels. This latter tie weighed 436 lb., including 374 lb. for the concrete.

Foreign Substitute Ties.—Steel ties are in extensive use in India, Africa, and South America, and to some extent in Europe, although the high price of steel in recent years has checked their further introduction.

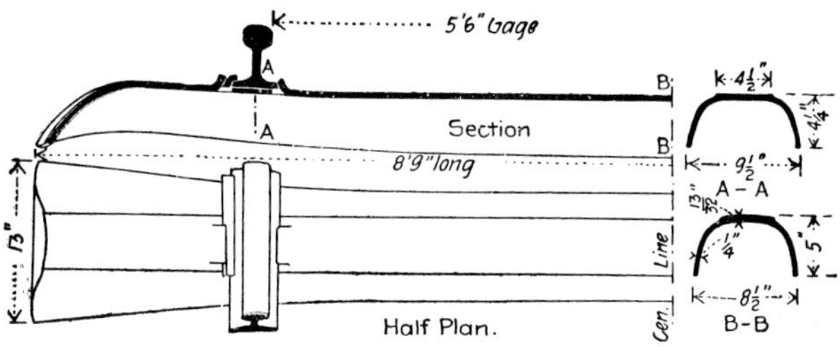

FIG. 19.—Rendel steel tie: Indian State Railways.

Most of the designs are variations of the inverted trough section invented by Vautherin, a French engineer, in 1864. As a rule, they are of shallow depth. An English tie widely used in India, Mexico, and other countries is a pressed-steel trough (Fig. 19), weighing 135 lb. for a length of 8 ft. 9 in. At each rail seat are two lugs stamped from the plate, one hooking over the rail base and a taper steel key being driven between the other lug and the rail. The Post tie, at one time used in Holland and France, was a trough of varying thickness and section, narrow and deep at the middle, but broad and shallow at the ends to give a bearing for the rail and ballast. Bolted clip fastenings had eccentric washers engaging the edge of rail base to allow for adjustment of gage. Ties 8 ft. 10 in. long, with metal ¼ to ½ in. thick, weighed 163 lb. The Haarmann tie, developed in Germany, was a 3-in. trough about 5 and 10½ in. wide at top and bottom, weighing about 195 lb. A distinctive type of substitute ties, largely used in India, consists of a pair of round, oval, or rectangular bowls or flat-bottomed plates of cast iron, cast steel, or reinforced concrete, connected by a steel tie bar which is put in after the bowls are placed. Reinforced-concrete ties have been used somewhat extensively in Europe, particularly in France and Italy, but usually on secondary railways.

CHAPTER IV

RAILS

Early History of Rails.—Practically the entire railway system of the world is laid with the T-rail or flange rail invented in this country in 1830 by Robert L. Stevens, chief engineer of the Camden & Amboy R. R., who also designed the hook-headed spike and flat splice bar or fish-plate joint. The first rails for this railroad, weighing 36 lb. per yard and in 18-ft. lengths, were ordered by him in England. In 1836, this type of rail was reinvented in England by Charles B. Vignoles, and in 1847 the fish-plate joint was reinvented there by W. Bridges Adams. A bar rail patented in England by Birkinshaw in 1820, having a thin web with top and bottom ribs, was the forerunner of the double-headed and bull-headed types of rails now used on English main line railways; these rails are not self-supporting, like the T-rail, but have to be carried in cast-iron chairs on the ties. The bridge rail, of inverted U-section, was designed in this country by Strickland in 1834 and in England by Brunel, in 1835. It is now practically obsolete, though used to a small extent in Europe on bridges and on underground railways, where it is always supported on longitudinal timbers.

Modern Developments.—A multiplicity and diversity of rail sections characterized railroad practice until about 1895, almost every engineer designing the rails for his own road. Consequently, large stocks of rolls had to be kept on hand by the mills, although many of the sections differed only slightly. To remedy this situation and to follow up its earlier investigations, the American Society of Civil Engineers in 1890 appointed a committee to design standard rail sections, which were adopted in 1893. In these sections, ranging from 40 to 120 lb. per yard, the width of base was equal to the height, the head was broad and relatively shallow, and the metal was distributed with 42 per cent in the head, 21 per cent in the web, and 37 per cent in the base. Some engineers have favored a smaller total height and a deeper head for the same weight of metal, but the overwhelming weight of opinion and experience endorses a high and stiff section with head broad enough to give a good bearing for the wheels and shallow enough to insure thorough working of the steel by the rolls.

These A. S. C. E. sections were adopted so extensively that in 1906 they represented 65 to 90 per cent of the output of the principal mills.

But as the heavier sections came into use they proved less satisfactory, and in 1906 a committee reported that they were not giving the service expected, even allowing for the fact that driving-wheel loads had increased 60 per cent while the maximum weight of rail had increased only 25 per cent, from 80 to 100 lb. per yard. The blame was laid on methods of rolling, but the manufacturers claimed that good rails could not be made in the heavy sections, partly on account of the wide and thin base. This form of base had been adopted to secure a broad bearing on the ties, metal tie-plates not having been developed at that time. In 1907 this committee recommended that for heavy sections the width should be ½ in. less than the height, that the metal in the base should be equal to, or slightly greater than, that in the head, and that the distribution of metal should vary with the weight.

In 1907 the American Railway Association proposed new sections for 60- to 100-lb. rails, but unfortunately it put out two distinct types for experimental use: *A*, a high and stiff section; *B*, a shallower and heavy-headed section of the same weight. As the latter was not used extensively, only a little temporary confusion resulted. The question of the respective merits of these sections was referred in 1908 to the A. R. E. A., which approved of the high and stiff type and in 1915 adopted the 90-lb. *A* section together with new sections for 100- to 120-lb. rails, which range has since been extended to 150 lb. In the A. R. E. A. sections the distribution of metal in head, web, and base varies with each weight of rail: from 38.2, 22.6, and 39.2 per cent in the 100-lb. rail to 34.7, 26.1, and 39.2 per cent in the 150-lb. rail. There are still several individual sections, but the adoption of the A. S. C. E. sections was so general that it appears practicable to secure a similar acceptance of the A. R. E. A. sections for heavy rails, especially in view of the marked tendency towards engineering standardization.

Apart from these two sets of standards there are the notable rail sections designed by P. H. Dudley for the New York Central R. R. To carry the increasing loads he considered it necessary to introduce heavier and stiffer rails as girders, with broad and well-worked heads to sustain the higher pressures per square inch of contact area. On the basis of this view and of rail inspection over some 25,000 mi. of track with his dynagraph car, Mr. Dudley in 1883 designed the first of a series of rail sections which now extends to a rail of 120 lb. per yard. The first 100-lb. rail in this country was designed by Mr. Dudley for the four-track line approaching the New York terminal of the New York Central R. R., in order to reduce the difficulty and expense of maintenance and renewals under heavy traffic on tracks partly in tunnel and partly in open cuts between retaining walls.

The relation of good, heavy rails to economy of operation is indicated by train resistance tests at 18 to 20 m. p. h., in which the resistance was 6 to

8 lb. per ton with trains of 25 to 30 cars (600 to 700 tons) on light rails and only 3 lb. with trains of 80 cars (3,428 tons) on Dudley 80-lb. $5\frac{1}{8}$-in. rails. In tire wear, also, it was found that while engines with 13,360-lb. wheel loads on 65-lb. narrow-headed rails averaged 19,400 mi. per $\frac{1}{16}$ in. wear of tire, engines with 23,000-lb. wheel loads averaged 19,300 mi. on the broad-headed 80-lb. rails and would make 150,000 to 190,000 mi. before the tires needed turning, so that tire wear no longer determined when the engines must be sent to the shops.

Modern Rails.—For present practice it may be said that the rail should have a relatively broad and shallow head, fairly sharp top corners, fishing angles not less than 13 deg. or 1 in 4, and a base narrower than the height, with fairly thick edges. In the A. S. C. E. sections the heads have vertical sides with $\frac{5}{16}$-in. top corners and 12-in. top radius. In the A. R. E. A. and Dudley sections the heads have a slight outward flare of 1 in 16

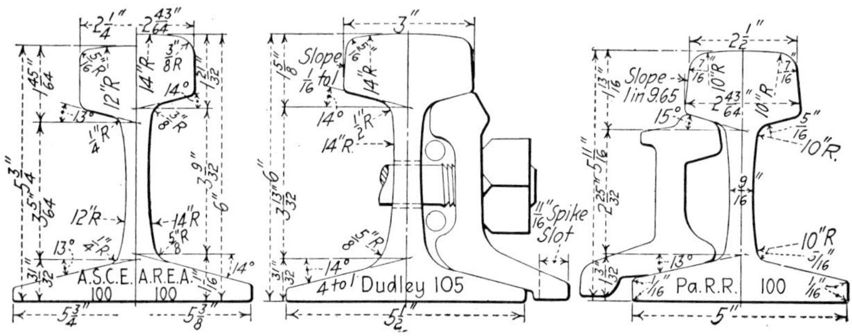

FIG. 20.—Rail sections.

and the top radius is increased to 14 in., while the A. R. E. A. also increases the corner radius to $\frac{3}{8}$ in. The Dudley sections of 90 lb. and over have uniform head width of 3 in. over the bottom, while in the A. R. E. A. sections this width increases from $2\frac{9}{16}$ in. in the 90-lb. to 3 in. in the 140-lb. section. Uniform width of head and base for several slightly varying weights of rail would simplify the tie-plate difficulty and need not affect the quality of the rail; at present the plates have to be punched with spike holes to fit every variation in rail. Round top corners of $\frac{1}{2}$- and $\frac{5}{8}$-in. radius are now practically obsolete. A top radius of 10 in. is retained by the Pennsylvania R. R. for rails up to 100 lb., above which it is 12 in. The Canadian Pacific Ry. had satisfactory results with a radius as sharp as 8 in., but abandoned this in order to adopt the A. R. E. A. sections in the interest of standardization. A 6-in. radius was used at one time by C. P. Sandberg, the British rail expert, but later he adopted a 12-in. radius. The web is made usually with curved sides and easy fillets to facilitate the draft of metal in the rolls and also to provide sufficient metal to prevent bending of the head over the web or of the web over the base, but many foreign rails have vertical sides for the web.

As to the base, a wide, thin section is difficult to roll, and a wide, thick section would be disproportionate, while in either case the extra width would not prevent cutting of the ties (since this is due to the motion rather than the pressure of the rail) and the metal in the edges would be practically useless between the ties. The logical base section is fairly thick with a width sufficient for stability but somewhat less than the height, relying on metal tie-plates to distribute the load and prevent cutting of the tie. In several foreign designs the top surface of the base or flange is composed of two slopes in order to increase the edge thickness. Thus the fishing slope of 1 in 4 extends only far enough to seat the splice bar, and is followed by a slope of 1 in 10 or 1 in 15 to the edge. This detail, however, was abandoned in the British standard sections in 1924, as there was rapid wear of the shaping rolls, involving frequent expense for dressing them to shape. Some typical rail sections are shown in Fig. 20, with dimensions and proportions in Table III.

Table III.—Proportions of Rail Sections

Weight, pounds per yard	Section	Height, inches	Base, inches	Proportions		
				Head, per cent	Web, per cent	Base, per cent
60	A. S. C. E.	$4\frac{1}{4}$	$4\frac{1}{4}$	42.0	21.0	37.0
80	A. S. C. E.	5	5	42.0	21.0	37.0
90	A. S. C. E.	$5\frac{3}{8}$	$5\frac{3}{8}$	42.0	21.0	37.0
90	Dudley	$5\frac{1}{2}$	5	38.7	22.6	38.7
100	A. R. E. A.	6	$5\frac{3}{8}$	38.2	22.6	39.2
100	Pennsylvania R. R.	$5\frac{11}{16}$	5	41.0	18.6	40.4
100	British	6	$5\frac{3}{4}$	43.56	21.07	35.37
105	Dudley	6	$5\frac{1}{2}$	40.9	24.0	35.1
115	Dudley	$6\frac{1}{2}$	$5\frac{1}{2}$	37.8	28.8	33.4
130	A. R. E. A.	$6\frac{3}{4}$	6	36.4	23.8	39.8
130	Pennsylvania R. R.	$6\frac{5}{8}$	$5\frac{1}{2}$	39.4	22.0	38.6
136	Lehigh Valley R. R.	7	$6\frac{1}{2}$	35.4	23.7	40.9
150	A. R. E. A.	$7\frac{3}{4}$	$6\frac{3}{4}$	34.7	26.1	39.2

Weight and Length.—Rails of 136 and 135 lb. per yard on the Lehigh Valley R. R. and Central R. R. of New Jersey, respectively, are the heaviest in use, but several railroads use 110- to 120-lb. rails on heavy-traffic lines. For ordinary main-line service, 80 lb. is about the minimum weight and 100 lb. rails are in general use. It has been estimated that traffic of 2,000,000 and 4,500,000 tons per mile per year warrants 90-lb. and 100-lb. rails, respectively; also that 130-lb. rails will give 50 to 100 per cent longer life than 100-lb. rails under the same traffic, while the latter will give nearly double the life of 85-lb. rails. To ascertain the weight of rail of a given section, steel is taken usually at 10.18 lb. for a 1-in.-square bar 3 ft. long, or 10.20 lb. for rails of less than 95 lb. per yard.

It is to be remembered that the rail is only one part of the track structure. Installing a heavier and stiffer rail gives a better load distribution and reduces cutting of ties, but it does not compensate for deficiencies in ties and ballast, nor does it warrant a decrease in the number of ties per rail. Heavy rails carelessly laid with uneven bearings on old ties may be damaged by becoming wavy or kinked in surface. Rails, joints, fastenings, tie-plates, ties, and ballast should be in proper relation as parts of a unit structure. Attention must be given to the maintenance of good line and surface on track with heavy rails, but such rails replacing lighter rails in ordinarily good track are found to reduce the maintenance work materially.

For many years the general length of rails was 30 ft., but for new rails the 33-ft. length is now used almost universally on main tracks of American railways, thus reducing the number of joints, which constitute weak or troublesome spots in the track. The A. R. E. A. specifications call for a 39-ft. length and the Southern Pacific Ry. is using 39-ft. rails of 90 and 110 lb. per yard, while the Philadelphia & Reading R. R. has some 45-ft. 90-lb. rails and finds them satisfactory. There is some trouble in handling them for renewals, as 20 men with tongs are required to handle a rail on the track. There is also more difficulty in keeping the rails in line, and on cinder-ballasted track more maintenance is required with 45-ft. rails than with 33-ft. rails. At one time 60-ft. rails were used to a limited extent, but some roads abandoned them on account of the difficulty in handling and the excessive width of contraction openings at joints. Rails 45 ft. long are used extensively in England and 78.72-ft. rails in France. Continuous rails, welded together in the track, are common on street railways, where the rails are embedded in the paving and thus protected from such rapid and extreme changes of temperature as may be experienced by rails in open track.

Short rails are required on the inside of curves to compensate for the difference in length of the two lines of rails and to keep the joints in proper relation on tangents. They are required also in making up turnouts and other special work. For this purpose the A. R. E. A. specifications provide for 11 per cent of the order to be in lengths varying by 1 ft. from 32 to 25 ft. (or 38 to 25 ft. for 39-ft. rails), these short rails being designated by painting the ends green. The New York Central R. R. specifies 10 per cent to be in lengths of 30, 28, 26, and 24 ft., while for every 100 tons there are to be three rails 32½ ft. long, the ends of which are painted red.

Rail Manufacture.—Early rails were of wrought iron, but the introduction of the bessemer steel process about 1865 made steel rails available at moderate cost and was an important factor in the rapid development of railroads. This process has been largely superseded by the open-hearth process, since the low-phosphorus ores specially suitable for the former are largely exhausted. Rails of open-hearth steel, in general, give a higher degree of wear and fewer fractures than bessemer rails.

In the bessemer process, molten metal from blast furnaces is delivered by 15- or 25-ton ladles to a 200- to 1,000-ton tank or mixer, from which other ladles charge converters of 12 to 25 tons capacity. Air at 25-lb. pressure blown in at the bottom passes up through the metal and burns out the carbon in a few minutes. A charge of ferro-manganese is added to give the required degree of carbon, and by tilting the converter the steel is poured into the casting ladle. With the open-hearth process, the metal from the mixers is charged into tilting furnaces of 50- to 200-tons capacity, and after about 2 hr. for burning out the carbon, the steel is recarburized to the required degree by adding liquid metal. Then the furnace is tilted to pour the steel into the casting ladle. In the duplex process, time is saved by treating the metal first in a converter and then in an open hearth furnace.

Steel from the casting ladle is poured into upright molds to form ingots 5 to 6½ ft. high, about 22 in. square at bottom and 21 in. at top, or 20 by 24 in. and 18¾ by 22 in., respectively. The ingots, weighing 6,000 to 8,000 lb., or even 14,000 lb., are taken from the molds and set upright in a furnace or soaking pit. In rolling, a blooming mill first reduces the ingot to a bloom or bar about 8 or 10 in. square and 15 to 25 ft. long. After cropping the ends to remove defective metal, the bloom is cut into three or four lengths, each making two or three rails according to weight. Those of the first bloom from the top are known as A-rails, the next are B-rails, and so on. Each bloom, with or without reheating in a furnace, goes to the rail mill to be rolled to final shape. There will be from 15 to 30 passes in the rolls, of which 6 to 19 are in the blooming mill and 9 to 13 in the rail mill. In rolling, the rails have the name of maker, date, weight, and other marks formed on the web.

The hot rails are cut by circular saws to such a length that they will be of the required length when cold, and then go to the hot beds for cooling. Bends or kinks are taken out in a cold-straightening press, the rail resting on supports 4 or 5 ft. apart. There should be as little of this work as possible, since the "gagging" tends to injure the steel and may cause incipient flaws which will develop into fractures when subjected to stress under traffic. It is desirable that the bend of the rail should not exceed 3 in. and some engineers consider that rails with an even swing or camber of 1 in. need not go to the press but can be straightened in track by spiking. After the burr left by the saws is chipped or ground off, the rails are measured for length, drilled for the joint bolts, and then placed on the shipping or inspection beds. A refinement for heavy rails, in view of the tendency to reduce the expansion spacing, is the milling of the ends to remove all burrs and make the ends absolutely square. A low finishing temperature was approved at one time, but with high-carbon steel this treatment may cause brittleness, though toughness might be restored by annealing, with extra expense. No requirements as to temperature are included in the A. R. E. A. specifications, but the New York Central R. R. specifies that the number of passes and the speed of rolls must be so regulated that on leaving the rolls the temperature of the rails will not exceed that which requires a shrinkage allowance at the hot saws of $6\frac{3}{4}$ in. for a 33-ft. 105-lb. rail and $\frac{1}{16}$ in. less for each 5 lb. decrease in weight. The rails must not be cooled by artificial means or by being held to cool before they go to the saws.

Second-class or No. 2 rails are those which vary slightly from the specifications but in a manner which does not affect their soundness or strength; such as surface imperfections, sharp kinks, or a camber of more than 4 in. in 33 ft. Such No. 2 rails are distinguished by painting the ends white.

Quality of Rail Steel.—While the quality of rails depends upon both the chemical composition and the treatment in manufacture, the latter is the more important. Much unsatisfactory service as to wear and fracture has been given by modern heavy rails, but unfortunately the manufacturers for several years took an arbitrary position, failing to cooperate in movements to secure better rails, and actively or passively opposing improved specifications prepared by railway engineers. This position is being modified, and in 1923 plans were made for a cooperative study of rail troubles by mill engineers and railway engineers. The manufacturers have blamed rail failures upon heavy traffic and wheel loads, but while these conditions certainly impose severe strains upon the rails, it appears from records and investigations that failures, as a rule, may be attributed to manufacture rather than service.

Thus, while breakages sometimes decrease with the introduction of heavier rail, there are many cases where there has been almost as much trouble as with the old and lighter rail. Even the failures from transverse fissures, so troublesome in recent years, are not due to traffic or service, as shown by the fact that they occur only in certain rails out of the same lot and carrying the same traffic. Furthermore, old rails with heavy traffic and high internal stress have been found to contain no fissures. Methods of producing rails of higher quality should not result in higher prices, since cost of manufacture has been reduced to a low figure and market price has no relation to quality or cost, but is fixed arbitrarily.

Chemical Composition.—In rail manufacture, the aim is to produce rails combining hardness (to resist wear or abrasion) with toughness (to resist fracture), and having a homogeneous composition without flaws or incipient cracks to develop into fractures when the rail is subjected to heavy stress in the track. Requirements for chemical composition are

TABLE IV.—CHEMICAL COMPOSITION FOR STEEL RAILS

Specification	Weight, pounds per yard	Carbon	Manganese	Silicon*	Phosphorus*
Open-hearth basic steel:					
A. R. E. A.	70 to 84	0.53 to 0.70	0.60 to 0.90	0.15	0.04
A. R. E. A.	85 to 100	0.62 to 0.77	0.60 to 0.90	0.15	0.04
A. R. E. A.	101 to 120	0.67 to 0.83	0.50 to 0.90	0.15	0.04
A. R. E. A.	121 to 140	0.72 to 0.89	0.50 to 0.90	0.15	0.04
A. S. C. E.	70 to 79	0.53 to 0.63	0.90	0.20	0.05
A. S. C. E.	80 to 89	0.58 to 0.68	0.90	0.20	0.05
A. S. C. E.	90 to 100	0.65 to 0.75	0.90	0.20	0.05
A. S. T. M.	50 to 69	0.50 to 0.63	0.60 to 0.90	0.20	0.04
A. S. T. M.	70 to 84	0.53 to 0.66	0.60 to 0.90	0.20	0.04
A. S. T. M.	85 to 120	0.62 to 0.75	0.60 to 0.90	0.20	0.04
Manufacturers Association	50 to 69	0.50 to 0.63	0.60 to 0.90	0.20	0.04
Manufacturers Association	70 to 84	0.53 to 0.66	0.60 to 0.90	0.20	0.04
Manufacturers Association	85 to 100	0.59 to 0.72	0.60 to 0.90	0.20	0.04
Manufacturers Association	101 to 120	0.62 to 0.75	0.60 to 0.90	0.20	0.04
Pennsylvania R. R.	100	0.62 to 0.75	0.60 to 0.90	0.10	0.04
Pennsylvania R. R.	130	0.68 to 0.87	0.60 to 0.90	0.10	0.04
New York Central R. R.	80	0.55 to 0.68	0.70 to 1.00	0.1 to 0.2	0.04
New York Central R. R.	90	0.60 to 0.73	0.70 to 1.00	0.1 to 0.2	0.04
New York Central R. R.	105, 115, 120	0.62 to 0.75	0.70 to 1.00	0.1 to 0.2	0.04
Bessemer steel:					
A. R. E. A.	70 to 84	0.40 to 0.50	0.80 to 1.10	0.10	0.10
A. R. E. A.	85 and over	0.45 to 0.55	0.80 to 1.10	0.10	0.10
A. S. C. E.	70 to 79	0.50 to 0.60	0.75 to 1.00	0.20	0.085
A. S. C. E.	80 to 89	0.53 to 0.63	0.80 to 1.05	0.20	0.085
A. S. C. E.	90 to 100	0.55 to 0.65	0.80 to 1.05	0.20	0.085
A. S. T. M.	50 to 69	0.37 to 0.47	0.80 to 1.10	0.20	0.10
A. S. T. M.	70 to 84	0.40 to 0.50	0.80 to 1.10	0.20	0.10
A. S. T. M.	85 to 120	0.45 to 0.55	0.80 to 1.10	0.20	0.10
Manufacturers Association	50 to 69	0.37 to 0.47	0.80 to 1.10	0.20	0.10
Manufacturers Association	70 to 84	0.39 to 0.49	0.80 to 1.10	0.20	0.10
Manufacturers Association	85 to 100	0.41 to 0.51	0.80 to 1.10	0.20	0.10
Manufacturers Association	101 to 120	0.43 to 0.53	0.80 to 1.10	0.20	0.10

* These percentages must not be exceeded.

given in Table IV. Carbon is one of the most important constituents affecting the strength, and its proportion is increased with weight of rail. A high carbon content is desirable to make the steel hard, but tends to make it brittle unless care is taken in proportioning the other elements and in the process of manufacture. Phosphorus makes the steel hard and brittle and tends to cause segregation in the metal. Silicon gives hardness with toughness, increases the density and prevents segregation of the elements in the metal. Manganese gives a dense metal with high resistance to abrasion. Sulphur tends to cause seams, but is not often mentioned in specifications, although those of the A. S. C. E. limit it to 0.06 per cent in open-hearth and 0.075 per cent in bessemer steel.

Special Steel for Rails.—Various special grades of steel have been tried to secure greater strength and wearing properties. Most of these are alloy steels having relatively high proportions of certain elements;

such as nickel, nickel chromium, silicon (0.25 to 0.50 per cent), and manganese (10 to 18 per cent). In others, elements are added to improve the quality of metal but do not appear in the finished steel; these include ferro-titanium, vanadium, and aluminum. Steel made in electric furnaces has been tried, as well as heat treatment of the finished rails, to raise the elastic limit without decreasing the ductility. A comparatively recent development is the Sandberg air-treated sorbitic surface-hardening process, in which the finished rail after cooling to a certain temperature is heated and quenched by special apparatus. This treatment has increased the life materially. It has been applied to rails in street railway track by means of a torch heater, and water jets mounted on a car which moves slowly along the track, but there has been some flaking or scaling of the hardened surface. All such special alloy-steel rails are costly, and for use only under special conditions, while general improvement of rails is to be looked for in careful or improved methods of manufacture rather than in special composition. Heat treatment of the rails as a means of improving the quality of the steel is a recent development with which experiments are being made on the Pennsylvania R. R.

Manganese steel has been used extensively on sharp curves on account of its wearing properties. At first these rails were cast, but now are rolled. With about 12 per cent manganese the steel gives a high resistance to abrasion and high tensile stress, but, on the other hand, it gives a low elastic limit, so that if unevenly supported, the rails may get bent or kinked instead of springing back to position. On 8-deg. curves where bessemer rails averaged a life of 1½ years, manganese rails have averaged 6 or 7 years. On curves of 90- to 500-ft. radius the Chicago Elevated Rys. use rolled manganese-steel rails: carbon, 0.95 to 1.35 per cent; manganese, 10.5 to 15; phosphorus, 0.10. These give from five to eight times the life of high-carbon, open-hearth rails having carbon 0.75 to 0.88; manganese, 0.60 to 0.90; phosphorus, not more than 0.04, and silicon not more than 0.20. On curves of larger radius these latter rails give such good service as not to warrant the use of manganese steel.

Open-hearth, high-carbon, ferro-titanium steel rails treated by the Sandberg sorbitic process noted above have been adopted by the Boston Elevated Ry. in view of the high first cost and renewal cost of manganese steel. In extended experience, this road has found cast manganese steel rails to have the objection of short length and an irregular section that prevented suitable splicing, so that the rails bend at the joints and do not make good track. Rolled manganese rails have a shorter life but they cost less, although the cost is still very high. The sorbitic steel rails are made to the following specifications: carbon, 0.8 to 0.9; manganese, 0.05 to 0.90; silicon, 0.10 to 0.20; metallic titanium, 0.10, and phosphorus not to exceed 0.04. The standard discard is 9 per cent. Test rails must bend cold to 80-ft. radius without sign of cracking and a 6-ft. piece on supports 3 ft. apart must stand a 15-ft. drop of a 2,000 tup. These rails are 33 ft. long, of the A. S. C. E. 85-lb. section, while the Chicago lines use the 90-lb. section. The large head is considered an advantage and the greater stiffness of other types of sections is not needed with the light equipment and rigid support of rapid-transit lines.

Rail Tests.—To determine the strength and quality of finished rails, a drop test is made, with a weight falling on a cropped piece of rail 4 to 6 ft. long, set on supports and having six 1-in. spaces marked along the head, at the middle, in order to show the elongation. The rail is usually tested with head down, so that the head will be in tension. The A.R.E.A. specifies a 2,000-lb. weight or tup and a 20,000-lb. anvil block on springs on a heavy foundation, with rail supports 3 ft. and 4 ft. apart for rails under and over 106 lb. per yard, standing (without fracture) one blow from the weight falling 18 ft. for rails up to 90 lb., 20 ft. for rails up to 120 lb., and 22 ft. for rails up to 140 lb. The New York Central (Dudley) specifications require that under a single blow of 18 ft. for 80- or 90-lb. rails, and 20 ft. for 105-lb. rails, the elongation must be at least 6 per cent for one 1-in. space or 5 per cent for two consecutive inches. The A. S. T. M. specifications provide for testing with head up and a drop of 15 ft. for rails up to 60 lb. and 21 ft. for those over 100 lb., giving the same results as the Dudley test. Rails which do not break under the drop test are nicked and broken to determine whether the metal is sound. An alternative quick-bend test specified by the A. R. E. A. is made with an hydraulic press of 350 tons capacity instead of a falling weight.

Expansion of Rails.—Since rails vary in length with changes in temperature, an open space must be left between adjacent rails. If this is not done, or if the joint bolts are screwed up so tight that the rails cannot slip in the splice bars, then in hot weather the rails in expanding may buckle the track either horizontally or vertically, creating a dangerous situation. The coefficient for linear expansion of steel for each change of 1°F. is 0.0000065, so that the expansion in inches for a 33-ft. rail in a 20° change of temperature would be $0.0000065 \times 33 \times 12 \times 20$, or 0.05148.

In laying rails, the proper width of joint opening is given by inserting strips of iron of the required thickness, these being removed after the joints are bolted up (see "Rail Joints"). Practice varies as to the width, but the tendency is to keep it small, and for 30- and 33-ft. rails a general rule is to make it $\frac{1}{4}$ in. for temperatures up to 25°F., $\frac{3}{16}$ in. up to 50°, $\frac{1}{8}$ in. up to 75°, and $\frac{1}{16}$ in. up to 100°F.; at higher temperatures the rails are laid in contact but without bumping them together. In tunnels, the width of opening may be $\frac{3}{16}$ in. up to 20°, $\frac{1}{8}$ in. up to 45° and $\frac{1}{16}$ in. up to 70°, beyond which the rails are laid in contact. Where heavy rails reach the same temperature as lighter rails, the expansion will be the same, but in practice less expansion spacing may be used for the heavier rail, since it takes a longer time to heat and the time of exposure to high temperature is short. The same applies to the 60- and 66-ft. heavy rails used in street railway track, where the rails are protected by being embedded in the paving. In Table V are given rules for expansion spacing.

TABLE V.—EXPANSION SPACING FOR RAILS

A. R. E. A.				Pennsylvania R. R.	
Temperature, degrees Fahrenheit	33-ft. rails, inches	39-ft. rails, inches	45-ft. rails, inches	Temperature, degrees Fahrenheit	60-ft. rails, inches
−20 to 0	5/16	3/8	7/16	To 25	7/16
0 to 25	1/4	9/32	11/32	25 to 37	3/8
25 to 50	3/16	7/32	1/4	37 to 50	5/16
50 to 75	1/8	1/8	5/32	50 to 62	1/4
75 to 100	1/16	1/16	1/16	62 to 75	3/16
Over 100	Close	Close	Close	75 to 87	1/8
				87 to 100	1/16

Creeping of Rails.—At many places the rails develop a tendency to creep or travel along the track, both up and down grade and generally in the direction of the heavier traffic on single track. This movement, which occurs particularly where the track is relatively light for the traffic, is due to a variety of influences, including the grades, direction of heaviest traffic, the action of braked wheels, the contraction and expansion of rails, and the wave motion of rails and track. It is a source of trouble at track crossings, turnouts, and drawbridges. The creeping of the entire track (rails and ties together) occurs sometimes on swampy country owing to the wave motion under traffic, but may be checked by the use of ties 10 to 12 ft. long with angle bars bolted to the middle of each rail and spiked to two ties to keep the rail from moving independently. The long ties may be anchored against piles in severe cases. Heavy track and roadbed construction is one of the best remedies for creeping. To prevent ordinary creeping, the flanges of angle bars have notches or holes to receive the spikes, but this use of the joint as an anchorage is being discouraged and special devices are being used extensively, (see "Rail Joints and Fastenings"). As a rule the joints are not slot spiked on bridges, the rails being allowed free movement in order to avoid undesirable strains in the structure. Where the creeping is continuous and excessive at bridges or trestles, an expansion joint made with switch rails is sometimes provided. The main rails are turned outward at the leaving end, and switch rails spiked against them make the connection with the main rails.

Rails for Curves.—To provide for the wear of the side of the rail head on curves, the Manning rail was designed with about $\frac{1}{4}$ in. additional metal on the gage side of the head, but extending only $\frac{1}{4}$ in. below the $\frac{5}{16}$-in. top corner so as not to give a full flange bearing when worn, as this might lead to derailments. Another special design is the "frictionless" rail for the inside of a curve, having a head only $1\frac{3}{4}$ in. wide in order to reduce friction and thus enable the inner wheel to slip backward in compensating for the unequal length of travel of the two wheels fixed to the same axle. While these rails were fairly successful, the result did not warrant the extra cost and complication involved. With ordinary rails, the worn outer or high rail is generally transferred to the low side, giving much the same result as the frictionless rail. Heavier rails are used sometimes, the Lehigh Valley R. R. using 110-lb. rails on curves where 100-lb. rails were laid on tangents. Special steels for curve rails and the use of short rails on curves have been mentioned.

Compound Rails.—Since a slight amount of actual wear in the head necessitates the removal of rails from the track, numerous designs have been made for compound rails in which the heads may be renewed, but they are complicated and costly and the parts would soon become loose

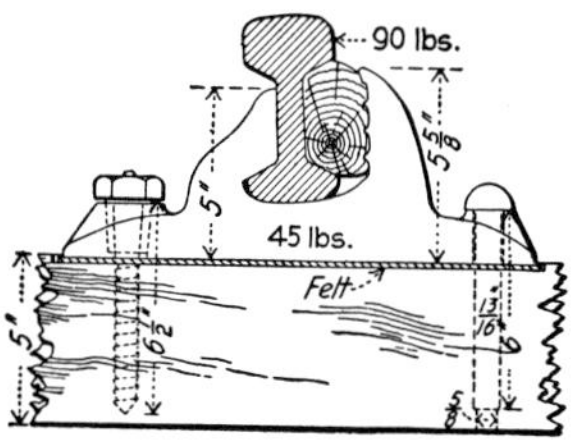

FIG. 21.—Double-head rail.

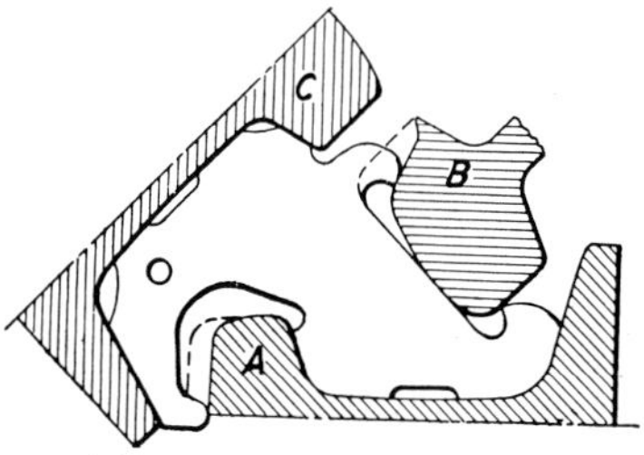

FIG. 22.—Gage for rail wear.

under the severe strains from heavy traffic. In some designs the rail was divided vertically, but, as a rule, a detachable head of flangeless T-section had the stem fitted into a grooved web. A compound rail used to a small extent for street railway track has a channel-shaped head with the flanges bent over the bulbed head of an inverted T forming the base. A powerful machine traveling on the track presses the flanges in, and, in making renewals, this machine cuts off the bent flanges to release the head.

Double-headed Rails.—The double-headed reversible rail, supported in chairs, was one of the earliest types in Europe, but as the indentations by the chairs made the reversed rails rough and liable to break, this form was superseded by the bull-headed rail, having the lower head only large enough to give a bearing in the chair and a hold for the "key" or wedge which is driven between the web and the chair to hold the rail in place. The bull-headed rail (Fig. 21) is the standard type for main lines in

England and is used to a limited extent in other countries. In the 100-lb. section the distribution of metal is approximately 50, 20, and 30 per cent for the head, web, and base, respectively.

Canting of Rails.—A practice which has come into use in this country in recent years, but which has long been general practice abroad, is that of giving the rails an inward cant or inclination so that their center lines will be approximately perpendicular to the inclined or coned surface of the wheels. In a majority of cases this inclination is 1 in 20, the same as that of the wheels, but a few roads use 1 in 38, 1 in 40, 1 in 44, or even 1 in 80. The cant should be given by means of metal tie-plates and not by adzing the ties, except that at switches and turnouts the ties may be adzed to bring the rails vertical. Investigation indicates that this practice reduces the lateral bending stresses in the rails on tangents as well as curves. On the other hand, wheels are soon worn to such an extent as to destroy the theoretical bearing or contact relations of wheels on rails. Rail canting is still experimental and its results are inconclusive, but it is claimed to have the following advantages: reduction of lateral stresses in rails, a better distribution of the load from wheel to track, a more central contact of wheel or rail, a more uniform wear of rail, and a reduction of maintenance expense by checking the tendency of rails to tilt outward and require periodical adjustment to proper gage.

Stresses in Rails.—Track is an elastic structure composed of flexible beams or rails on flexible cross-beams or ties resting on a yielding bed of ballast which affords varying degrees of support. A wheel load on the rails is distributed over several ties due to the stiffness of the rails and the yielding support, the length covered by this distribution increasing with the stiffness of the rail. The spacing of wheels also has an influence upon the load distribution. Probably the maximum driving axle loads are 75,000 lb. and 72,600 lb. for freight and passenger engines, respectively, on the Pennsylvania R. R., but the Baltimore & Ohio R. R. has 68,000 lb., while axle loads of 60,000 to 65,000 lb. are fairly common. In addition to the vertical static load there are dynamic effects at low joints and other irregularities, besides lateral thrust on tangent as well as curved track. Investigations by the Committee on Stresses in Railway Track indicate the following conclusions:

1. Stresses in rails are governed by the general track depression rather than the span between ties.
2. These stresses are less in heavy rails than in light rails.
3. They increase with speed of trains, this increase perhaps exceeding 50 per cent.
4. Both rail stress and track depression increase in direct proportion to the load.
5. Both stress and depression are at a minimum with a deep and well-tamped bed of stone ballast.

6. The rail is subjected to transverse bending, and the stress in the outer flange is usually much higher than that in the inner flange. Owing to the lack of uniform support by ties and ballast, the rails are subjected to severe and varying strains beyond those due to the rails acting as girders between adjacent ties. In fact, the rail may rather be considered as a bridge between driving wheels as piers, with upward pressure from the ties as loads.

Electric locomotives are generally claimed to be easier on the track than steam locomotives, owing to the lighter wheel loads of the former and the absence of reciprocating parts and wheel counterbalances. On curves, however, electric locomotives are found to develop excessively high stresses in the rails, due largely to the pivoting action of the engines.

For the entire length of an engine and tender, the track will be depressed slightly below the normal surface of the rail head, with an extra depression under each wheel. With the heavy loads of modern engines this depression may be $\frac{1}{8}$ to $\frac{3}{8}$ in. Under a heavily loaded wheel, the compression of tie and its bending to a full bearing may be 0.05 in., the compression of 24-in. stone ballast, 0.15 in., and the compression of roadbed, 0.15 in., or a total of 0.35 in. The deflection of an 85-lb. rail between two ties 22 in. c. to c. will not exceed 0.01 in. In front of the engine truck the rail is slightly above normal level, due to the wave motion of rail and track. With a rail carrying a moving load, the part under each wheel has compression in the head, tension in the base, and a shearing stress across the rail section at or near the ties which are carrying the load. At a little distance on either side of the wheel, the stresses are reversed, the head being in tension and the base in compression. These continuously reversing stresses help to develop any incipient flaws or defects in the rail. An important step in comparing the efficiency of different rail sections and determining the stresses in rails under load is the introduction of the stremmatograph, invented by P. H. Dudley and used in his extensive investigations as well as in the present investigations of the Committee on Stresses in Railway Track. This instrument records the stresses under varying conditions of load, which is a matter beyond mathematical analysis.

Wheels and Rails.—The conical form of wheel tires, already mentioned, tends to steady the position of the wheels upon the rails and keep them in line, but it has little effect in reducing the slip of wheels in passing around curves, although it has been considered theoretically to have such an effect due to the tendency of the outer wheel to crowd the rail and thus ride on a larger diameter. Flat-topped rails and cylindrical wheels have been proposed, but there are difficulties in rolling such rails with the desired pressure to compact the metal, and experience indicates that the rails would wear to a convex surface and the wheels to a concave surface, as with the ordinary form of construction. The A. R. A. standard section for wheel tires is an inclination of 1 in 20 for the tread (changed in 1907 from 1 in 25) and this is followed almost universally; among exceptions is the New York Central R. R., which has adopted a slope of 1 in 38 to increase the contact area, while the Baltimore & Ohio R. R. has tried the steeper slope of 1 in 13 with a view to reducing the flange wear by keeping the wheel flanges away from the rail heads. The taper of the wheel has little relation to the spreading of rails and widening of gage by traffic.

Contact areas of wheels on rails are approximately ovals $\frac{3}{4}$ by 1 in. to 1 by $1\frac{1}{4}$ in., with areas of 0.40 to 0.6 sq. in. or as much as 1.07 sq. in. on light rails, but the shape and area vary with the condition of wheel and rail and the deflection of the rail.

Driving wheel tires, usually $5\frac{1}{2}$ in. wide, have a flange $1\frac{1}{8}$ in. deep and $1\frac{1}{4}$ in. thick at the throat, with a fillet or curve of $\frac{5}{8}$-in. radius running into the slope of the tread. In worn wheels the depth of flange must not exceed $1\frac{1}{2}$ in. and the thickness must not be less than $\frac{15}{16}$ in. For car wheels, the flange is 1 in. deep and 1 in. thick (at $\frac{5}{8}$ in. below top of rail) with fillet of $\frac{7}{8}$-in. radius. The wheel gage or distance back to back of tires is from $53\frac{1}{8}$ to $53\frac{1}{4}$ in. On tangent track the form of the wheel tires and flange has little relation to wear of wheel or rail, but on curved track it exerts an influence, since the wheel flange cuts and grinds the side of the head of the outer rail. At switches and frogs, tires worn hollow, or with outside false flanges, may cause both wear and damage, while wheel flanges worn thin or sharp may lead to derailment. Any damage to track by defective wheels should be reported promptly by the section foremen.

Severe stresses in rails are caused at times by the counterbalancing, wheel spacing, and load distribution in the locomotives. The effect of normal load on the wheels is increased to a serious extent by the weights added to the driving wheels to counterbalance the rotating and reciprocating parts of the machinery. This counterbalancing is a difficult problem, and its solution is largely a compromise which results seriously to the track, as shown by the investigations on stresses in track. In fact, direct injury, vertical or lateral bending, or even fracture, of rails may be caused by faulty counterbalancing or by "dead" engines hauled at high speed with rods disconnected, so that the balance weights in the wheels have no counterbalance and exert powerful hammer blows upon the rails. It is desirable that motive-power engineers should give more attention to the track aspect of the counterbalance, and cooperate more effectively with engineers of the track department, in the interests of the railroad.

Wear and Life of Rails.—The life limit of rails in track is sometimes determined by depth of wear of rail head or by the amount of tonnage carried, but surface wear or abrasion is rarely the cause of removal. In practice, rail renewal on main track is governed largely by crushing or deformation of the heads under heavy wheel loads; by uneven surface or battered and depressed ends, making a rough-riding track, or by the desirability of laying heavier rails on account of increased loads or traffic. A wear of $\frac{1}{8}$ in. in a length of 24 in. may be the limit for end wear. There is always demand for the heavier "relayer" rails from the main track to replace old light rails on branches and lines of lesser traffic, so that rails are likely to be removed from important tracks long before they have become worn, in order to serve this purpose and also to enable main tracks to be kept in first-class condition with the new rail. A heavy rail which has lasted 10 or 15 years in the main track may give the same or greater length of service on branch lines and then perhaps be in condition for yard or sidetrack use.

Head wear and abrasion of 60- to 80-lb. rails on tangents is given in Wellington's "Economic Theory of Railway Location" (1887) as $\frac{1}{16}$ in. per 14,000,000 to 15,000,000 tons, with an allowable depth of wear of $\frac{3}{8}$ to $\frac{5}{8}$ in., or a total tonnage of 150,000,000 to 200,000,000 tons. There is modern record of 18,400,000 tons of traffic per $\frac{1}{16}$ in. of wear on 100-lb. rails, while some of the early high-carbon 6-in. 100-lb. rails of the New York Central R. R. carried 250,000,000 tons with a wear of only $\frac{1}{16}$ in. On the other hand, it has been estimated that with modern conditions of traffic and heavy wheel loads the allowable wear limit would be reached with 60,000,000, to 120,000,000 tons. Traffic figures are not generally available, however, owing to the great variation in density of traffic on different divisions or engine districts and the impracticability of keeping tonnage records for the traffic over various kinds of rails.

In tunnels, there is likely to be abnormal wear due to dampness, use of sand on damp rails, and also to the gases from the engines, unless the rails are protected by painting. Corrosion may occur locally due to the dripping of moisture from coal, ore, and refrigerator cars. A peculiar character of rail wear, which occurs more particularly on electric railways, is the formation of transverse corrugations on the head, causing a rough and noisy track. These corrugations are about 1.18 to 1.57 in. apart and 0.0078 to 0.0118 in. deep. The cause has never been determined definitely, but it has been attributed to the vibration of the rolls of the rolling mill as the rails pass through, and also to uneven or unbalanced action of the wheels. Tight gage, with rails bedded in concrete, is considered a probable cause on street railways, where it is general practice to smooth or face the rails in track with grinding machines. Besides the effects of locomotives improperly counterbalanced, as noted above, rails may be injured by wear or "burning" of the steel when driving wheels slip badly in starting trains.

Curve Wear of Rails.—On curves, the side wear of the outer rail is often the limiting factor and this varies with the intensity of traffic, quality of steel, degree of curve, types of locomotives, and especially the speed of trains. The rail should be removed before this side wear reaches the bottom corner of rail head, for if this point is reached, the wheel flanges will strike the splice bars. If the top of the outer part of the rail head is not badly worn, the rail may be used on unimportant lines or sidetracks, being turned so that the unworn side becomes the gage side of the rail.

Templets for permissible limit of wear are used by the Southern Pacific Ry., the Chicago, Milwaukee & St. Paul Ry., and the Pennsylvania R. R., this last being shown in Fig. 22. A caliper gage for side wear is used by the Nashville, Chattanooga & St. Louis Ry., and outside rails must be removed when the width of head is only 2 in. at $\frac{5}{8}$ in. below top of rail (or $1\frac{3}{4}$ in. wide on branch lines). Rails on the inside of curves must be removed when the thickness or depth of head is reduced to 1 in. or when the joints are badly battered. The Norfolk & Western R. R. uses a gage which is seated on the rail base and enables the top and side wear to be measured; for a 100-lb. rail the allowable depth of wear is $\frac{7}{16}$ in. on tangent and $\frac{3}{8}$ in. on the low rail of curves; on the high rail, the vertical limit is $\frac{1}{4}$ in. measured at center line of rail, with $\frac{3}{4}$ in. side wear measured at the gage line $\frac{5}{8}$ in. below original top of rail. The Northern Pacific Ry. allows $\frac{3}{16}$ in. wear of 90-lb. rails on the low side of curves. Such specific limits are only for a general guide, with sound rails, so that roadway officials may remove rails whenever they judge this to be desirable for safety or to improve traffic conditions.

Rail Failures.—Defects in the steel are usually the primary causes of rails failing in the track, and such failures may be cracks, excessive wear, or actual fractures. Apart from rails broken and injured by train wrecks or broken wheels, the failures are classed as follows, the figures representing the average percentages of the different classes: (1) ordinary break, complete fracture without internal defect, 31 per cent; (2) break due to internal transverse fissure, 20 per cent; (3) split web, 21 per cent; (4) split head, with or without internal defect, 17 per cent; (5) crushed head, but showing no cracks on the underside, 7 per cent; (6) flowed head, with the top metal rolling out to the sides but without indication of crushing of the metal, 3 per cent; (7) broken base, with cracks or portions split out, 1 per cent.

Failures in bessemer steel rails are principally split heads and web and base failures, which are likely to be detected before breakage or total failure of the rail. A serious trouble with modern open-hearth rails is failure due to transverse fissures, which are interior defects and difficult to discover. It has been claimed by manufacturers that these fissures result from overstrain by heavy traffic and wheel loads; but this is not sustained by the records, which show that of many similar rails few develop transverse fissures, while rails that do fail usually contain several such defects. Further, rails from reheated blooms seldom fail in this way. The indications are that failed rails have interior defects which form the starting points for progressive fracture under stress of traffic loads. Approximately 50 per cent of the failures occur in the *A* rails, or rails made from the top of the ingot. To reduce these troubles, some roads specify 25 per cent discard from the top of the ingot, which is effective but involves extra cost. Other methods are to roll the *A* metal into tie-plates, if the mill is equipped to do this, or to have a nick-and-break test on a piece from the top end of each *A* rail.

Reconditioning Old Rails.—Rails are removed from track on account of general wear or of depressed and battered ends. Those of the first class may be made available by rerolling to a lighter section, the rails being heated only to such a temperature as not to affect the chemical composition of the steel. When elongated by rolling, they are sawed to length and drilled for the bolts. Thus an old 110- or 100-lb. rail may make a new 90- or 95-lb. rail, or a 90-lb. may be reduced to an 80-lb. section. Owing to the high cost of treatment, including transportation to and from the mill, the railroads have not used this plan extensively, but with the increasing use of rails weighing more than 100 lb. per yard it is likely to be more generally adopted, since it produces new rails of weight suitable for lines which do not need the heaviest rails. Rerolling mills do considerable business in reducing old rails to very light sections for mine, industrial, and contractors' tracks. A notable application of the autogenous welding process is in the building up of battered or broken rail ends and frog points in the track, thus avoiding the expense of removal, sawing, drilling, transportation, and replacement.

But the cropping of rails with worn ends is practiced extensively and is done by railway forces. It may be done at points where reclamation plants are established, but several roads use portable plants which can be sent to various division points and thus avoid transportation of rails. In the latter case, also, the rails are likely to be practically uniform in size and wear, requiring little or none of the sorting necessary for a miscellaneous collection of rails. A crop of 12 to 18 in. is usually sufficient to include all bolt holes. For a portable plant, a 60-ft. steel car is equipped with engine and boiler, 44-in. disk saw, two drills with two or three spindles, roller supports for rails, and a pneumatic hoist for handling rails. The stationary plant may have also a shear, a forge, and a rail bender. With a force of 20 men from 6,000 to 10,000 ft. of rail may be sawed and drilled in an 8-hr. day.

Old rails shipped to the reclamation plant or the stores department may be classified as follows for use after sawing or reconditioning: (1) relayer rails for main running and passing tracks; (2) for secondary running and passing tracks and heavy-traffic branches; (3) for light traffic branches; (4) for sidetracks and yards; (5) for frog and switch repairs, (6) scrap.

Rails for the first three classes should be not less than 20 to 24 ft. long, free from kinks and distortion, and sorted or matched as to height, varying by $\frac{1}{32}$ in. For sidetracks and yards, 15-ft. lengths may be used. It is generally considered that rails from curves should not be used on tangents or on curves of greater radius than the original. An economic practice in reconditioning rails while in the track is the raising of low joints by means of new and heavier splices.

CHAPTER V

RAIL JOINTS AND RAIL FASTENINGS

Requirements of Rail Joints.—Since early days the rail joint has been a weak point and a source of trouble as well as a cause of continual labor cost in track maintenance, but a brief consideration of the work imposed upon it will indicate why it is difficult to design an effective or permanent rail joint. This device must provide for longitudinal movement of the rails due to expansion and contraction under temperature changes, but at the same time it must hold in proper alinement, both vertically and longitudinally, the free ends of two independent rails, and must transmit a continual succession of loads of varying intensity, with reversal of stress between loads.

On one hand are the rapidly applied loads from fast passenger trains, with the severe hammer blows from driving wheels carrying static loads of 18,000 to 30,000 lb., followed by car wheel loads of 6,000 to 18,000 lb. On the other hand is the heavy pounding of slower freight trains with driving wheel loads of 25,000 to 30,000 and even 36,000 lb., followed by wheel loads from 7,500 lb. for empty cars to 15,000 or 30,000 lb. for loaded cars. These loads may be aggravated by impact due to worn or loose parts, by faulty counterbalancing of the engines, or by flat spots on worn car wheels. In addition there are lateral blows and pressures due to the swing of cars and engines on tangents and their action in passing around curves.

For a perfect joint and a smoothly riding track, the splicing must be such as to make the joint as strong and stiff as the solid rail; but it must also be as flexible as the rail, so as to return to normal position after the deflection under each of the innumerable loads, and thus carry the wave motion of the rail uniformly along the track. The joint also must be simple in construction and of relatively low cost, although an apparently expensive joint may be economical by reason of long service and low maintenance. An obvious way of reducing joint troubles is in increasing the length of rails. Tight bolting is necessary to maintain good joints and track, and it also increases the lateral stability of the joints. But excessive tightness may cause the splice bars to grip the rails so hard as to prevent their proper slip with temperature changes; the expansion opening will then concentrate at relatively loose joints instead of being distributed uniformly. Even lubricating the contact surfaces, as is now done frequently in applying new splices, may not prevent this gripping of the rail.

At each rail joint when under load there are two beams fixed at one end and loaded at the other, each carrying half the load. The deflection of each beam is nearly ten times as great as that of the solid rail carrying the same total load and considered as a beam. The strength and stiffness of the rail itself counts for little at the joint, and does not distribute the load over several ties, as it does in the solid rail. Consequently, the joint ties have to take practically the full impact of the load, besides being subjected to a rocking or pumping motion. For this reason, these ties are likely to go down more than the others, causing low joints. Tamping the joint or shoulder ties to a firm bearing will not compensate for weak splices, so that extra cost for bars of sufficient strength and stiffness is more economical than continual labor cost for raising and tamping low joints, besides producing a better condition of track. When joints begin to go down and the contact surfaces get worn, great improvement may be effected by the prompt application of new splices, which in some cases will bring the rail ends up again. This practice also tends to increase the life of the rails materially.

Expansion Spacing and Rail Wear.—Much wear of the rails at the joints is caused by the wheels jumping over the deflected rail ends and cutting the receiving rail joint beyond the joint. Thus the end of the receiving rail tends to be permanently depressed, its incline up to the normal level being flatter than that of the leaving rail. It has been shown that the wheel, in spite of springs and vertical play, does not follow the deflection of the rail ends, but jumps the depression (especially as the curves are convex) and strikes the receiving rail a few inches from the end, causing the rail to wear or cut at that point. The drop at the expansion gap between the rail ends is a negligible factor as to shock, for it would be almost imperceptible. Heavy rails and stiff joints will minimize the deflection and consequently the pounding, noise, and wear at the joint.

Expansion movements of rails in the joint are by no means uniform, owing largely to varying tightness of the splices, since it is possible to make the bolts so tight that movement of the rail is restricted. As a result, the movement travels until a relatively loose joint permits free movement, or the force becomes so great as to buckle the rails or joints. With a cumulative contraction movement there will be an excessive width of gap at the loose joint. Difficulty with tight joints may be reduced by lubricating the contact surfaces (a practice which is on the increase) and by providing for a bolt tension which will maintain the strength of the joint and still allow the rails to slip. Tests indicate that, with a bolt tension of about 13,000 lb., a temperature change of 12°F. should develop force enough to slide the rails in the splice bars.

Splice Joint.—The type of rail joint in practically universal use in all countries is the splice joint which was invented by R. L. Stevens in 1830 and began to come into general use about 1855. It consisted originally of two fish plates or flat bars fitted between the inclined surface of the head and base of the rails and bolted through the webs. With the inclined fishing surfaces or contact surfaces, the tightening of the bolts wedged the plates firmly into position. About 1868 the angle splice bar was devised, having a flange which extends over the rail base to give lateral strength, and this flange also forms a means of anchoring the rail by driving spikes into notches in the flange of the bar. If this flange projects only slightly beyond the rail base, the spikes may be crowded aside by creeping rails, so that in some bars the flange is wide enough for punched spike holes. The flange may either stand clear of the tie or take a bearing upon it. There is a growing opinion that the rail joint should not

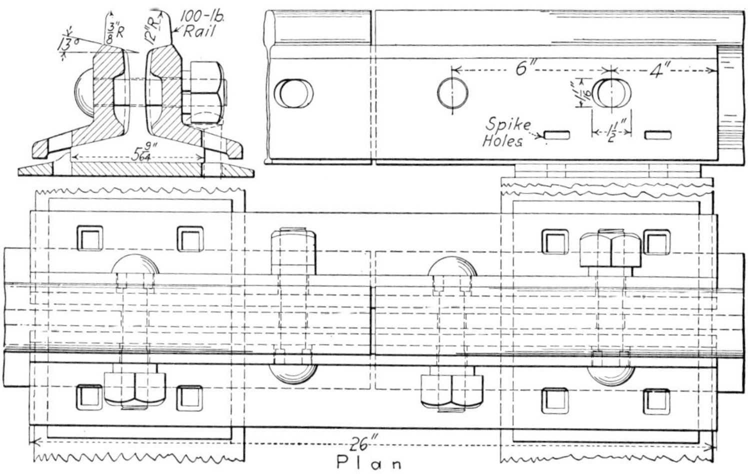

FIG. 23.—Rail joint; Chicago & Northwestern Ry.

thus be made to serve for an anchorage, as noted further on, and that no part of the joint should project down between the ties.

A difficulty in joint design is in making it equal to the solid rail in strength, stiffness, deflection, and elasticity. If deficient in strength, it soon becomes a depressed or low joint; if too strong, it forms an anvil upon which the rail heads are battered. Increased strength is provided sometimes by increasing the thickness of the bar or by forming heavy horizontal ribs along the top and bottom of the web. In the New York Central (Dudley) angle-bar section, the metal is distributed with a special view to combining stiffness with sufficient elasticity to give long service before the bar takes a permanent set, while the web is thin enough to stand clean punching without distortion or warping. Varying ideas in splice bar section are shown in Figs. 20, 23 and 24.

Splice bars of Z-section have been used in a number of joints, an example of which is the Bonzano joint (Fig. 24), in which the broad flange of the bar has its ends resting on the ties but is bent down to a vertical position between the ties. Such bars have been used more particularly in Europe, sometimes with bolts put through the lower webs to cause the flanges of the bars to grip the rail base. There is some objection to having any part of the joint extending below the base of the rail, and ample stiffness can be secured without the depending flanges, if desired. Of the innumerable joints invented, many are quite impracticable, and few of those tried in service have been found superior to a good spliced joint. Trials of experimental joints may fail to give reliable results, as such joints are often put on new rails or when the track is surfaced, and are assembled with special care. An ordinary splice joint applied with equal care and under the same conditions might show results even better than those of the trial devices.

Steel for Splice Bars.—Splice bars made of soft steel having only 0.15 to 0.20 per cent carbon have given trouble by lack of elasticity resulting in depressed or low joints with heavy and continual expense for tamping. The present tendency is to use high carbon bars, often with heat treatment to increase the efficiency. With such bars secured by bolts of high tensile strength, rail joints can be made practically equal to the solid rail and capable of standing severe service. Heat treatment is effective principally on steel of not less than 0.5 carbon. That marked increase in strength of joints may be effected by the use of heat-treated bars is indicated by tests made on the Pennsylvania R. R. in 1915 with joints placed on supports 20 in. apart and subjected to blows from a 2,000-lb. tup. The results are summarized in Table VI. These tests also showed little advantage of six-bolt over four-bolt joints in bars of about the same length, while the latter had the advantage of fewer parts. Similar tests on the Santa Fe Ry. showed that the strength and rigidity of the joint depends upon the strength of the bar, and that the strength of the steel increases with the carbon content, when the other elements are constant, and the heat treatment and quenching are the same.

For splice bars, the New York Central R. R. and the A. R. E. A. specifications call for open-hearth steel with carbon 0.35 to 0.60 per cent and not more than 0.8 manganese and 0.04 phosphorus; the bars to be punched, slotted, and shaped or straightened at not less than 1,382°F. (1,470° for A. R. E. A.) and then oil quenched. Test specimens must show a minimum tensile strength of 100,000 lb. and a minimum elongation of 12 per cent in 2 in., and must bend cold 90 deg. around a pin whose diameter is three times the thickness of the piece, without cracking on the outside of the bend. Or a full section of the bar may be bent cold around a pin whose diameter is three times that of the greatest thickness of the bar. In section, the bars vary considerably, as will be seen by the illustrations, but many are reinforced for stiffness by heavy ribs. The 24-in. bars for the 130-lb. rails of the Pennsylvania R. R. have an area of 13.68 sq. in. per pair and weigh 3.88 lb. per inch, while the finished weight is about $89\frac{1}{4}$ lb. For the 100-lb. rails these figures are 9.30 sq. in. and 2.64 and 61 lb., respectively. For the 38-in. bars of the New York Central R. R. 105-lb. rails, the area and weight per pair are 9.08 sq. in. and 95.4 lb.

TABLE VI.—TESTS OF RAIL JOINTS

Splice bars	Number of bolts	Steel	Elastic limit of joint, head up	Ratio of elastic limit of joint to elastic limit of 100-lb. rail
Plain	4	Untreated	50,000	29
Plain	4	Treated	125,000	71
Plain	6	Untreated	70,000	40
Plain	6	Treated	125,000	71
Continuous	4	Untreated	130,000	74
Continuous	4	Treated	210,000	120
Bonzano	4	Untreated	120,000	69
Bonzano	4	Treated	255,000	120
Plain, with base plate	4	Treated	240,000	137
Plain, with base plate	4	Treated	255,000	160
100-lb. solid rail	..	Untreated	175,000	
100-lb. solid rail	..	Treated	280,000	

Weight of bars per pair: plain, 69 lb. and 79 lb. for four-and six-bolt; Continuous, four-bolt, 94 lb.; Bonzano, four-bolt, 75 lb.

Breakage and Wear of Angle Bars.—Former serious troubles from broken splice bars have been reduced greatly by the general use of heavier splicing; and may be kept low by maintaining well-tamped joint and shoulder ties so as to prevent low rail joints. Breakage may occur in different ways, owing to varying conditions of load and of support in the ballast, while the continual transfer of load from one rail to the other causes continual reversal of strains in the splice bars as each wheel passes. There is an upward movement of the rail ends with a wheel approaching the joint, followed by deflection with the wheel crossing the joint, and then a second upward movement after the wheel has passed the joint tie. As a result, fractures begin mainly at the top of the bar due to the tension caused by these upward movements. In some rail sections the width of fishing surface does not admit of sufficient vertical metal in the splice bar section to make the strength of splice equal to that of the rail, so that deflection at the joints is greater than in the solid rail. The result is not only to bend the bar downward but to cause wear of the contact surfaces immediately under the ends of the rails; the bars may even be nicked or notched by the sharp ends of the rail heads, thus promoting cracking when the top of the bar is put in tension by the upward movement of the rails. To eliminate this particular kind of wear, the Abbott bar is made with a slight depression in the middle of the top, so as to clear the rail heads, while in the "head-free" joint the splice bars have no contact with the underside of the rail heads. This last is a modification of the Continuous joint, shown in Fig. 24, in which the shape of the head-free bar is shown by dotted lines.

Respacing Joint Ties.—In track maintenance and rail renewals it has been general practice to respace the joint and shoulder ties to restore them to the original position at joints which have shifted by creeping rails, or to adjust them to joints in the new rails. This involves an immense amount of manual labor, a rail renewal gang of 158 men, sometimes having 68 assigned to the respacing of ties. Apart from the time and expense involved, this shifting disturbs the well-tamped ballast beds under the ties. It is doubtful if this work is necessary, especially with heavy rails and splices, and several railways pay no attention to the position of the ties in relation to the joints, after the track is once built. The results appear to warrant this practice.

Length of Bars and Spacing of Bolts.—Although there is great diversity of practice in joint details, the general tendency is toward short splice bars with four closely spaced bolts, but six-bolt bars are used to some extent. Long bars are represented by the 38-in. three-tie joints of the New York Central R. R., having holes spaced uniformly 5.6 in. c. to c., but bars 40 to 48 in. long and with bolt spacings of 8 to 12 in. are obsolete. Six-bolt joints with 27- and 26-in. bars are used by the Delaware, Lackawanna & Western R. R. and the Lehigh Valley R. R., respectively, while four-bolt joints with 26- and 24-in. bars are standard on the Chicago & Northwestern Ry. and the Pennsylvania R. R., respectively. A standard bolt spacing of $5\frac{1}{2}$ in. for either six or four bolts is recommended by the A. R. E. A. While the rails have round holes for the bolts, the splice bars have round and oval (or oblong) holes, the latter to fit necks on the bolts which prevent them from turning. Formerly, all round holes were in one bar and the oval holes in the other, but present practice is to alternate the holes in each bar and thus have half the bolt heads inside and the others outside the rail, in order to give a better distribution of pressure in making the joint tight. Particulars of some splice joints are as follows:

	Chicago & Northwestern Ry.	Pennsylvania R. R.	Lehigh Valley R. R.	Delaware, Lackawanna & Western R. R.	New York Central R. R.
Number of bolts	4	4	6	6	6
Weight of rail, lb. per yard	100	130	136	118	115
Length of bars, inches	26	24	26	27	38
Spacing, middle bolts, in.	6	$5\frac{1}{2}$	4	$4\frac{1}{8}$	5.6
Spacing, outer bolts, inches	6	$7\frac{1}{2}$	4	$4\frac{3}{4}$	5.6

Supported, Suspended, and Bridged Joints.—Rail joints may be classed as follows: (1) a supported joint resting on a tie; (2) a suspended joint between two ties and held up by the splice bars; (3) a bridged joint between two ties but seated on a bridge plate whose ends rest on these ties. The plain one-tie supported joint is little used, since it does not afford sufficient bearing for heavy loads, but a modified three-tie arrangement is the standard on the New York Central R. R. for main tracks carrying heavy traffic and wheel loads. In this case, the three-joint ties are only 5 in. apart, or $14\frac{1}{2}$ in. c. to c., and the 38-in. six-bolt splice bars extend over the three ties. The great majority of track has either the suspended or the bridged joint, the ties being then 6 to 10 in. apart or 14 to 20 in. c. to c.

Bridged joints, which are used mainly for heavy traffic lines, are of two distinct types: (1) with the bridge plate held to the rail only by the spikes; (2) with this plate bolted or clamped directly to the rail. Of the

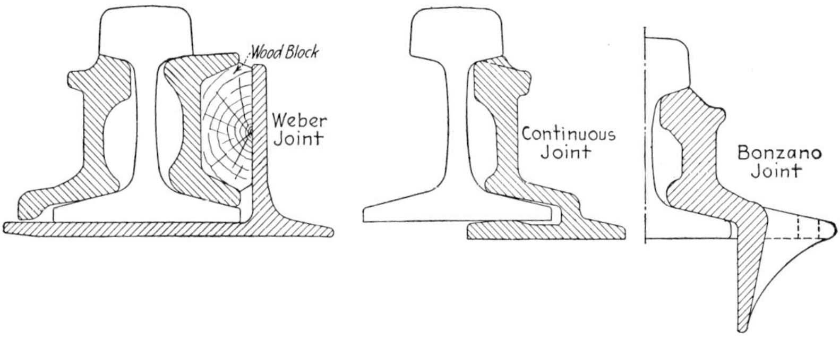

FIG. 24.—Special forms of rail joints.

former type, the Neafie joint on the Delaware, Lackawanna & Western R. R. has a grooved plate, and the Abbott joint has a flat plate with the edges bent downward to form stiffening ribs; the Carter joint, formerly used on the Chicago & Northwestern Ry., had a plate 26 in. long with a rib along each edge and the outer edge of rail base laid against the rib. Of the latter type, which is more effective and much more largely employed, there are two principal representatives—the Continuous and the Weber, (Fig. 24). In the Continuous joint, angle bars with double flanges engage both the top and bottom of the rail base; their bearing does not extend to the extreme edge of the rail but leaves this free, to avoid pinching or possible fracture. In the Weber joint, the rail ends rest on an L-shaped bridge plate with its web on the outer side of the rail and having a channel-shaped splice bar and wooden filler between this web and the rail; an angle bar is used on the inside of the rail. Both of these joints are made mainly in short lengths and with four bolts.

Broken and Even Joints.—In general practice, the rails are laid with broken joints; that is, the joint is opposite the middle of the opposite rail. Even or square joints, directly opposite each other, are used to some extent on tangents and on new track not fully ballasted. With broken joints, the pounding which tends to cause low joints is better distributed, especially when the track is well ballasted, so that maintenance work is reduced and the cars ride smoothly. Tracklaying proceeds more quickly with broken joints, since on curves the joints can run a little distance ahead of the middle of the opposite rail before it becomes necessary to insert a short rail in order to keep the joints in proper relation. For this reason, broken joints have been used on curves even when square joints were used on tangents. But to insure good riding track, this distance of joint from the center of opposite rail should not exceed 6 to 12 in. on both tangents and curves.

Miscellaneous Types of Joints.—Numerous forms of joints designed to eliminate the drilling of rails for bolts have been tried, usually with some arrangement of wedges, but none of these have come into extensive use. Several special forms of joints have flat bars or fish plates instead of the familiar angle bar; such a design is used on the Delaware, Lackawanna & Western R. R., in which the plates have heavy ribs at top and bottom to form a dumb-bell section which is reversible, the outside of the bar becoming the inside and giving unworn contact surfaces. The idea that shock at the rail joints is due to the expansion spacing of the rails led to early experiments with miter joints, in which the rail ends were cut at angles of 45 to 60 deg. This practice was followed on the Lehigh Valley R. R. until 1895, when it was abandoned on account of many rails breaking near the ends. This mitering of the rails does not affect the depression of the rail ends, which is the main cause of the shock. Another experiment was with the rail ends halved, placed side by side and bolted together. As used in Germany, this scarf joint was 8½ in. long with 26-in. splice bars and four bolts, the two middle bolts passing through both rails. To avoid weakening the rails by thus cutting the webs so thin, a rail was designed with head eccentric to the web so that only the head and base were cut away for the scarf, leaving the two webs of full section. As the bolt holes must be large enough to allow for expansion, the joint was helped only by the friction between the webs, which depended upon the tightness of the bolts.

Since the deflection of the rail ends is one of the troubles with rail joints, numerous designs of non-deflecting joints have been made. In England, some railways place a single cast-iron chair as a bridge resting on the two joint ties and giving a continuous support to the ends of the two rails, while other roads use a single long chair resting on one tie, as a supported joint. Both of these designs are used with bull-head rails. Another variation consists of two joint ties with connecting bridges to carry the rails, all made in one "twin" steel casting. Rail deflection is eliminated also in the supported joints of the Chicago Elevated Railways, with a tie under the joint, shoulder ties with only ½ in. spacing, steel tie-plates on all ties, and the whole supported rigidly by the steel girders. In a form proposed by Sir Bradford Leslie, an English engineer, wood joint ties held together by tie rods have two longitudinal blocks framed between them and having long base plates to receive the ends of the two rails, while ordinary tie-plates were placed on the ties.

Welding of rail joints is rarely applied in steam railway track, but it has been used in building up the worn ends of rails. A notable piece of work is the welding of three pairs of 66-ft. rails to form continuous rail lengths of 198 ft. through a long and heavily traveled street crossing on the Chicago & Northwestern Ry. In this way and at small expense, maintenance work at the crossing was reduced to a minimum.

Compromise or Step Joints.—Where rails of different heights meet, either in open track or where sidings have lighter rails than the main track, the splice must be so formed as to hold the tops of the rail heads in the same plane. Bent or cranked splice bars may suffice where there is little traffic, but it is better to have special joints in which half the length is shaped to fit one rail and the remainder to fit the other rail. Malleable and forged-iron splices are used for this purpose, with grooves shaped to engage the edges of the rail base tightly and prevent any independent vertical motion. Forged or cast-steel taper rails 8 ft. long with opposite ends shaped to the respective sections have been used less extensively. They have the disadvantage of requiring two joints close together, but the Southern Pacific Ry. avoids this objection by making the taper rails 16½ ft. long, two being planed from a 33-ft. rail.

Bridge and Expansion Joints.—Swing bridges usually have the end rails, 10 to 15 ft. long, projecting about 12 or 24 in. to take a bearing on the abutment and pivoted at the heel so as to be lifted clear of the fixed rails when the bridge opens. The rail ends may be cut square or may have a miter cut about 6 or 8 in. long, the miter joint making the easier-riding track. Provision must be made for lateral movement of the bridge, which may warp as much as ½ in., with possibility of the rail ends interfering as the bridge closes. As the lift rails cannot be spiked to the ties, each rail lies in a channel-shaped rest plate with flaring sides which guide the rail into alinement and hold it laterally when lowered into position. The joint may then be lined and locked by wedges sliding longitudinally between the rail webs and lugs on the channel and the abutment. To insure safety, the raising and lowering mechanism is interlocked with the bridge-locking gear, so that the lift rails must be pulled down into proper position before the signals can be released.

In a sliding-rail arrangement introduced on the Nashville, Chattanooga & St. Louis Ry., the 16½-ft. end rails move longitudinally, the rear end of each rail being bent outwards and fitted against a standard switch rail spiked to the ties. When the bridge is closed, the outer end of the rail enters a sleeve or socket which holds the end of the fixed rail on the abutment. With bascule or vertical-lift bridges, the meeting ends of the rail may be mitered and so arranged that the fixed rail laps over the bridge floor in the first case and the bridge rail laps over the abutment in the second case. Where the rail ends are not mitered to form a continuous bearing, they should be connected by sliding bars or easer rails on the outside, in order to give a bearing for the rims of wheels in crossing the gap.

On long-span steel bridges, expansion joints in the rails must be provided to take up the longitudinal movement of the structure. At the several expansion joints on the Quebec cantilever bridge, a taper connection is used for the track rails, as shown in Fig. 25. The tapering end of one rail is laid against a long casting of corresponding shape which

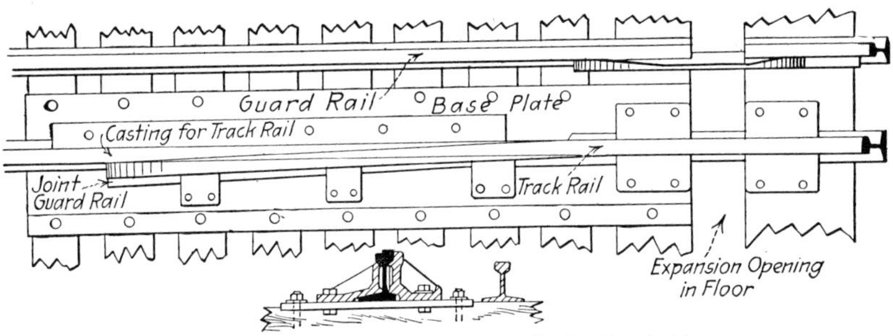

FIG. 25.—Expansion joint in track; Quebec bridge.

is formed integrally with a heavy base plate carrying the tapered rail. This latter is held against the casting by an outside rail but is free to slide longitudinally. The next main rail is spliced to the end of the cast manganese steel rail. The track guard rail is not extended across the gap in the floor, but has a steel casting laid against it and bridging the gap; one end of the casting is bolted to one rail and the other end is free to slide against the other rail, being held in position by guides.

On the Thebes cantilever bridge, a 7-in. open joint is left between the rails at the ends of the anchor arms, and the wheels are carried across the gap by an easer rail or bar, 4 ft. long, of rectangular section, laid against the outside of the two track rails; this bar is bolted to one rail and slides against the other, being held in position by guides on the base plate. To insure a good wheel bearing, the outside of the rail head is planed away to a width of about $\frac{3}{8}$ in. and guard rails with $1\frac{3}{4}$-in. flangeway hold the wheels over so that they engage the easer rail properly. On the Memphis bridge, there are three sets of expansion joints formed by pairs of short switch rails resting on the base plate and held in line by guides. The joint is $3\frac{1}{2}$ ft. long when closed and provides for a movement of about $1\frac{1}{2}$ ft. In another design, the rail ends are halved for a length of 12 in. and rest on a base plate bolted to the ties and having riveted side plates extending over the rail base. Neither rail has the web cut, but one is bent slightly to allow the web to overlap, the line on the gage side being maintained by planing the inside of the head of the bent rail.

Insulated and Bonded Joints.—Where track circuits are used for automatic block signals and at interlocking plants, the rail joints at the end of the block sections must be insulated to prevent the passage of the electric current. This is effected by means of a ½-in. filler or end post of vulcanized fiber or treated wood inserted between the rail ends, strips of fiber or other insulating material between the contact surfaces of the rails and splice bars, fiber bushings on the bolts, and fiber washers between the nuts and the bars. If the rails have a base support, this also must be insulated from one or both rails.

To provide free passage for the current within the insulated section, the rail joints are electrically bonded, usually by wires attached to copper studs set in holes drilled in the webs or bases of the rails. When these wires are laid between the rail and the splice bar, they sometimes get between the fishing surfaces; this may be avoided by placing them outside of the splice bar and holding them in position by thin washers on two of the track bolts. In some cases, particularly on electric railways and the electrified sections of steam railways where heavier currents for traction have to be carried, copper bonds are welded to the sides of the rail heads or webs.

Bolts and Nuts.—Track bolts for rail joints are generally of open-hearth or bessemer steel of medium carbon, having a tensile strength of about 55,000 lb. and a minimum elongation of 20 per cent in 2 in. There is increasing use of bolts of quenched carbon and alloy steel having carbon not less than 0.3 per cent, a tensile strength of 100,000 and 110,000 lb., respectively, and a minimum elongation of 12 per cent. These high-tension bolts are being used because those of ordinary steel may be stretched by trackmen applying excessive force on a long wrench, and also because the lateral force on the joint under load (due to the wedging action of the fishing surfaces) may reach 45,000 to 48,000 lb. per square inch. Track bolts are shown in Fig. 26 and in some of the illustrations of rail joints.

Whatever kind of steel is used for the bolts, the nuts are usually of soft, untreated steel of sufficient strength to develop the ultimate breaking stress of the bolts. Square nuts are commonly used, and those for bolts of heat-treated steel are ¼ in. longer on account of the greater strength required. It is usually required that the nut should go on for a few turns by hand and then with a 10-in. wrench, while the force required to screw it up tight with a 24-in. wrench should not be more than 60 lb. or less than 40 lb. Experiments on the Chicago, Burlington & Quincy R. R. indicate that with high-pressure nutlocks (noted below) it may be practicable to use nuts which can be screwed up entirely by hand, thus effecting a large saving of time and labor cost in the removal and application of joints.

In the Harvey grip-thread bolt, the thread is deeper than in the U. S. standard and is of the ratchet type, having the face at an angle of 10 to 15 deg. while the back of the thread is at about 45 deg. By using a nut in which the angle of the threads is slightly different from that of the bolt, a bending and upsetting or distorting action is secured which produces a very tight fit and resists the backing or slacking of the nut. When a nut is unscrewed, however, little damage is done to the thread. The hole in the nut is enlarged at the bearing face so as not to engage the

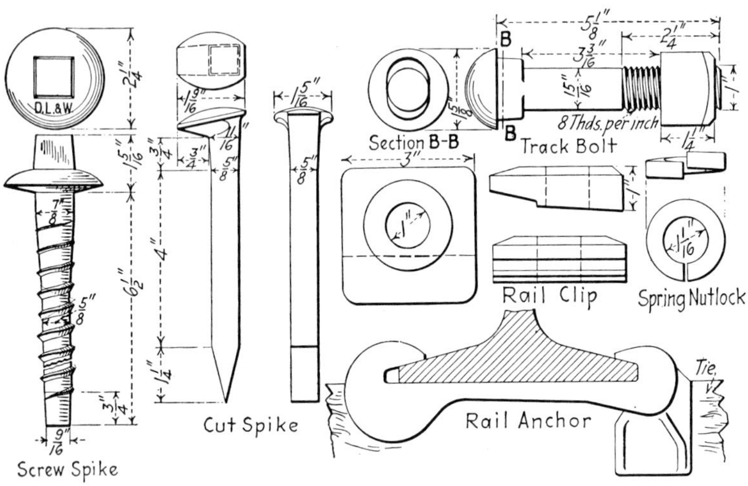

FIG. 26.—Rail fastenings.

bolt, thus insuring unaltered threads on which to screw the nut when taking up any slack.

Heavy rails may have 1-in. or $1\frac{1}{8}$-in. bolts in holes of $\frac{1}{16}$-in. greater diameter; $\frac{3}{4}$- and $\frac{7}{8}$-in. bolts are used with lighter rails and 1- to $1\frac{1}{4}$-in. bolts for frog and switch work. An oval neck fitting an oval hole in one splice bar prevents the bolt from turning with the nut, but the New York Central R. R. has used a square neck and bolt hole. It is general practice to put the bolts with heads and nuts on opposite sides alternately, so as to distribute the pressure and to prevent all the bolts being broken by derailed wheels. This plan cannot be followed with light or shallow rails, as the flanges of worn wheels might come in contact with the nuts. Bolts are packed usually in 200-lb. kegs, the number varying with size and type of nut; bolts $\frac{3}{4}$ by $3\frac{1}{2}$ in. average 240 to 260 lb. per keg with square nuts, or 260 to 280 lb. with hexagonal nuts; bolts 1 by $4\frac{1}{2}$ in. with square nuts average 115 to 130 per keg.

Nut Locks.—Loose bolts are a cause of wear in both rails and ties, so that economy in material and maintenance cost results from keeping the bolts tight. Spring washers or nut locks are usually applied between the nuts and splice bars to prevent the nuts from working loose under the influence of vibration. They are usually of the spiral type, flattened more or less by the nut so as to exert a pressure against it and so hold its threads tightly in contact with those of the bolt. But vibration and movement in the joint may cause the nut to slacken, thus reducing the pressure and efficiency of the nut lock; periodical tightening is required, therefore, to check the wear at loose joints.

In order to obtain and maintain a permanently tight joint, high-pressure washers or spiral-spring nut locks have been devised of such power as to maintain a constant high tension. The A. R. E. A. specifications require that after ordinary spring washers have been loaded three times sufficiently to close them flat, they must stand a pressure of 1,500 lb. for 1-in. bolts or 2,000 lb. for 1⅛-in. bolts without being compressed to within 0.01 in. of being flat or solid. But for high-pressure washers, the three preliminary applications are three loads of 28,000 lb. for 1-in. bolts, 36,000 lb. for 1⅛-in. bolts, and 45,000 lb. for 1¼-in. bolts, with subsequent test pressures of 8,000, 10,000, and 12,000 lb., respectively. One washer of this class is bent so that instead of having a single spiral turn, it has three bearing points and four curves, thus developing high pressure in being flattened under the nut.

A nut lock of a different type consists of a flat bar long enough for two bolts and bent so as to be concave to the splice bar at each bolt and convex between them. As the nuts are screwed up and tend to flatten the curves, the plate develops a resisting pressure of 16,000 to 26,000 lb. on each bolt. The bar also tends to equalize the pressure of the bolts, as compared with individual nut locks. Lock nuts or self-locking nuts, to eliminate the necessity of spring washers, are used to some extent, as for the grip-thread bolt noted above. But with heavy rails, the spring washers are usually employed.

Tests of Rail Joints.—To determine the strength of rail joints, a test recommended by the A. R. E. A. provides for a joint bolted up complete, with ⅜-in. expansion space and bolts not in contact with the rails. This joint is set on supports 48 in. c. to c. and loaded till the rail ends meet, the test being made with head up and head down. The results are compared with those of a continuous piece of rail of the same section and with the same span, the efficiency of the joint being the relation of its elasticity to that of the rail taken at 100. The rigidity is expressed as the percentage ratio of the deflection of the joint under the load which develops its elastic limit, to the deflection of the continuous rail under the same load.

Typical Rail Joints.—The standard joint of the Chicago & Northwestern Ry., with its 100-lb. rail, is of the suspended type, on ties 18 in. c. to c., and spliced with 26-in., four-bolt, angle bars which have the unusual feature of punched holes instead of slots for the spikes, the flanges of the bars extending about $1\frac{1}{2}$ in. beyond the rail (Fig. 23). These holes are 1 by $\frac{15}{16}$ in. and register with the spike holes of the tie-plates. Four 1-in. bolts with necks 1 by $1\frac{3}{8}$ in. are placed with heads alternating on opposite sides of the rail, and have square nuts with spiral washers or nut locks. The bolts are of quenched carbon steel with rolled threads, eight to the inch. In the rail, the bolt holes are $1\frac{1}{16}$ in. in diameter; in the bars, they are $1\frac{1}{8}$ in. and $1\frac{1}{8}$ in. by $1\frac{1}{2}$ in. For the three-tie joint of the New York Central R. R., with 105-lb. rails, the 38-in. bars are of oil-quenched, high-carbon steel with flanges slotted for spikes and $\frac{1}{16}$ in. clear above the bottom of the rail. The bolt holes are $1\frac{1}{16}$ in. in the rail and alternately 1 in. and 1 by $1\frac{1}{4}$ in. in the bars, all spaced 5.6 in c. to c.

With 130-lb. rails, the four-bolt suspended joint of the Pennsylvania R. R. has 24-in. splice bars weighing about $89\frac{1}{4}$ lb. per pair, with the middle bolts $5\frac{1}{2}$ in. and the outer ones $7\frac{1}{2}$ in. c. to c. The bolt holes in the rail are $1\frac{5}{16}$ in. in diameter and $1\frac{3}{16}$ by $1\frac{1}{2}$ in. These bars are slotted for the rails. The Lehigh Valley R. R. also uses the suspended type of joint for its 136-lb., 7-in. rails, with six $1\frac{1}{16}$-in. bolts, 4 in. c. to c. In the rails, the bolt holes are $1\frac{5}{16}$ in., while in the bars they are alternately $1\frac{3}{16}$ and $1\frac{3}{16}$ by $1\frac{1}{2}$ in. A base-supported, six-bolt joint with 118-lb. rails on the Delaware, Lackawanna & Western Ry. has 27-in. reversible splice bars of dumb-bell section, the joint not being utilized as an anchor for the rails. The middle bolts are $4\frac{1}{8}$ in. and the others $4\frac{3}{4}$ in. c. to c. In the outer bar, the end holes are $1\frac{1}{8}$ in. in diameter and the others $1\frac{1}{8}$ by $1\frac{5}{16}$ in., these positions being reversed in the inner bar. A double-shoulder or grooved base plate 27 in. long and $12\frac{3}{14}$ in. wide is fastened to the shoulder ties by screw spikes; the rails are held by cut spikes, but the shoulders of the plate are so made that the spike heads are kept $\frac{1}{8}$ in. clear from the top of rail base. The 1-in. bolts, $5\frac{3}{4}$ in. long under the $1\frac{1}{16}$-in. heads, have oval necks 1 by $1\frac{5}{16}$ in., with eight rolled threads per inch and square nuts with a recess in the contact face.

RAIL FASTENINGS

Track Spikes.—Two types of rail fastenings in general use are the cut or drive spike and the screw spike, of which the former is almost universal while the latter is used to some extent on heavy-traffic main track and on rapid-transit lines. The former is held by friction only, while in the latter this friction is supplemented by the mechanical bond of the spike thread in the solid wood. Both types of spikes are subjected to three forces, which by innumerable and continual repetitions tend to loosen the hold of the spike in the tie: (1) a vertical pull due to the wave motion of the rail; (2) an outward thrust due to lateral pressures and blows of the wheel flanges against the rail heads, both on tangents and curves; (3) an overturning tendency of the rail on curves, the rail tilting on its outer edge, drawing up the inner spike, or twisting its head and forcing the outer spike back into the wood. In addition, the neck of the spike is likely to be abraded or worn by the edge of the rail base, unless shoulder tie-plates are used, in which case the spike may be placed with its body or shank $\frac{1}{8}$ in. clear of the rail.

The continually repeated vertical force finally starts the spike and then gradually pulls it, while the lateral force enlarges the spike hole, the general effect being to loosen the spike and promote decay of the wood. Continual redriving and respiking results in "spike killing" the tie by the destruction and decay of the fibers around the spike, so that in badly maintained track the spikes can sometimes be pulled out by hand. A comparatively modern practice, employed mainly with screw spikes but to some extent also with cut spikes, is to keep the spike head about $\frac{1}{8}$ in. clear of the top of the rail flange, in order to allow free vertical movement of the rail and thus largely relieve the spike from the vertical pull. Tie-plates are sometimes made with stops to prevent the spike head from coming in contact with the rail.

Any lateral movement of the rail is objectionable, for it not only loosens the grip of the spike and damages the tie, but also causes a widening of gage, with consequent necessity for adjustment and respiking. Tie-plates help the spikes to resist lateral force by causing the inner and outer spikes to act together; by preventing the rail base from cutting into the tie, they also resist the overturning tendency of the rail.

Drive and Screw Spike.—For screw spikes, it is necessary to bore holes in the ties, this boring being done usually by machine at the tie yard, as the use of hand-boring appliances in the track is too slow and costly, except in renewing turnouts or other special work. The holes also aid the penetration of preservative during treatment. Two holes at each end are usually bored; sometimes there are four holes, but consideration must be given to the weakening of the tie as a beam, especially if additional holes should be bored for respiking. Boring the ties has been recommended also for cut spikes in treated ties, as the holes not only assist the penetration of the preservative but also prevent the crushing and tearing of the wood by the spike, while if the diameter is about $\frac{1}{16}$ in. less than the width of the spike the holding power will be increased. One difficulty is that the cut spike sometimes has a tendency to diverge and cut into the solid wood. However, at present, ties are rarely bored for cut spikes.

The main advantage of screw spikes is in the mechanical bond of the projecting thread, which cuts its way to a bearing in the solid wood. This bond depends in part upon the density of the wood, but there has been much trouble from trackmen hammering the spikes down or overdriving them with power spike setters. In the first case the wood is crushed and torn, while in the second case the threads cut and chop the wood and destroy the solid bearing. Screw spikes will work loose, especially in soft wood ties, even when properly placed, and then it is difficult to make them tight; occasionally they may become so rusted in the wood that they cannot be withdrawn.

In first cost, cost of application, and cost of maintenance the screw spike has shown at a disadvantage in some cases. Thus the cost per tie in track, including all labor and material, has been given as $1.778 with tie-plates and cut spikes; and $2.155 with tie-plates, screw spikes, and spike-hole lining; or $1,044 and $1,782, respectively, per 1,000 ft. of track. Annual cost of maintenance per foot of track with these two types of fastenings is given as 31.59 and 81.96 cts., respectively, with treated red oak ties, and 23.69 and 65.66 cts., respectively, with treated pine ties. Another record gives $338 and $593, respectively, per 1,000 ft. of track. Greater time is consumed, also, in placing and renewing screw spikes.

Favorable and unfavorable results with screw spikes are reported by different railroads. Thus they have given excellent service on the Delaware, Lackawanna & Western R. R. and the Lehigh Valley R. R., but have proved less satisfactory on parts of the Pennsylvania R. R. having somewhat heavier traffic, although this last road has used special devices to increase the holding power of the spikes. Their efficiency is greater with heavy rails and hardwood ties on well-tamped ballast, and with the unyielding base of track on elevated and subway lines. It can be said safely that screw spikes have proved good and economical rail fastenings in some roads, and that probably their efficiency and economy will be further developed.

Resistance of Spikes.—Numerous experiments have shown that there is great variation in the resistance or holding power of spikes in ties, due to difference in the species and quality of the wood. There seems to be little difference, however, between treated and untreated wood. Results obtained in testing machines are not directly indicative of track conditions. In the machine, the tie is held firmly and there is a direct and steady pull; but in the track, there is an irregular and continually repeated pull combined with lateral thrust, while the tie may have more or less vertical play in the ballast. It appears that as soon as the first resistance has been overcome, or the spike started, a much lower force is required to pull it; and if it is redriven, the initial resistance is not again obtained. If the spike head is kept clear of the rail instead of being driven tightly upon it, there should be less trouble with the loosening of spikes, as noted above.

As an average, a $\frac{9}{16}$-in. spike newly driven in a good oak or pine tie may have an initial resistance of 3,000 to 3,500 lb., while in inferior woods it may be only 1,800 to 2,000 lb. Resistances up to 7,000 lb. have been recorded but cannot be accepted as occurring in ordinary track. Screw spikes may average 5,000 to 9,000 lb. in oak and 4,000 to 7,000 lb. in pine. This resistance is maintained for a longer time than with a cut

Table VII.—Resistance of Spikes

Wood	Resistance to vertical pull		Movement in lateral impact tests	
	Cut spikes, pounds	Screw spikes, pounds	Cut spikes, inches	Screw spikes, inches
White oak	6,950	12,026	0.70	0.39
Red oak	6,299 to 8,935	10,962 to 14,686	0.71	0.43
Short leaf pine	4,038 to 4,620	9,182 to 9,805		
Long leaf pine	3,334 to 4,465	8,474 to 9,170	0.69	0.79
Red gum	2,540 to 4,383	9,205 to 10,000		
Chestnut	2,980	9,418	1.23	0.56
Beech	....	....	0.84	0.51

spike, owing to the mechanical bond of the spike thread in the wood. But under combined repetition of heavy wheel loads the threads pull through the wood, crushing and distorting the fiber and thus permanently reducing the hold of the spike. Test records with screw spikes have run as high as 12,000 to 14,000 lb.

In some of these tests the cut spike showed little slip until the maximum load was nearly reached, while the screw spike began to slip at about 66 per cent of its maximum load and did not reach this load until the thread had pulled out of its original seat. Further, the cut spike held about 75 per cent of the screw spike load up to the point where slip began and about 50 per cent of the maximum load on the screw spike. In lateral resistance tests, cut spikes have shown a resistance of 1,619 lb. in pine to 2,026 lb. in red oak; screw spikes ranked 20 to 80 per cent higher. In lateral impact tests made with a 100-lb. pendulum weight falling $1\frac{1}{2}$ ft. and striking five blows on a bar resting against the neck of the spike, all the spikes were bent and the cut spikes were partly pulled. Results of tests on cut and screw spikes are given in Table VII.

Cut or Drive Spikes.—The common track spike (Fig. 26) is $\frac{5}{8}$ to $\frac{9}{16}$ in. square and $5\frac{1}{2}$ to 6 in. long under the head, which is oval on top, about $1\frac{1}{2}$ by $1\frac{1}{4}$ in., with a projection of $\frac{3}{4}$ in. beveled 1 in 4 on the underside to fit the slope of the rail base. The lower end is tapered for about $1\frac{1}{4}$ in. to a blunt chisel edge. The A. R. E. A. specifications require bessemer or open-hearth steel of 55,000 lb. tensile strength, with an elongation of 25 per cent in 2 in.; a test spike must bend cold on itself without cracking on the bent portion and the head must bend upward in line with the face of the spike without cracking. Planks at road crossings may be fastened to the ties by crossing spikes $\frac{1}{2}$ in. square and 8 in. long under a $\frac{3}{4}$-in. square head; $\frac{3}{8}$ in. boat spikes 7 to 8 in. long may also be used. Track spikes are put up in kegs of 200 lb., averaging 490 spikes $\frac{1}{2}$ by 5 in., 335 spikes $\frac{9}{16}$ by $5\frac{1}{2}$ in., or 260 spikes $\frac{9}{16}$ by 6 in.

Experimental spikes devised to increase the hold in the wood have had twisted, grooved, and toothed or jagged surfaces, but none of these have come into use. The Goldie spike, with sharp diamond point, was used extensively at one time but has declined on account of the extra cost. A special 6-in. spike used by the Delaware, Lackawanna & Western R. R. for attaching tie-plates has a rounded point, a large head, and a body $\frac{3}{4}$ in. square with grooved faces and with corners projecting diagonally $\frac{1}{8}$ in. from the body.

Screw Spikes.—Good design and manufacture are essential to the success of screw spikes. In the A. R. E. A. design (Fig. 26), the shank or body is $\frac{7}{8}$ in. in diameter under the head for a sufficient distance to include the thickness of rail base and tie-plate; then it tapers to $\frac{5}{8}$ in., or $\frac{7}{8}$ in. over the $\frac{1}{8}$-in. thread, which has a pitch of $\frac{13}{32}$ in. The thread dies out $\frac{3}{4}$ in. from the end of the spike, and the body tapers to $\frac{9}{16}$ in. at the end. This smooth, tapered end is to allow the spike to be set firmly in the bored hole with a light tap, as hammering the threaded portion into the hole crushes and tears the wood fibers. A $2\frac{1}{4}$-in. mushroom head has the underside beveled 13 deg. and has a top or rectangular projection about 1 by $\frac{3}{4}$ in. and $\frac{3}{4}$ in. high for the socket wrench. The effective length in the tie is $5\frac{1}{2}$ to 6 in. Test spikes must show a tensile strength of 60,000 lb., with 20 per cent elongation in 2 in., and must bend cold through 180 deg. without cracking.

To insure the most efficient service, it is recommended by the A. R. E. A. that the spike head should not be in contact with the rail, but $\frac{1}{8}$ in. clear with rails of 100 lb. and over, or $\frac{1}{16}$ in. for specially stiff and substantial track. As a rule, the rail is held directly by the spike head, but in some designs the spike passes through a washer or clip which bears on the tie and rail; this clip relieves the spike from abrasion by the rail, and the hole may be placed eccentrically, so that by turning the clip, the width of track gage may be adjusted. Spike holes are bored usually by machine at the tie-treating plants, both for economy and to assist impregnation of the wood. If bored in the track, either by hand or by power devices, as may be necessary at turnouts or other special work, creosote should be poured into the holes before the spikes are placed.

Two devices used extensively in France to increase the hold and efficiency of the screw spike are the Thiollier and Lakhovsky fastenings. For the first, the bored hole is tapped with a spiral thread into which is screwed a steel helix of greater pitch so that the helix is compressed when in place. The screw spike is then inserted in such a way that its threads cut into the wood between the coils of the helix. Under conditions of American track this has not proved successful, and if the spike is so driven that its threads follow those of the helix, the friction is so reduced that vibration causes the spike to work loose. The Lakhovsky device requires the spike hole to be enlarged for 75 per cent of its depth to receive a split sleeve having a thread formed on the outside and a conical nut placed inside. This sleeve is dropped into the hole, and as the spike is screwed down, it draws up the nut, forcing the threaded faces of the sleeve outward into the wood. The extra cost of material and labor make such devices impracticable for general use, but the indications are that the screw spike alone may be made a sufficiently effective fastening.

Bolts and Wedges.—Bolts with clips or clamps are used largely for fastening rails to steel and concrete ties, as well as to unballasted steel decks on bridges. The clips are usually made to allow for adjustment of gage, and in some cases the rear end enters the bolt hole so as to take some of the lateral strain on the bolt. Twin or U-bolts, with the legs on opposite sides of the rail, are sometimes used. In foreign practice through bolts in wood ties are used to some extent, and the chairs for double-headed rails are fastened by round spikes, screw spikes, or wood trenails. In these chairs the rails are secured by wood or steel spring keys, driven in as wedges, the jaw on the chair being at an angle with the rail web. With some forms of steel ties, flange rails are secured by steel wedges or taper keys which engage the side and top of the rail base and are driven between the rail and a lug on the tie (see Fig. 19).

Tie Bars.—At switches and sharp curves in yards, the rails are sometimes held to gage by flat tie bars having the ends bent up or lugs formed on the ends to engage the rail bases. Such attachments might be useful on sharp curves in the main track in order to relieve the rail fastenings. A block between the lug and the web of the rail would help to resist overturning. Adjustment of gage may be made by fillers in the lugs, by adjustable lugs, or by making the bar in two pieces with the adjacent ends forged and threaded for a turnbuckle. Tie rods passed through the rail webs are common in street railway track, being covered by the paving; but in ordinary track, such bars would be objectionable, as likely to trip employees and to be damaged by loose brakebeams or other parts hanging from trains.

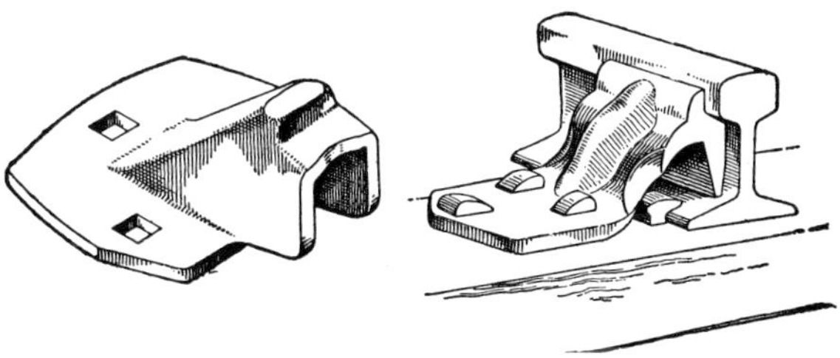

FIG. 27.—Rail braces.

Rail Braces.—On account of the lateral thrust on the rails at curves and turnouts, the spikes on the outside are reinforced usually by braces (Fig. 27) which bear against the web and head of the rail and are spiked to the tie, although shoulder tie-plates are sometimes considered to give sufficient lateral support on curves. Long tie-plates should be used to hold the rails and braces (and are sometimes made integral with the braces), as otherwise the value of the brace may be reduced by the outer edge of rail cutting into the tie, thus throwing the load on the top of the brace and tending to loosen it by pulling the spikes. In general, such braces are placed on every third or fourth tie on curves of 4 to 6 deg. and every third or second tie on sharper curves, but the Southern Pacific Ry. places them on every third tie for curves up to 2½ deg. and on every other tie for sharper curves. Both rails should be braced, as the inner or low rail receives a severe thrust from slow trains; the two braces should be on the same tie. If frog guard rails are not bolted or clamped to the track rails, they should be backed up by two or four rail braces, according to length, seated on tie-plates carrying the two rails.

Rail Anchors.—Creeping of rails along the track, as mentioned under "Rails," is the cause of considerable trouble and extra maintenance work, particularly at turnouts and track crossings. At the latter, especially, it results in excessive wear and a rough-riding track through the crossing being out of line. At switches and interlocking plants long anchor bars are frequently used, having one end bent up and bolted against the rail web, while the flat portion lies parallel with the rails and is spiked to several ties. Although it is still general practice to spike the splice bars to the ties in order to anchor the rails, many roads prefer to leave the joints free and to anchor the rails in some other way (see "Rail Joints").

One method is to anchor each rail at the middle, allowing it to expand in either direction, so that there is no cumulative action. For this purpose, the Chicago, Burlington & Quincy R. R. has used a pair of angle bars 5 in. long, bolted to the middle of the rail and slot spiked to the tie; this design was modified by making the flange long enough to rest on two ties, to which it was spiked. As extra bolt holes in the rails may be undesirable, there has developed an extensive use of rail anchors which are clamped or wedged on the rail base and bear against the side of the tie. Such a device is shown in Fig. 26. With any tendency of the rail to travel, the fastening wedges to a tighter grip. Since the longitudinal force may be sufficient to move the ties in the ballast, enough anchors must be used to absorb or resist the force developed at any particular point. Another device, which has been used particularly on elevated railways, is a tie-plate having a lug which laps over the rail base and has a setscrew to exert a powerful pressure on the rail.

An example of the economic importance of preventing rail creep is afforded by a double-track crossing where the two parallel tracks of one line moved 2½ in. per month in opposite directions. Each adjustment or straightening of the crossing cost about $95 for labor. But after rail anchors had been applied at a cost of about $6, the movement was permanently stopped, eliminating the periodical expensive maintenance charge.

CHAPTER VI

SWITCHES AND FROGS

As railway wheels have flanges on the inside to hold them on the rails, it is necessary to provide special appliances to enable trains to pass from one track to another. This diversion is effected by a short, diagonal, curved track or turnout (Fig. 28), connecting the two tracks. The turnout consists of a switch with movable rails to direct the train along one or other of the tracks as desired, a frog having grooves to enable the wheel flanges to cross the rail intersection, and lead rails connecting the switch and frog. Further details of turnouts are given in Chap. XXI. Switches are of two types: split and stub, of which the former is now universally used, while the latter is obsolete.

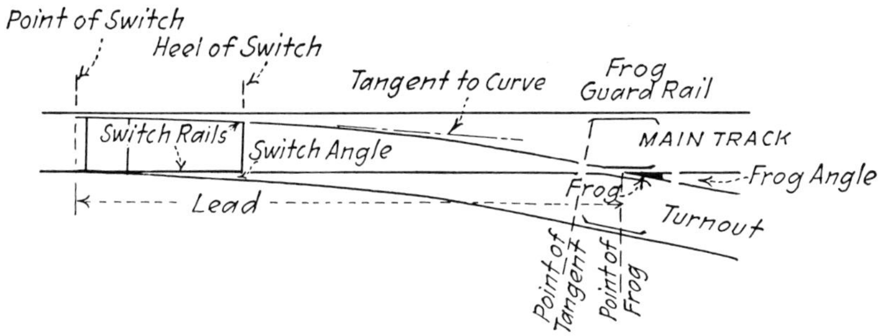

FIG. 28.—Turnout with split switch.

Split Switch.—In the split or point switch (Fig. 29), there are two movable tapering rails having their rear ends spliced to the lead or track rails, while their opposite ends are planed to a feather edge so as to fit close against the main or stock rails. The two switch or point rails are connected to move as a unit and have a throw or lateral movement of $3\frac{1}{2}$ to $5\frac{1}{2}$ in. When the switch is set for the main track, one switch rail lies against the turnout stock rail, so that wheels cannot follow the curved turnout rail; at the same time, the opposite switch rail stands clear of the main track stock rail, leaving an opening for the flanges of wheels traveling along the main rail. With the switch set for the turnout, one point rail lies against the main track rail so as to engage the wheel flanges and divert the wheels onto the turnout lead, while the other rail stands clear of the turnout stock rail, allowing passage of the flanges of wheels along that rail.

A kink or bend in the turnout stock rail is formed 8 to 12 in. ahead of the point rail by means of a rail bender, so that when the switch is set for the main track, the gage sides of the stock rail and switch rail will be in exact line. At the heel of the switch, the distance between gage lines of stock and switch rails is 5 to 6½ in., with a clearance of 3 or 3½ in. between the rail heads. The switch angle is the angle included between the stock and switch rails when in contact. A facing switch is one in which a train approaches the points, so that its route will be governed by the position of the switch rails. A trailing switch is one approached from the heel. Thus any switch may be either facing or trailing according to the direction

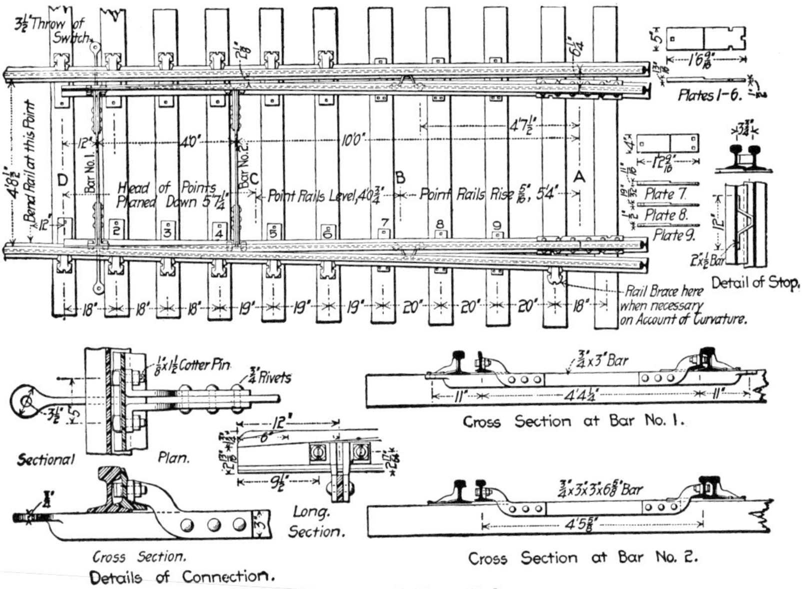

FIG. 29.—Split switch.

of the train. A switch is also right- or left-handed according as the turnout leads to the right or left of a man standing on the main track ahead of the switch and facing the turnout.

Switch Rails.—For main track turnouts the length of switch rail is usually 15 or 16½ ft., sometimes 18 ft. for turnouts with frogs of No. 6 to No. 10, and 30 or 33 ft. for turnouts of higher angles. Yard switches may have 10- to 12-ft. switch rails. The thin tapering portion of the rail is generally reinforced by flat bars bolted to both sides of the web or by a T-bar bolted to the gage side. At the extreme end of the rail, the part bearing against the rail head is planed or chamfered on the outside for 18 or 20 in. to give an end thickness of ⅛ in., and has the top also planed away on an incline so that the extreme end or point is 1½ in. below the top of the stock rail. With this arrangement, wheel flanges will not strike the point and will not engage the side of the switch rail until they reach its heavier and stronger section.

To protect the point further, a beveled block or pad is sometimes bolted to the gage side of the stock rail just ahead of the switch, so that wheel flanges that may be hugging the rail will be pushed out far enough to clear the switch point. Fixed guard rails placed ahead of the switch to line up the wheels so that they will take the turnout properly are now little used, except for switches on curves. For turnouts carrying heavy traffic, the switch rails are frequently made of manganese steel on account of its high wearing properties; or a point of cast manganese steel 24 to 30 in. long may be spliced to the blunter portion of the switch rail, especially for switches on curves or where side wear is excessive. For turnouts from curves the switch rails are straight, except at very sharp curves, as on industry tracks. In such cases, and for sharp turnout curves on rapid-transit lines, the switch rail on the turnout side is often about 18 in. longer than the other, with a guard rail opposite extending almost to the point of the switch rail. At switches on right-angle curves of the Chicago Elevated Rys., a cast manganese steel housing protects the switch point; when the switch is set for the curve, the open switch rail lies within this housing.

At the heel joint, a block or filler is placed between the switch and stock rails and secured to them by through bolts, which also pass through the angle splice bar on the inside of the joint between the switch rail and lead rail. This is more convenient than making an ordinary spliced joint at the heel. Since long and heavy switch rails would not move freely if bolted tight at the heel, the bolt holes in this rail may be reamed out to take a gas-pipe thimble or bushing placed on the bolt and bearing against the heel block and angle bar, so that the bolts may be drawn tight without gripping the rails. This practice is sometimes considered objectionable on account of the loose splice joint. On the Pittsburgh & Lake Erie R. R., therefore, the heel joints are bolted up tight, but the lead rail is so spiked that it swings or bends at a point $7\frac{1}{2}$ ft. back of the heel of the switch. In this way there is the flexibility of a 24-ft. rail, but with the expense of switch points only $16\frac{1}{2}$ ft. long. Stops or lugs riveted to the web of the switch rail and bearing against the stock rail support the former when subjected to pressure in guiding the wheel flanges. With switch rails of less than 12-ft. length there will be one stop about $3\frac{1}{2}$ ft. from the heel joint; longer rails may have the first stop about 5 ft. from the joint and one or two others at 4-ft. spacing.

In order to carry the false flanges of worn wheels, the switch rail is usually slightly higher than the stock rail. Thus, with a $16\frac{1}{2}$-ft. switch rail the rise begins at about 4 ft. from the point and reaches $\frac{1}{4}$ in. at 7 ft.; this elevation continues to the heel and then runs out in about 5 ft. on the lead rail. The elevation is given by the varying thickness of the slide plates or riser plates which carry the switch and stock rails.

Switch Plates and Tie Rods.—On all ties from heel to point of switch the stock and switch rails on each side rest on tie-plates or slide plates, while the stock rails should have rail braces on the outside to resist lateral pressure. These plates are of varying thickness under the switch rail in order to give it the elevation noted above. On the head tie or head block, a single long plate may be used, extending under the points of both the switch rails. A similar plate may be placed on the tie ahead of the points to insure exact width of gage. Tie-plates should be placed on all ties under the lead rails, except that at the less important sidetrack and yard turnouts these plates may be used only at, and opposite, the rail joints.

The switch rails are connected near the points by a tie rod or head rod to which is attached the operating rod from the switchstand, this latter rod passing under the rails. One or two other tie rods will be used in ordinary turnouts; sometimes three or four at high-speed turnouts or yard switches having heavy traffic, while the less important yard switches may have only the head rod. Flat bars are generally used, with riveted lugs which are bolted to the web or base of the rail. The bars should lie between ties closely spaced, so as to be protected against derailed wheels and from loose parts dragging from trains. At sharp turnouts, as on elevated and other rapid-transit lines, a tie rod is sometimes attached to the extreme ends of the switch rails to insure against any play or bending of these thin blades.

To permit adjustment of gage, the tie bars may be made in two parts, with the adjacent ends forged round and threaded to fit a turnbuckle; an alternative plan is to use eccentric clips on the bolts which attach the rail lugs to the rods. Where the use of track circuits makes it necessary to insulate the tie rods, they may be in two parts, the adjacent ends being bent up and bolted together, with a fiber insulating block between them, and with fiber bushings on the bolts. Another method is to lap or butt-strap the ends of the flat bars, fiber insulating strips being placed in the lap or between the bars and the splicing straps.

Example of Switch.—In the typical switch shown in Fig. 29, the switch rails are bolted to a heel block, and connected by the head rod and one tie rod. Opposite *B*, are the stops which bear against the stock rails. A gage plate is placed under the points, and each pair of rails rests on a tie-plate. The stock rails are supported against lateral pressure by rail braces. The switch rails, reinforced by two bars, have the points chamfered for 18 in. and are planed to $\frac{1}{8}$ in. thickness at the end.

Switch Locks.—Switches at interlocking plants are mechanically locked in position. At non-interlocked switches it is sometimes desirable to insure that the point rails are in exact position and also to prevent their movement after the switch has been set. For this purpose main-track facing switches are sometimes equipped with switch locks, usually operated in connection with the switchstand. One device consists of a rod parallel with the head rod and having lugs to engage the bases of the switch and stock rails. After the switch is thrown, the rod is revolved automatically on its axis so that the lugs stand above the level of the base of the rail and prevent lateral movement. Before the position of the switch can be changed, the rod must be revolved, which may be effected by a treadle on the switchstand, thus lowering the lugs clear of the rail. In another design, a locking bar parallel with the switch connecting rod has two holes and slides in a sleeve having a locking bolt or plunger moving parallel with the rails. This bolt, operated by a bell-crank connection, must be withdrawn from one hole in the locking rod before the switch lever can be shifted; it must also engage the other hole in the rod before the switch lever can be placed in its final position. The lock may be connected electrically with a distant signal.

Spring Switches.—If an engine or train on one track trails through a switch that is closed, or set for the other track, it is likely to damage the

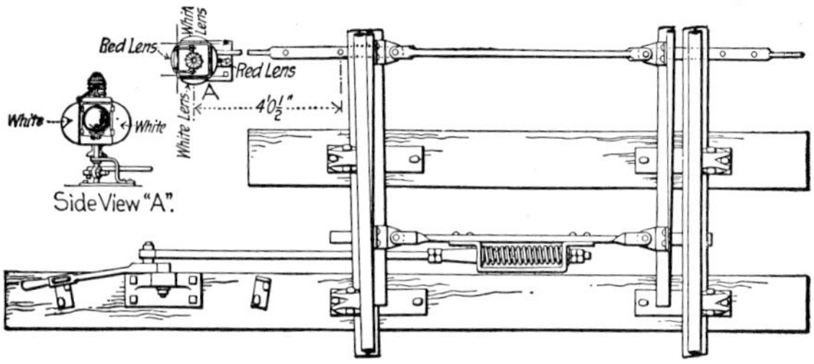

FIG. 30.—Spring switch with low switchstand.

switch and may be derailed. Although such movement is rarely permissible, it may be provided for by a spring switch or spring switchstand. In any case, the engine or train must pass entirely through the switch before backing, as reverse movement would otherwise get wheels on both tracks, resulting in derailment.

With the spring switch (Fig. 30), the connecting rod from the switch stand is attached to a spiral spring in a frame or stirrup riveted to the head rod, the spring being stiff enough to serve as a rigid connection in ordinary operation by the switchstand lever. But if a train trails through the closed switch, the wheel flanges wedging between the switch rail and stock rail will force the former outward by compressing the spring. This spring throws the rail back as soon as the pressure is released, the switch being left in its original position after the train has passed. Such repeated snapping or sudden movement of the switch rail under long trains, as each wheel or truck passes, is destructive to all moving parts. Consequently, the spring switch is now little used except in industrial plants and on electric interurban lines, where trains consist of only one or two cars.

A special application of the spring switch, however, is at the departure ends of passing sidings or the departure tracks of freight yards, to avoid the expense of providing switchmen or the alternative delay and cost of stopping a train after pulling out while the brakeman closes the switch and rejoins the train. As used for this purpose on the Santa Fe Ry., the spring is on the switch connecting rod and its strap is riveted to the switchstand rod. In order to reduce the racking due to frequent rapid operation by the wheels, the head rod of the switch has attached to it a rod connected to an oil buffer or dashpot on the side opposite to the switchstand. Signals control the movement of trains at these switches. In some cases, the siding switch is operated electrically from a distance, thus eliminating the alternatives of a switchman or a spring switch (see "Sidetracks").

Continuous-rail Switches.—To eliminate the interruption or break in the main rail at switch and frog, special devices have been designed for the purpose of having both the main rails continuous. The principle of operation is that with the switch set for the turnout, inclined rails will engage the wheels and elevate them so that their flanges will pass over the main rail (see "Frog Substitutes"). Owing to the sharp elevation of the turnout rails, these devices are not adapted for main-line or high-speed turnouts, but are used to a small extent at passing sidings, where speeds would not exceed 20 m.p.h. In the Wharton switch, there is one inside switch rail and one outside switch rail; the latter, on the turnout side, is shaped to engage the treads of the wheels and is on such an incline that the wheel flanges are carried over the stock rail. The McPherson switch, of generally similar type, was used extensively at one time on the Canadian Pacific Ry.

Slip Switches.—It is sometimes desirable to connect a number of parallel tracks with a diagonal ladder track where there is no room to

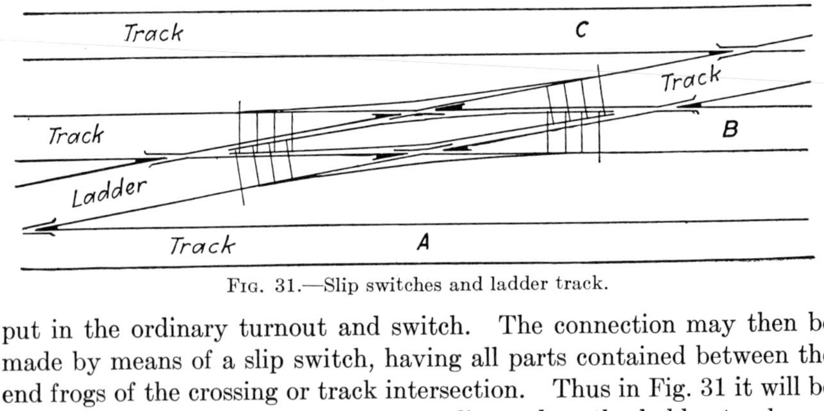

FIG. 31.—Slip switches and ladder track.

put in the ordinary turnout and switch. The connection may then be made by means of a slip switch, having all parts contained between the end frogs of the crossing or track intersection. Thus in Fig. 31 it will be seen that a train on track *A* may be diverted to the ladder track and thence to either of the parallel tracks *B* and *C*.

The slip switch is used extensively in track approaches to large passenger stations. At the south approach to the Chicago Union Station, where 6 main tracks connect with 14 station tracks, there are 33 double slips with No. 8 and No. 9 movable-point frogs. In the No. 8 slip switches, the length is 54 ft. 7 in. between switch points and 76 ft. 2 in. between the frog points of end frogs; switch rails are 15 ft. long, with $4\frac{1}{2}$-in. throw, and the movable-point frogs are 12 ft. long with 4-in. throw. At the St. Paul Union Station the slip switches have No. 4 frogs. Slip switches may be used also at junctions, but must then provide for higher speeds. At the Germantown junction of the Pennsylvania R. R., where a four-track line divides into two double-track lines, there were slip switches 115 ft. long, having No. 15 frogs and 28-ft. switch rails to insure easy passage of high-speed trains.

Three-throw Switches.—In special cases, two turnouts may diverge in opposite directions from the same point, requiring a frog (crotch frog) at the intersection of the curved lead rails at the middle of the main track

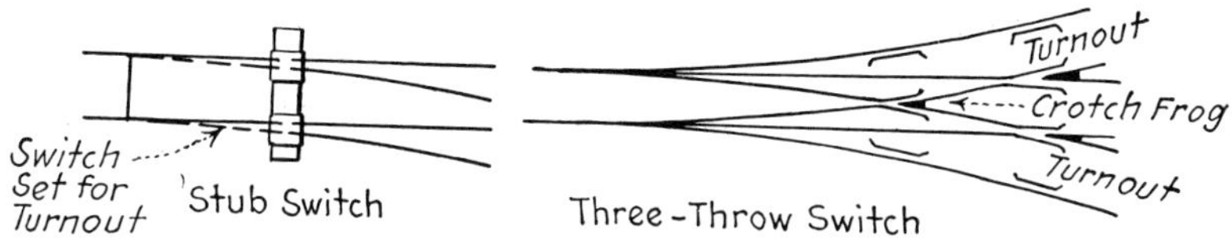

FIG. 32.—Stub switch and three-throw switch.

(see Fig. 32). It is advisable to set one pair of switch rails slightly in advance of the other, in order to keep the points distinct, although this plan throws the crotch frog off the center line of the main track.

Stub Switch.—In this old device (Fig. 32), now obsolete except for occasional use in industry or minor yard tracks, there are no tapered point rails. Two of the track rails have their heels (or the ends farthest from the turnout) spliced to the adjacent rails, while the other ends are free to slide laterally so as to line up with either the main track or turnout rails. Thus the switch rails are pivoted at the facing end, instead of at the trailing end as in a split switch. They are fastened together by tie rods so as to move together and are usually spiked to the ties for a few feet from the heel, so that they will be sprung to an approximate curve when the switch is set for the turnout. The free ends of the switch rails slide between stops on a base plate or head chair, while the ends of the main and turnout rails are rigidly fastened to the chair. This form of switch is not safe, efficient, or economical for carrying trains; its parts are subject to excessive and rapid wear and battering, and it is severe upon the rolling stock and wheels which pass over it.

FROGS

At the intersection of the main and turnout rails beyond the switch, there is installed a device termed a frog, which provides the gaps or openings necessary for the passage of wheel flanges, forms a support or path for the wheel treads, and guides the wheel to the proper connecting rails. In the typical frog (Fig. 33) the main parts consist of a point, V-shaped in plan, and two wing rails. The lead rails from the switch are connected to the wing rails, while a main track and a turnout rail are connected to the diverging ends of the frog point. The end of the frog nearest the switch is called the toe, while the rear end, behind the point, is the heel. Wheel flanges ride through the open spaces or flangeways between the wing rails and the frog point. The narrowest part of the flangeway, just ahead of the point, is the throat. The standard width of flangeways is $1\frac{3}{4}$ in.,

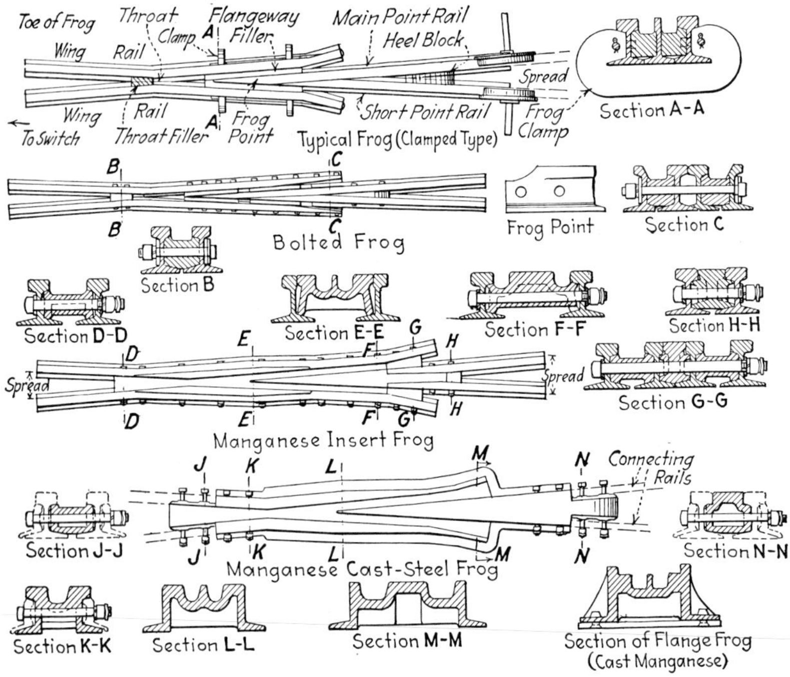

FIG. 33.—Types of frogs.

with $1\frac{7}{8}$ to $2\frac{1}{8}$ in. at the throat, while the depth should be not less than $1\frac{7}{8}$ in. at the fillers placed at the throat and between the frog point and wing rails to maintain the proper width. The actual point, about $\frac{1}{2}$ in. wide, is a few inches back of the "theoretical point" at the intersection of gage lines.

It will be seen that a trailing wheel on the main track will ride along the frog point, cross the flangeway, and continue along the wing rail and lead rail to the switch. In the same way, a wheel on the turnout will ride along the opposite side of the point to the other wing rail and lead rail to the switch. Frogs may be made right and left, with the lead rail wing sufficiently longer than the other wing to bring the switch points opposite each other without cutting rails.

Rail Frogs.—In the ordinary frog, built up of track rails, the point is planed to shape from the main rail, and the turnout rail is shaped to dovetail into the point. Fillers are placed between the rails, as noted, and in the V at the heel is a heel block or raising block which has an inclined surface at the rear so as to engage and raise the false flanges of hollow-worn wheels so that they will not wedge into the V and tend to burst or split the frog point. This block may be spiked to the ties to anchor the frog in place. It is sometimes made to extend beyond the rails of the frog point to form a support against which the track rails may be bolted. A filler block at the toe may also be extended beyond the wing rails to form a similar attachment for the lead rails. To give a flaring entrance to the flangeways at the heel, the ends of the wing rails may be bent outward or may have the inside of the heads planed away. Several types of frogs are shown in Fig. 33.

Rail frogs are usually of the bolted type, put together with long, through bolts 1 to $1\frac{3}{8}$ in. in diameter according to the weight and size of rails used. In clamped frogs, the point and wing rails are held together by two transverse clamps or yokes about $1\frac{3}{4}$ in. thick, and 3 in. deep, set edgewise, and shaped to fit against the outside of the wing rails. These clamps are driven up tight upon the frog and are secured in position either by stay rods passing through them, parallel with the rails, or by a stop bolted to the rail web. Riveted frogs, having the rails riveted to a large base plate, and generally without fillers, are little used and only for light rails. Tie-plates are placed on every tie under the frog, with one plate on a tie directly under the point. For heavy traffic, a base plate about 4 ft. long may be placed under the frog, this being secured by bolts having clips on the rails and countersunk heads in the plate. These tie-plates and base plates are $\frac{1}{2}$ to $\frac{3}{4}$ in. thick.

Manganese Frogs.—Owing to the rapid wear of rail frogs under heavy traffic and the consequent expense for both material and maintenance labor, there is extensive use of frogs made of cast manganese steel, which will outlast five to ten ordinary frogs. In rail-bound manganese frogs, the casting forms the point, flangeways, fillers, gage sides of wing rails, throat, and heel block. The wing rails are track rails bent to embrace the castings, and the rails at the heel are fitted to its heel block, all being heavily bolted together. In solid manganese frogs, the entire frog is a single casting, having extensions at toe and heel which are drilled for the rail connections, angle splice bars being used on the outside of the joint.

Frog Guard Rails.—To hold the wheels in line so that they will take the proper flangeway and will not strike the frog point, a guard rail in each track is placed against the rail opposite the frog, leaving a flangeway for the wheels. As a rule, the guard rail (Fig. 34) is a piece of an ordinary rail 8 to 15 ft. long, spiked to the ties and secured to the track rail by bolts or clamps, the width of flangeway being secured by means of spacing blocks or fillers. Tie-plates should extend under both rails in order to prevent cutting of the ties and consequent displacement of the guard rail. As a rule, the guard rail is placed with its center about a foot ahead of the frog point, but at spring-rail frogs it must extend sufficiently far to protect the movable part of the wing rail. Its ends are bent outward to give a flaring entrance, guiding the wheel flanges into the flangeway.

Practice varies widely as to length of guard rails. The A. R. E. A. standards include three lengths; 8 ft. 3 in., 11 ft., and 16 ft. 6 in., with the straight portion 4 ft. 3 in., 5 ft., and 7 ft. 6 in., respectively. Their ends flare to give a clear entrance of 4 in., but in the two longer rails there is first a flare to $2\frac{1}{2}$ in. in 20 and 38 in., respectively. The Chicago & Northwestern Ry. uses an 8-ft. rail straight for 5 ft.; the Illinois

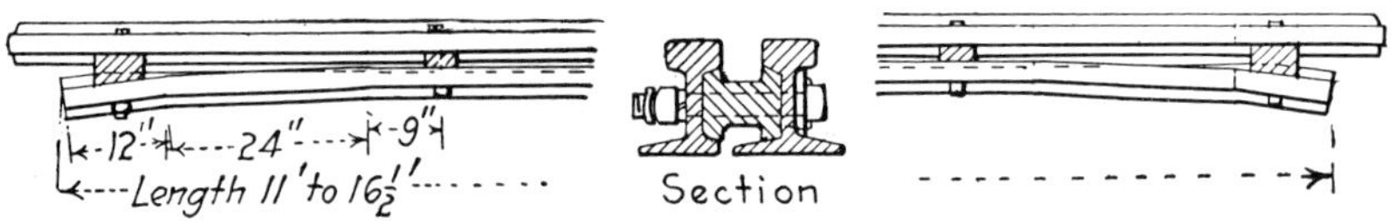

FIG. 34.—Frog guard rail.

Central R. R. and Pittsburgh & Lake Erie Ry. use 11 ft., with the straight portion 5 ft. and 7 ft. long, respectively. The Norfolk & Western Ry. uses 15 ft. for turnouts up to No. 9, $16\frac{1}{2}$ ft. for No. 10, and increasing lengths up to 33 ft. for a No. 29 frog; this last is held to the track rail by four clamps or yokes. For yards, a length of 7 ft. is often preferred, but guard rails are frequently omitted at yard switches.

A rail laid flatwise or upright and supported on chairs makes a good guard rail, the flat rail being laid with its head forming the flangeway. This arrangement enables light rails to be utilized and provides openings which facilitate the cleaning of snow from the flangeway. A specially rolled guard rail has its flange on the inner side depressed so as to fit under the base of the track rail, which thus holds the guard in position. Cast manganese-steel guard rails are used also, and may be formed with projections serving as tie-plates and rail braces. Some of these cast guards are about 1 in. higher than the running rails in order to guide driving wheels which have flat or blind tires.

For guard-rail flangeways the standard width is $1\frac{3}{4}$ in. on tangents, giving 4 ft. $6\frac{3}{4}$ in. between the gage sides of frog point and guard rail. On curves, the width will be increased by the amount of widening of track gage for the curve, the distance of 4 ft. $6\frac{3}{4}$ in. remaining constant. Easy riding of trains and wear of frogs depends largely upon maintaining the guard rail in proper position, with the proper width of flangeway.

Flange Frogs.—It will be seen that, with the arrangement as described above, the wheel passing through the frog is kept in position to take the proper path by its mate wheel on the other end of the axle engaging the guard rail. With a special type of frog, known as a flange frog, the wheel is guided by its rim, or the outer edge of the tread, engaging a rib or flange along the wing of the frog and extending about $1\frac{1}{4}$ in. above the level of the rail head (Fig. 33). Thus each flange frog eliminates the two guard rails, their attachments, and, consequently, the periodical work of spiking and adjusting the guard rail to maintain the proper width of flangeway. The ribs flare away from the gage lines of the frog at the throat, in order to give an easier entrance angle and less shock to the wheels. Frogs of this type are used principally in yards and passenger station approaches, where switching movements are numerous.

Foot Guards.—To guard against men getting their feet caught under the rail heads in the angles of frogs and switches and at the flaring ends of guard rails, these spaces should be closed with foot guards, as men thus trapped may be run over before they can get free. Wood fillers may be wedge-shaped blocks driven in, or blocks of triangular section, bolted to the rail web and having one side sloped from the bottom corner of the rail head to the edge of the rail base. Metal guards are preferable, either cast-iron blocks which serve also as spacers, or steel straps bent into a long U or stirrup and driven between the rail webs, where they are held in place by their spring action. At frogs, the heel block serves as a foot guard in the angle back of the frog point.

Frog Numbers.—Frogs are classified by numbers indicating the spread of the angle formed by the frog point, the spread of a No. 8 frog being 1 in 8. This number may be determined in different ways: (1) measure the distance, in inches, between the points where the width over gage lines is 2 in. and 3 in., this distance giving the frog number; (2) divide the length from heel to true point, on the gage line, by the width over gage lines at the heel; (3) take the sum of the width over rail heads at the heel and between rail heads at the toe, and divide this into the length from toe to heel, on gage lines.

A right-hand frog is for a turnout to the right, as seen by a man standing at the switch. A crotch frog is placed at the intersection of the lead rails in a double turnout (with three-throw switch) and its number is equal to that of the main frog multiplied by 0.707. The frog angle is that included by the point rails. Table VIII gives the angles of the frogs generally used, together with the degree of turnout curve from a tangent (see also "Switch Work").

It is desirable to have as few sizes of frogs as possible in regular service: No. 4 may be used for sharp turnouts on industry spurs into warehouses, but can be used only by switch engines of short wheelbase; Nos. 5 and 6 are used in yards and are about the sharpest for ordinary practice; No. 7 or No. 8 is used for ladder track connections, so as to occupy as little track space as possible, while Nos. 8 and 9 are used also in yard layouts;

Nos. 9 and 10 are the minimum for main line turnouts and crossovers; Nos. 12 to 20 are used for more important turnouts; and Nos. 14 to 24 or even No. 30 at junctions and the connections of double track to single track. Long frogs make a better-riding track and are more easily connected to the rails. Yard frogs may be 8 to 10 ft. long, the length increasing with the frog number. A little more than half the length is behind the point. Thus in a No. 10 frog the toe length may be 6 ft. 5 in., and heel length, 10 ft. 1 in.; while for No. 16 the lengths will be 8 ft. 8 in. and 15 ft. 4 in., respectively.

TABLE VIII.—FROG NUMBERS AND ANGLES

Frog number	Frog angle		Turnout curve		Frog number	Frog angle		Turnout curve	
	Degrees	Minutes	Degrees	Minutes		Degrees	Minutes	Degrees	Minutes
5	11	25	24	03	11	5	12	4	58
6	9	32	16	40	12	4	46	4	10
7	8	10	12	17	14	4	05	3	04
8	7	10	9	23	15	3	49	2	40
9	6	22	7	25	18	3	11	1	52
10	5	44	6	01	20	2	52	1	28

Spring-rail Frogs.—To avoid the shocks caused by wheels riding across the frog flangeway, light-traffic turnouts from main lines carrying heavy traffic and fast trains are very generally equipped with spring-rail

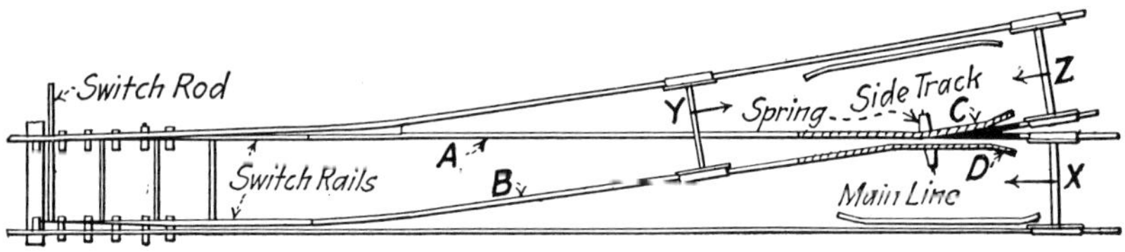

FIG. 35.—Turnout with spring-rail frog.

frogs which give a continuous rail for the main track. Crossovers also may have spring frogs. In this construction, shown in Fig. 35, *A* and *B* are the lead rails. The turnout wing rail, *C*, is hinged at the heel and is free to slide outward, but is normally held in close contact with the frog point by strong springs, so that the flangeway is closed. Thus wheels on the main rail have a continuous bearing through the frog. The opposite wing rail, *D*, is secured in the usual position, giving an open flangeway. With a wheel entering the turnout from the switch, its flange will strike the spring wing rail and force it back so that the wheel flange can pass between it and the frog point. The guard rail opposite the frog pulls the wheel over so that it cannot strike the frog point.

With a wheel on the turnout approaching the heel of the frog, the flange will wedge between the frog point and spring rail, forcing the latter back so as to allow the wheel flange to pass. Objection is made sometimes to this type of frog on the ground that a joint loose enough to allow proper movement of the spring rail is not strong enough to carry heavy loads, so that it will become a bad spot in track and necessitate early renewal of the frog. Detail construction of a spring-rail frog is shown in Fig. 36.

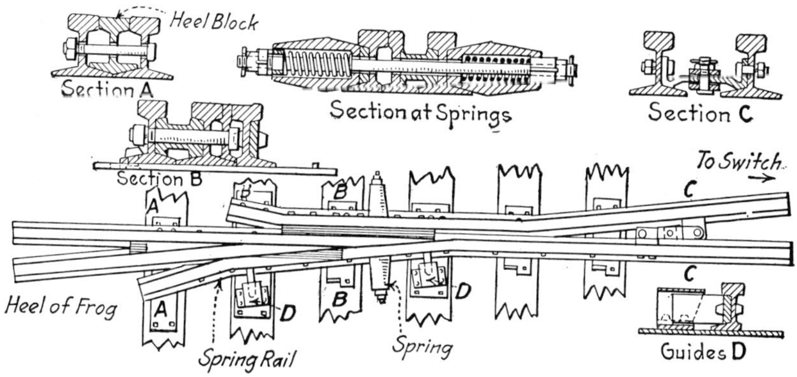

Fig. 36.—Spring-rail frog.

Frog Substitutes.—To provide actually continuous main-track rails at turnouts, several devices have been invented and tried to eliminate the frog. In most cases the turnout lead rail rises on an incline so as to raise wheels high enough for their flanges to pass over the main rail. A gap in the lead rail is left to allow passage for wheels on main track, but when the switch is set for the turnout, this gap is closed by a swinging rail or casting operated in connection with the switch, thus making the turnout rail continuous. In the Roach device, however, there was no elevation of the turnout rail, but the action was somewhat similar to that of a stub switch. The frog proper was a cast-steel bar or tongue pivoted at the heel and moving laterally on a base plate so as to coincide with one or other of the lead rails according to the position of the switch. While none of these devices is used to any extent, and many of them are obsolete, they indicate a desirable line of development in track design.

Crossing Frogs.—At track grade crossings, each of the four rail intersections has a crossing frog to give a flangeway in both directions. In crossings of 25 to 90 deg. the frog wing rails are usually extended to form a continuous guard rail and flangeway along the four sides of the crossing (Fig. 37). Where traffic is heavy it is usual, also, to place an easer rail in contact with the outside of the track rail, in order to carry the false flanges of hollow-worn wheels and so prevent them from battering the rails at the intersections. The ends of these easer rails, on the outside of the crossing, are planed to an incline so as to engage the worn wheels without severe shock.

All these parts need to be heavily built, bolted together, and provided with base plates under each corner or extending under each side of the crossing. Where the angle of crossing is less than 8 deg. or where one or both tracks are curved, the fixed frog may be replaced by pairs of

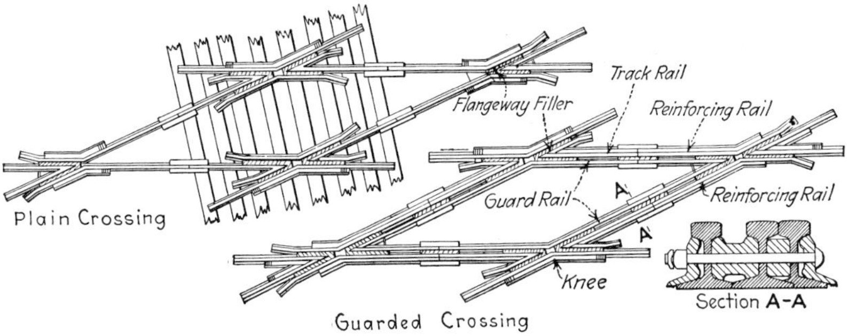

Fig. 37.—Track crossings.

movable-point frogs which are placed face to face and operated simultaneously in connection with the switch (Fig. 38). This arrangement may be applied also at track crossings fitted with slip switches.

These crossing frogs may be built up of steel rails or formed by manganese steel castings. In the latter, the four intersections may be separate castings bolted to connecting rails, as in the rail-bound, cast-

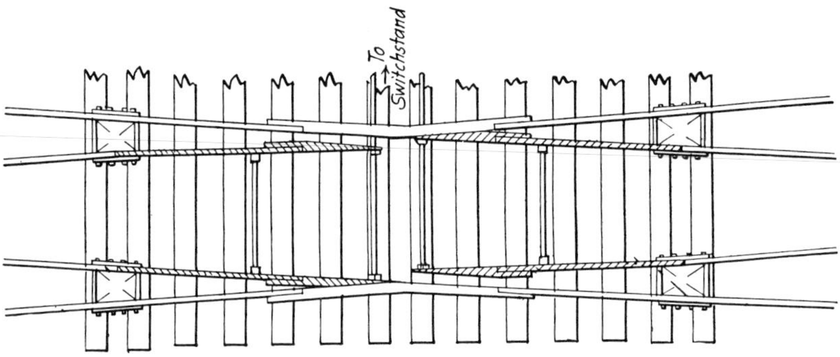

Fig. 38.—Movable-point crossing.

steel frogs already noted; or the entire crossing may consist of a single casting or of two castings with a bolted joint in the center line of one track. Joints in the sides of a crossing are undesirable, and a single large casting is likely to wear unevenly, besides being difficult to handle. To avoid the trouble with numerous bolted parts and also with large castings, the Balkwill construction has the crossing composed of four crossings, U-shaped in plan, to form the four sides of the crossing (the one long leg being grooved for the flangeway); four outside angle-shaped corner pieces are fitted between the legs of the adjacent U-castings, the joints coming at the intersections of the flangeways. Angle braces are placed inside the corners and the parts are heavily bolted together, but worn parts are readily renewed.

Devices giving continuous rails have been tried on a small scale in order to eliminate the violent shocks of wheels passing over the gaps at the intersections of the flangeways. In the Leighton-Haskell device, four grooved, cast-steel blocks move diagonally in the corners of the crossing. The Eymon device has two diagonal rods each operating two triangular sliding blocks in opposite corners of the crossing; the rods move simultaneously in opposite directions, one rod closing the two flangeways in one track and the other opening the flangeways in the other track. In both devices the mechanism is interlocked with that of the signals, so that continuous rails are formed for the track which has a clear signal.

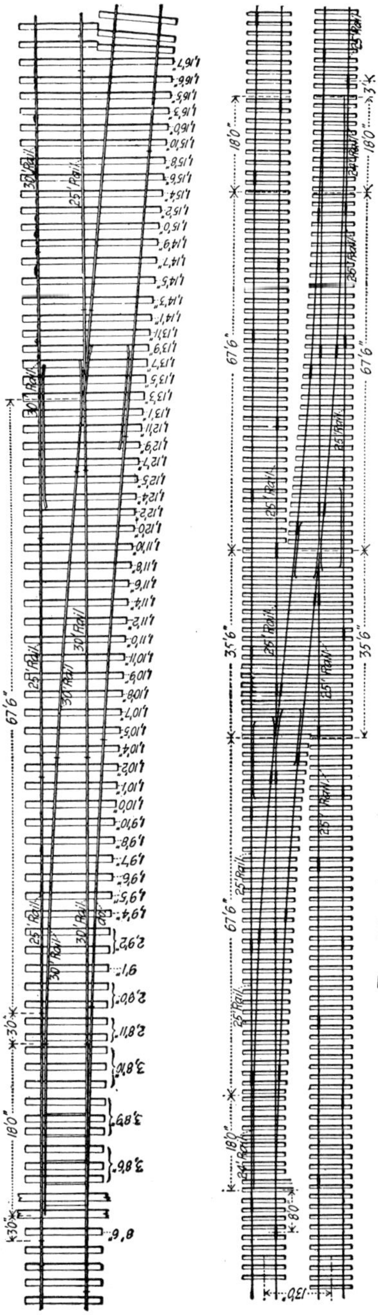

FIG. 39.—Arrangement of switch timbers at a turnout and a crossover.

Switch Ties and Timbers.—To carry the diverging rails of the main track and turnout, switch timbers or ties of varying length are generally used to support the four rails. An alternative plan used occasionally is to place ordinary ties, alternated for the two tracks, with six or eight long timbers under the frog. Long timbers are somewhat difficult to renew, but they give a more substantial and uniform support. They are spaced 16 to 24 in. c. to c., but usually not less than 7 in. apart in the clear, to allow for tamping; the spacing must be adjusted as required, especially in getting a tie under the frog point. The first two ties at the switch are the head blocks, usually 12 to 14 ft. long to carry the switchstand, but a single head block is sometimes used. Where main track switches are close together, the length of head block may be 14, 16, and 20 ft. for the first, second, and third, with ordinary, intermediate, and high switch stands, respectively. At interlocked or mechanically operated switches, where movements are controlled by signals and no switchstands are required, the head blocks are the same length as the ties.

Switch ties are commonly specified to be 7 by 9 in. in section, sometimes 7 by 10 in. To ascertain the required lengths of the several timbers, take the difference in length (in inches) between the last standard tie at the switch and the last long tie beyond the frog; divide this by the number of ties to be placed between them; add the result to the length of each tie in succession, beginning at the heel of the switch. As a rule, however, the timbers are arranged in groups of equal length, increasing by 6 in. successively, as indicated in Table IX. Their ends are all in line along the main track rail. These timbers, sawed to length, are stacked separately at the tie yard and made up into sets as required. In the No. 10 turnout of road *B*, the first 13½-ft. tie is under the frog point, 80 ft. 3½ in. from the point of the switch. The arrangements of ties and timbers at a turnout and a crossover are shown in Fig. 39.

TABLE IX.—SETS OF SWITCH TIMBERS

	Length of switch timbers, feet / Number of switch timbers																		Total number	Total linear feet	Total feet b.m.
	8	8½	9	9½	10	10½	11	11½	12	12½	13	13½	14	14½	15	15½	16	16½			
Road *A*:																					
Turnout No. 8	..	..	10	6	6	4	2	4	3	2	4	2	4	3	3	2	3	2	60	713½	3,746
Turnout No. 10	..	..	11	9	6	5	4	4	3	4	2	3	6	4	2	3	4	2	72	855	4,489
Turnout No. 16	..	..	11	8	10	7	6	7	6	5	6	5	7	5	5	5	5	4	102	1,244	6,531
Turnout No. 8	8	..	20	12	12	14	10	8	6	4	14	8	4	..	..	..	..	..	120	1,295	6,799
Turnout No. 10	8	..	22	18	21	10	8	17	6	8	12	8	4	..	..	..	..	..	142	1,523½	7,998
Single slip switch No. 8	..	..	..	..	..	..	9	7	7	9	4	6	3	2	1	..	..	..	48	595	3,124
Double slip switch No. 8	..	..	..	..	..	..	9	8	6	8	4	8	4	2	..	..	..	..	49	608	3,192
Road *B*:																					
Turnout* No. 7†	..	8	6	5	4	3	2	2	4	2	2	2	2	..	..	..	..	..	42	......	2,310
Turnout No. 10‡	..	7	8	8	3	4	4	3	6	3	4	4	2	..	..	..	..	..	58	......	3,305

* 15-ft. switch rail. †16½-ft. switch rail.
‡ 16½-ft. switch rail.

For turnouts, the longest timber is usually 14 to 17 ft., ordinary ties being then used. For crossovers, the maximum is frequently 14 ft., but will be 21½ ft. if single timbers carry tracks which are spaced 13 ft. c. to c. and are laid with 8½-ft. ties. This latter plan is not usual. The heaviest wear comes upon the ties at the switch and frog, in spite of tie-plates, and these ties may have to be renewed before the others. But where turnouts under heavy traffic need renewal of 30 to 40 per cent of the ties, it is generally preferable to renew the entire set at one time.

Switchstands.—For switches operated by hand, the operating mechanism is contained in a switchstand mounted on the long ties or head blocks at the point of switch. A typical construction consists of a frame bolted to these ties and having a vertical shaft with target and lamp holder at the top, and a crank at the bottom to carry the connecting rod which leads to the head rod of the switch. The shaft is revolved on its

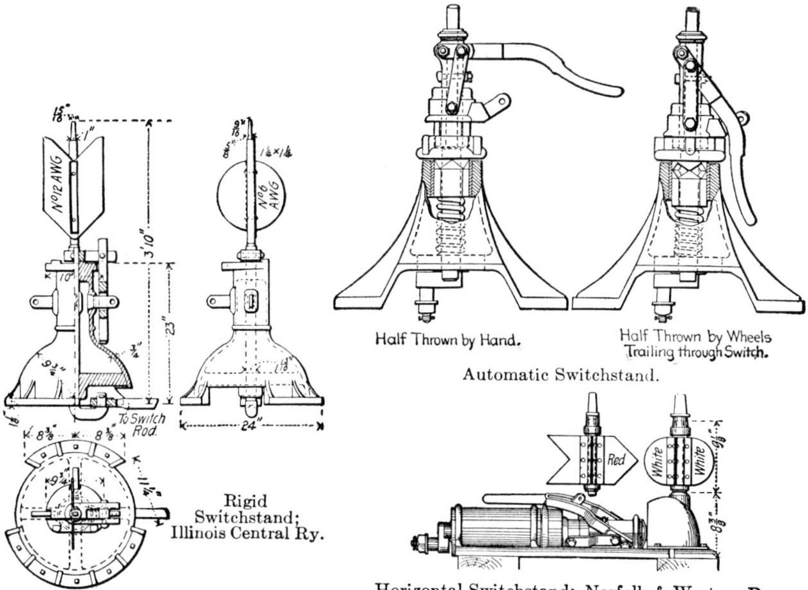

Rigid Switchstand; Illinois Central Ry.

Automatic Switchstand.

Horizontal Switchstand; Norfolk & Western Ry.

FIG. 40.—Switchstands.

axis by means of a lever which is hinged to it, and which normally hangs vertical and is secured in a slot or to a lug on the switchstand frame. To throw the switch, the lever is raised to a horizontal position, clear of the lug, so that it can be swung round, turning the shaft and crank and thus shifting the switch rails. For yard work, there are low switchstands with the lever operating in a vertical plane, parallel with the rails. There is an increasing use of interlocking switch and signal plants at main-line turnouts and yard entrances, the switches being operated by rods from levers in a tower, but only hand-operated switchstands are mentioned here. Main-track switchstands may be arranged to operate distant signals by wire or electric connections.

Switchstands at main track turnouts are usually 7 to 8 ft. from the rail, or at a less distance in yards. For the former, the height to target may be 14 to 18 ft. above the tie. Where two or three switchstands come in line, close together, as at station approaches, the first may be about 4 ft. high and the others 6 and 8 ft., so that the lights and targets will all be visible from an approaching engine. Another plan is to put them at varying distances from the track (see "Switch Timbers"). Yard switchstands may be 1 to 2 ft. high. Different designs of switchstands are shown in Fig. 40.

Automatic or Spring Switchstands.—To meet the condition of trains trailing through closed switches, as noted under "Spring Switches," switchstands may have spring connections by which the switch will be thrown to the open or opposite position by the action of the first wheel wedging its flange between the stock rail and the closed switch rail. These automatic switchstands are used mainly at yards. With normal hand operation the connections are rigid, the raising of the lever to turn the switch disconnecting the automatic mechanism. By this method, the switch is operated only once and not by each wheel as with spring switches; instead of being returned to the original position when the train has passed, it remains in its new or open position, which is indicated by the target and lamp. If it is desired to return the switch to its original position after having been opened by a train, the spring connection in the switchstand can be so made as merely to yield with the pressure from each wheel, as in the spring switch. This latter arrangement is employed mainly, however, at the sidings of electric railways. In both kinds of switchstands the target is rigidly connected, so as to show the position of the switch rails at all times (see Fig. 40).

Derails and Derailing Switches.—To prevent a car or train on a sidetrack from standing foul of, or running onto, the main track, a derailing switch with stub or point rail may be placed in the outer rail of the siding. This derail is interlocked with the switch and is opened (so as to send a car off the track) when the switch is set for the main track. The derailing point rail inside one track rail may be supplemented by a 12-ft. lifting rail outside the other rail. When the point is set to derail one wheel, the lifting rail is against its corresponding track rail so as to engage the wheel treads. This lifting rail, extending 2 ft. ahead of the point, is bent on an incline and supported on blocks on the ties, rising $1\frac{1}{2}$ in. above the track rail in a length of 4 ft. This incline and the fact that it is equivalent to superelevation on the inside of what corresponds to a turnout curve combine to exert considerable resistance to the movement of the car. A simpler device is a hinged block which may be laid upon the rail and is so shaped that if a car moves over it, the block will raise the wheel and carry it across the rail so as to drop on the ties. This derail may be operated by hand or by mechanism from the switch stand or interlocking tower. Derails of both types are shown in Fig. 41, in position to throw a car off the track. To allow a car to pass to the main track, the block at the left is revolved so as to stand clear of the rail, while the switch rail at the right is shifted to lie against the derailing rail.

At the interlocking plants of junctions and grade crossings, it is usual to install derailing switches to prevent trains which disregard signals from reaching the danger point when the signals are set for the other track. The distance from derail to danger point will depend upon traffic conditions and speed of trains. In some cases, however, the derailed train has been wrecked or has actually run into the other train at the crossing. The signal section of the A. R. A. has taken the position that derails should not be used in main tracks, except where required by law or by order of national or state authorities. It recommends, however, that on heavy grades deflecting tracks may be used to check runaway cars or trains.

FIG. 41.—Derails in track.
At left: derailing block. At right: derail with switch rail.

Switch Targets and Lamps.—Two sheet-iron targets are usually attached to the top of the switchstand shaft, distinctive shapes and colors being used to indicate for which track the switch is set. Common combinations include a square, diamond, or circular target for one position, and an oval or narrow upright target for the other position. Red and white are the usual colors at main track switches, sometimes with stripes or other designs to make them conspicuous. For yard switches, the colors are frequently red and green or yellow and green. In any case, the white (or green) is for the superior track. Low switchstands may have targets on top, or the height may be reduced by placing 10-in. annular targets around the lenses of the lamp case.

Lamps are provided usually on all switchstands on single track lines, and at all facing point switches on double- or multiple-track lines. They are mounted on top of the shaft, which is made of such shape that the lamp cannot be placed in the wrong position. The lenses are usually 4½ or 5½ in. in diameter, having a spherical face, and the back either flat or with concentric corrugations, the latter being the more general form. The colors are usually red and green for main line turnouts and crossovers, yellow and green for yard turnouts and crossovers; sometimes red and purple for a middle siding, or yellow and purple for a derail siding from the main track. Section foremen are usually responsible for cleaning, trimming, filling, and lighting switch lamps, except in block signal territory. This work is done daily for ordinary lamps, but those with long-time burners, which burn continually, are only inspected daily and trimmed when necessary. Such lamps have founts for 30 to 40 oz. and burn for 5 to 7 days, but 1-gal. founts have been used and lamps kept burning 28 to 30 days. The oil founts should be emptied, washed, and refilled once a month. Trackwalkers may be required to inspect, light, and extinguish lamps and report every lamp failure daily to the foreman, who in turn makes a similar report to the roadmaster.

There is an increasing use of electrically lighted switch lamps. The case and attachment are so made that no focusing or adjustment is required when installing the original lamps or making renewals. The bulbs used may be guaranteed for a life of 1,200 hr. at rated voltage; they may be 2-candlepower lamps for 110-volt current or the lamp case may have a transformer for a 10-volt, $2\frac{1}{2}$-watt lamp, but this increases the cost. For signals, the 10-volt, $2\frac{1}{2}$-watt lamp is so arranged that in case of failure of the supply line the lights are switched automatically to a storage battery which is kept charged from the supply line.

CHAPTER VII

FENCES AND GRADE CROSSINGS

FENCING

Types of Fencing.—Right-of-way fences are of three types: woven wire, strand wire, and board—the first being most extensively used. In some states the design and height of fences and gates are prescribed by law, but the usual height is $4\frac{1}{2}$ ft. The panel length is usually 8 ft. for board fences and 16 ft. for wire, although for the latter it is sometimes 20 or even 25 ft.

Woven-wire fencing consists of longitudinal wires with vertical stay wires woven or welded to them, the longitudinal wires formed with bends or tension curves to take up expansion and contraction. All the wires are galvanized, but as the coiling of the stay wires may crack the zinc coating, recent practice provides for passing the woven fence through a bath of zinc and then through an annealing furnace. To reduce trouble from rusting, experiments are being made with steel wire containing a proportion of copper. On irregular ground, where woven fencing cannot follow the sags and depressions, the fence is built of separate wires or strands; a vertical wood stay, about 1 by 3 in., may be placed at the middle of each panel to maintain the spacing and prevent sag of the wires. With five strands, the bottom wire may be 6 or 8 in. above the ground and the others spaced 8, 12, and 14 in. or uniformly 12 in. apart. With either kind of fence on wood posts the wires are secured by staples driven into the posts.

Two styles of wire fencing used on the Chicago, St. Paul, Minneapolis & Omaha R. R. are shown in Fig 42, both with $16\frac{1}{2}$-ft. panels and having a fir or white pine stay at the middle to which the wires are stapled to prevent sagging. The posts are of white cedar, 7 ft. long for line posts and 8 ft. for end, corner, and straining posts. The woven-wire fence of the Michigan Central R. R., shown in the same illustration, is 55 in. high, with eleven No. 9 wires spaced from 3 in. at bottom to 9 in. at top and having No. 12 stay wires 12 in. apart. It is placed with the bottom wire close to the ground. Various styles of wire fences are in use, but there is a movement to reduce these to a limited number of standards, and the A. R. E. A. specifies four classes, as follows: (1) Woven fence of nine smooth wires spaced 4, $4\frac{1}{2}$, 5, $5\frac{1}{2}$, 6, 7, 8, and 9 in., and stay wires 12 in.; bottom wire 5 in. above ground, with a barbed wire $2\frac{1}{2}$ in. below it if a hog-tight fence is desired. All wires are of No. 9 gage, except No. 7 for top and bottom. (2) Woven fence of seven No. 9 smooth wires spaced

$6\frac{1}{2}$, 7, $7\frac{1}{2}$, 8, $8\frac{1}{2}$, and 9 in., and No. 12 stay wires 12 in. apart; bottom wire 7 in. from the ground. (3) Combination fence with seven-wire woven fencing having the bottom 2 in. from the ground and wires spaced 3, $3\frac{1}{2}$, 4, $4\frac{1}{2}$, 5, and $5\frac{1}{2}$ in.; No. 9 stay wires 12 in. apart. Above this, three strands of barbed wire spaced $4\frac{1}{2}$, 10, and 12 in. (4) Strand fence with five strands of round or ribbon wire, smooth or barbed, with a spacing of 10 in. from ground and 10, 10, 12, and 12 in.

Board fences have usually four or five lines of 1- or $1\frac{1}{4}$-in. boards, 6 to 8 in. wide and 5 to 10 in. apart, in 16-ft. lengths for 8-ft. panels, breaking joints on alternate posts. A top board or a cap board (laid flat) is sometimes used with wire fences to prevent damage to animals

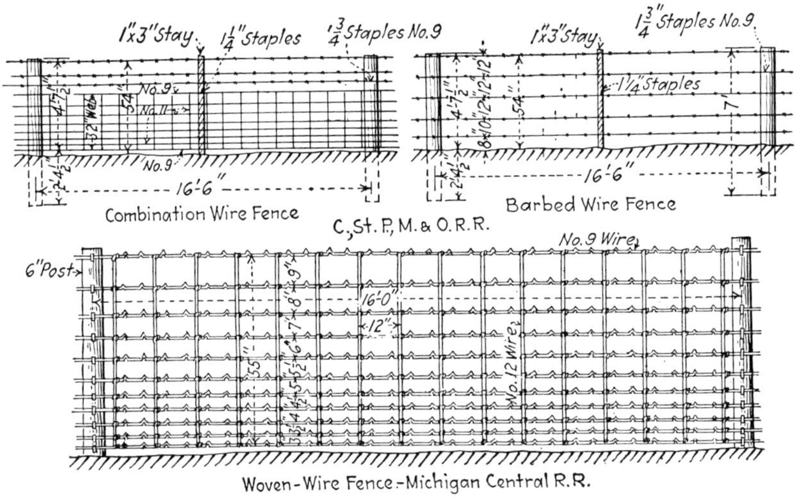

FIG. 42.—Right-of-way fences.

where barbed wire is used. To make a fence hog tight, a barbed wire may be placed about 2 in. from the ground, or a board in contact with the ground. Boards and wires are placed on the field side of the posts, except that on curves the wires are strung on the convex side.

Fence Posts.—Wood posts are usually of durable local timber, the principal species (with average life) being as follows: bois d'arc, 30; creosoted pine, 25; cedar, 24; locust, walnut, and white pine, 21; chestnut and catalpa, 20; cypress, 12; oak, 8. Anchor or braced posts should be provided at intervals of not more than $\frac{1}{4}$ mi. Line posts should be 7 ft. long, 4 in. in diameter at the small end and set 28 in. in the ground; in sags or depressions they should be anchored by two cleats spiked to the sides near the bottom. On rocky ground, the post may be set on the end of a 6- by 6-in. sill $3\frac{1}{2}$ to 4 ft. long, and supported by one or two braces spiked to the sides of the post and sill. All posts are set with the large end down, and the top may be sawed on a slope of 1 in 4 to shed water.

Anchor, corner, end, and gate posts, set not more than 10 ft. from the adjacent line posts, should be about 8 ft. long, 8 in. in diameter at the small end, and set 40 in. in the ground. They should be anchored by having two cleats, 2 by 6 in. and 16 to 24 in. long, spiked in gains on opposite sides of the post near the bottom, the cleats being at right angles with the fence. Every such post should be braced by a diagonal strut about 4 by 4 in. set in a gain 12 in. below top of post and 12 in. from the ground on the adjacent line post. This panel is also cross-braced by a cable of two No. 9 wires looped around the anchor post near the ground and 12 in. below the top of the line post, the cable being twisted tight (see Fig. 45).

Concrete and Steel Posts.—Concrete posts, which are used to an increasing extent, are usually of circular, semicircular, or square section. Line posts, as recommended by the A. R. E. A., are 7 ft. long, tapering from $4\frac{1}{2}$ to 3 in. square or 5 to $3\frac{1}{2}$ in. in diameter, and reinforced by four No. 8 wires or $\frac{3}{16}$-in. rods. Their weight is about 100 lb. Such posts are estimated to last 50 years. Anchor and other special posts would be of larger section and perhaps set in concrete footings. Braces may be of wood, concrete bars, or steel tubes, placed horizontally or diagonally, but in combination with the wire cross-brace noted above. A concrete mix of 1:4 is general, using stone or gravel, and care must be taken that all the steel is well covered. The posts should be left in the forms for 24 hr. at least, and then stacked upright in a shaded place and kept wet for 10 days. They should not be used or shipped in less than 90 days. Fencing may be attached to inserts in the posts or secured by tie wires around them.

Steel posts, usually of tubular or T-section, are estimated to last 22 years, but only 5 to 8 years in wet or swampy ground. Slots or lips to hold the wires may be formed in tube posts, while the back rib of the T-post may be notched so that tie wires will not slip. These posts can be driven with a sledge hammer.

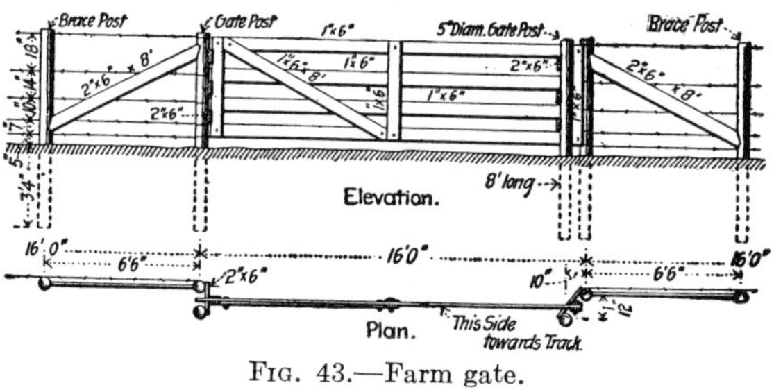

FIG. 43.—Farm gate.

Gates.—Farm gates in right-of-way fences are usually 4 ft. high and 12 to 18 ft. long, or as required by law, and sometimes 20 ft. to pass agricultural machinery. They should open away from the track. Hinged gates should be long enough to lap the post so that they cannot be swung the wrong way, and should be so hung as to close by gravity, since farmers are frequently careless about closing gates. A sliding type of gate which is used extensively (Fig. 43) moves between two posts at the rear end, placed diagonally, and about 12 in. c. to c., with cleats between them supporting the boards of the gate. The gate may slide back for its full length, parallel with the fence; but usually it slides back only a short distance and is then swung open as permitted by the position of the two rear posts.

To prevent drooping of the gate, its hinged end post may be about 7 ft. high with a brace extending from top of post to the bottom of the gate at the other end. The gate post is then of the same height, with the upper hinge near the top and a stay to the bottom of the next line post. A more ordinary gate consists of four or five planks, 16 or 18 ft. long, with the ends sandwiched between two 4-ft. vertical planks, the whole stiffened by a vertical at the middle with two diagonals from its top to the ends of the bottom plank. All pieces are 1 by 6 in. Metal gates, which are used extensively, have a rectangular frame of 1½-in. tubing, divided into three panels by vertical tubes and stiffened by a diagonal stay wire across the gate or by a diagonal tube in the middle panel. Over this frame is stretched woven-wire fencing. The gate fastening should be strong and simple, and should allow for considerable distortion or displacement without becoming ineffective. Gates not used frequently may be secured with chain and padlock.

Yard and Station Fences.—To prevent trespassing at yards and shops, tight board fences are frequently used, with round or 6 by 6-in. posts 6 to 8 ft. apart, having two 2 by 4 in. girts mortised into the face; vertical planks 6 to 8 ft. high are nailed to the girts and may be finished with a cap board and strand of barbed wire on top. Expanded metal or heavy wire mesh may also be used, attached to tall steel posts having their tops inclined inward to carry three or four lines of barbed wire, making the fence practically unclimbable. Picket fences or iron railings may be placed between the tracks at stations to keep passengers from crossing except at specified points. In one design, posts 4½ in. square have iron rests for the rails of 9-ft. panels which are removable to facilitate attention to tracks or trains. The 3 by 3-in. rails have 1½-in. pickets 4 ft. long and 6 in. apart. Ornamental grounds at stations may be enclosed by ribbon wire or woven fencing on steel angle or T-posts; or lines of 2-in. pipe may be set in wood or iron posts 6 to 8 ft. apart. Hedges may also be used at such grounds.

Snow Fences (Fig. 44).—Trouble with snow occurs principally at cuts, and snow fences are built to prevent drifting snow from reaching the right-of-way, the fence checking the wind and allowing the snow to settle. Permanent fences along the right-of-way may consist of 12½-ft. posts, 8½ ft. above the ground, spaced 8 ft. c. to c., with 1 by 6-in. boards about 6 in.

apart if the fence is 100 ft. or more from the track (allowing the drift to accumulate on both sides), or closely spaced if it has to be within 50 ft. of the track. In either case, a 12-in. opening is left along the bottom. Hedges may be used as permanent snow fences where local and economic conditions permit. Temporary worm fences may be built of ties, but such use of ties is not desirable, except in emergency.

More extensive use is made of portable snow fences which can be erected outside the right-of-way, a small rental for this occupancy being paid in cultivated sections. These fences are set up early in the winter

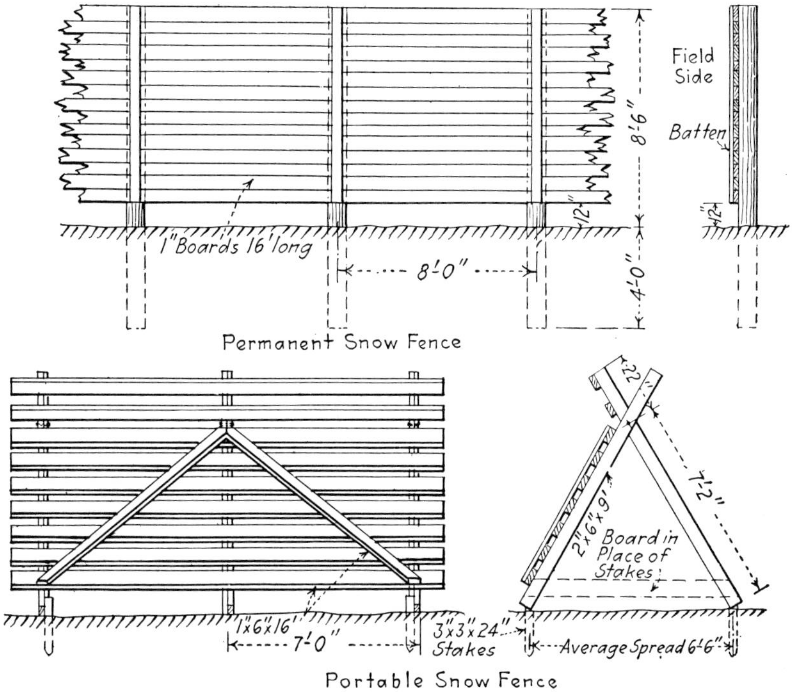

FIG. 44.—Snow fences.

and removed when the snow period is over. Ordinarily a single line is sufficient, placed about 100 ft. from the track or right-of-way, on the windward side and approximately at right angles to the prevailing winds, or in the most effective position to meet the local conditions. In exceptional cases two or more lines of the same spacing may be necessary.

Portable fences are usually made in 16-ft. sections, with horizontal planks spiked to three folding bents or A-frames. These sections may be placed continuously or about 16 ft. apart; with the latter arrangement the intervals are closed by panels laid against the back braces. In the A. R. E. A. design (Fig. 44), the two 9-ft. posts are put together with a bolt about 2 ft. from the top, but with posts up to about 8 ft. long, this hinge may be at the top. If stakes cannot be driven at the feet of the posts, a plank may be spiked across the bottom of each bent and loaded with field stone or old ties to hold the fence in position. A permanent improvement which is desirable where the snow season is long, is the widening of cuts by flattening the slopes to about 1 on 4, thus giving space for drifting snow to accumulate outside of the roadbed.

Wing and Apron Fences.—At road crossings, a wing fence is built on each side of the road from the edge of roadbed to the right-of-way fence. To exclude small animals the track ditch may be closed by stakes driven into the ground and having their tops nailed to a board between the fence posts. At the track end of the wing fence is a rectan-

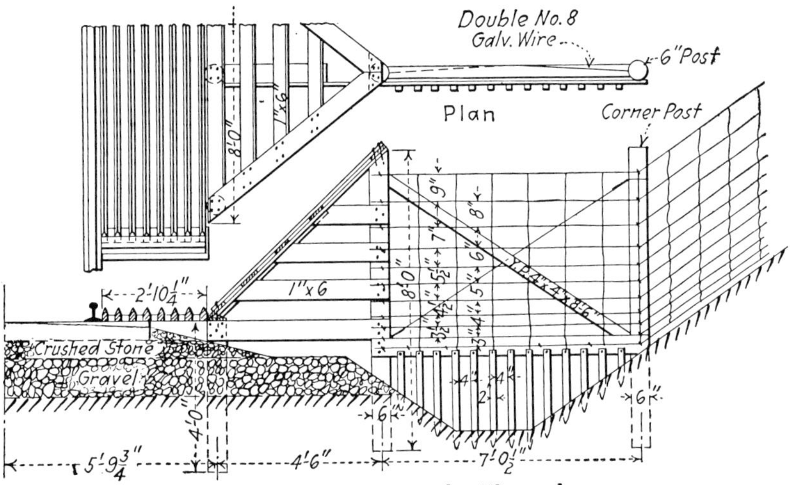

Fig. 45.—Wing fence and cattleguard.

gular or triangular fence panel parallel with the track. This is sometimes vertical, but a safer plan is to place it sloping away from the track (Fig. 45) in order to be clear of employees on car steps or freight-car ladders. The foot of the apron panel is about 3 ft. from the rail. Where there is much snow, the aprons and ends of wing fences may be removed in winter in order to give clearance for snow plows. Aprons and board wing fences should be kept whitewashed.

Cattleguards.—At grade crossings of railways and highways, cattleguards are usually placed in the tracks, at each side of the highway, and for the entire width of the roadbed, to prevent cattle from straying onto the railway. Wing or lateral fences extend from the roadbed to the right-of-way fences, as noted above. Cattleguards are also placed sometimes at the approaches to bridges, tunnels, and deep cuts. Ordinary farm cattle rarely give much trouble, but some of the animals are inveterate wanderers, and when straying on the road they will make repeated attempts to cross the guards in order to reach some tempting feed. When being driven they may run blindly into the cattleguard, which must be long enough to prevent them from jumping across it. Range cattle are hard to turn, while hogs and sheep are persistent in trying to reach forbidden ground.

Surface cattleguards, now used almost exclusively, are made of wood or metal, and consist of a series of slats parallel with the rails and so shaped and spaced as to afford no foothold. A typical design of wooden construction is that of the Michigan Central R. R. (Fig. 46); this cattleguard is built in sections 9 ft. long, laid between and outside of the rails, and secured to the ties by lag screws in the cross-pieces at

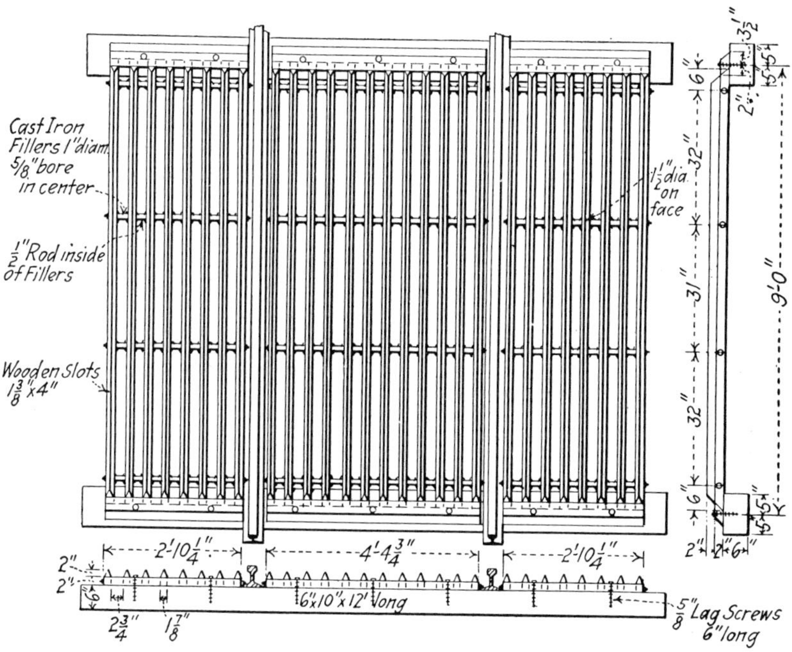

FIG. 46.—Surface cattleguard.

each end. Another design has planks $1\frac{1}{4}$ by 5 in. set on edge, but in an inclined position, being held in inclined slots in the end pieces. Various forms of metal cattleguards have slats of flat plates, angles, and T-bars, generally in an inclined position, while the small angles and T-bars are carried in cross-pieces to bring them nearly level with the rail heads. Burned-clay blocks of triangular section, as used on some electric railways, are held in position by end pieces spiked to the ties.

For pit cattleguards, now practically obsolete, an excavation 5 to 10 ft. wide and 2 to 3 ft. deep below base of rail was made across the roadbed and walled with timber or stone. The rails were carried across this pit on stringers or upon ties laid on the stringers, the edges of the ties being beveled, except at the rails. Such a pit holds moisture and refuse, makes a bad-riding spot, causes heaving of track in winter, and is liable to be burned and thus make a dangerous trap for trains. Furthermore, many accidents have been caused by animals getting caught in the pit and causing trains to be derailed.

GRADE CROSSINGS: CONSTRUCTION AND PROTECTION

Highway Crossings.—Railway grade crossings of public highways, which are a necessity in many cases, especially in thinly settled districts, may become danger spots to both methods of transportation as districts become settled and both railway and highway traffic increases. There is continual agitation for separation of grades at various points, and a great deal of such work has been done, often under considerable difficulties and generally at large expense. But the number of crossings is so great (more than 356,300 in 1923) and the cost of their elimination by separating the grades is so great (estimated at more than $18,000,000,000) that only slow progress can be made, especially in view of the many other costly works required on the railways. There are possibilities in eliminating many crossings by relocation, and new crossings should be established only where real necessity exists.

The dangers of grade crossings are due largely to the rapid increase in the number of automobiles and to the reckless driving which is characteristic of this traffic, with speed as a main consideration and general disregard or defiance of traffic rules, signals, and warnings. From 75 to 90 per cent of crossing accidents are attributed to this recklessness, against which there is no effective protection. Laws requiring all vehicles to stop before crossing a railroad are objectionable, because they cannot be enforced and because they are likely to cause obstructions and dangers in busy highways. If there is no train approaching, the stop is a useless interference with traffic; and if a train is approaching, the reckless driver is likely to try to cross ahead of it, even if he does first come to a stop. In view of these conditions, the education and control of those who drive automobiles and the punishment of those who drive recklessly are of much more importance than the planning for the general elimination of grade crossings. Since such elimination is more particularly in the interest of the general public, its cost should be shared by the public, through the municipal or highway authorities. Some states have laws apportioning the expense between the railways and the state, county, municipal, or other local authorities, but the laws are not uniform.

Construction of Crossings.—The crossing should be substantially built, well drained, and kept in good repair, not only for safety but because defective or unsightly crossings are frequently a cause of ill feeling on the part of the public. The highway approaches, also, should be well built and maintained, with track ditches carried through them by iron pipe drains, or clay pipe if the earth cover is thick enough to protect it. The ends of the drains may be laid in small, dry walls and covered with gratings. A 20-ft. width of railway roadbed should be the minimum. The A. R. E. A. recommends that the crossing should be level with the rail heads for at least 1 ft. beyond the rails, 3 in. lower at 10 ft. and not more than 9 in. lower or 3 in. higher at a distance of 30 ft. Steep approaches are undesirable, whether at cuts or fills, and one state specifies that the grade must not exceed 1 per cent for 25 ft. from the track or 3 per cent on the right of way. To avoid rail joints in the crossing, 60-ft. rails may be used, or the rails may be welded together for the necessary distance. A 6- or 8-in. concrete slab foundation has been used in some cases, under a good bed of stone ballast, in order to prevent settlement and to reduce the work of surfacing. In such construction at concrete-paved street crossings, the Southern Pacific Ry. builds a vertical wall to connect the slab with the ends of the concrete paving. It also places tie rods through the rails, 6 ft. apart, to hold the gage.

Special care should be taken to have the highway approaches direct and as free as possible from trees, bushes, or advertising signs which obstruct the view of the track. The construction of humps and depressions in the paving, or curves in the road to check and warn drivers, is rarely advisable or effective, but good results have been obtained from a central curb in the road for some distance from the crossing, with a warning sign at the end, so that drivers will slow down for the narrow roadway and will be prevented from dodging in and out of the traffic to get ahead of other machines.

Paving Road Crossings.—Since railway tracks require periodical surfacing and other maintenance work, a permanent or continuous paving

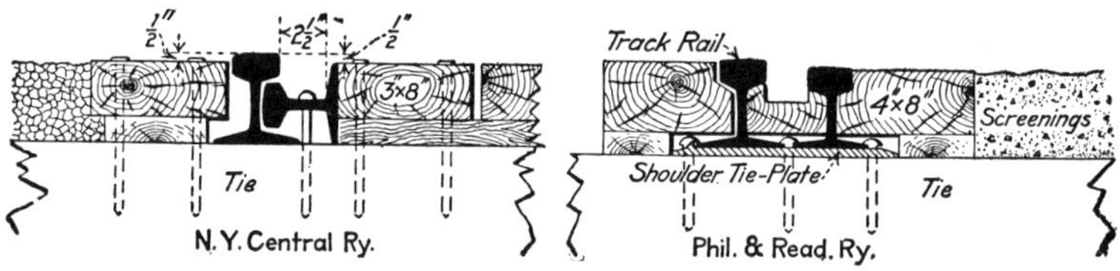

FIG. 47.—Planked grade crossing.

is rarely used, except occasionally for city streets. Country road crossings are paved usually with 3- or 4-in. plank, 8 to 12 in. wide; or for light traffic there may be a plank on each side of each rail, with gravel or cinders filled between them and retained by short cross-planks. This paving is from 12 to 18 ft. wide for farm and side roads, and for the full-paved width of main roads. Its length extends usually about 2 ft. beyond the outer rails. Different styles of paving are shown in Figs. 47 and 48.

Any available timber that does not split or splinter readily is suitable for the planking, but planks grooved or milled to fit under the rail and form a flangeway are likely to split. Where traffic is heavy, hard woods like oak, birch, and gum may be used; but at the less important crossings, soft but durable woods like cedar, pine, and cypress are suitable, as they will fail by wear and decay at about the same time. Treated timber may be economical in some cases. Long spikes fasten the planks to the ties, wood strips being nailed on the ties if necessary to bring the planks nearly level with the rail heads. The ends of the planks are beveled for 8 to 12 in. so as not to catch any loose parts hanging from trains. Where track is liable to heave with frost, the planks may be removed from unimportant crossings in winter, so as not to foul engine pilots or snow plows. The Ohio highway department has tried 2-in. planks laid on edge and spiked together, fastened to the ties with lag screws and covered with an inch of tar and gravel composition.

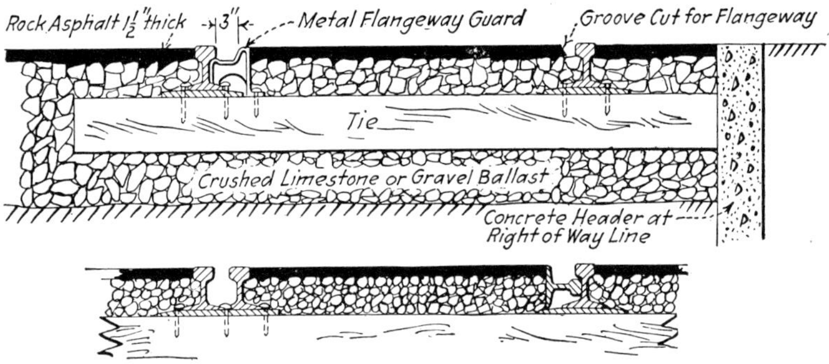

FIG. 48.—Paved grade crossings.

A flangeway, $2\frac{1}{2}$ in. wide and at least $1\frac{7}{8}$ in. deep, is left along the inside of each rail and may be provided by laying the plank that distance from the rail head. To prevent the calks of horses' hoofs catching on the rails, however, the flangeway should be filled to the underside of the rail head by a wood strip or a layer of concrete. The flangeway is frequently formed by an old rail which may be set upright with a filler between the rail webs; more usually this guard rail is laid flat, with its head against the web of the track rail and its base protecting the edge of the planking or gravel filling. If these rails are lighter and of less height than the track rail, they may be placed on small wood or iron blocks. At the sides of the crossing, the guard rails should be flared out to a clearance of 4 in. so that wheel flanges will not strike their ends, these openings being fitted with foot guards similar to those used at frog guard rails.

Concrete and Other Paving.—Reinforced-concrete slabs are satisfactory if the track and ballast give a firm foundation, but are likely to crack and break if uneven settlement occurs. The slabs are usually 18 to 25 in. wide and in lengths of 5 to 8 ft., with the edge at the flangeway faced with a steel angle or curb. Cored holes provide for spikes or lag screws and for attachments in handling the slabs. The thickness is usually 4 or 5 in., wood shims on the ties being used if necessary to bring the slabs level with the rails. The end slabs are beveled. Concrete planks, $4\frac{3}{4}$ by 8 in. in 8-ft. lengths, have been used on the Chicago, Burlington & Quincy R. R. Poured concrete slabs extending from bottom of the tie to top of rail have been tried experimentally; also rough-surface sheets of steel and cast iron.

For bituminous macadam paving, which is used extensively and may be applied hot or cold, the old ballast is replaced with clean crushed stone, drainage is provided, and the track is put in good line and surface, after which the new ballast is tamped and dressed to a level 4 in. below the top of rail. On this is laid a binder course (about $\frac{1}{2}$ gal. bitumen to 1 cu. ft. of ballast), tamped to a thickness of $2\frac{1}{2}$ in., and then the $1\frac{1}{2}$-in. wearing surface (1 gal. to 1 cu. ft. of $\frac{1}{2}$- to $\frac{3}{4}$-in. stone). The top course is either finished level with the rail by a steam roller, or is tamped and left about $\frac{1}{2}$ in. high to allow for compacting by the wheels of vehicles. Finally the surface is given a thin seal coat of $1\frac{1}{2}$ to 2 gal. bitumen to 1 cu. ft. wet sand, or the bitumen is spread at the rate of $\frac{1}{4}$ gal. per square yard and then covered with clean gravel or fine stone chips. The figures given are approximate, as the mixtures and proportions vary in different designs and under different conditions.

Flangeways in this paving are usually formed by steel rails; they are sometimes left to be formed by the wheels of trains, compressing the soft material, but this may result in disintegration of the paving along the rail. If there is no guard rail, a flangeway with a slope of 45 deg. may be formed with a hot iron. Such paving may crack or disintegrate due to vibration and slight settlement, but some mixtures are self-healing. Bituminous paving 1 to 2 in. thick, or rock asphalt 2 in. thick is sometimes laid on planking. In other cases, the well-tamped ballast is flushed with a bituminous composition (about $\frac{1}{2}$ gal. per square yard) and covered with a 1-in. sand cushion for paving brick.

Protection of Road Crossings.—To warn drivers and passengers to look out for trains, signs are erected near the crossing, and in some states additional signs are required at a distance of about 300 ft. These latter must be provided by the highway authority (see "Track Signs"). Where there is heavy traffic on the railway or highway, or both, the signs are supplemented by devices to indicate the approach of trains. These are usually visual signals, either flashing lights or swinging targets, stationary signs not being sufficiently conspicuous. Gongs or bells are sometimes used, but unless they are loud, they are not heard at a distance and loud ringing may be a nuisance. Curbs and guards in the highway approaches have been mentioned already.

The warning devices come into action automatically by means of a track circuit when a train approaches within a certain distance or will take not less than 20 sec. to reach the crossing, and they continue in operation until the train has cleared the crossing. The signal section of the A. R. A. has a standard 20-in. swinging target painted white, with a black border and cross, and a red lens or roundel at the center. In swinging, the disk has a stroke of $2\frac{1}{2}$ ft. and makes 30 to 45 strokes per minute. For the flashing lights, two lamps are recommended, $2\frac{1}{2}$ ft. c. to c. and 9 ft. above the road, fitted with lenses $5\frac{3}{8}$ to $8\frac{3}{8}$ in. in diameter and hooded so as to be prominent even in daylight at a distance of 300 ft. Each light makes 30 to 45 flashes per minute.

Crossing Gates.—Gates and watchmen are generally provided at crossings in suburban, residential, and city districts and also at those within switching limits. The gates consist usually of light wooden arms attached to castings pivoted in iron posts at the curb line and about 10 ft. from the center of track. As a rule, there are two arms meeting at the middle of the road and two shorter arms extending across the sidewalk. Swinging rods support the longer arms when lowered. These gates may be operated by a crank handle attached to gearing in one post, with wire and chain connections to the mechanism in the other posts.

More generally they are operated by compressed air by means of a hand-operated plunger air pump in the gateman's cabin, a 7-lb. pressure being sufficient. One or both posts on each side of the street contain a cylinder and piston or an iron case with rubber diaphragm, connected by piping to the pump. Two strokes of the pump lever usually operate the gates on both sides of the track. For electric operation, a controller in the cabin or the telegraph office at the station regulates the movement of motors in one post on each side of the track. With either pneumatic or electric operation, the power mechanism on one gate usually operates the other gate by wire and chain connections led through pipes under the road. If the gateman's cabin is elevated, he may operate power gates at two or three crossings, provided he has a clear view of each crossing and its approaches. Only careful and sober men should be intrusted with the care of crossings and gates.

To make the gates conspicuous they are often painted black and white in 12-in. diagonal stripes and may carry a green or yellow flag by day and lamps at night. A red lamp is not advisable as automobilists are apt to mistake it for the rear light of a standing car and may try to pass around it, with more or less disastrous results. There has been, however, a movement to use yellow instead of red for automobile tail lamps. The gates described are merely warnings and are frequently damaged by automobiles being run into or through them. At some crossings where automobile speeding is particularly troublesome, heavy timbers mounted in guides have been installed effectively to stop cars which may try to cross in front of an approaching train.

Street crossings at yards and stations, where there are numerous tracks with frequent trains and switching movements, are often protected by flagmen and should have, near the middle, an open, paved refuge space for teams. Such crossings are dangerous, particularly if street cars are included, and especially in dark or stormy weather when the view of drivers and flagmen is obstructed by smoke, steam, snow, or fog.

TRACK CROSSINGS

Railroad Grade Crossings.—Track crossings or intersections of railways at grades are numerous, especially in flat country, but few new crossings are built, owing to the expense of interlocking signal equipment and its operation. There is also the delay, damage, and expense incident to the stopping and starting of heavy trains, besides which accidents do occur at crossings in spite of all safety equipment. At "know-nothing" crossings, which have no such protection, all trains must stop, but interlocking plants are usually compulsory.

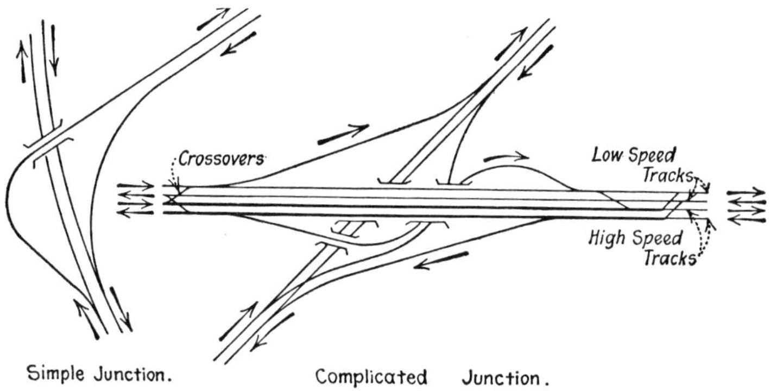

FIG. 49.—Junctions without track crossings.

In many cases such crossings can be avoided or eliminated with permanent operating and economic advantages to each road, and much work has been done in eliminating single or grouped crossings in and near large cities. The crossing of trains at busy junctions, especially on four-track lines, has been avoided in some cases by a "flying junction," carrying one line over or under the normal grade (Fig. 49). This plan avoids all danger and interruption to traffic, as well as the expense involved in the construction and maintenance of costly special track work.

At track intersections, the crossing frogs (see Chap. VI) and connecting rails may be supported in different ways: (1) by four longitudinal timbers about 12 ft. long, halved together at the intersections (Fig. 50); (2) by long timbers extending diagonally under the four rails (Fig. 37); (3) by ordinary ties placed at such angles as to afford the best support to both tracks. The first arrangement is used very generally for angles of 75 to 90 deg., but a modification with timbers for the heavy-traffic track, and ties for the other, is used for angles of 45 to 90 deg. In all cases the rails should be placed on tie-plates. Steel longitudinals have been used, built up of 16-in. steel channels placed 16 in. back to back, with concrete filling between and continuous plates on top. The second arrangement is that most generally used for angles of less than 80 deg., while the third is least satisfactory.

Since track crossings sustain very severe service, due to the pounding and shock from heavy loads and the blows from innumerable wheels, their maintenance is troublesome and expensive, both as to wear of parts and general distortion. Creeping of one or both tracks throws the crossing out of line and thus increases the wear, besides making a

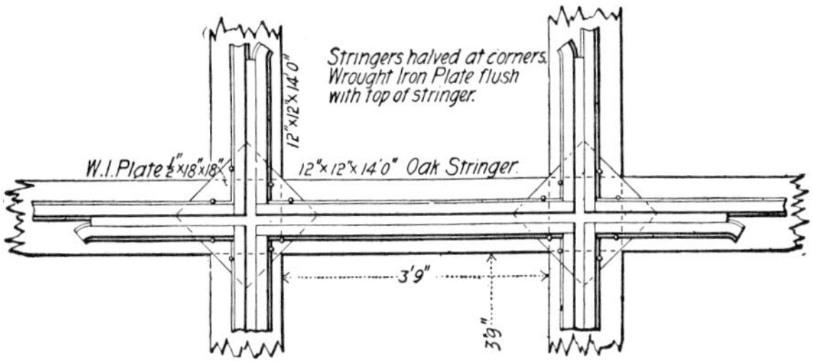

FIG. 50.—Track crossing on framed timbers.

hard-riding spot in the track (see "Creeping"). A good foundation is essential for good service, and in some cases stone ballast is placed for a depth of 20 to 30 in., well drained by tile or perforated cast-iron pipe. A concrete base under 10 to 20 in. of ballast has given good results, preventing all settlement or sinking of the ballast into the subgrade. But if the ties are laid directly upon the slab, as occasionally in paved streets, the crossing is likely to be a hard spot unless the adjacent tracks are heavily ballasted.

Electric Railway Crossings.—Where electric and steam railways intersect, either on country roads or in paved streets, the track construction is similar to that for steam railway crossings (Chap. VI), except that in the electric track the outside easer rail to engage worn wheels is often omitted. The crossing frogs are usually of the built-up type for lines of lighter traffic, and of cast manganese steel for heavier traffic and at paved crossings.

On high-speed interurban lines the crossings have usually the regular interlocking signal equipment, but with the high cost of construction and operation of such plants the railway may often find it economical to avoid grade crossings by means of diversions, subways, or viaducts. At unprotected crossings in city streets it is the rule for the electric car to stop while the conductor walks across to see whether trains are approaching. This precaution is open to the objection that it takes the conductor away from the car while it moves over the crossing, at which time there is the chance of the trolley jumping off the wire and leaving the car stalled on the track. At such crossings, therefore, the trolley wire should be carried in an inverted metal trough or guard which is electrically connected, so that it would not only catch the trolley but also maintain the supply of current to the car.

Crossings of electric lines over busy terminal railroad tracks with numerous trains and switching movements are an element of danger and should be avoided or eliminated where possible. At all points where electric railways operated by overhead wires cross steam railways, precautions must be taken to guard against the danger of charged wires falling upon the railroad cars or striking men whose duties require them to stand on the cars. Specifications drafted by the A. R. E. A. require a clear headroom above the rails of 22 ft. for trolley wires and 25 or 30 ft. for feeders of less or more than 750 volts, respectively.

Tracks in Paved Streets.—Tracks crossing city streets are usually paved like road crossings, but concrete paving or brick on a concrete base is sometimes laid. Small towns sometimes have main tracks laid lengthwise in a street, while at large ports there are usually tracks extending along the waterfront streets. In the latter cases the paving is generally brick or stone block, the portion between the rails being sometimes level with the bottom of the rail heads to form a flangeway. A line of heavy plank is sometimes laid against the outside of the rail head.

Track construction of this kind employed by the Public Works Department at Philadelphia for slow freight traffic consists of 9-in. 159-lb. grooved girder rails on creosoted wood blocks, with tie bars through the rails at 6-ft. intervals to hold the gage. The wood blocks, 7 by 9 by 24 in., are embedded in a concrete base, the top of which is flush with the 6-in. street base, but has grooves 24 in. wide to permit of placing the rails on the blocks. The grooves are then filled with concrete. On the concrete base is a 1-in. sand cushion for 5-in. granite-block paving laid close against the rails and grouted with cement. The guard side of the grooved rail head is slightly lower than the running head, except that on curves the guard is made very thick and of full height. This construction was found preferable to that with 9-in. 143-lb. T-girder rails on cross-ties. For similar tracks along city streets, the New York Central R. R. has used ordinary rails supported on iron chairs 3 in. high, spiked to the ties, which are set 3 in. lower than usual. The flangeway is formed by a Z-bar of special section bolted to the web of the rail. Concrete filled between the ties is covered with 2 in. of sand on which the stone paving blocks are laid. For 4-in. brick paving with a 2-in. sand layer, the rails are laid on base plates of such thickness as to give a height of 6 in. from the top of the tie to the top of the rail. In this case the flangeway is formed by a steel rail laid on its side. These details are worked out jointly by the railway and municipal engineers.

CHAPTER VIII

TRACK ON BRIDGES AND ELEVATED STRUCTURES

Bridge Floors.—In the design of bridges and trestles, the deck and track construction are closely related, and this construction follows two general types: (1) an open floor, usually with ties laid across stringers or girders of the structure; (2) a solid or closed floor carrying ordinary ballasted track or some special design of track.

The track on bridges and approaches should be kept in good line and surface, and at the ends of approaches it should be well bedded and tamped so as to avoid shocks when trains pass on and off the structure at high speed. Long trestles and bridges should have a plank walk at one side or between the tracks for the use of trackmen and other employees. Refuges for track gangs and section cars may also be provided, the latter only on very long structures. The smaller refuges can be placed conveniently at points where fire-protection water barrels are placed, or at intervals of not more than 100 ft. Refuge platforms on deck structures of the Southern Pacific Ry. are 4 ft. 8 in. long, carried by three long ties alternating with the regular ties, their length on the refuge side being 9 ft. 6 in. from center of track. The floor is of 2-in. plank and at the end is a hand railing.

Ballasted Floors.—Ballasted floors have advantages in regard to safety, permanence, and smooth riding, and while the first cost is greater, there is a decided economy in bridge and track maintenance. The extra dead load due to the floor and ballast may require a somewhat heavier and more substantial structure, but within certain limits this is economical, as it gives increased life and decreases the maintenance cost, while the ballast bed practically eliminates impact stresses and greatly reduces vibration. For long-span bridges, however, the extra material and weight may be excessive, so that open-floor construction, or at least an unballasted floor, is desirable. With a ballasted floor, the track on the bridge is the same as on any other part of the roadbed, to be examined daily by the section gang and any necessary maintenance or repair attended to at once, as part of the routine work.

With open-floor construction, however, the section foreman is not permitted to undertake track work on the bridge. If any defect is noted, he reports it, and the carpenter gang, or bridge and building gang, may have to make a special trip for a very minor repair job, with resulting delay and high labor cost. Adjustment of surface is effected by shimming between the caps and stringers of trestles or between the stringers and ties; for lining, the spikes must be drawn, rails adjusted and spikes re-driven. These operations are more difficult and take more time (especially for fine adjustments) than raising or shifting ties in ballast, while much respiking shortens the life of ties. There is often difficulty in maintaining good surface at the end of a structure, where the rails pass from ties on ballast to ties on the rigid foundation of a steel deck or of a trestle which may settle slightly. This particular difficulty is readily avoided with ballasted track.

Open-floor Construction.—Typical open-deck construction on steel bridges (Fig. 51) consists essentially of sawed ties laid upon track stringers or across the top chords of deck spans. Rails are laid and spiked in place in the usual way, but are supplemented by guard rails and guard timbers

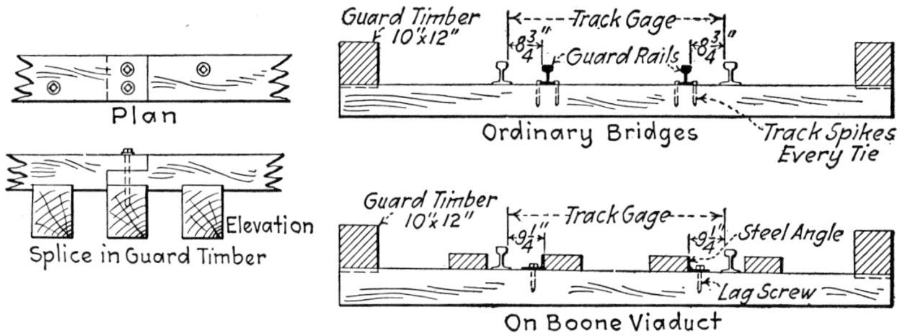

FIG. 51.—Open floor construction on bridges: Chicago & Northwestern Ry.

to keep derailed cars on the track. Every third or fourth tie is fastened to the structure by hook bolts engaging the flanges of the stringers or girders. Bridge ties are usually of yellow pine (creosoted or untreated) on account of its having little tendency to warp; fir and white oak are also used, while some electric roads employ red oak, black walnut, and white elm ties. Their length is usually 9 or 10 ft., sometimes 12 or 14 ft., and every third or fourth tie may be 2 ft. longer to carry a plank walk or the supports of an electric third rail.

In section, the ties vary from 8 by 8 to 10 by 12 in., or sometimes 8 by 16 in., according to the spacing of supports and length of ties. On the Southern Pacific Ry., the minimum size is 8 by 10 in. for ties 10 ft. long on stringers 7 ft. c. to c., while ties 8 by 16 in., 14 ft. long, are used where stringers are 12 ft. apart. For curves, the superelevation is given by large ties having the top cut at an incline which will give the desired elevation; the minimum depth of 10 in. is retained at the low end, while the maximum at the high end is 13 in. On other roads the height is reduced at the low end, 7- by 9-in. ties for tangents being replaced on curves by ties 7 by 7 in. at the low end and up to 14 in. high at the upper end to give a maximum elevation of $3\frac{3}{4}$ in. for the high rail. On double track, separate ties may be used, or timbers about 24 ft. long to carry both tracks. Ties and timbers should be not more than 4 in. apart and held securely against displacement.

Sawed ties are used on account of their uniform thickness, as they are carried on rigid supports. They are usually boxed or dapped about 1 in. to fit over the flanges of girders or stringers, and also cut to compensate for the thickness of cover plates, besides being grooved to fit over the rivet heads. On the Baltimore & Ohio R. R., however, this grooving is omitted where there are four or more lines of rivets; the ties are first set in place on the rivets, and after traffic has indented the wood, the ties are removed and shallow holes are bored at the indented points. The dapping of the ties may be made sufficient to compensate for the camber of the span, so that the track is level when there is no load on the bridge. The Southern Pacific Ry. specifies that if such cutting would reduce the depth of ordinary ties to less than $8\frac{1}{2}$ in., deeper ties must be used at the ends of the bridge, where the depth of cut would be greatest. All cutting and boring should be done before the ties are treated, and, where any cutting is done after treatment, the fresh surfaces should be well coated with creosote oil.

Trestle Floors.—For trestles, the open-floor construction (Fig. 52) is practically the same as on bridges, except that the ties are carried by two rows of packed stringers laid along the caps. On account of this length

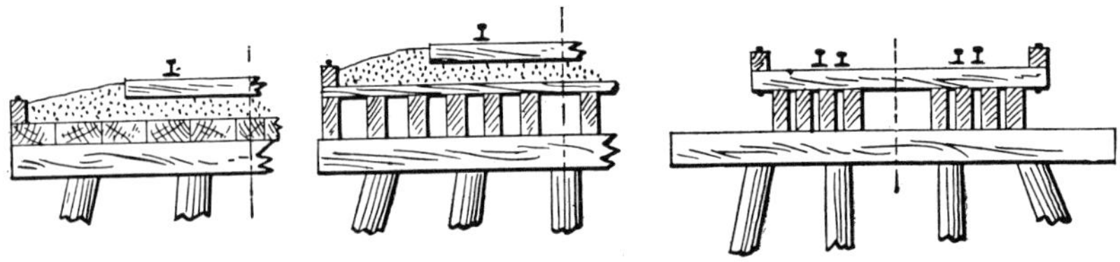

FIG. 52.—Trestle floors; ballasted and open deck.

of bearing, the ties are larger than 8 by 8 in. and are not dapped, but every third or fourth tie is fastened to the outer stringers by screw bolts or drift bolts. Open-deck culverts and short trestles over roads and cattle passes may often be replaced to advantage by solid embankments, with pipes or arched or box openings, thus materially reducing maintenance work and the danger of fire or washout.

On industry tracks which cross gutters for surface water, pairs of old rails are sometimes used as stringers, fastened together by bolts having spacing blocks on which the track rails are laid. At small irrigation ditches, each rail may be carried in a trough span formed by a pair of steel channels set back to back and having saddles riveted between them to support a timber stringer for the rail. A special construction employed under conditions requiring a very thin floor consists of transverse I-beams carrying two $\frac{1}{2}$-in. plates 30 in. wide, on which are shallow cushion timbers giving continuous bearing for the rails (see "Track Elevation Bridges").

Solid Floors: Ballasted.—For ballasted-floor bridges outside of towns and cities the deck is often a concrete slab supported on stringers or transverse framing. If the concrete extends the full width of a through-girder bridge, it will be shaped up against the webs and furnished with flashing to prevent water from entering between the steel and concrete, but in many cases the slab is only wide enough for the track (about 14 ft.) and has its sides formed as curbs 16 in. high to retain the ballast. Drain holes are formed in the bottom. This latter plan reduces the weight of floor and enables all steel to be inspected and painted. Other designs have a steel-plate deck on the floor framing.

The Rock Island Lines have used a solid floor of 14-ft. ties, 8 by 10 in., boxed $\frac{1}{4}$ in. over the deck girders and set on edge, close together, with every fourth tie fastened by hook bolts. Stone ballast 8 in. deep under the ties is retained by curb timbers which are bolted to the floor and are held against overturning by inclined braces having one end spiked to the top of the curb and the other end spiked to the floor. Inside 80-lb. guard rails are spiked to the ties. The top chords of the girders are coated with asphalt and covered with sheets of zinc bent down over the edges.

For ballasted-floor trestles (Fig. 52) the solid floor construction may be of three types: (1) with stringers or joists carrying transverse planking; (2) with ties laid close together; or (3) with longitudinal stringers or heavy planks placed close together. The first plan is the best and is most generally used. It permits inspection of all parts and enables new stringers to be placed without disturbing the deck, such stringers being one span in length and having the top at one end beveled so that it can be inserted between the cap and deck. In the A. R. E. A. recommended design there are twelve stringers 7 by 16 in. covered by 4-in. planks 14 ft. long, with 8 by 8-in. curb timbers bolted through the outer stringers, but raised 1 in. above the deck by separators on the bolts, so as to allow for drainage. The ties rest on 9 in. of ballast and have inside guard rails as protection in case of derailment.

Untreated floor timbers are sometimes protected with a layer of roofing paper laid in hot asphalt, but this covering is likely to be damaged in track work, and it is more economical to use treated timber throughout. The ballasted-deck trestle affords better protection against loss and traffic delay due to fire; it also gives better-riding track, greater safety and greatly reduced maintenance expense. It will be the more economical when the difference between the annual investment cost of the open-floor and the ballasted-deck types is equal to, or only slightly greater than, the annual cost of maintenance of the former.

Over narrow and shallow drainage openings, the ballast may be carried by a span of I-beams or old rails embedded in concrete, or carrying a concrete slab or a deck of transverse planks. On the Southern Pacific

Ry. (Fig. 52), ballasted-floor pile culverts with spans not exceeding 10 ft. have caps 15 ft. long carrying a solid deck of longitudinal creosoted timbers 8 by 17 in. (8 by 10 in. for middle timber), laid flat and drift bolted to the caps. This deck carries ballast 8 in. deep under the ties and retained by a curb timber along each side of the floor, secured by bolts with 1-in. separators to form a drainage opening under the curb.

Solid Floors: Unballasted.—Special deck and track construction has been used where limit of distance between street or water level and railway grade necessitates a floor of minimum thickness. This occurs most frequently for track-elevation bridges, noted below. For trestles, a floor has been made of ties laid close together with tie-plates spiked to alternate ties. The floor is given a coating of hot tar or pitch and is then covered with 3 in. of washed gravel as a protection against fire caused by sparks or engine cinders. Such construction, however, presents difficulties in track work.

Track-elevation Bridges.—On the solid-floor bridges required to carry elevated tracks across streets, both ballasted and unballasted tracks have been used, but the latter is exceptional. Both through- and plate-girder spans are employed, largely according to headroom conditions, but through girders between tracks cannot be used where turnouts and cross-overs have to be provided. Short spans may be composed of I-beams about 18 in. c. to c., embedded in concrete which is covered with a water-proofing course and then about 5 or 6 in. of reinforced concrete as a base for the ballast. Longer spans may have plate girders with a deck of transverse troughs partly filled with asphalt-concrete, or a deck of transverse I-beams covered with a steel deck plate, a concrete slab or 3-in. creosoted planking.

Self-supporting floors, consisting of a series of longitudinal troughs carried by the abutments and by intermediate girders on bents in the street, have been used in several cases, the ballast being retained by facia or parapet girders along each side. On the Chicago & Northwestern Ry. this deck has been found satisfactory and more economical than transverse troughs between through girders, the latter design being used where a shallow floor is necessary. The longitudinal troughs are approximately 18 in. deep and 15 in. wide, in spans of 12 to 30 ft. Asphalt concrete is used for the filling, the metal being heated to insure a close adhesion of the hot mixture. If ballast filling is used, drain holes must be provided, with gutters leading to downspouts. A more recent deck construction, now used extensively, consists of large precast concrete slabs which form self-supporting spans and are laid side by side, with ends resting on the abutments and on intermediate bents of steel or concrete. The outer slabs are L-shaped to form ballast curbs or parapet walls.

Unballasted solid floors usually have a steel deck plate, but to reduce the weight a tight plank floor may be laid. A design used by the Chicago,

Milwaukee & St. Paul Ry. for shallow-floor construction consists of 15-in. I-beams of 15-in. spacing laid across the bottom chords of through girders and covered with a ½-in. deck plate. Two double rows of steel angles riveted to the plate form troughs 18¾ in. wide for 3-in. wood cushion blocks forming a continuous bearing to which the rails are fastened by clips on bolts which pass through the deck plate. This design, but with a plate 30 in. wide for each trough, instead of a complete deck plate, has been used for open-floor construction where a shallow floor was necessary.

There are difficulties in keeping the bolts tight, in making a watertight deck over streets, and in protecting the steel against corrosion by brine dropping from refrigerator cars, while the cushion blocks require renewal in 4 or 5 years. A modified design has the troughs formed by 15-in. steel channels with wood cushions or planks 1½ to 2 in. thick. Concrete is not often used without ballast, but a design employed by the Big Four Ry. has transverse I-beams completely embedded in a floor slab, while I-beam steel ties placed between them have their tops projecting sufficiently to permit of placing the rail attachment bolts in the flanges.

Open-floor Elevated Railways.—For elevated rapid-transit lines, open-deck construction is usually employed, similar to that on bridges, but solid floors with and without ballast are also in use. A typical open-floor design is that of the Chicago Rapid Transit Co. (Fig. 53), having 90-lb. high-carbon rails of A. S. C. E. section fastened by spikes (screw spikes on curves) to longleaf yellow pine ties 6 by 8 in., 8 ft. long, spaced 18 in. c. to c. (15 in. on curves). With a solid bed of three ties, ½ in. apart, under each joint, this arrangement averages 40 ties to a 60-ft. rail. Every fourth tie is 9 ft. long to carry the insulated support of the third rail. Steel tie-plates, 7 by 9½ in., with seats inclined, are placed on every tie. Alternate ties are fastened to the girders by hook bolts, except that on curves every tie is fastened; the inside guard timbers are bolted to the same ties. The tracks are spaced 12 ft. c. to c. on tangents and 13 ft. on sharp curves. Four lines of guard timbers are used ordinarily, with inside guards 6 by 6 in., placed 6 in. from the gage side of the rail head (for wheels 5½ in. wide). The outer guards are 6 by 8 in., on edge, placed 10⅝ in. from the gage side of the rail and bolted to those ties to which the inner guard rail is not fastened. On curves, a steel restraining or guard rail is used on the low side. This is an 80-lb. rail with 1¼ in. sheared off the base: it is laid on its side, on malleable-iron chairs riveted to a forged plate, leaving a 2⅛-in. flangeway between the guard-rail head and the track rail. Shims of proper elevation are placed on the ties on the high side of curves to give superelevation.

The third-rail conductor has its head 6½ in. above the running rail and its center line 20⅛ in. from the gage side of the track rail. Between the tracks are 4 by 6 in. transverse timbers, 6 ft. apart, to carry the conduit or trunk for the electric cables, which forms a walk 38 in. wide. Where no conduit is required, a 30-in. central walk is installed, and in some locations there is an outer walk protected with railings. These walks are for the convenience of trackmen and others. Station platforms are 3 ft. 5 in. above the rail heads, with the edge 4 ft. 7 in. from center of track.

Two lines of guard timbers to each track rail are commonly used on elevated railways, the outside guard being for extra precaution and to

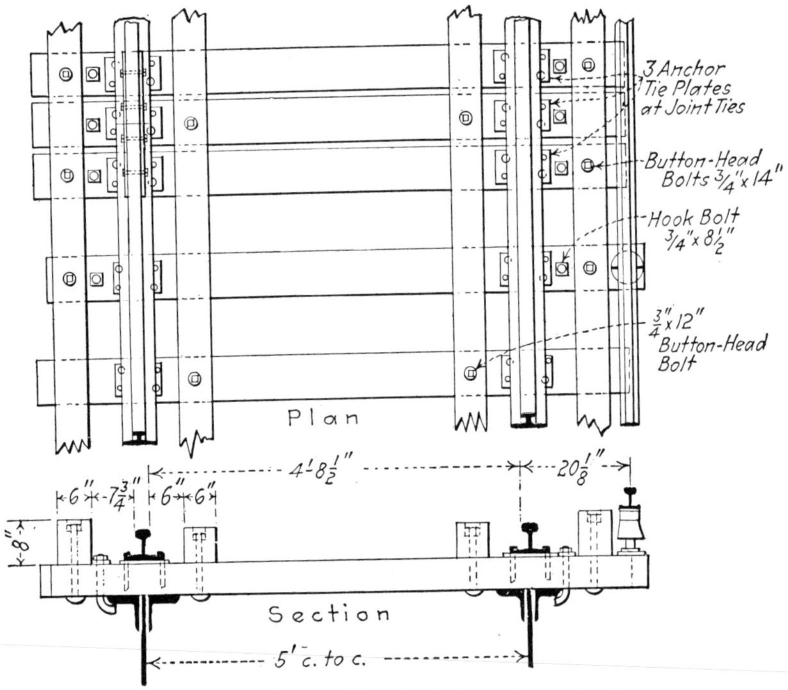

Fig. 53.—Open-floor and track; Chicago elevated railways.

hold the ties from bunching. The inner guard is more effective, as it is less likely to be climbed by the flange of the wheel, with its larger diameter. On curves, it may be replaced by a steel guard rail set at the close flangeway provided for curved track, the guard rail flaring out beyond the curve to the wider tangent spacing.

Solid-floor Elevated Railway.—A solid-floor construction used at Philadelphia is carried by transverse girders of 24 ft. span between 50-ft. longitudinal trusses (Fig. 54). In the earlier design, the deck was built up of plates and Z-bars to form a series of shallow longitudinal channels or troughs 18 in. wide and 5½ in. deep, this deck being covered with reinforced concrete shaped to form a central drain covered with removable concrete slabs. Upon the concrete is broken stone ballast with ties

6 by 8 in. and 8 ft. long. The track rails have inside guard rails of ordinary section and $8\frac{1}{2}$-in. spacing, or special guard rails with the web sloping inward so as to give a 4-in. flangeway. These special rails are fastened to the track rails by bolts with spacing blocks. The third rails, of the under-running type, are not carried by the ties but are suspended from the framing of an elevated walk along each side of the structure. Water is found to seep through the concrete, travel along the troughs, and drip to the bottom flanges of the transverse girders, causing corrosion of steel and deterioration of paint.

On the newer Frankford line, three lines of half-through longitudinal girders have transverse 12-in. I-beams framed between them at intervals of 5 ft. 3 in. and embedded in concrete jack arches, the surface of the

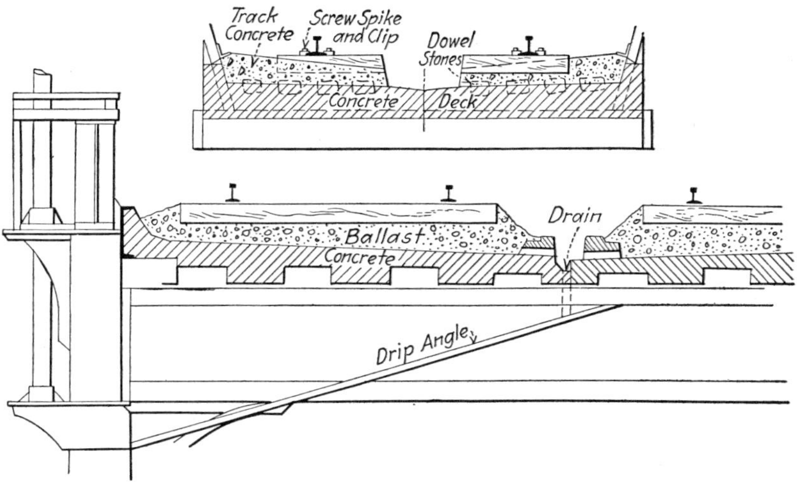

FIG. 54.—Ballasted decks; Philadelphia Elevated Ry.

concrete being sloped in drainage planes. This construction has eliminated the dripping of water, with consequent reduction in maintenance expense. On the concrete is the ballasted track. An unballasted construction used at stations and on a part of the open line (Fig. 54) has for each track two narrow concrete slabs which are anchored to the deck slab by lines of old granite paving, the blocks being half embedded in the slab so as to form dowels. A drainage channel is left between the two top slabs. Pockets formed in the slabs contain tie blocks $5\frac{1}{2}$ by $9\frac{1}{2}$ in., 34 in. long, to carry the track rails. The 90-lb. A. S. C. E. rails are secured by screw spikes with clips.

Station Floors and Tracks.—In large stations having the track floor above the street level a substantial and watertight construction is essential. In the Indianapolis Union Station a concrete track deck is formed with pockets for creosoted blocks 7 by 9 in., 36 in. long, with steel tie plates for 100-lb. rails secured by screw spikes. This bare floor with central drain can be thoroughly washed and flushed. At the St. Paul Union Station, a concrete deck carries washed-gravel ballast with creosoted ties and 90-lb. rails.

Longitudinals of Oregon fir, $7\frac{1}{2}$ by $15\frac{1}{2}$ in., resting on the top chords of plate-girder stringers, are used in the Chicago terminal station of the Chicago & Northwestern Ry. These timbers, secured by hook bolts, lie in troughs in the concrete floor, the spaces at the sides being packed with asphalt and covered with flashing. Rails of 100-lb. A. S. C. E. section are laid on tie-plates 8 by 8 by $\frac{1}{2}$ in., 24 in. apart, with two screw spikes in each plate. Both timber and steel longitudinals are used in the Pittsburgh terminal station of the Pittsburgh & West Virginia Ry. The older construction consists of longitudinal timbers 7 by 14 in. on 8 by 10-in. ties 30 in. apart, carrying a sheet-steel deck with asphalt coating. A 15-in. steel channel is laid upon each timber, the sides of which are

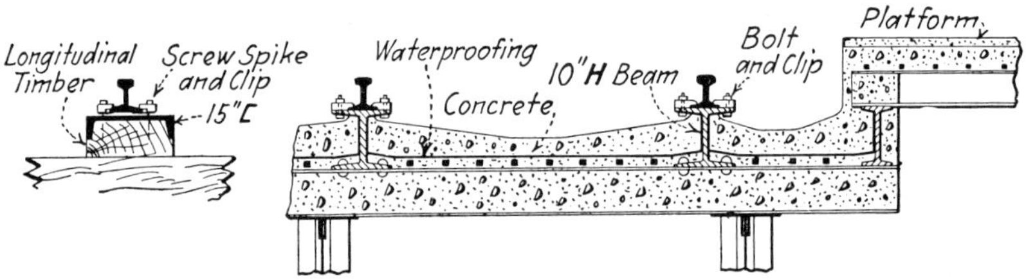

FIG. 55.—Station deck with longitudinal supports under rails.

shaped to fit, and to this the 90-lb. rails are fastened by screw spikes with clips. In a partial reconstruction, the wooden ties were replaced by 8-in. I-beams carrying longitudinals of 10-in. H-beams to which the rails are secured by clips on bolts passing through the flanges of the beams. A reinforced-concrete floor, with waterproofing course, is filled between the steel framing and sloped to form drains. Both types of deck and track construction in the Pittsburgh station are shown in Fig. 55.

Waterproofing and Fireproofing.—With bridges over streets, the engineer has a difficult problem in making and maintaining watertight decks, to prevent dripping and also unsightly damp patches and streaks on the abutments. In general, a waterproofing course is placed, usually consisting of layers of felt mopped with hot pitch, but vibration and expansion movements may rupture or displace this course, and water will find its way through the finest cracks or openings. Complicated flashing may be required also where concrete is laid against vertical steel surfaces. Periodical renewal of this waterproofing must be expected on either steel or concrete floors. Protection of steel decks against corrosive liquids dripping from refrigerator and coal cars is another problem. Where steel or concrete is subjected to the sand-blast effect of engine exhausts, as in low-headway bridges over tracks, it is protected usually by blast plates of cast iron or other material placed over the middle of the track, or by a tight plank ceiling beneath the floor. Pockets where engine gases may collect should be avoided or filled.

For fireproofing wooden structures, the use of solid floors with ballast or a thin protective bed of gravel provides good protection to the deck, while open-floor structures sometimes have sheet iron laid over the caps and stringers, the sides of the plates being bent down on an incline along each side of the timber. Water barrels are placed usually at intervals along trestles. These decks need careful watching, and at long or important structures watchmen may be assigned to follow every train and look out for signs of fire. Grass and weeds must be kept cut and cleared from around timber structures.

Examples of Open-deck Bridge Tracks.—On the Quebec cantilever bridge over the St. Lawrence, 12-ft. fir ties 8 by 10 in., laid on edge, are dapped ½ in. for the stringers and spaced 4 in. apart by combined tie plates and spacing blocks. The track rails are of 85-lb. and the guard rails of 56-lb. section, with 13½-in. spacing from the gage side of track rail to the center of the guard rail. No outside guards are placed. On the Thebes cantilever bridge over the Mississippi, ties 10 by 10 in., 5 in. apart, carry 8 by 8-in. outside guard timbers, dapped for the ties and bolted to alternate ties. The inside guards are heavy steel angles, with the vertical leg facing the rail and 8 in. from it. The standard construction on the New York, New Haven & Hartford Ry. has ties 8 by 8 in., 11 ft. long for single track and 24 ft. for double track, all spaced 8 in. apart, which is wider than desirable. Outside guard timbers 6 by 8 in., laid flat, are dapped ¾ in. for the ties and placed 10½ in. clear from the track rails. The inside guard rails, 8 in. clear from track rails, are extended straight on the approach for 15 ft. and then inclined inward on a flat curve for 60 ft. until they meet in the center line of the track.

Typical floor construction on the new Ohio River bridge of the Kentucky & Indiana Terminal R. R., at Louisville, Ky., consists of four plate-girder track stringers 6½ ft. c. to c., with ties 7 by 9 in., 11 ft. long, spaced 13 to 14 in. c. to c. and dapped ½ in. for the stringer chords. Every third tie is fastened to the outer flanges of the stringers by a ¾-in. hook bolt. Across the two inner lines of stringers are laid 2 by 9-in. joists, 8 ft. long, 4 ft. 4 in. apart, carrying a walk of three 2 by 8-in. planks. Some of the deck spans, with five lines of stringers, have ties 8 by 12 in., 24½ to 28 ft. long, carrying both tracks. This bridge also serves an electric railway of 5 ft. ½ in. gage, and on part of the structure the two tracks on each side are gantletted 8¾ in., so that each provides a guard rail for the other. The steam tracks, 13 ft. c. to c., have 85-lb. rails on tie-plates, every third tie having a combination tie-plate and rail anchor. The electric tracks have 75-lb. rails and no tie-plates. Guard timbers 6 by 8 in. are placed 10 in.

clear from the outside of the track rails, dapped 1 in. for the ties and secured to every tie by a lag screw $\frac{3}{4}$ by 9 in. These timbers have halved joints over the ties (Fig. 51). On curves, beveled ties are used, 7 by 7 in. at the smaller end and up to 7 by 14 in. at the larger end, giving a maximum superelevation of $3\frac{3}{4}$ in. at the outer rail.

On the Boone viaduct of the Chicago & Northwestern Ry. (Fig. 51), four lines of deck girders 6 ft. 6 in. c. to c. carry ties 8 by 8 in., 12 ft. long, 12 in. c. to c. Inside of each track is a guard composed of a steel angle 6 by $3\frac{1}{2}$ by $\frac{1}{2}$ in., with its 6-in. horizontal leg towards the rail and fastened to every third tie by a 6-in. lag screw. This angle is backed by a 4 by 10-in. plank spiked to the ties, and a similar plank is laid against the outside flange of the rail. In addition there are 10 by 12-in. guard timbers bolted along the ends of the ties. This construction is used also for steel bridges on curves. Bridges on tangents have similar outside guards, but the inside guards are steel rails spiked to every tie and brought together at 50 ft. beyond the ends of the bridge. The distance from gage side of track rail is $8\frac{3}{4}$ in. to inside guard rails and $9\frac{1}{4}$ in. to the guard angles. Particulars of practice as to bridge floors and guards are given in Table X.

TABLE X.—BRIDGE FLOOR-CONSTRUCTION

Railroad	Ties			Guard rail spacing		
	Wood	Size, inches	Length, feet	Spacing center to center, inches	Inside, inches	Outside, inches
Baltimore & Ohio	Oak	8 by 8	9	14	10	14
Chicago, Burlington & Quincy R. R. (A)	Douglas fir	8 by 8	10	14	$9\frac{1}{2}$	18
" " (B)	Douglas fir	8 by 10 8 by 12	10	12	$9\frac{1}{2}$	18
Chicago & Eastern Illinois (A).	Longleaf yellow pine	8 by 8	9 to 12	12	10	On ends of ties
" " (B)	Creosoted pine	8 by 10 8 by 12	9 to 12	12	10	
Michigan Central	Southern pine	7 by 9	12	13	12	27
Missouri Pacific (A)	Fir; treated and	7 by 8	9	12	8	Varies
Missouri Pacific (B)	untreated pine	8 by 10 8 by 12	10	12	8	Varies
Southern Pacific (C)	Douglas fir and	10 by 10	Varies	4 clear	8 minimum	8 minimum
Southern Pacific (D)	longleaf pine	8 by 10	Varies	4 clear	8 minimum	8 minimum
Wabash	Pine	8 by 8	9	12	8	....
Lake Shore Electric	White oak, red elm, black walnut	7 by 8	9	16	8	....
Michigan Electric	Longleaf yellow pine	6 by 8 8 by 8	10 (12 for third rail)	14	6	4 inches from end of tie
Quebec bridge	Douglas fir	8 by 10	12	4 clear	$13\frac{1}{2}$	none

A, trestles; B, bridges; C, main lines; D, branches.

Guard Rails and Rerailing Devices.—Provision must be made for safety in case of derailment on bridges, and effective protection, which need involve little extra expense, may save many lives and thousands of dollars in preventing an accident. As a rule, guards are placed parallel with the rails, and on open floors means are taken to prevent the ties from being shifted or bunched by wheels riding on them. In rare cases, through-truss bridges have horizontal skid girders built into the trusses to prevent car bodies from striking the members and perhaps causing collapse of the structure. A special feature of the Boone viaduct, noted above, is a heavy railing which was designed to hold the body of a derailed car. On the outside girders are deep triangular brackets connected across the structure by transverse angles, and each bracket carries a 10-in. vertical channel set edgewise to the track. These posts are riveted to the bracket webs and have outside braces attached to the ends of the transverse angles. The posts are capped with a heavy bulb angle, 4 ft. 6 in. above the rails. Through girders are likely to resist such glancing blows, but parapets and heavy railings on masonry structures and deck spans are of doubtful value as safeguards.

In general, reliance is placed upon inside guard rails and outside guard timbers laid along the ties to keep derailed wheels on the track, although the timbers are sometimes spaced so far from the rails that they serve principally to hold the ties in place. These timbers are placed only on open-deck construction, but the inside guards are used on practically all structures and also to some extent on high fills. The inside rails are the more effective, since the retarding action of the back of a wheel riding against the guard rail tends to hold the truck in line, while if the face of a wheel engages the outside guard timber, it tends to swing the truck out of line and thus make the wheel more likely to climb or break the guard.

Outside Guards.—Guard timbers are usually of pine or oak, 6 by 8 in. (laid flat) being the average size. Smaller sizes are undesirable, but there is some use of timbers 8 by 10 in. or even 10 by 12 in. on large bridges. They are secured to all ties by lag screws or to every third or fourth tie by through bolts having ordinary washers for the nuts or a cup washer socketed in the timber, the latter being sometimes used to keep the bolts below the top of the guard timber where snow plows and flangers are operated. The joints are scarfed or halved and bolted through a tie under each joint, although a few roads make a vertical diagonal scarf fastened by two horizontal bolts. These timbers are usually boxed or dapped 1 to $1\frac{1}{2}$ in. to fit over the ties and thus form spacers. As the pieces between daps sometimes split off, an alternative arrangement less generally used has the timbers left full size and fastened to every tie with a long lag screw. Metal spacing blocks sometimes placed between the ties are usually a combination of tie-plate and spacer, as on the Quebec bridge, noted above. Another practice of limited application is to reinforce the guard with a steel angle about 3 by 3 in. on the top inner corner.

In present practice the guard timbers do not extend beyond the ends of the structure, but formerly it was usual to extend them 50 to 100 ft. along the approach, flaring outward on long ties to a width of about 10 ft. in order to catch derailed wheels and guide them onto the bridge. At the ends of these flaring guards, bumping posts were sometimes placed in line with the bridge trusses, so that a car far enough from the rails to strike the truss would have its trucks stripped from under it before reaching the structure. The bumpers were composed of three piles bound together with cable or a 16 by 16-in. timber 10 to 12 ft. long having its top about 4 ft. above subgrade. With the heavier bridge and track construction of the present day, the use of extended outside guards and rerailing devices at bridges has become almost obsolete.

Guard Rails.—For the inside guards, steel rails of slightly smaller section than the track rails are placed 8 to 10 in. from the latter and spiked to the ties. They may be reinforced also by rail braces on every second or third tie. To catch "wandering" derailed wheels approaching the bridge, the guard rails are extended about 50 ft. beyond the structure, being inclined inward to meet on the center line, where they are bolted to a casting or frog point; this piece may be either beveled or bent down between the ties so as not to catch loose brake beams or other parts dragging from trains. The Canadian Pacific Ry. uses either two guard rails with their centers 18 in. from gage side of track rails, or three rails with one on the center line and the others $13\frac{1}{2}$ in. from the track rails. Steel angles instead of rails are sometimes used, as on the Boone viaduct.

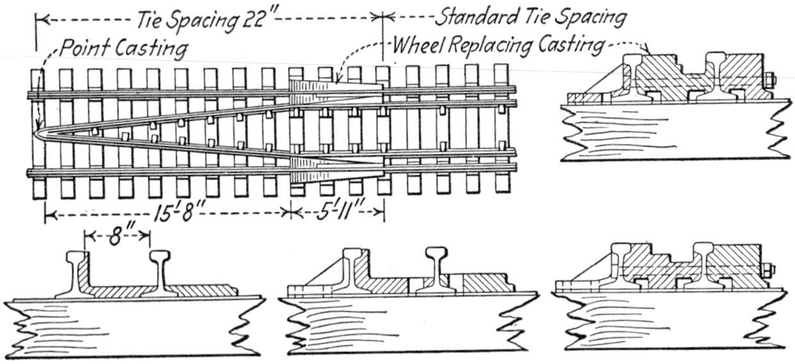

FIG. 56.—Rerailing device for bridges; Southern Pacific Ry.

Rerailing Devices.—Devices to guide derailed wheels onto the rails before reaching the bridge are now rarely used. They consist essentially of castings forming inclined planes, placed on both sides of the track rails and so arranged that wheels riding along the ties would run up the inclines and at the same time be guided inward, so that the outside wheel would ride across the rail head and drop into place with its flange on the inside of the rail. Such rerailers are used on the Southern Pacific Ry. at trestles more than 150 ft. long and at truss bridges; also at tunnels and on sharp curves where considered desirable. They are placed 30 to 50 ft. from the bridge or other point guarded. In this design (Fig. 56) a casting 6 ft. long is placed on each side of each rail, the inner casting fitting between the track and guard rails, which are 8 in. apart. Beyond the rerailers the guard rails swing inward and meet in a distance of about 16 ft. For operation in both directions, two sets of castings are placed end to end, forming a double incline about 12 ft. long.

CHAPTER IX

TRACK SIGNS AND ACCESSORIES

Track Signs.—Numerous marker and cautionary signs are required along a railroad, to give information to trackmen and trainmen, to warn the public, and to indicate property lines or political boundaries. For effectiveness and economy it is desirable to have few and uniform designs, using the same design for various purposes, with appropriate lettering. All signs should be of simple construction, durable and economical, while those for the guidance of trainmen should be specially conspicuous; they should have large, plain lettering, as little lettering as possible, and little attempt at ornament. Caution boards with long inscriptions are of little value, but may be required by law. Where state laws or the regulations of state authorities prescribe the design, wording, or positions for railroad signs, these requirements must be followed.

In signs conveying orders or information to trainmen, such as "slow" and "whistle" signs, the lettering faces down the track. They are placed on the engineman's or right-hand side of the track, with their posts 8 or $8\frac{1}{2}$ ft. from the rail or 10 ft. from the center of the track, but no part of the sign should come within the standard clearance limit of 8 ft. from the center of the track. Signs which are for the information of trackmen may face the track.

Boards or targets on posts constitute the majority of signs, but for many purposes the post alone is sufficient. Wood posts are usually of oak, cedar, or chestnut, set 3 ft. deep, or more if the post is high and carries a large board; they should be painted with, or dipped in, a preservative to about 12 in. above the ground line, while the top may be pointed or rounded to shed water. Ornamental turned work is undesirable, but square posts may have the edges chamfered. Concrete is used to some extent for post signs, usually with a painted surface or patch for the lettering, but if made with white cement and white aggregate, only the letters or figures need be painted. These posts may be made economically by railways which have plants for producing such concrete parts as culvert pipe, bridge slabs, and fence posts. Stone posts and slabs are used to a limited extent, mainly for mile posts. Cast-iron plates with raised letters and rim serve for many of the smaller signs, but cast-iron posts with disk or target tops are likely to be broken and cannot be repaired.

Steel boards or targets of $\frac{3}{16}$ or $\frac{1}{4}$-in. plate are cheap and effective, and are adapted to a variety of purposes. They may be bolted or screwed to wood posts or riveted to steel posts made of angles, T-bars, rails, or old boiler tubes filled with cement grout and having the top flattened. These posts may be set in precast-concrete anchor blocks, but if simply set in the ground, there should be a cross-rod through the bottom to prevent the post from being turned. Scrap material may often be utilized economically for signs of this kind. A special type has the letters or figures punched out instead of painted, thus requiring no skilled work in painting. The use of old rails for marker posts depends upon the scrap value of rails.

Black letters and figures on a white ground, or white figures on black, are most conspicuous; post signs are often brown, with a white patch or panel for black lettering. All the A. R. E. A. signs are white with a black border and letters or figures. Brown or black may be used also to make the rear of a sign inconspicuous. Red and green (or yellow) are used in some cases, principally for "stop" and "slow" signs, respectively. Large malleable-iron letters and figures nailed in place simplify the work of painting and renewal by unskilled labor. Varnish has given good results in making the colors last longer. Broken stone or small field stone laid around the base of the post gives a neat appearance, keeps weeds away, and protects the post from grass fires. Signs should be kept in good repair and are usually painted every year or two years by a small gang from the regular painting force, these men working at other times on bridges and buildings.

TYPICAL RAILROAD SIGNS

Some of the principal railroad signs are noted below, with particulars of different types, including those recommended by the A. R. E. A. Typical examples are shown in Figs. 57, 58, and 59.

Boundary Signs.—A state-line marker, *A*, is an oak post set 13 ft. from the rail and facing the track. Another design has the state names on opposite sides of a 6- by 8-in. post with an 8-in. horizontal board 5 ft. long, lettered "State Line." County and township lines and city limits may be marked by 6-in. posts 8 ft. long, with the upper part $5\frac{1}{2}$ in. square and lettered on the sides; a cast-iron sign has a 36-in. circle with cross-arm for the name and a socket at the bottom to fit the post. Railway divisions and subdivisions may be marked by 6-in. posts or by targets on iron tube posts, these signs being placed 30 or 40 ft. from the track or at the right-of-way line.

Bridge Signs.—Bridges, trestles, culverts, and other openings, however small, are numbered consecutively, either from a terminal or division point or from each mile post (as 25-*A*, 25-*B*, etc.). New structures may have fractional or decimal numbers. For steel through bridges, the number may be painted on the end post or end of girder, sometimes with the name of the stream. For deck bridges and other structures on which the number cannot be painted, the A. R. E. A. sign is an octagonal plate 24 by 24 in. riveted to the flattened end of an 8-ft. boiler tube. A simple design is a short block of triangular section spiked on top of the end of a tie and having the number painted on one face. For drawbridges, there are usually three warning signs, each with a tall post and a board lettered "Drawbridge," and with "One Mile," "1000 Feet," and "Stop," consecutively.

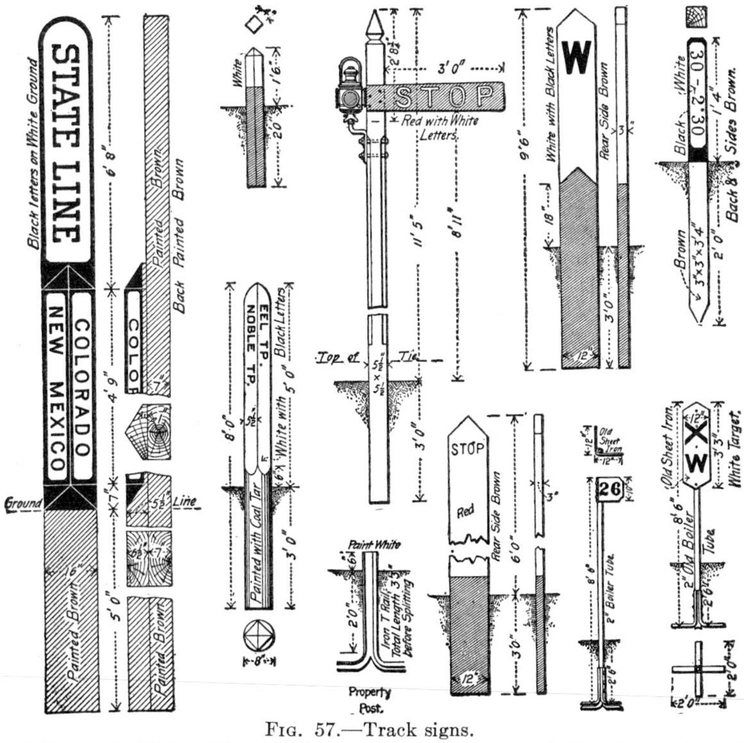

FIG. 57.—Track signs.

From left to right: State and township boundary posts; clearance post (above); stop signs; whistle posts; section number sign; curve elevation post (above).

Clearance Signs.—To indicate the nearest point at which the end of a car on a siding may stand without fouling cars on another track, a post or target may be placed on the outer side of the track, 8 ft. from the center line. The clearance point is usually where the distance between gage lines of rails of the two tracks is 8 ft. at main-line turnouts (tracks 13 ft. c. to c.) or sometimes 7 ft. at yards and house tracks. In yards, the ties at the clearance point may have the ends painted white, but low stakes set between the tracks are objectionable, as men may fall over them.

Crossing Signs (Highway).—To warn persons on public roads that they are approaching a grade crossing, a conspicuous sign is placed 10 to

20 ft. from the center of the track and in good view along the road. The style and lettering are sometimes prescribed in state laws or regulations. The A. R. E. A. design has an 8-in. post 16 ft. long, 12 ft. above the ground, carrying two 12-in. boards 8 ft. long which cross at 50 deg. and form an X. A simple sign consists of a long cross-arm or a rectangular board on a post. Another design has a square frame of four

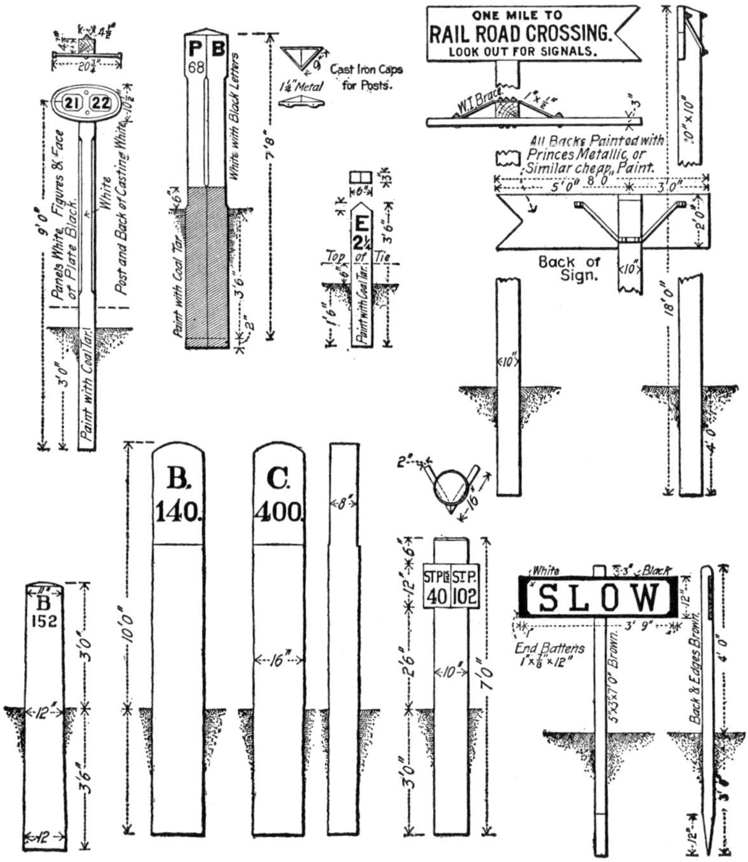

FIG. 58.—Track signs.

Above: section number sign; mile post; curve elevation post, crossing sign. Below: stone mile posts; wood mile post; slow sign.

boards about 5 ft. long, placed in diamond form between two 3 by 8-in. planks which form the post. A cross-arm on the post and lettered "Danger" is sometimes required; and where two parallel railroads are less than 400 ft. apart a similar cross-arm may be lettered "Two Crossings." Where a distant sign is required by law, the A. R. E. A. standard is a post 4 to 5 ft. high carrying a 24-in. white disk with black rim and cross, and marked "R. R." in the upper quadrants.

Crossing Signs (Railway).—These are set at a half-mile or a mile from a crossing to warn enginemen. The sign may be a board or a pair of cross-arms at 45 deg. mounted on a post and lettered "Railroad Crossing, One Mile," or with the distance marked on the post. At non-interlocked crossings there will be an additional sign, "Stop; Railroad Crossing, 200 Feet."

Curve Signs.—Where curves are numerous, they are sometimes numbered (from each mile post) and marked by small signs giving the degree for the information of trackmen in adjusting the elevation and bending rails. For transition curves, signs or posts are placed opposite the point of curve and point of spiral (see "Curves"). The signs may be attached to telegraph poles or to pipe posts. In the A. R. E. A. recommended practice, a 4 by 4-in. post, 3 ft. 10 in. long, set 2 ft. in the ground, is placed 7 ft. from the gage side of the rail at the point of runoff on tangent for circular curves or the beginning of easement for transition curves; this is lettered "E. 0." (for "Elevation, Zero"). A similar post at the end of the easement or point of full elevation on a circular curve has the elevation marked on the side facing trains approaching the curve; on the face is the elevation (E-2½) and on the opposite side is the number of the curve. The posts are black to 3 in. above the ground, and then white with black lettering.

Flanger Signs.—These are temporary signs, put up early in winter to indicate where the blades of snowplows and flanger cars must be raised to clear obstructions, as at switches, bridges, and tunnels. The A. R. E. A. design is a post set 60 ft. in advance of the point it protects and carrying a 12-in. triangular white target, painted black with a white spot. Other designs have a target, painted black with a white spot, or with two 6-in. holes; or yellow with a black spot. More conspicuous is a 10-in. black blade about 2½ ft. long, sloping upward and outward from a post 6 ft. high set 10 ft. from the track center and 50 to 200 ft. in advance of the obstruction.

Mile Posts.—Wood posts about 8 in. in diameter or 10 by 10 in. are commonly used; sometimes with simply the mile painted, or with the initial of the terminal and the distance on opposite sides. The A. R. E. A. standard is a 9 ft. reinforced-concrete post of triangular section with a 17-in. base and 12-in. sides having recessed figures 6 in. high. Concrete posts on the Chicago & Eastern Illinois R. R. are 8 by 8 in., 8 ft. long, with a painted black band for white figures. These are made of 1:1:2 concrete and weigh about 500 lb. Flat posts 16 by 4 in., of white concrete (made with white cement, sand, and gravel), used on the New York Central R. R., are 9 ft. long, 5½ ft. above ground, with 6-in. recessed letters (for terminal) and figures; these recessed marks, 1 in. wide and ⅜ in. deep, are painted black. Stone posts, used to a limited extent, are usually of granite, flat or square, with a painted panel for the figures.

A cheap and simple V-shaped sign consists of two boards or a bent steel plate forming panels about 9 by 15 in. placed on the telegraph pole, the pole being painted white above ground. Their positions are only approximate. On some roads the half-mile is indicated by a 24-in. white band on the nearest pole.

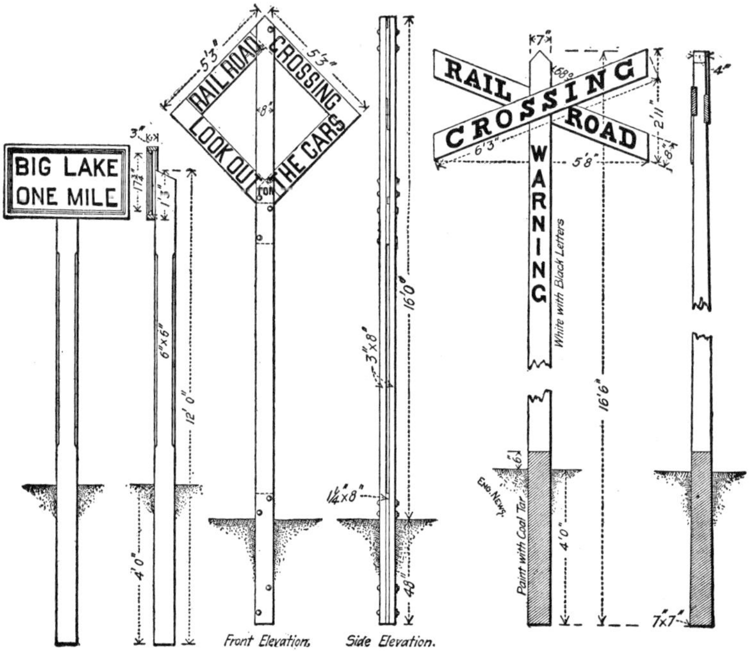

FIG. 59.—Track signs.
Station sign; highway crossing signs.

Property Post.—Right-of-way lines, station grounds, and other property lines should be carefully marked by permanent monuments, especially at corners and angles. If the post cannot be set at the exact spot, it should be placed on line, with the distance painted on the side facing the corner. The A. R. E. A. design is a 5-ft. post of reinforced concrete, standing 2 ft. above ground and tapering from 7 to 5 in. square, with a cross-cut in the top to mark the exact corner. On the side facing the track it is marked "Property Line," with the initials of the railroad on another side, all lettering being recessed. These posts are to be set on all property corners, at 1,000-ft. intervals on tangents and at the P.C. and P.T. of all curves. Cast-iron monuments or 3-ft. pieces of rail, having the bottom split for 3 in. and opened up like an inverted T, are sometimes placed at important land corners, with intermediate wood posts 500 ft. to 1,000 ft. apart. These posts are on the railroad property, with the outer face on line, and are usually white with black lettering.

Section Signs.—These signs, which mark the limits of track sections, need not be conspicuous and are generally placed at a distance from the track or near the right-of-way line. The A. R. E. A. design is an oval steel target 18 by 30 in., on a tube post; it is white with a black border and vertical stripe forming two panels for 6-in. figures. Frequently, a post about 2 by 6 in. is set edgewise to the track, 7 or 8 ft. from the rail, with the section numbers on opposite sides. The simplest form is a board or iron plate fastened to a telegraph pole.

Sidetrack Limit Sign.—Sidings long enough for two trains may have the middle indicated by a post or target, usually without lettering.

Slow and Stop Signs.—These are to notify enginemen of approaches to track crossings, drawbridges, etc. The A. R. E. A. design is a 12 by 32-in. arm on one side of a post 6 to 7 ft. high, the arm being lettered "Slow" or "Stop" in 6-in. letters. Flat posts about 3 by 12 in. and large board signs are also used. In some cases they conform to signal indications by having a 3-ft. horizontal red arm lettered "Stop" or an inclined green (or yellow) arm marked "Slow." Red and green (or yellow) switch lamps may be placed on the signs at night. The "slow" board is placed usually at 2,000 to 3,000 ft. and the "stop" at 200 ft. from the point protected.

Speed Limit Signs.—At sharp curves and old bridges and at points where track or bridge repair work will last for more than 3 or 4 days, signs or slow boards are set about 3,000 ft. in advance of the working limit to notify enginemen to reduce speed. An ordinary "slow" sign will then be placed 50 ft. in advance of the limit. The A. R. E. A. sign is a 12-in. board at an angle of 45 deg. upward from a post and lettered with the speed required. At night a lamp is arranged to throw a yellow light on the sign. At 150 ft. beyond the working limit is placed a post with a vertical board 10 by 22 in. lettered "R. S." for "Resume Speed." Another design is a green board 15 by 36 in. lettered in white, "Reduce Speed, —— Miles per Hour," a pocket holding a sheet-iron plate with the number. With this, a green lamp is used. At 500 ft. beyond the working limit is a similar sign, but painted white and lettered "Resume Speed" in black. If these signs must be placed between the tracks, the board is about 12 by 18 in. and 6 in. above the ballast. For sharp curves, a target may be placed facing the traffic and lettered "Curve; Speed Limit—— Miles."

Station Signs.—As an indication to enginemen, a sign is usually placed a mile from each station and a whistle post at ½ or ¼ mi., measured from the outer switch or the yard limits. The A. R. E. A. sign is a 12 by 42-in. cross-arm 18 in. below the top of an 8-in. post rising 6 to 7 ft. above the base of the rail. The board has a black border and "Station" in 6-in. letters, with "1 Mile" on the post above the board. A similar sign board of suitable length for the lettering is used in advance of yards, crossings, junctions, and drawbridges. A common form of station sign is a board about 18 by 30 or 24 by 36 in., lettered with the station name and "One Mile," but the A. R. E. A. sign is uniform for all stations.

Station Name Signs.—For the convenience of passengers on trains the name boards at stations should be large, distinctive, and conspicuously placed, with no advertising signs near them. Small boards on the ends of station buildings or platform shelters may not be seen readily. Boards placed on posts on the platforms should be at some distance from the track, and have lamps placed so as to illuminate them at night. Such signs are sometimes placed a little distance beyond the platform and at a slight angle with the track so as to face the trains. The board should be not less than 18 in. wide. White or yellow plain block letters on a black or dark blue ground are prominent; also black letters on white or buff. Enameled iron signs in blue and white are used generally on suburban and electric lines. On many main lines the distances to the two main terminals are given at opposite ends of the sign. Glass slips in the station windows and lamps are useful, and it is a good plan to mark the name in large letters of white stones or shells on a strip of grass or cinders behind the platform. Experiments have been made with distinctive signs, unusual in shape or design, such as a disk or circle with the name board across it.

Whistle and Ring Signs.—To notify enginemen where to sound the whistle or engine bell, these signs are placed usually ¼ mi. on each side of stations and road crossings. Concrete posts 9½ ft. long, 6 ft. above the rail, as used by the New York Central R. R., are made from a 1:2:4 mix of white portland cement, white sand, and ¾-in. white gravel. The post is 6 by 4 in., widening to 12 in. at top to form a panel for an 8-in. recessed "W," painted black. For highway crossings the A. R. E. A. design has a board 14 by 10 in. on a post, with a 6-in. "W." Steel-plate targets on tube or pipe posts are also used, but the usual style is a wood post 10 by 4 or 8 by 8 in., 5 to 8 ft. high, lettered "W" or "R," or both. Sometimes "X-W" indicates a crossing and "S-W" a station. Where the bell is to be rung to warn track or bridge gangs, a "whistle and ring" post ("W-R") is set 1,000 ft. from the point to be protected.

Yard Limit Signs.—These denote the limits covered by yard gangs and switching engines, within which trains must be run under control. The A. R. E. A. design has two 12-in. boards framed at right angles and placed to form a V on top of a 6 or 7 ft. post. The arms are lettered "Yard Limit." A board or target is often used, sometimes with the name of the station in addition.

Miscellaneous Signs.—Various signs required for special purposes may generally be uniform in style, differing only in the lettering. This gives economy of manufacture. Two such styles recommended by the A. R. E. A. are: (1) a steel-plate target on a pipe post; and (2) a 12-in., horizontal board across a post 6 or 7 ft. high and extending 18 in. above the board; both with black letters and border on a white ground, with black pipe post, and white or black wood post. Some of these signs are as follows: "Block Section Limit;" "Clean Ashpan Here;" "Dump Ashes Here;" "Beginning (and end) of Double Track" (placed opposite the clearance points); "Blind Siding," and "Passing Siding" (placed opposite the switch); "Shut This Gate" (at farm crossings); "Derail" (at the head block of turnouts where the derail is not interlocked); also "Crossing," "Junction," "Station," "Yard," "Drawbridge," "Water Station," and "Coaling Station," all of which are also marked "One Mile."

MISCELLANEOUS ACCESSORIES

Bridge Warning or Tell-tale.—To warn men who may be on the tops of box cars that the train is approaching a structure (bridge or tunnel) where the headroom above the rails is less than $22\frac{1}{2}$ ft., a tell-tale or tickler is placed 100 to 300 ft. in advance on the main line, and 50 to 100 ft. on yard tracks, and for both tracks on double-track lines. This device consists of a post with top arm extending across the track and carrying a row of cords or ropes which will strike a man standing on a box car and are so hung that the wind cannot blow them up over the cross-arm.

On the Rock Island Lines, a wood or concrete post set in concrete carries a 13-ft. arm with knee brace and is held by a wire guy or backstay anchored in the ground. To the arm, which is $25\frac{1}{2}$ ft. clear above the rails, is attached a 3 by 7-ft. panel of 1-in. wire mesh in a frame of 1-in. round bars; from the lower side of the frame are hung $\frac{3}{8}$-in. cords 42 in. long and 6 in. apart, having their ends bound with copper wire. The post is set with its center line $9\frac{1}{2}$ ft. from the center of the track; on curves, this distance is $10\frac{1}{2}$ ft., with a 14-ft. top arm. The Chicago & Eastern Illinois R. R. bridge warning (Fig. 60) consists of a 4- and 3-in. pipe post on a concrete footing and having a top arm of 2-in. pipe extending 13 ft. across the track and 4 ft. to the rear, the rear end being anchored to the post by a long diagonal tie rod. To prevent sagging, the pipe arm has a 15-in. strut on top just back of the post, with a truss rod attached to both ends of the arm and passing over the strut. Eyebolts 6 in. c. to c. in the arm, which is 24 ft. above the rails, carry $\frac{3}{8}$-in. rods 36 in. long, having looped ends. To each bottom loop is attached a $\frac{3}{8}$-in. rope or sash cord 42 in. long, having the ends wrapped with copper wire for 6 in. On multiple-track lines and at yards, the panels or arms carrying the rope ticklers may be suspended from a taut $\frac{3}{8}$- or $\frac{7}{8}$-in. wire cable on posts held by braces or guys. This cable is kept horizontal by hangers from a catenary cable attached to the tops of the posts.

Mail Cranes.—For hanging mail bags at stations, to be taken off by the mail car of a moving train, a mail crane is erected consisting of a post with two pivoted arms. In the typical design of the Railway Mail Service Bureau of the U. S. Post Office Department (Fig. 61) the post is seated on a frame carried by three 16-ft. ties. At the top, about 10 ft. above the rail, is a cross-arm pivoted at the middle and balanced so that normally it stands vertically against the back of the post; about 3 ft. below this is a short arm which normally hangs vertically against the face of the post. These arms are horizontal when a mail bag is attached to them by its top and bottom straps engaging spring clips on the arms. As the "catcher" arm on the mail car picks off the bag, the arms of the crane drop back against the post. The crane is best carried on two or three long ties so that it will maintain a uniform relation to the track.

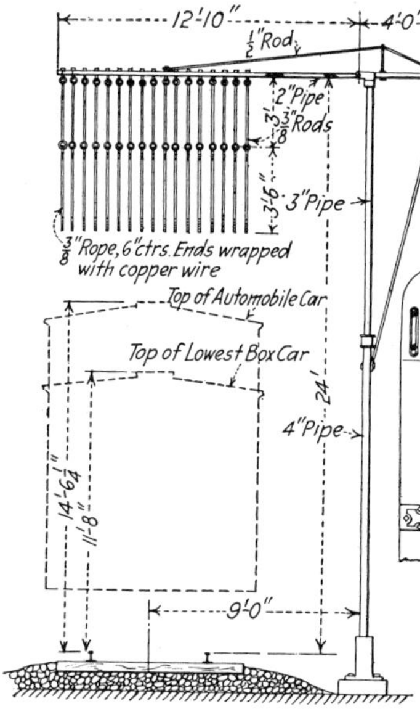

FIG. 60.—Bridge warning.

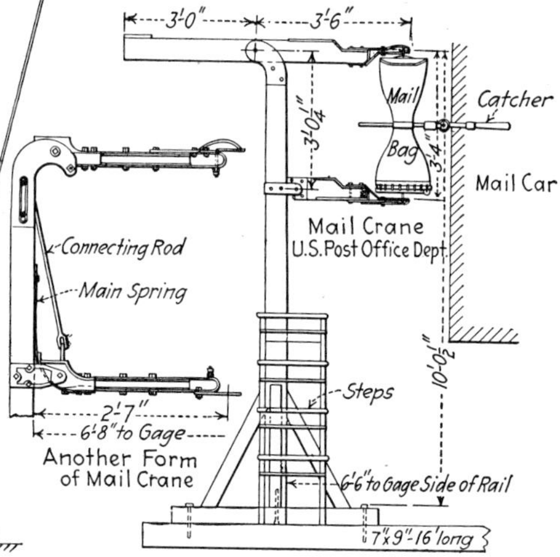

FIG. 61.—Mail cranes.

An alternative design, also shown in Fig. 61, has two short arms connected by a link so that as the lower arm drops, the upper one is swung upward, the action being assisted by springs. Iron mail cranes for four-track lines where passenger trains use the inner tracks have fixed arms on a post which revolves so that the arms ordinarily stand parallel with the rails. Mail is delivered from moving trains by throwing the bags from the car. Mechanical devices to deliver and receive the bags have been tried, but without much success. Since the introduction of the parcel-post service, trains stop at least once a day at all stations having post offices, thus facilitating the exchange of mail so that the ordinary mail cranes and catchers meet the requirements of exchange from moving trains.

Bumping Posts.—Bumpers or bumping posts provided at the ends of stub tracks may be high enough to engage the car coupler or platform or simply high enough to engage the wheels. A typical high bumper has a heavy timber post seated on a sill parallel with the rails, the post being braced to the rear end of the sill by an inclined timber and tied to the forward end by rods. Two such frames side by side and connected by cross-pieces (Fig. 62) make an effective bumper. Across the face of the posts is a cross-timber at the height of the car coupler and carrying the striking block. With a 7-ft. post, the sills will be bedded deep, the center line of the coupler being about 40 in. above the rails. Numerous bumpers are made with steel rails bolted to the track rails and inclined inward and upward in V-shape, the apex being bolted to a casting on a heavy post. Against the face of this casting rests a 12 by 12-in. post with a striking

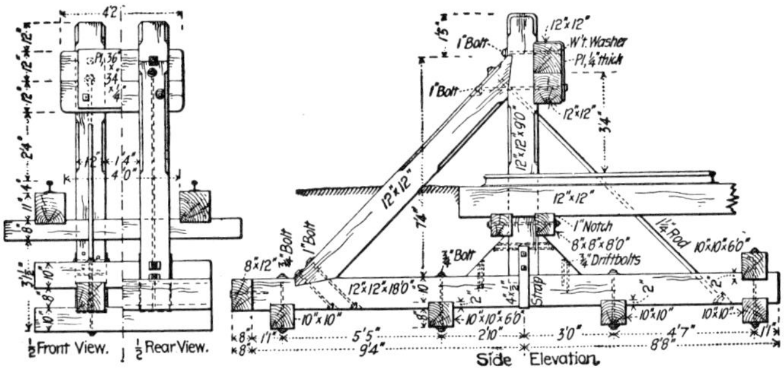

Fig. 62.—High bumping post.

block bolted to it. In some designs the apex and block are supported by a similar V-frame as a back brace, no post being used. In passenger stations having elevated tracks, the bumpers may be of steel-plate or reinforced-concrete construction built into floor framing of the respective materials.

Large reinforced-concrete blocks are frequently used in yards and sometimes at terminals, but it is not unusual to find them displaced by severe blows. Such blocks at the Camden terminal of the Philadelphia & Reading R. R. are 11 by $8\frac{1}{2}$ ft. for 5 ft. from the base up to rail level and then taper for 5 ft. to a top $6\frac{1}{2}$ ft. square, the face having a bench for the 12 by 16-in. back timber of the striking block. Damage to cars and bumpers may be reduced by making the striking block of two timbers with nests of coiled car springs between them.

Some large English terminals have hydraulic bumpers consisting of a timber frame carrying a pair of 10-in. horizontal cylinders perforated with $\frac{1}{2}$-in. holes and each enclosed in a larger cylinder, leaving a $2\frac{1}{2}$-in. annular space. The piston of the inner cylinder is normally at the outer end of its stroke, and the head of the piston rod forms the striking block. The cylinder, filled with a mixture of petroleum, soap, and water, is connected to an air chamber in which a pressure of 40 to 45 lb. is maintained by occasional charging. In the earlier bumpers the cylinders had a stroke of 2 ft., but those of recent construction have a stroke or travel of 7 ft. When a train strikes the bumper, the car or engine buffers drive back the piston rods, the pistons forcing the fluid out through the holes against the air pressure, and the resistance increasing as the number of holes for escape becomes less. Other methods are to check the motion of the train by sliding shoes or stop blocks on the rails, as noted below, or by placing each rail in a trough filled with sand; but all these methods require a certain length of track, which perhaps cannot be spared.

Low bumpers (Fig. 63) are frequently heavy timbers laid across the rails and backed by posts or by brackets bolted to the rails; they may be bolted to the ties or held by tie rods or straps bolted to the rails ahead of

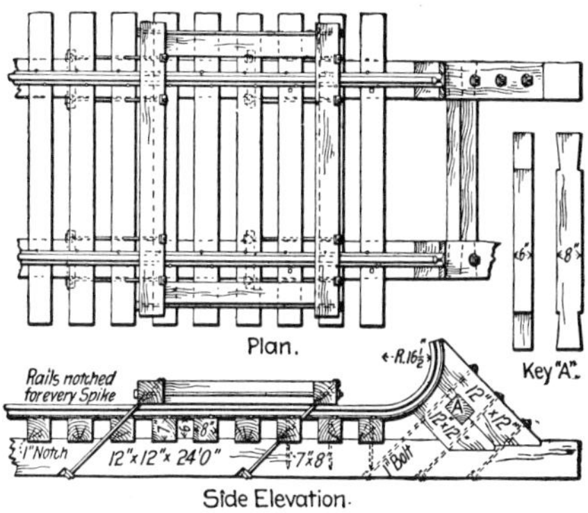

FIG. 63.—Low car bumper.

the bumper. As a single timber may be jumped rather easily, the bumper is more effective if made of two timbers separated by spacing blocks 2 or 3 ft. long or by nests of car springs. A simple low bumper consists of a casting about 16 in. high bolted to each rail and having the face shaped to the radius of a car wheel.

Where yard space permits, it is often desirable to absorb the momentum of the car by a heap of earth, which at stations may be dressed to a pyramid 3 or 4 ft. high and 5 to 8 ft. square on top, with slopes of 1 on $2\frac{1}{2}$ or 3. A modified design has a 2- to 3-ft. fill retained in a three-sided box about 12 ft. long, built of planks or old ties with outside posts.

Stop Blocks.—A stop used on enginehouse tracks consists of a block 22 in. long bolted to the end of each rail and having its face curved to 10 in. radius; its lowest point is $2\frac{1}{2}$ in. below the rail head, so that a wheel will drop into a pocket. The top of the block is only 3 in. above the rail head, so as to clear the engine pilots. A portable stop block clamped to the rail serves to hold cars in position on shop or yard tracks and may be fitted with a padlock.

A movable or sliding stop to check cars on yard or station tracks consists of a feather-edge shoe lying on the rail head and held in place by side lugs. As the wheel rides on the shoe, it is raised on a slight incline and then drops into a pocket, the rear of which is curved to fit the wheel. The momentum of the car causes the shoe to slide along the rail, the friction quickly checking the speed. In another design the drop pocket is omitted, but at the rear of the shoe is a grooved roller; when the wheel rides up on the shoe and pushes it along, it presses a hinged block against the roller as a brake. Such stops are set 10 to 30 ft. from the end of the track. Portable sliding stops, used to check the movement of cars in gravity yards, are shoes like those described above but without the rollers. A track brake used for the same purpose on the Indiana Harbor R. R. has two rails which are parallel with, and above, the track rails and may be moved inward so as to bear against the rims and backs of the car wheels. These brakes are operated by compressed air under the control of men in towers along the yard (see "Gravity Switching").

Sand Track.—To stop runaway cars on heavy grades a sand track 100 to 200 ft. long may be provided, gantletted with the main track and connected to the latter by switches. In this device (Fig. 64), the

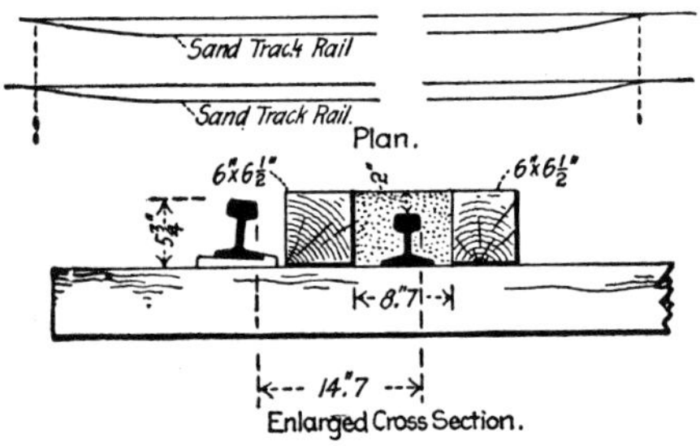

FIG. 64.—Sand track.

sand-track rails, about 15 in. from, and 1 in. lower than, the others, are each flanked by timbers forming a trough 6 to 9 in. wide and deep enough for 2 to 2½ in. of sand over the railhead. To catch cars which break away from an ascending train, the upper end of the sand track will have a spring switch set normally for this track so as to divert the cars into the gantlet. To check cars running too fast on the inclines of a gravity yard, some European roads have installed a sand track with a facing switch to be operated by a yardman on emergency, since many of their cars have no brakes at all or brakes operated only by a man standing on the ground. Sand tracks may also be formed at the ends of stub tracks, or the rails may be simply covered with sand for 20 ft. or more.

Spare Rail Racks.—For prompt renewal of broken rails, spare rails are placed at intervals. There may be a rail and a pair of splice bars near each mile post (or two rails on double track), but for greater distances two or three rails may be placed at each point. A common support consists of three stakes or posts about 21 in. high, made from old car sills or trestle stringers and stepped to form seats for two or three rails. Smaller stakes may have the top slotted to receive a single inverted rail. Stakes of old rails with riveted brackets are also used. Instead of posts, the Union Pacific Ry. uses blocks (Fig. 65), each consisting of two 4-ft. lengths of old ties driftbolted together, the lower one being half embedded in the roadbed and their inner faces being 12 ft. from center of

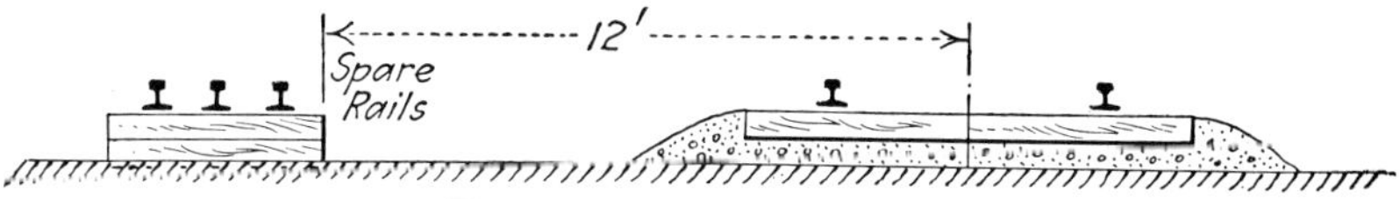

FIG. 65.—Spare-rails on track.

track. A pair of blocks carries two or three spare rails, which are painted occasionally to prevent rusting. At these rests, which are located where the cut or fill is low, the roadbed is widened to 17 ft. from center of track for a length of 40 ft.; this space is kept smooth and neat, and the wood blocks are painted white.

Section Car Turnouts.—For convenient placing of hand and motor section cars while the men are at work, turnouts or takeoffs are provided about $\frac{1}{2}$-mi. apart, and are on each side on double track. To form such a turnout the Union Pacific Ry. widens the roadbed to $19\frac{1}{2}$ ft. from outer rail for a length of about 24 ft., while the ballast is widened to 14 ft. from the rail for a distance of 12 ft.

CHAPTER X

RAILWAY BUILDINGS

Many minor buildings of a railroad come under the care of the engineering, maintenance-of-way, or bridge and building department. These include section tool houses, dwellings for section forces, cabins for switchmen, telegraph and signal towers, small stations, and also the portable buildings and camp cars used by maintenance or special forces on such work as double tracking and bridge repairs. The buildings are generally of wood, but brick and concrete are often used, the latter either monolithic or in precast hollow blocks. Stucco on metal lath with paper or felt sheathing over a wood frame is also used. Materials and style of construction depend largely on local and climatic conditions. An attractive appearance can be obtained at little cost, and all buildings should be kept neat, clean, in sanitary condition, and in good repair. Buildings and other structures should be numbered for convenience of record and reference, and should have the number painted at some designated place.

Painting.—In painting wood buildings, a shade darker than the walls is usually employed for belt rails and other trim, and sometimes the walls below the sills are darker than the upper part. Effective combinations are tuscan red with a darker brown, two shades of stone color or green, or the colonial yellow and white. Stock pens, sheds, and fences at road crossings should be kept well whitewashed. Both paint and whitewash can be applied economically as a spray by compressed air, the former being mixed thinner than when used with a brush. A convenient equipment for this work consists of a car fitted with a compressor (or air-brake pump), air tank, paint and whitewash tanks, and the necessary hose lines and nozzles. The work is rapid and economical, especially on rough lumber. Such a car may be part of the equipment of the painting gang which goes over the road at intervals.

Section Tool House.—Every track section has a building to accommodate the hand or motor cars, push cars, tools, and supplies of the section gang and to give room for the men to work at cutting shims, sharpening tools, or sorting small scrap in bad weather. For important lines, the A. R. E. A. recommends a house 14 by 20 or 12 by 18 ft., with its longer side parallel with the track and having an 8-ft. sliding door at the end of the track side; for less important lines, a house 10 by 14 ft. would have the shorter side parallel with the track and double swinging doors on the track side.

The standard section tool house of the Cleveland, Cincinnati, Chicago & St. Louis Ry. (Fig. 66), is 12 by 16 ft., with its front at least 20 ft. from the nearest rail and having double swinging doors. A track for the section cars is formed of two 2 by 4-in. sticks spiked to the 2-in. floor planks. The walls are of drop siding and the roof of shingles dipped in preservative. A smaller tool house of the Great Northern Ry., 8 by 12 ft., has its

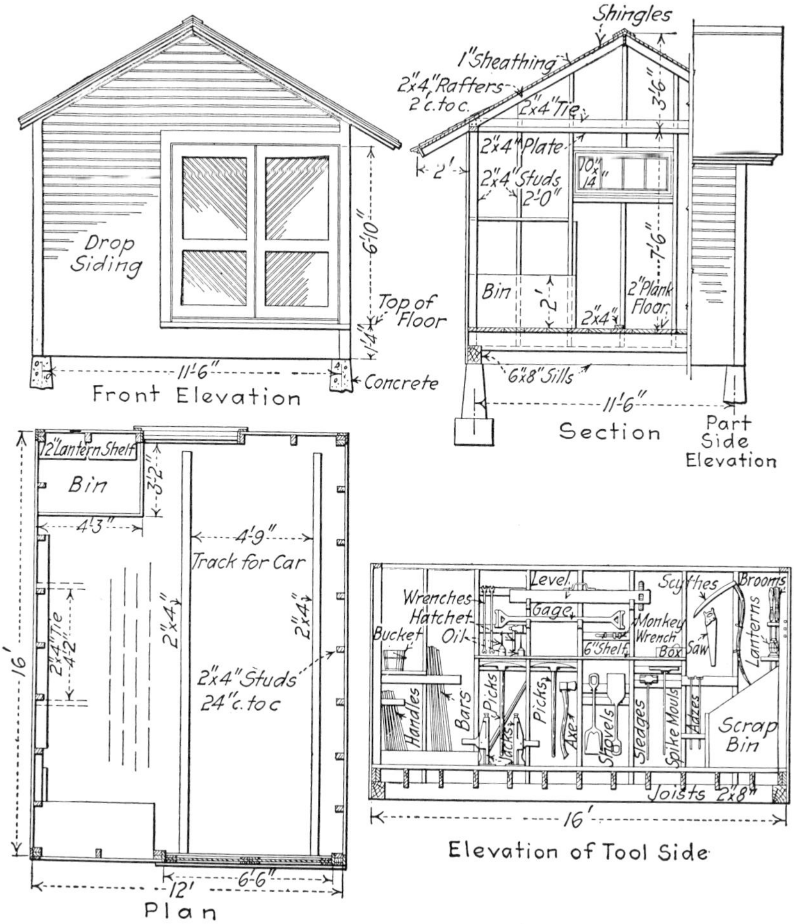

FIG. 66.—Section tool house; C., C., C. & St. L. Ry.

narrower side, with 6-ft. doorway, 16 ft. from the nearest rail. A tool rack is at the rear and a rack for bars on one side. This building, an outdoor track for the push car, and a platform of old ties 12 by 4 ft. for piling scrap, are placed side by side on a cinder bed 25 ft. square, surrounded by 3 ft. of gravel.

In frame buildings, the walls, about 7 ft. high, stand usually on four sills supported by short wood posts or concrete pedestals. Joists on these sills carry the plank floor, and a track of 2 by 4-in. timbers or old rails on the floor is sometimes extended to the ballast line. Plank floors, however, soon wear out, while concrete is usually too expensive, but a satisfactory floor is sometimes made by leveling and tamping the earth and covering it with 3 or 4 in. of fine crushed stone or screenings. The building should be at least 22½ ft. from center of main track, giving ample room for the cars to stand between the house and the track. Where the headquarters of two sections are at the same place, a double house may be built.

Along one side of the larger buildings may be the tool racks, shelves, and boxes for bolts, spikes, and small scrap. On the other side will be a work bench, part of which may be covered with a zinc tray to serve for filling and cleaning lamps. A shelf desk is convenient for the foreman in writing up time books, requisitions, and reports. Sometimes the rear or one side is partitioned to form a room 5 to 8 ft. wide. A locker may be provided for special tools and edge tool. Bars and long-handled tools may stand on end, with their tops resting in horizontal hooks or inclined notches in a shelf. Oil cans may be set in a shallow pan filled with sand, to prevent the floor from becoming oil soaked. Gasoline for the motor car may be stored in an underground tank and handled by a pump inside the house. Miscellaneous supplies, ladders, and spare lumber are sometimes stored in the attic or upon the tie timbers of the roof, the attic being accessible by a fixed or swinging stairway, but material thus stored is likely to be forgotten and neglected.

Section Houses.—At points remote from towns, quarters must be provided for the section foreman and his gang, the former frequently boarding the men and having the house rent free, but being required to keep it in proper condition. These buildings are usually of frame construction and vary greatly in design. A neat one-story frame building for section laborers, as used on the Cleveland, Cincinnati, Chicago & St. Louis Ry., (Fig. 67), is 25 by 28 ft., divided into four rooms. Its walls are of 1-in. sheathing, covered with roofing paper and then an exterior covering of barn siding. Red cedar shingles are used for the roof. In warm regions the kitchen may be separate from the house and connected with it by a covered platform or porch. A one-story building on this plan has a 30-ft. front, facing the track, with living and bedroom each 15 by 15 ft.; a bedroom 12 by 12 ft. forms an L and opens onto a 9-ft. rear porch extending 18 ft. to a separate kitchen 10 by 12 ft.

Where some of the section men live with the foreman, a two-story building is often used; the first floor has living room, kitchen, and one or two bedrooms, while above are two or three bedrooms. In general, however, there is a separate bunk house for the men, and the foreman has his house to himself. A section foreman's house, as used on the Louisville & Nashville Ry. is shown in Fig. 68. On the Great Northern Ry., a one-story house for foremen is 22 by 24 ft., with living room 12 ft. 9 in.

square, a bedroom 9 ft. 3 in. square at the front, and a kitchen and a bedroom at the rear. Walls 9 ft. high carry a 7-ft. pyramidal roof, with chimney at the center.

Cement-stucco houses on the Nashville, Chattanooga & St. Louis R. R. have a wood frame covered with building paper or felt and then with wire mesh to which the stucco is applied in two coats by a cement-gun. A double house for two colored families is 27 by 31 ft., each with an 11-ft. kitchen and a 15-ft. living room. The foreman has a house to himself.

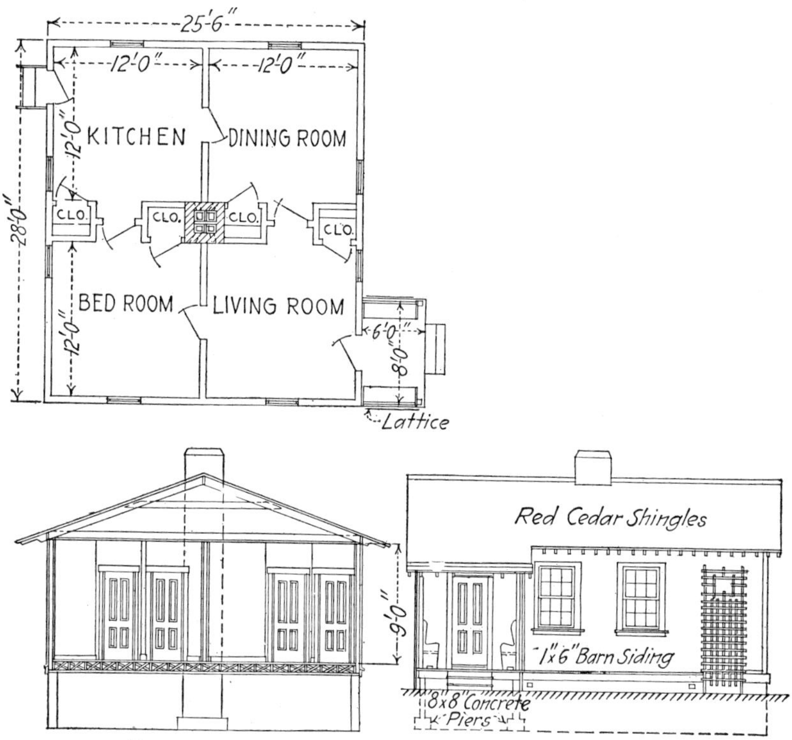

FIG. 67.—Section laborer's house; C., C., C. & St. L. Ry.

This house is identical in plan with the others, but has doorways in the central partition. The cost is little more than for all wood construction and the cost of periodical painting is eliminated.

Concrete bunk houses for section men on the Santa Fe Ry. are built of hollow concrete blocks, which keep the building warm in winter and cool in summer. A 10-room house 103 by 12 ft. has two end rooms 19 ft. wide with a 6-ft. porch between them. There is a concrete floor and a ceiling of metal lath and plaster under the wood roof trusses, which have $\frac{7}{8}$-in. plank sheathing and composition shingles. Each room is for not more than two men. This arrangement is usually preferable to an open bunk room. On the Great Northern Ry., a frame bunk house, 11 by

36 ft., for 12 men, has a 16-ft. sitting room at one end (with stove and wash bench) and a 20-ft. bunk room with six double-deck steel bunks. It is 7 ft. high with horizontal rafters 2 by 8 in. dressed to 2 by 4 in. at the ends and having 1 by 6-in. sheathing above and below, with three-ply roofing on top. Two large ventilators are placed over the bunk room. A similar building for eight men, on the Michigan Central R. R., is 10 by 26 ft. with an 8-ft. sitting room at one end and two double-deck steel bunks on each side of the bunk room.

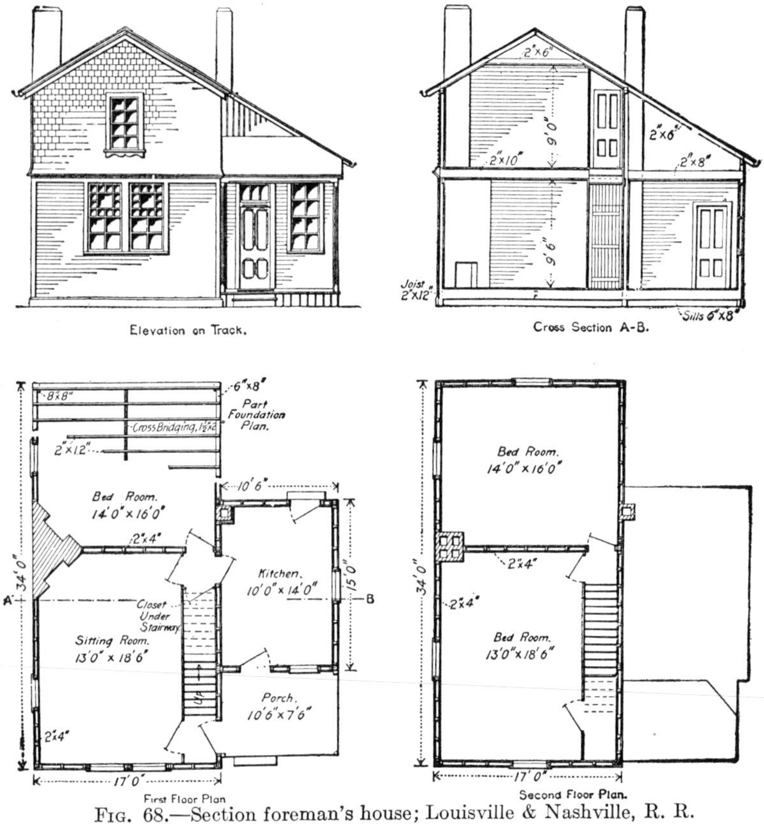

FIG. 68.—Section foreman's house; Louisville & Nashville, R. R.

Cabins.—Cabins for gatemen, watchmen, switchmen, and yardmen should be not less than 6 ft. square or in diameter, $7\frac{1}{2}$ to 8 ft. high, of neat design, and having a seat, a stove, and a locker for lamps. For frame cabins where winters are severe, an inner lining and ceiling or a layer of building paper or felt may be used. A coal bin adjacent to the building should have a strong cover and padlock. Two-story towers for gatemen at crossings, with the operating room 12 to 15 ft. above the rails, may be 9 ft. square, with interior or exterior stairs and a coal bin in the lower room. A design frequently used in towns and suburbs has one or two posts of steel or wood 16 to 20 ft. high, with caps or side brackets carrying a cabin 6 by 8 ft. approached by steps or a ladder. The posts are set 6 ft. in the ground. Small cabins, telephone booths and other structures along the railway are sometimes made of precast concrete, ready to be shipped on cars and set in place with a crane or derrick.

Toilet Buildings.—A sanitary closet and septic tank used by the Nashville, Chattanooga & St. Louis Ry. at section headquarters, camps and small stations has a concrete tank 7 by 4 ft. and 4 ft. deep, with two baffle walls and a top outlet of 4-in. tile leading to a bed of cinders. When completed, the tank is filled with water, and five shovels of well-rotted manure are added to start the bacterial action. About 4 gal. of water are to be poured in daily through the seat. For temporary use by bridge and extra gangs, a dry closet is used with double seat, in a folding shanty 5 by $3\frac{1}{2}$ ft. and 8 ft. high.

For sanitary disposal at bunk houses and small stations, several railways use a septic tank composed of three or more sections of 24- to 36-in. sewer pipe, laid horizontally and fitted with plates closing the joints and ends. In the top of each end plate is a hole for 4-in. pipe inlet and outlet; the inner plates are baffles, with holes at top and bottom, alternately. The outlet may connect with lines of tile laid in fan shape and 10 to 18 in. deep, with a grade of 8 in. per 100 ft. in sandy soil or 2 to 4 in. in clay.

Station Platforms.—Passenger platforms are usually 6 to 8 in. above the rail heads and 4 ft. 9 in. to 5 ft. from center of track. As a rule, the width should be not less than 8 or 10 ft. Platforms between tracks are usually 12 to 16 ft. wide, with a crown not exceeding $1\frac{1}{2}$ in. Side platforms have a slope of $\frac{1}{8}$ or $\frac{1}{4}$ in. per foot towards the track and are usually one or two steps below the station floor. A platform about 3 ft. 3 in. high, with ramp approach, is convenient for handling freight to and from wagons and trucks. At the Chicago Union Station, passenger platforms 13 ft. $9\frac{1}{2}$ in. wide and $7\frac{3}{4}$ in. above the rails alternate with baggage platforms 10 ft. 9 in. wide and 20 in. high. On elevated and subway lines and at a few terminal stations, the platforms are about 39 in. above the rail heads and level with the car floors, as is the usual European practice. The edges of such platforms should project beyond the supports, so that a person falling from the platform may roll clear of trains. High platforms are used also at freight stations.

Concrete, brick, and wood plank paving are commonly used, with asphalt at some large terminal stations. Brick on a sand cushion over a bed of slag or gravel makes a good platform, but should have a concrete curb wall, built monolithically or of precast slabs. In one design these slabs are 24 in. deep and 42 in. long, 6 and 4 in. thick at top and bottom, with a vertical rib at one end and a groove in the other end. Wood platforms have planks 3 by 6 or 2 by 4 in. spiked to joists about 3 by 10 in.

spaced 12 to 16 in. apart, laid parallel with the rails and supported on wood sills or steel caps carried by footing blocks, posts, or small piers. An open space under the planking gives air circulation and prevents rapid decay. Where no paving is used, gravel or coarse cinders may be filled between wood or concrete curbs. At points where passengers and baggage trucks cross the tracks, there should be planking level with the rail heads, as at road crossings. At small stations the platforms extend usually about 100 ft. beyond the ends of the building.

Trainsheds and Platform Shelters.—The lofty and long-span steel trainsheds formerly favored for large stations are almost obsolete, modern practice being to use low roofs in relatively short spans and having a longitudinal opening or smoke duct over each track. Vertical slabs along these ducts and projecting above and below the roof serve to protect the platforms from driving rain and snow. Flat skylights are provided over the platforms, but both light and ventilation are improved in some cases by triangular skylights over the middle of each platform and rising above the main roof. In the Indianapolis Union Station the spans are 39 ft. 3 in. c. to c., with a central headroom of 18½ ft. above the rails. The smoke ducts are 36½ and 19 in. wide at top and bottom, with sides 63½ in. high and rising about 2 ft. above the top surface of the roof.

At many large intermediate stations, no trainshed is used, but each platform has a shelter roof about level with the tops of the cars. This roof is carried usually by a central line of posts with cross-arms at the top supported by kneebraces. An "umbrella" roof slopes down from a central ridge to the eaves and gutters, but a "butterfly" roof slopes down from the eaves to a central gutter having downspouts against the posts. The posts and framing may be of wood, steel, or concrete, with roof of plank and metal or composition covering; precast-concrete roof slabs are also used.

Clearance Limits.—Posts, buildings, and other structures beside the track are required to be at least 7 or 7½ ft. from center of track, but there are usually places at which there is less than the standard clearance. To avoid trouble with shipments of large machinery, steam shovels, or bridge steel, it is desirable to have an easily accessible record of all limiting obstructions such as buildings, low or narrow tunnels, and through bridges. Overhead structures must be 22 ft. above the rails, in order to clear men standing on the top of box cars; where this height cannot be obtained, as in tunnels and at some highway bridges, warnings must be erected, as already noted in regard to "Bridge Warnings." In the standard tunnel section of the Central Pacific Ry., the height is 14 ft. 3 in. from top of rail to springing line, with a roof radius of 8 ft., giving a clear width of 16 ft. and a total height of 22 ft. 3 in. On curves of more than 2 deg., the height to springing line is 13 ft. 9 in. and the roof radius is 8½ ft., giving a clear width of 17 ft. Bridge clearances are noted in the chapter on "Bridge Work."

CHAPTER XI

WATER AND FUEL STATIONS; TURNTABLES; SCALES

Railway Water Supply.—For delivery of water to the locomotives at terminals and intermediate points there are about 15,000 water stations averaging 20 mi. apart, and railway water supply ranges from 750 to 900 billion gal. annually, an average of 8,000 gal. per mile per day, or 10,000 and 2,000 gal. for main and branch lines, respectively. Its cost averages 5 to 10 cts. per 1,000 gal., but treated water may cost as high as 25 to 30 cts. From 75 to 80 per cent is supplied by the railroads, the remainder being purchased from waterworks plants. About 75 per cent of the total is for the locomotives, 12 per cent for washing out and filling the boilers, 8 per cent for sanitary and drinking supply at shops, terminals, and stations, and 5 per cent for stationary boilers. Water supply for the locomotives has an important bearing on efficiency and economy of train service, the expense of providing good water in ample quantity being small in relation to that of boiler repairs and train delays due to deficient or poor supply. For this reason, and in view of the many chances of waste and inefficient use at a multitude of water stations, it is desirable for large railroads to have a complete water service organization, headed by an engineer or superintendent and a chemist.

Water is taken usually from wells and streams and pumped to elevated tanks, but at towns it is sometimes obtained from the local waterworks. Storage or impounding reservoirs of 100 to 1,000 million gal. capacity are sometimes provided to conserve small streams, to provide a large reserve supply, and to secure the advantages of sedimentation. Intakes for pumping stations on streams carrying much drift should have a chamber divided by a screen, the rear portion serving the pump suction and the forward section having an intake pipe with valve on the outer end, which is enclosed by a cage or grating.

Pumping Plants.—Both steam and oil engines are used for operating reciprocating and centrifugal pumps, usually by belting. For plants up to 6 hp. the choice may depend on whether 100 lb. of coal or 1 gal. of oil is the cheaper, delivered, or which plant has the lower interest cost. At some points, steam may be available. Electric motors are being used to an increasing extent, and where current is available or can be purchased at a low rate, they have many advantages, especially as they require less attention or labor cost, while the pumps can be controlled from distant points or automatically by the varying water level in the tank. This system of pumping may be the cheapest where electric control enables pumpmen to be dispensed with. Windmill pumps, used to some extent by railroads crossing great plains where winds are practically constant, are operated usually by 14-ft. wooden wheels with direct drive instead of gearing. Hydraulic rams may serve economically as pumps where local conditions permit their use. In view of short hours, high wages, and overtime pay, ample storage capacity at water stations is an economic necessity in order that the required amount may be pumped during working hours. Where the probable daily supply does not exceed 500,000 gal., the source should be sufficient to allow the total amount to be obtained in 7 hr. at terminals, or 4 hr. at intermediate water stations. The pump capacity in gallons per minute may be determined by dividing the quantity required in 24 hr. by the number of hours for pumping and then by 60.

Location of intermediate water stations must be governed by traffic conditions. A station should not be placed in a sag of grade lines simply because a stream exists at that point, since the cost of piping the water to a more suitable location at the top of the grade will be negligible in comparison with the continual expense incident to the starting of heavy trains on grades. At times, the location may have to be changed on account of changes in operating conditions, as where divisions are shortened to enable train crews to make the run in working hours, or where the engines make unusually long runs, with different crews.

Water has to be supplied also at stockyards where animals are kept awaiting shipment or where they are let out of trains on account of the law which limits the time which they may be kept in cars. At the smaller yards, with summer use only, hand pumps may be used or small windmills serving galvanized iron tanks of 1,000 to 2,000 gal. Elevated tanks serve the larger yards. Concrete drinking troughs, which are frequently used, are economical and sanitary. Another phase of water service is the supply of drinking water at stations and on cars, as governed by the Standard Railway Sanitary Code. Its rules relate to the quality and purity of the water and the icing, filling, and cleaning of the cooler tanks.

Tanks and Standpipes.—Flat-bottomed wood tanks, used very extensively, are built up of vertical staves hooped with flat steel bands or circular or half-round rods having adjustable fastenings. The rods are generally preferred, as being less liable to corrosion, more easily inspected, and facilitating the painting of the wood. Creosoted timber of cheaper grades will give as good service as the best untreated timber; with all sapwood, the empty-cell process giving 5 lb. of creosote per cubic foot will suffice, but without this sap restriction the full-cell process is preferred, with 15 lb. per cubic foot. Two common sizes are 50,000-gal. tanks 24 ft. in diameter and 16 ft. high and 100,000-gal. tanks 30 by 30 ft., the latter being about the largest made in wood. These tanks should be kept filled to prevent decay, shrinkage, and leakage. They are mounted on towers of timber or steel, usually of the 12-post type, but timber is not used for heights of more than about 40 ft. Concrete towers used by the Chicago & Northwestern Ry. are 10 or 15 ft. square for the two sizes of tanks, with a circular top deck 24 or 30 ft. in diameter. Figure 69 shows this railroad's standard water station with a 30-ft. tank.

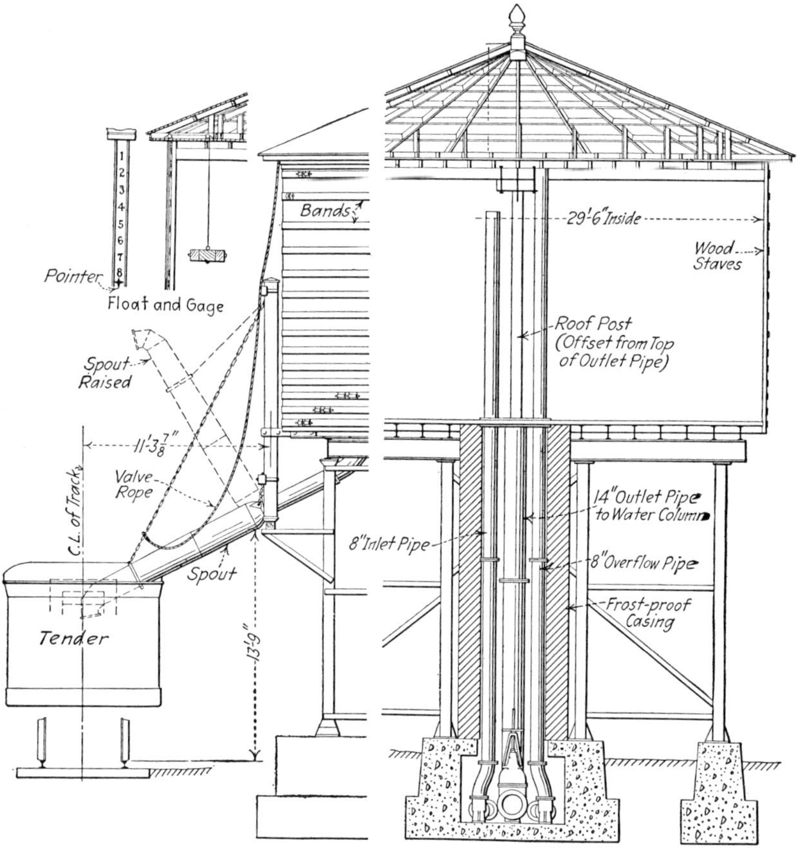

FIG. 69.—Water station and tank; Chicago & Northwestern Ry.

Steel tanks almost invariably have conical or hemispherical bottoms and are carried by four columns riveted to the shell. They are made in sizes of 100,000 or 500,000 gal. capacity. Steel tanks are generally preferred to wood in hot, dry regions. They show little corrosion if kept full, except at the fluctuating water level, but they should be thoroughly painted before being put in use, as they are not likely to be repainted for a long time, this work requiring the tank to be out of use for a week or more for the drying of the surface and the paint. Concrete tanks are used to a limited extent. The average life of railway tanks has been estimated at 30 years for wood, 40 to 50 years for steel, and 75 for concrete.

Supply, delivery, and overflow pipes for wood tanks are enclosed in a central rectangular trunk for protection, the trunk being of frostproof construction in cold regions. Steel tanks usually have a large central water leg or riser attached to the bottom and extending to the ground; this, with the deep bottom, assists in sedimentation, the sludge being blown out at intervals. Within this riser is the discharge or delivery pipe, extended high enough to draw off the clear water in the upper part of the tank.

As a rule, the steel riser is of such size that the water will not freeze, but it may be sheathed if necessary. Even in the northern states, protection against freezing is confined to enclosing the pipes and roofing the tank, although in some cases a stove, heater, or steam pipe is placed in the trunk or in a chamber formed by enclosing the tower. The pumps also may be placed in this chamber. On the Canadian Pacific Ry., both the tank and its supports are completely enclosed in a wood tower, which is about 28½ ft. in diameter for a 24-ft. tank, thus permiting inspection of the tank hoops. With a stove in this tower there is no trouble with ice in the tank or clogging the valves. To indicate the water level, a float operates either a pointer sliding on a scale attached to the side of the tank, or a ball sliding on a staff above the roof, but these devices may be omitted if a regulating valve is used to maintain a constant water level.

Steel standpipes, or tanks on ground foundations instead of upon towers, are used only to a limited extent in railway work. They are generally about 24 ft. in diameter and 30, 40, or 60 ft. high, with capacities of 60,000 to 200,000 gal. If they supply water directly to the engines by spouts, the available capacity above the level of the spout is 70 to 80 per cent of the total capacity, but, as a rule, these standpipes serve water columns.

Where the tank is to deliver water directly to the engine tenders, it is placed sufficiently near the track to allow a hinged spout to be lowered to the tender manhole, with a pipe leading from a bottom valve to the spout (Fig. 69). The fireman standing on the tender pulls down the spout by a chain or handle and operates the valve by another chain or a lever. More generally, however, the tank is placed further away and serves water columns placed beside the track at convenient points. The height from top of rail to bottom of tank for direct delivery is 16 to 20 ft. with 8 or 10 in. discharge pipe to the spout. For serving water columns, the height is 25 to 30 ft. according to length and size of water main.

Water Columns (Fig. 70).—The water column consists of a vertical pipe with horizontal arm at the top, the pipe revolving on its vertical axis in a pedestal casting and being connected to the main from an elevated tank. A valve pit houses a gate valve and regulating valve on this main. When the column is swung to its normal position, with spout parallel to the rails, it is latched automatically so that the wind cannot blow it round to foul the track. The horizontal pipe arm may have a flexible joint to allow it to be depressed to the tender manhole, or it may be a short pipe discharging into an independent adjustable spout. Lateral play of the spout is desirable to avoid the necessity of spotting the engine in exact position and also to prevent injury if the spout should be struck by an engine moving to or from the water column. It is sometimes required that engines of heavy trains must be uncoupled and proceed alone to the water

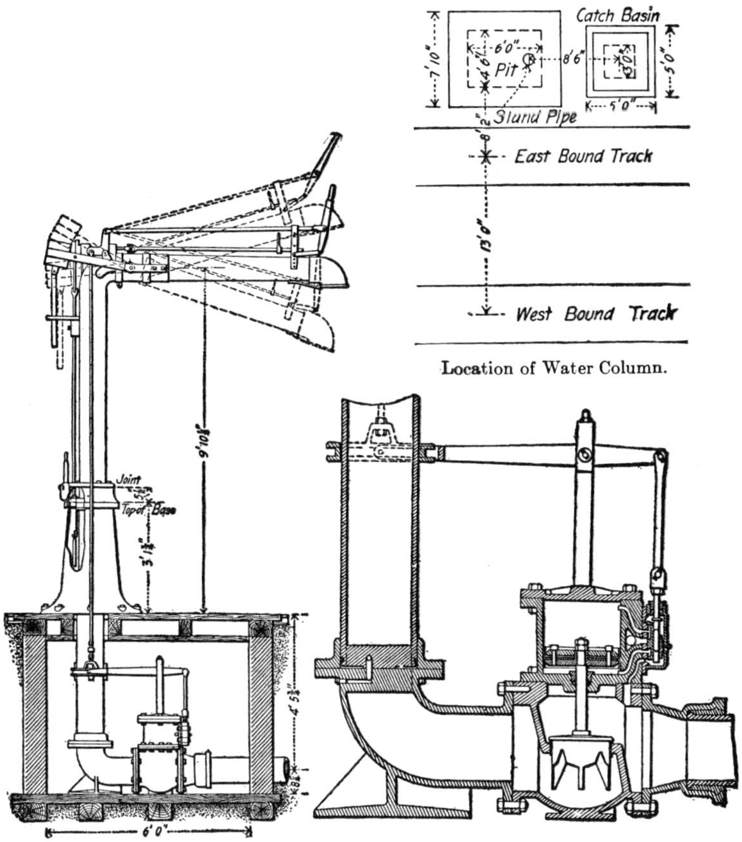

FIG. 70.—Water column and valve.

column, but this may involve more delay and trouble than spotting a train under control of the brakes.

A hand lever or a hand wheel and screw on the arm of the water column may operate the valve in the pit. A balanced valve or an air cushion prevents shock or water hammer if the valve is closed suddenly, but slow closing is insured by the screw gear. Water held in the vertical pipe by the closing of the pipe escapes by a drip valve, the pit having a drain to carry off this water. To fill large tenders rapidly and thus reduce the delay to trains in taking water, high head and large mains are desirable, the main being somewhat larger than the water column. Most water columns are 10 to 12 in. in diameter, but 12-in. columns with 14 in. mains have been used, delivering 4,000 or 4,500 gal. per minute from tanks 20 and 25 ft. high, respectively.

Frost protection is rarely provided, except that the valve pit may be insulated and made with double bottom and cover. Track drains must be provided where engines take water, so that the roadbed will not be kept wet by leakage and spilling. If leakage or drip results in ice on the track, the pump man or other attendant must keep flangeways open and see that the ice is not high enough to catch engine pilots. During extremely cold weather, the roadmaster may direct the section foreman to furnish a man to look after the track at water stations. All tanks, pumps, piping, valves, and other equipment should be examined at least once a year to see that it is in proper condition and to provide against leakage; a special inspection may be made before winter to look for leakage and to see that there is proper protection against freezing.

Track Tanks.—To enable fast trains to make long runs without stopping, the engines may be equipped to take water from long, shallow tanks laid between the rails. In the tender is a vertical pipe having an elbow at

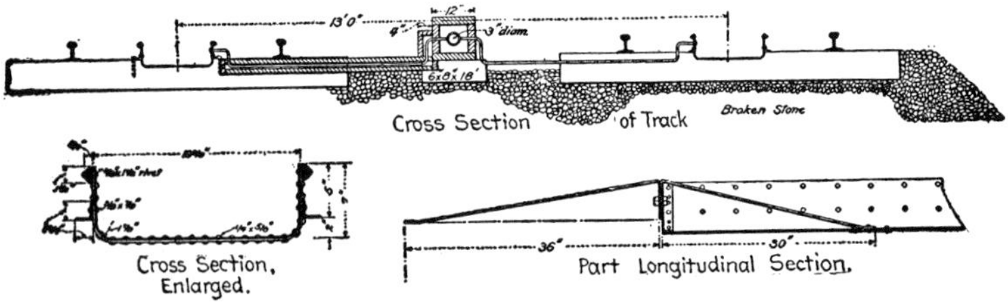

Fig. 71.—Track tanks; Michigan Central R. R.

the top while the lower end extends through the bottom and slopes forward, having at its end a hinged and counterbalanced scoop which is operated by the fireman by means of a lever or a compressed-air device. With this scoop lowered into the track tank, the water is forced up the pipe and into the tender by the speed of the train, the most efficient action being at 40 to 50 m.p.h. This system, which in general is used only by passenger trains, was invented in England in 1861 by John Ramsbottom. An installation on the Michigan Central R. R. is shown in Fig. 71.

Track tanks of the Baltimore & Ohio R. R., between Philadelphia and Baltimore, are 1,200 ft. long and about 30 mi. apart. They are 19 in. wide and 8 in. deep, with outside shelf angles resting on 8 by 9-in. white oak sawed ties which are boxed out 1 in. to receive the tank. Track spikes hold the tank in line, but allow for expansion and contraction, while at midlength the tank is anchored by being bolted to the ties. At each end is an inclined plate $13\frac{1}{2}$ ft. long, rising from the bottom to pass over the end plate and then descending to the ties. This incline raises the scoop if it is lowered too soon or if not raised before the engine passes off the tank. A 50,000 gal. tank 28 ft. above the rails, with an 8-in. main, serves a 6-in. pipe laid along the ties and having three $3\frac{1}{2}$-in. connections to the track tank, at the middle and at 200 ft. from each end.

Automatic valves may maintain the water level, but before each passenger train is due, the pump man should see that the tank is full, as the current of air from other trains may blow out some of the water. Ample roadbed drainage and a heavy bed of stone ballast should be provided, while the ballast slopes are sometimes protected against wash by loose slabs of stone or concrete. To reduce the splashing and spilling of water, the Great Western Ry., England, bends the tops of its pressed-steel tanks inward to form flanges about $1\frac{1}{2}$ in. wide.

To prevent freezing, a $2\frac{1}{2}$-in. pipe from the boiler at the pump house is connected to a 2-in. pipe laid along the ties and having 1-in. branches 45 ft. apart which are led through the side of the tank. Each branch ends in a nipple having a $\frac{1}{8}$-in. hole inclined downward, a check valve behind the nipple excluding water when steam is shut off. These pipes are enclosed in wood trunking. With 80-lb. pressure such a system has been found effective in very cold weather. Ice must be kept cleared from the tank and also from the track and rails, but to avoid this trouble and expense the New York Central R. R. has used steam heating by means of a 2-in. pipe laid along each end of the ties, the pipe being covered with a steel shield to confine the air and prevent the spilled water from coming in contact with the pipes. Expansion joints divide the pipe into 200-ft. lengths, each anchored at the middle and having a trap to remove the condensation. With zero temperature, the steam consumption is equivalent to about 50 hp. for each 2,000-ft. tank, with 8 tons of extra coal daily, requiring one extra fireman. The cost of operation and maintenance (exclusive of fixed charges) has been about half the former annual expense for ice clearing by hand.

In renewing track tanks, with an uninterrupted traffic of 75 to 100 trains daily, the new 30-ft. sections were distributed along the track and the old tank was cut apart at 30-ft. intervals with a torch. A derrick car removed each old section and placed a new one, two or three riveting gangs following to make the joints.

Water Treatment.—Good feed water is an important element in fuel consumption, boiler cleaning, repairs, and steaming efficiency of the locomotives, and particularly in the elimination of delays resulting from engine failures on the road. Poor water is a leading cause of engine failures due to leaky flues and cracked firebox plates, while it accounts for 12 to 20 per cent of the loss of service from engines sent to the shops for boiler repairs. Natural supplies of clean, soft water are exceptional, but available waters may be treated chemically to give the required quality. They vary in character so widely that each supply must be investigated as to the proper treatment; furthermore, the character is likely to change materially from time to time, so that periodical tests are necessary as a check. In spite of the great economic and practical value of such treatment, it is applied probably to not more than 10 or 12 per cent of the waters which could be treated with advantage.

Certain peculiar waters may be beyond treatment, as on part of the El Paso & Southwestern Ry., where it was found economical to abandon attempts to treat local well water and to bring a good supply from the mountains, distributing it by a pipe line laid along the track for about 120 mi. Troubles due to bad water are the deposit of scale, foaming in the boiler, or corrosion and burning of the plates. Treatment may be required to clarify muddy or slimy water, to neutralize acid or corrosive water, or to soften hard water. This last is most general. Soft scale or mud may be blown out, but hard scale collecting on the plates and tubes reduces the transmission of heat and may result in the burning or leakage of tubes, necessitating repairs or perhaps causing an engine failure. It is usually advisable to apply treatment at all the water stations on a bad-water division, so that the engines will have a uniform supply of good water.

Chemical Treatment.—Water hardness may be due to sulphates or carbonates of both lime and magnesia. The sulphates are the more objectionable since they form hard scale, while carbonates (in the absence of sulphates) deposit soft scale or mud. For removal of the sulphates, the water is treated usually with a solution of soda ash (sodium carbonate), but barium hydrate has been used instead, as less likely to cause foaming. For removal of the carbonates, the water is treated also with milk of lime and as the small remaining amount of carbonate of lime may cause clogging of the pipes and injectors, a final treatment may be given with sulphate of iron, sulphate of alumina, or sodium aluminate as coagulants to promote sedimentation, for which ample time must be allowed.

It is usually beneficial to treat any water containing more than 12 grains of hardening matter to the gallon, or even less if there is high proportion of sulphate of lime. The dose must be adapted to the various waters and changed as the character of the raw water changes from time to time. Treatment may be applied continuously or intermittently. In one intermittent plant, two standpipes of 400,000 gal. capacity are used alternately, the raw water being treated with a soda-ash and lime solution in the inlet pipe. Simple tests of the raw and treated water are made daily by the pump man, the amount of chemical required being then determined from standard lists. With an 8-hr. pumping period, about

16 ft. are allowed for chemical reaction and sedimentation before the water is used. For continuous treatment in bad-water districts on the Chicago, Milwaukee & St.Paul Ry., the raw water receives continuous streams of milk of lime, soda ash (sodium carbonate) and sulphate of iron in a mixing tank. From this tank the water flows to the bottom of the settling tank, the overflow from which is pumped to the storage tank. After 40 min. flow through the mixing tank, with slow, continuous stirring to cause complete chemical action, the water remains about 3 hr. in the settling tank.

Coaling Stations.—For the delivery of coal to locomotives at terminals and at intermediate points approximately 50 mi. apart, various methods are employed, according to local conditions. Where only a few engines daily are to be supplied, coal may be handled from cars or a stock pile by a fixed or locomotive crane, and dumped either into the tender or into buckets or small cars on a coaling platform. But most coaling stations have gravity delivery from elevated pockets or bins, such stations being of two types: (1) the trestle or pocket type; and (2) the so-called mechanical type, which reduces the labor factor to a minimum.

In the first type, a trestle about 25 ft. high has on one or both sides a series of pockets of 10 to 20 tons capacity, each with a hinged chute that can be lowered to the engine tender. Different quantities of coal are placed in the pockets, so that an engine can take any required amount. The coal is dumped or shoveled from cars on a track level with the tops of the pockets and reached by a grade of about 5 per cent, if the cars are pushed up by a locomotive, or 20 per cent if they are pulled up by a cable and hoisting engine. An inclined conveyor taking coal from a track hopper or a stock pile on the ground has also been used, the supply to the pockets being regulated by transverse troughs and gates. A coaling station of this type occupies considerable area and can serve only two tracks.

Mechanical Coaling Stations (Fig. 72).—Here the typical features are a track hopper into which coal is dumped from cars, an elevating conveyor, and an overhead bin of 50 to 1,000 tons capacity, having gravity chutes with regulating gates to shut off the flow when the tender is filled. The bin may have two or more compartments for different grades of coal and to supply suitable grades for hand-fired locomotives and those equipped with mechanical stokers.

Two general types of elevators are used: (1) an endless chain with flights or small buckets; and (2) a bucket of $\frac{1}{2}$- to 2-yd. capacity traveling in a tower, the bucket being filled from the track hopper by an automatic loader and tripped at the top of its travel so to discharge the coal into the bin. Two buckets in balance may be used. With electric operation in the larger plants, the machine may be started by a push button and will continue to operate until stopped in the same way, while in the smaller plants with buckets, a round trip is made for each push of the button, the bucket coming to rest in its loading position. These plants can handle 75 to 100 tons per hour and an engine can take 15 tons in about 45 sec.

For the smaller stations, the bin is on a tower at one side of the engine track. With a larger bin mounted over the coal-car track, it may serve one or two engine tracks on each side. An alternative arrangement has the coal-car track at one side and two engine tracks under the bin, with perhaps one or two other tracks on the opposite side. The spouts or chutes and the regulating gates may be handled by the fireman or by a

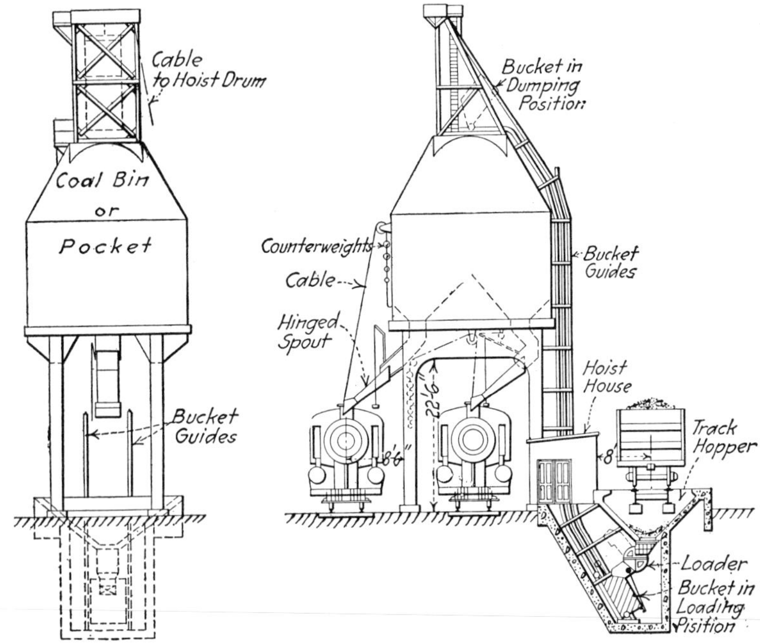

FIG. 72.—Coaling station.

station attendant. Automatic weighing or recording devices to show the amount of coal delivered to each engine are very desirable in the interests of fuel economy and should be more extensively used.

A storage supply of coal must be kept on hand to provide against delay or interruption of the regular delivery, and may be sufficient for 1 to 5 days, according to distance from mines. Very generally, storage coal is held in cars on sidetracks; cars sufficient for a day's supply are placed on a siding of about 1 per cent grade, so arranged that they can be moved by gravity to the track hopper as required, the empty cars running then to another sidetrack.

Since car storage is expensive and keeps cars out of service, it is usual to provide a stock pile where large amounts of coal are required. For stock piles of less than 7,500 tons it may be economical to reclaim the coal by such available equipment as a locomotive crane, steam shovel, or ditching machine, but special equipment may be desirable for larger storage. At the Richmond terminal of the Richmond, Fredericksburg & Potomac Ry., where 80 engines daily take an average of 7 tons each, there is ground storage for 25,000 tons, piled 15 ft. high, sufficient for 2 months. Coal for storage is dumped into the track hopper of the coaling station and handled by the conveyor, but instead of going to the 1,000-ton elevated bin it is delivered to the stock pile by a chute. A cable-operated drag scraper spreads the coal over the storage area and also reclaims the coal by delivering it from the pile to the track hopper to replenish the elevated bin.

Fuel-oil Stations.—Where the locomotives use oil fuel, the oil is discharged from tank cars on a special sidetrack into a continuous concrete trough between the rails or into separate pits or hoppers which are spaced about 40 ft. apart to suit the distance between the discharge openings of

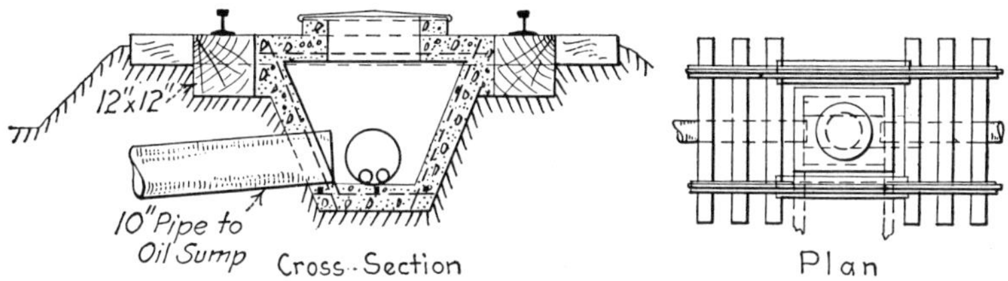

FIG. 73.—Fuel-oil unloading track.

adjacent cars. The pits, of steel or concrete, are about 8 ft. long, 24 to 30 in. wide, and 21 to 24 in. deep, with lateral drain pipes connected to a main leading to an underground collecting or storage tank or a pump pit.

In a somewhat different plan shown in Fig. 73, closed hoppers are used, 36 in. square on top and 14 in. at the bottom, 30 in. deep below base of rail and having an 18-in. manhole in the top to receive the oil flowing from the valve on the car. These hoppers are connected by a 10-in. pipe having two $1\frac{1}{4}$-in. heater pipes inside, while the last and lowest hopper has a pipe leading to the storage tank. Some fuel stations have the hoppers outside the track, a portable pipe or chute being laid from the car discharge opening, but this plan is likely to cause more leakage and dirt. With the oil track on a trestle, the receiving trough or pits may discharge the oil directly through the roof of a storage tank on the surface, but usually the track is on the surface and the storage tank is underground, as described.

From the collecting tank or sump, the oil is pumped usually to steel storage tanks of 20,000 to 60,000 bbl. capacity on the surface of the ground. To supply the locomotives, the oil is pumped either directly to the oil columns (which resemble the water columns already mentioned) or to elevated tanks of 20,000 to 100,000 gal. capacity, from which it flows to the columns by gravity. These oil columns are usually 6 or 8 in. in diameter, with the horizontal arm hinged for depression to the tender manhole or fitted with a telescopic drop spout. Where this arm is arranged to swing parallel with the track when not in use, it stands normally a little above the horizontal, so as to drain back to the main pipe of the column. In some designs, however, the arm swings vertically and stands normally latched in a vertical position. Where cool weather occurs, a steam pipe may be placed in the column, so that the oil will be warm enough to flow freely. The discharge is usually 250 to 400 gal. per minute from 6-in. and 8-in. pipes. Tender tanks hold 2,500 to 3,000 gal. of oil.

At the oil trough or pits, each track rail is laid directly on a longitudinal timber (Fig. 73), or on short blocks placed across a pair of stringers. The rails on this track and at the oil columns need to be carefully bonded to prevent stray electric currents from flowing to the pipes and causing sparks. These fuel-oil stations must have fire protection apparatus, and the surface storage tanks are usually surrounded by earth banks to retain the oil in pools in case of bursting or leakage of the tanks. Fuel oil weighs approximately 7.5 lb. per gal., and about 1,200 lb. of oil are equivalent to a ton of coal.

Sand Supply.—Dry sand for the engine sandboxes is supplied usually at the fuel stations. Raw sand dumped into a track hopper is elevated to a bin from which it flows to a drier heated by a furnace or composed of coils of steam pipes, the dried sand passing over screens to a drum or tank. When the drum is filled, the inlet is closed and compressed air at 30-lb. pressure blows the sand through a 4-in. pipe to an elevated bin, which may be lined with hollow tile to provide ventilation and thus keep the sand dry. Other 4-in. pipes leading downward are fitted with hinged spouts and cutoff gates to regulate the delivery to the engines on the coaling tracks. A coaling station with 300-ton coal bin may have storage for 100 tons of wet sand and 30 tons of dry sand. Where a separate sandhouse is provided, the wet sand is usually shoveled from cars to a bin and thence to the drier and screen. The dried sand flows to a cylinder set under the floor and having a $2\frac{1}{2}$-in. or 3-in. pipe for compressed-air delivery to a small elevated hopper with swinging spout ending in a rubber hose.

Ashpits.—Provision for dumping and cleaning the fires of locomotives must be made at engine terminals and occasionally at intermediate points, the latter usually for freight engines. Of numerous types of ashpits and ash-handling plants employed under varying conditions of climate, area available, track layout, and daily number of engines (or rather the maximum number per hour), the principal types may be classed as follows: (1) shallow track pits between the rails; (2) depressed track pits; (3)

water-filled pits; (4) long track pits having small cars or buckets; (5) short track pits with skips on lateral inclined tracks; (6) track hoppers with underground removal; (7) steam-jet pits. An extended report on ashpits is given in the "Proceedings" of the A. R. E. A. for 1922. Ash-handling plants should have such storage capacity that the pits will need to be cleaned only during working hours.

Where few engines are handled or where engines on the main track clean fires while taking water, a pit $3\frac{1}{2}$ to 4 ft. wide and $1\frac{1}{2}$ to $2\frac{1}{2}$ ft. deep, between the rails, may suffice. A common form to serve both few and many engines is the depressed-track ashpit, 50 to 100 ft. long, having two parallel engine tracks. In this arrangement, shown at the right in Fig. 74, each engine track is over a pit about 3 ft. deep and 9 ft. wide, with its floor sloping towards a central pit $5\frac{1}{2}$ to 7 ft. deep in which is the ash-car

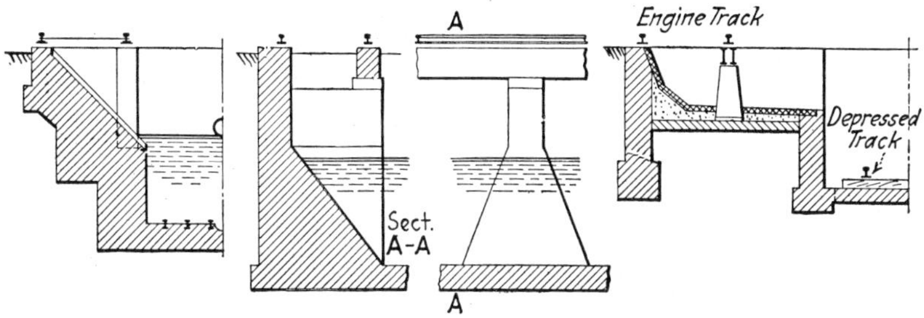

FIG. 74.—Types of ashpits.

track, approached by an incline at one or both ends. Ashes dumped onto the paved floor are quenched with water and shoveled into the ash cars. One rail of the engine track is on the pit wall and the other on girders of 6 to 12 ft. span supported by pedestals of masonry, steel, or cast iron. This plan involves considerable labor and the work is hard and dirty, especially in bad weather.

Mechanically operated ashpits are more generally used, in order to reduce the labor factor. Water-filled pits, of which two designs are shown in Fig. 74, are usually 25 to 45 ft. wide and 10 to 16 ft. deep, with vertical or inclined walls and having along each side an engine track supported in the same way as at a depressed-track pit, so that the ashes fall into the water. An ash track may be laid on the ground beside the pit, or on girders and piers along the middle of the pit. For safety and convenience, there should be a platform or walk outside of each engine track. This type of pit, 50 to 350 ft. long, is used extensively where from 50 to 300 engines are handled daily. A smaller type to serve 40 to 100 engines has a pit between the track rails and 3 to 4 ft. deep. Removal of ashes from the small pit is effected by a locomotive crane with grab bucket; for the larger pit, the bucket may be handled in the same way or by a gantry or an overhead traveling crane spanning the pit and the ash track.

Long track pits, between the rails, may be fitted with 2-yd. cars running on rails in the pit to a fixed hoist, or it may have buckets which are lifted and dumped by a gantry or overhead crane. A pit on this latter plan on the Chicago, Burlington & Quincy R. R. at Lincoln, Neb., to serve 150 engines daily, has two pits 245 ft. long, each filled with 35 hopper-shaped, bottom-discharge, 2-yd. buckets 7 ft. long and 3 ft. deep. Each engine drops an average of $1\frac{3}{4}$ yd. of ashes. Short pits 5 to 8 ft. long, with $1\frac{1}{2}$- to 3-yd. skips standing on tracks at right angles to the engine track, are used to serve 25 to 100 engines daily, several units being required. The track rises on a steep incline which extends above the ground and across the ash track, an electric or air hoist hauling up the skip, which is dumped automatically.

With hopper ashpits on the Pittsburgh & Lake Erie R. R., four tracks have each a hopper 14 ft. long and 8 ft. deep, of 225 cu. ft. capacity, with bottom gates discharging into an electrically operated car of 75 cu. ft. capacity. This car runs in a transverse tunnel and dumps into the bucket of a vertical hoist, or it may run onto the hoist platform; this hoist serves an elevated bin from which the ashes are spouted to an ash car. The tracks are 15 and 20 ft. c. to c., the former being the minimum to allow for safe and ready work in cleaning fires. About 12 engines per hour can be cleaned, averaging 3 to 4 cu. yd. each. In the steam-jet system, an 8-in. pipe extends across the floor of a rectangular pit and has an opening to receive the ashes. One end of the pipe has an air inlet; the other end connects with a vertical pipe and horizontal arm leading to a hopper with a dropspout over the ash car. A steam jet in the pipe acts as an ejector.

Long ashpit tracks have the objection that engines needed promptly may be blocked on the pit. Track connections and crossovers should be arranged to enable engines to go to the pit without waiting their regular turn, and also to enable switching or outbound engines to move away from the ashpit without using the turntable. Approach tracks should be of sufficient capacity to accommodate engines which arrive in groups and cannot all be cleaned at once.

Engine Pits.—In the roundhouse, each engine stands over a pit between the rails to enable the men to get under it to inspect and repair the machinery. The pit is usually of concrete with a convex floor so as to drain to the side and keep the surface dry. To warm the enginehouse, and to thaw snow and ice from the engines, heater coils are arranged along the sides of the pit, sometimes placed in recesses or protected from falling materials by overhanging curbs. Electric lights may also be placed in pockets in the walls, and there will be attachments for the cords of portable lights. Track rails are usually laid on tie-plates or base plates and secured by nuts and clips on anchor bolts embedded in the side walls of the pit. To provide a foundation for jacks used in raising the engines, the pit walls are made very wide for the full length or are widened at certain points. A concrete floor is sometimes laid against the rails, but more generally there are lines of planks to serve as supports for the jacks. At many modern engine terminals the incoming locomotives are inspected at similar pits about 100 ft. long and 16 ft. c. to c., placed near the entrance to the terminal. Ample track space should be provided for engines waiting their turn at inspection and then waiting for admission to the roundhouse. These pits are connected by a subway, with stairs to the inspector's office, which is usually arranged in a rectangular building or shed which covers the pits and is open at both ends. This office is connected with the roundhouse foreman's office by telephone and by pneumatic tubes for transmitting the instructions as to repair work required.

Turntables.—To reverse the engines at terminals and to serve enginehouses having radial tracks, a turntable is used, carrying a single track and revolving in a circular pit. In turntables of the center-bearing type, the entire load is carried by a central pivot and pedestal, the table taking an end bearing only as it tilts under the load of an engine moving on or off. The resistance is low, but the engine must be spotted or balanced so that the table swings free.

With the great increase in weight and length of engines and the application of tractors or motor-driven end wheels to operate turntables, the tendency has been to adopt tables having both end and center bearings to distribute the load. The ends are supported by wheels riding on a circular rail laid on a bench of the pit wall and carrying approximately 36 per cent of both dead and live load. Since the engine has not to be spotted, the table need be long enough only for the maximum wheelbase of engine and tender, thus saving length and also eliminating the time taken to spot an engine in balancing position. These turntables may be of the continuous-girder or twin-span types, the latter resembling two adjacent bridge spans connected on a center pier.

Turntables are 80 to 120 ft. long, according to the longest engine and tender to be handled. Generally they are of deck-girder construction, but sometimes with through girders to reduce the depth of pit, and occasionally with through trusses, as in 120-ft. turntables of the Santa Fe Ry. and 100-ft. tables of the Southern Pacific. Long tables are best driven from both ends, and in any case the operation is controlled from a cabin on the table. Electric, gasoline, or compressed-air motors may be used. Current is taken usually from an overhead wire with connection to an arched frame at the center, but it may be supplied by a cable extending up through the pivot pier. Such a tower must give ample clearance for the engines (see Fig. 102). End latches are required to hold center-bearing tables in position, but not for other types.

The side wall, center pier, and floor of the pit are usually of concrete, the wall having a step or bench for the circular rail, which is fastened to wood blocks or to metal castings embedded in the concrete. The pit should be well drained and kept clean; in cold climates it may be equipped with steam pipes to prevent clogging by snow, in which case snow cleared from the yard may be dumped in the pit to be melted. In northern Canada, the turntable is sometimes enclosed in an extension of the enginehouse to obviate the troubles due to snow and severe cold.

Some passenger stations in England have turntables pivoted at the end and swinging in quadrant-shaped pits; these connect with lateral spurs to mail and express platforms.

Transfer Tables.—While a turntable serves radial tracks, a series of parallel tracks may be served by a single-track transfer table traveling in a straight pit at right angles to the tracks, so that an engine or car run onto the table from one track can be transferred to any other track on the same or opposite side of the pit. This arrangement is used principally at shops, as with a locomotive and boiler shop on opposite sides, or with a long car shop having the repair department in one end and the painting department in the other end. But the transfer-table arrangement is also used occasionally to serve rectangular enginehouses having parallel tracks.

The table has a steel frame carried by several pairs of wheels on rails 8 to12 ft. apart, spiked to ties, short blocks, or longitudinals. To reduce the depth of the pit, usually 16 to 24 in., the axles may be mounted on top of the table frame. All the axles on one side are usually connected by shafts and are thus driven from a motor of 15 to 25 hp. The length of the table, usually 75 to 100 ft., depends upon the longest engine or car to be handled; it is usually 80 ft. for cars. A 120-ft. transfer table of 427 tons capacity is installed at the San Bernardino shops of the Santa Fe Ry., but at shops the engine and tender are generally handled separately. A transfer table travels far enough beyond the building to serve at least one yard track. Its ordinary speed is about 125 and 300 ft. per minute, loaded and empty, respectively, but for a table with the unusual travel of 1,000 ft. at the car shops of the Louisville & Nashville R. R., the light speed is 850 ft. per minute. For hauling cars and dead engines on and off the table, there is usually an electric capstan giving a speed of 125 to 150 ft. per minute for the former and 100 to 125 ft. for the latter.

Y-tracks.—To provide against accident to the turntable, it should be supplemented by a Y-track on which engines may be turned if necessary. Such tracks are frequently used also for turning cars and complete trains. Thus all passenger and sleeping cars using the Peoria Union Station of the Peoria & Pekin Union Ry. are turned on a Y formed by a double track connection of two divisions, the cars being made up in cuts or trains of 8 to 15 cars handled by switch engines. At a terminal on the Denver & Rio Grande Western Ry., which has a rectangular enginehouse, a Y-track with 10-deg. curves is provided for turning the engines in order to avoid troubles with a turntable and pit in a district of heavy snow.

Track Scales.—For weighing cars at yards, a track is provided with a scale 50 to 65 ft. long, with capacity of 100 to 200 tons, or a sectional capacity for 75 to 100 tons on each pair of levers without producing excessive strain in the mechanism. The scale pit, 7 ft. deep, is protected against leakage along the platform and should be well drained, ventilated, and lighted to keep the parts in proper conditions and to facilitate inspection. Short easer rails with slip joints form the connection between the approach track and the scale table, and prevent impact on the scale mechanism as heavy cars pass from the approach onto the scale table.

If the scale track is not used exclusively for cars to be weighed, a dead-rail track is gantletted with it and for 30 to 100 ft. from each end of the scale, the rails being 8 to 12 in. apart. Where a large proportion of the cars are to be weighed, the scale track should be straight and the dead track offset. One of the dead rails is carried on the wall of the pit and the other on the stringers supported by pedestals in the pit, the scale platform being then made in two sections. A relieving gear may be used, operated by hand or compressed air, which causes the scale bridge to be raised vertically off the knife-edges a sufficient distance to overcome the deflection in the bridge under maximum loads, so that there will be no pressure or wear on the knife-edges during the passage of locomotives and trains not to be weighed. This arrangement facilitates movement between the receiving and classification yards, as it eliminates the throwing of switches by the yard force in order to pass trains over the scale; the control of the device is in the hands of the weighmaster and is set by him for the passage of engines and non-weigh trains when the scale is not in use for weighing. Mechanically operated semaphore signals at the ends of the scale indicate whether the scale is set for weighing or for the passage of engines and non-weigh trains.

An alternative arrangement is to provide a runaround track for switching movements, leaving the scale track for weighing exclusively. Where cars are uncoupled, the scale track need be only long enough to clear the switches; but if cuts of cars are weighed, the track must be long enough for a complete cut in advance of and beyond the scale, so as not to block the runaround track. At yards operated by drilling or poling, the scale should not be placed in a main drilling lead unless the movements over the lead are few, or unless more than 60 per cent of the cars are to be weighed. In such yards, the scale should be raised above the general surface sufficiently to give drainage away from it and to give a grade on which a car will run by gravity from the scale.

For cars weighed in motion, the speed should not exceed 4 m. p. h. and each car must be alone and entirely upon the scale for at least 3 sec. In a gravity yard, the grade must be such that cars will not exceed this speed over the scale or in running as far as desired on the classification tracks. On the scale itself the grade should not exceed 1 per cent (the mechanism being placed level) and the same grade should continue for 45 ft. above and below if cars are handled singly, or 100 to 200 ft. if cars are handled in cuts. With a recording beam, when the weight is obtained, a pressure on the latch causes this weight to be stamped on a card or tape. For busy yards, especially where cuts of cars are weighed in motion, an automatic device on the scale gives a printed record of the weight of each car.

Truck and Wagon Scales.—For weighing wagons and motor trucks at freight houses and team yards, scales of 5 to 25 tons capacity are provided, the latter being necessary in view of the extremely heavy loads now carried by motor trucks. These scales have platforms 8 or 9 ft. wide and 15 to 24 ft. long. The scale beams may be in a housing at the side of the scale, but it is better to arrange them in a cabin or in the freighthouse, where they are better protected and under better supervision. Both track and wagon scales need to be tested with known loads at intervals to maintain their accuracy.

CHAPTER XII

TERMINALS, YARDS, AND SIDINGS

The word "terminal" covers all the trackage, facilities, and property at the end of a railway or an operating division. It is applied also to different classes of terminals and integral parts of terminals; as passenger, freight, line, division, branch, or district terminals, rail-and-water terminals, and engine terminals. The word "yard" denotes the system of tracks for the switching and storage of cars at terminals, division points, and large stations. It is applied also to individual groups of tracks in the yard. Thus a freight yard will include receiving, classification, departure, and other yards; and a large passenger yard will have coach and coach-cleaning yards.

Freight Yards.—Yard operations have a most important bearing upon the cost and speed of freight transportation. Delay in freight yards beyond the reasonable time for separating and assembling trains interferes seriously with the forward movement of cars, and as cars on long journeys have to pass through several yards, the delays are cumulative. For this reason, there is an increasing use of solid or "main tracker" trains composed of cars for the same distant destination, so that at intermediate yards the trains require only inspection of cars and change of engines and crews. But such trains compose only a small part of the total traffic. At both small and large yards there may be delays due to defects in arrangement or operation, while at large cities there is often extended delay in transferring cars from one railroad to another.

A freight yard is simply one of many facilities in the railway transportation business, its main purpose being the sorting and distribution of cars and the forwarding of traffic. The more quickly each car is passed through the yard, from the time of its arrival off the main track until it is delivered at its local unloading point or started out to continue its journey, the greater will be the operating efficiency and economy of both the yard service and the transportation service as a whole. Furthermore, if cars are handled promptly, they will make more mileage and do more work, and consequently the capacity of the car equipment will be increased. On nearly every railway, it will pay to study the design and operation of each yard with a view to effecting improvements to secure increased efficiency and economy, and such improvements may often be obtained at relatively small cost. In many cases, the traffic has outgrown the yards and one of the most notable developments during recent years has been in providing more adequate facilities for this important part of the railway business.

Freight Yard Design.—Freight trains ordinarily are composed of cars for various destinations, so that at terminals and division points the inbound trains have to be broken up and the cars separated and classified. Thus cars may have to be set out for unloading, forwarded to local points on the next division, sent through to the next division point, transferred to connecting roads, held for repairs, or held for orders. Outbound trains may be composed of through and local cars from the previous division, cars loaded locally, and cars from connecting roads. Outbound local trains must be made up in station order for distribution of the cars along the division. Inbound cars terminating at the division point may have to be distributed to freighthouses, local yards or team tracks, warehouses, grain elevators, coal and ore piers, and industrial plants. Empty cars must be distributed to these various places, and when loaded, must be brought to the main yard to be placed in outbound trains. Upon the efficiency of handling these complicated movements will largely depend the economy and efficiency of the transportation service, as noted above.

A main or general yard consists of two units for traffic in opposite directions; each unit is subdivided into interconnected groups of tracks, each termed a yard, such as receiving, separating, classification, departure, and repair yards. The layout or plan must provide for continuous forward movements of the cars in the several switching operations, as reverse movements detract from the efficiency.

In relation to the main line, the yard may be placed on one or both sides of the main tracks or between them. With the first plan, trains in one direction must cross the other main track in entering and leaving the yard. If the main tracks separate the eastbound and westbound portion of the yard, there will be considerable crossing of the main tracks by light engines, yard engines, caboose cars, and interchange cars. While both of these arrangements permit of free extension and enlargement of the yard, costly interlocking plants at the main-track connections may be required for safety and to minimize the interference with main-line traffic. With the main tracks spread to pass around opposite sides of the yard district, the traffic problem is simplified, but after a few years it is likely to be found that insufficient room is provided for enlarging the yard, so that additional facilities can be obtained only at the expense of considerable main-line reconstruction. In addition, the separation of maintenance and signaling of the two main tracks adds to the expense. The arrangement, however, is logical on a four-track line having the two freight tracks in the middle. Each plan has its advantages and drawbacks, so that the selection in any particular case will depend largely upon local traffic and topographical conditions, and upon the opinion and experience of the officers making the selection.

In addition to the engine terminal, the yard must include facilities for changing engines and cabooses, inspecting and repairing cars, and supplying ice to refrigerator cars and water to stock cars. Stockyards for cattle and spur tracks to industries, piers, etc., may also be required. Two essential features of a general yard are the receiving and classification yards. The latter may have certain tracks assigned for making up outbound trains, but if sufficient number and length of tracks cannot be thus provided there will be also a departure yard. Auxiliary to these may be a separating yard between the receiving and classification yards for preliminary classification, and a sorting yard between the classification and departure yards to facilitate the final arrangement for trains. But these auxiliaries are not generally provided.

Each end of the yard is connected with the main track by a "lead" track which should be long enough for one or two trains and which may extend to the next telegraph station, being then operated as part of the main line and enabling trains to enter and leave the yard without interfering with the main tracks. At busy points, the switches between the main tracks and the receiving yard and departure yard may be operated by switch tenders or from interlocking towers, in order to avoid the delays due to stopping trains while their crews operate the switches.

The "body" tracks are the parallel tracks (in groups) upon which cars are switched or stored, and at each end these tracks open from a diagonal "ladder" track, whose angle is usually that of the frogs. To interconnect a series of long, parallel, body tracks a "ladder" may be run diagonally across them, connecting with each one by slip switches (see Fig. 31). Where the body tracks are numerous, as in a classification yard, the entering end may be wedge shaped, with two diverging ladder tracks, in order to concentrate the switches and to lessen the wear of frogs and switches. One body track in each group, or a drill track parallel with the outer body track and connected with the ladder, is kept clear of cars to allow of movements from one group to another. Running or thoroughfare tracks are kept open to enable yard engines to move from place to place and to enable both yard and road engines to move to and from the engine terminal facilities.

Body tracks are usually 13 or 14 ft. c. to c.; sometimes alternately 13 and 15 ft. The first yard track should be 15 to 20 ft. from the main track, and ladder tracks should be 15 to 20 ft. from parallel tracks, in order to give ample room for switchmen. Repair tracks may be spaced 18 ft. c. to c. or alternately 16 to 18 ft. and 24 to 30 ft., with a narrow- or standard-gage track for material in each wider space. In rating the car capacity of tracks, measured between clearance points, 42 ft. per car is usually assumed; for car repair tracks this distance is taken as 50 ft. to allow for work on the ends of the cars. Compressed-air piping may be provided at receiving yards for testing brakes and detecting the necessity of light repairs; at departure yards for testing the brakes and charging the brake system before the road engine is attached, and at repair yards to operate tools and appliances.

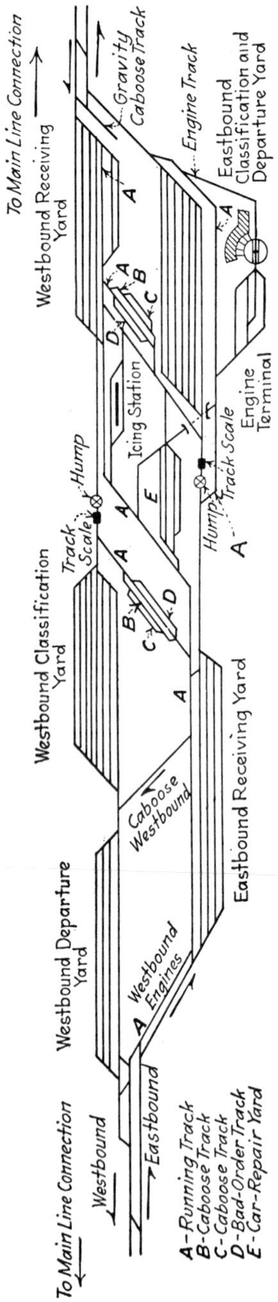

FIG. 75.—Typical layout for a division freight yard.

Where much freight is transferred between cars, a covered platform of convenient length will be provided, about 24 ft. wide and 40 in. above rail level, with two tracks 12 ft. c. to c. on each side. For filling the ice bunkers of refrigerator cars there may be a high platform 15 ft. high connected with the ice house and delivering block ice to the roof hatches of the cars; such a platform may be 16 ft. wide between two tracks 29 ft. c. to c. An upper platform about 23 ft. above the rail provides for chuting crushed ice to the hatches.

In the typical plan of the layout for a division yard, Fig. 75, two distinct arrangements are shown. For westbound traffic, there are separate receiving, classification and departure yards. But for eastbound traffic, the classification and departure tracks are combined in one group or yard. At the east end is shown a gravity track for caboose storage. The caboose of a train in the westbound receiving yard can be switched to this track and be ready to be run to the rear of an eastbound train for return over the division. This yard is assumed to be located on a loop from the main line, with a double-track connection at each end, so that the position of the main tracks in relation to the yard is not a factor in the design or extension of the yard.

Operation of a Freight Yard.—When a train has entered the receiving yard, the engine and caboose are cut off and the cars are inspected and marked for handling, or a cut list is made out for the guidance of the switching crew or the men at the switching hump. In exceptional cases, the cars are then switched to a separating yard, where those with the same destination or commodity are placed on the same track. A separating yard at the Galesburg yard of the Chicago, Burlington & Quincy R. R. provides for cuts of empty cars, which are switched into this yard before the remainder of the train goes to the switching hump. Solid trains of empty cars may then be run around the hump to the departure yard.

From the receiving yard (or the separating yard) the cars are switched to the classification yard, where they are classified or grouped as required for making up trains, which are then placed in the departure yard ready for the caboose and road engine. Where the classification yard can have tracks of sufficient length and number to permit some to be assigned for making up trains, a departure yard may be omitted. Cars held for orders go to the "hold" tracks or storage yard; those needing repairs go to the repair or "cripple" track, whence they are shifted to the repair yard. Empty cars may be placed on special tracks to be clear of yard movements, and fast-freight cars may also be placed on special tracks at the departure end of the yard to facilitate the forwarding of fast-freight trains.

Length of Yard Tracks.—In receiving and departure yards, the tracks should be long enough for trains of maximum length, including the engine and caboose. Where the departure yard is omitted, the classification tracks assigned for outgoing trains should be long enough to hold full trains, as otherwise the trains must be doubled out of the yard, which is an undesirable and expensive practice. Where this condition obtains, there is the alternative of lengthening the classification yard or providing a separate departure yard. Where a departure yard is to be provided, there is considerable flexibility as to the length of the classification tracks, which in extreme cases may be sufficient to hold only a few cuts of cars, the cuts being assembled on the departure tracks. As a rule, however, it is desirable to have at least some of the classification tracks long enough for full trains.

In classification yards, the number of tracks must be ample for the number of classifications required, with a few extra tracks for use in case one classification track is filled temporarily and cannot hold all the cars destined to it. The length usually need not be sufficient for a full train, a half-train length being ordinarily sufficient. If, however, certain tracks in this yard are assigned as departure tracks these should be of full train length, as noted. Even where there is a separate departure yard it will often be convenient to have some classification tracks of full train length, especially where a second hump switching movement or a long transfer movement would otherwise be necessary. The desirability of long tracks increases as the number of classifications decreases, since each track must then hold more cars. If the length is too great, the cars must be switched at high speed, increasing the liability of damage to cars and freight.

The hold or storage tracks should have a ladder at each end, or there may be a tendency to neglect the cars at the stub end. They should also be connected with the classification yard, to allow cars to be shifted readily for forwarding. Caboose tracks should be arranged for convenient switching to and from trains; the outbound track may have a grade of 1 to $1\frac{1}{2}$ per cent to facilitate moving the caboose to its train.

Team Yards.—In team yards, where freight is transferred directly to and from wagons, the tracks are usually in pairs alternately 12 ft. and 40 to 50 ft. c. to c., with paved driveways in the wider spaces. They may be parallel with the main track or at an angle with it, according to the site. For convenience of shifting cars the track length should be limited to about 20 cars. The yard should have a wagon scale and a crane for handling heavy or bulky freight.

Yard Tracks.—Tracks in yards should be of only little lighter construction than on the main line and should be well maintained, as this facilitates the yard work. Lightly built track will soon get in bad condition and require excessive maintenance, especially at switches, besides retarding the car movements. Yard engines are sometimes operated with badly worn tires that are very severe on the rails, while the heavy engines now used in large yards call for substantial track construction. The yard should be well ballasted and drained, and kept neat and free from rubbish. Concrete ties are said to be specially adapted for yards where renewals are difficult and costly. Rails are very generally of 80 to 90 lb. per yard, with No. 8 and No. 9 frogs. The diverging angle of the body track may be made as large as possible and the curve continued beyond the frog to make up the total central angle.

Curves in yards should rarely exceed 9 deg., or 8 deg. on all tracks carrying full trains. Complicated track arrangements should be avoided; slip switches are convenient and sometimes necessary, but are expensive in first cost and maintenance. Switchstands may have the track numbers painted on them as a help to the yard men. Usually the switches are operated independently, but at some large and busy yards all the switches of a long ladder track are operated by levers concentrated in a tower but not interlocked. Power operation of switches in gravity yards is noted under "Yard Switching."

Since much yard work must be done at night, efficient lighting is desirable but is a difficult problem owing to the cars on the tracks. One method is to place lamps of about 2,000 cp. on 30-ft. poles spaced about 150 ft. apart along the gravity track or ladder and 300 ft. along classification or body tracks. Towers 100 to 150 ft. high, with clusters of lamps, have been used for general yard lighting, but the most modern practice is to use flood lighting, in which powerful lamps throw the light horizontally across the yard. This plan gives a better diffusion of light, which not only facilitates the yard work but also results in less damage to cars and freight and less theft from cars. Groups of four to eight lamps of 1,000 watts may be placed on steel poles 60 to 70 ft. high or on the roofs of coaling stations and other convenient buildings.

Yard Switching.—For classification work the cars may be handled in different ways. In drilling, the train is pushed and pulled to and fro by a switch engine, each car or "cut" (group) of cars being uncoupled in turn from the rear end. The train is pushed down the drill track and then stopped, leaving the uncoupled car to run by its own momentum into the desired track. In poling, the engine runs on a track parallel with the train and by means of a pole attached to the engine or a poling car it pushes each car or cut far enough to cause it to run to the desired track. This is preferable to handling the entire train for each movement. The poling track should extend parallel with the ladder track, so that cars which may fail to reach the proper switches can be given an extra push. These methods are for flat switching, although a grade of 0.3 to 0.5 per cent on the ladder track will facilitate the car movement.

Gravity Switching.—In gravity switching, the train is pushed to the summit of a hump, and as each car or "cut" passes over, it is uncoupled and runs down a grade which gives it sufficient speed to travel along the ladder and into its destined body track. Switches on the ladder tracks may be operated by switchmen, but in large gravity yards they are usually operated by electric or pneumatic mechanism, controlled electrically from an elevated tower at the hump by means of push buttons or small levers. In a few cases the slope of the ground can be utilized, but, as a rule, a switching hump is built and trains are pushed from the receiving tracks up an easy grade to the summit. This method of switching is used extensively in large yards, and with proper operation there should be no more damage to cars or freight than by other methods. As each car starts, a brakeman mounts it to regulate the speed and stop it at the proper point on the classification track. A motor car on a special track through the yard returns the men to the hump.

The grades for gravity switching will vary with local conditions; steeper starting grades may be required in winter, when the lubrication of car journals is cold, and also where there is a large proportion of empty cars, as these run less easily than loaded cars. Grades may be adjusted by ballasting or by means of an adjustable table at the summit. On the ascending side the grade may be 0.5 to 1 per cent, increased to 1 or $1\frac{1}{2}$ per cent for about 75 ft. at the top to close the cars together and thus facilitate uncoupling. On the descending side the starting grade may be 2 to 4 per cent (maximum 5 per cent), usually followed by a somewhat flatter grade and then 0.5 to $1\frac{1}{2}$ per cent along the ladder and into the body tracks.

Where the cars are to be weighed as they run down, the grade over the scale must not exceed 1 per cent (see "Track Scales") and the profile is, therefore, broken by a flat portion just below the starting grade. Examples of grades for gravity switching are given in Table XI. It is generally advisable to have two hump tracks to prevent blockade in case of derailment or other accident. If only a small percentage of the cars are to be weighed, the track scale may be placed on the reserve track. A typical profile of a switching hump is shown in Fig. 76, being that of the Denison yard of the Missouri-Kansas-Texas R. R.

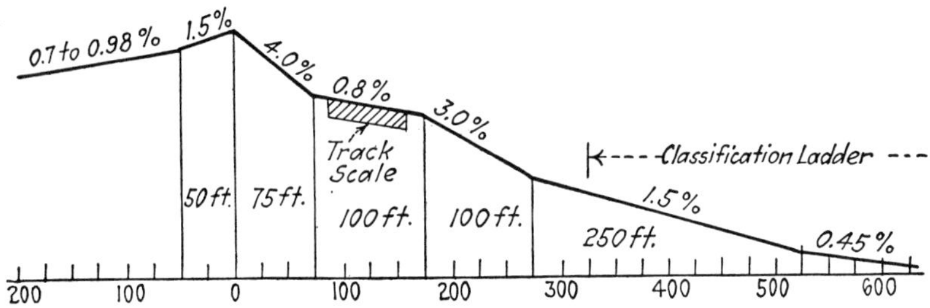

FIG. 76.—Profile of switching hump; Missouri-Kansas-Texas R. R.

TABLE XI.—GRADES OF GRAVITY SWITCHING HUMPS

		Grades						
		Starting		Over scale		To ladder		On ladder
		Per cent	Feet	Per cent	Feet	Per cent	Feet	Per cent
A. R. E. A.	Cold climate	4	50	1	100	4	100	1.5
A. R. E. A.	Moderate	4	37½	0.8	89	3	100	1.25
A. R. E. A.	Warm	4	26	0.6	77	2½	100	1.00
Pennsylvania-Detroit Ry.	Detroit	1 to 3	21	0.8	200	3	100	1.00
Missouri-Kansas-Texas Ry.	Denison	4	50 to 100	0.8	100	3	100	1.5 and 1.25
Pennsylvania R. R.	Indianapolis	5	100	no scale		3	100	1.25 and 1.00
Michigan Central Ry.	Niles	2	60	no scale		4	50	1 and 0.5
Wheeling and Lake Erie Ry.	Canton	2½	25	0.6	50	2½	50	0.38 and 0.24

Yard Brakes.—Track brakes used experimentally to eliminate the use of car riders promise effective results in controlling the speed of cars. The first brake is installed on the lower side of the hump, and others are on the ladders leading to the classification tracks, one or more brakes being operated from an elevated tower by means of compressed air.

Each brake consists of two pair of bars 20 to 40 ft. long, parallel with the rails and slightly above them, so that when the bars of each pair are pressed together, they will bear against the face and back of the car wheels. These bars are faced with renewable blocks, and behind each bar is a heavy angle iron connected to it by nests of coiled springs. The operating levers which close and open the brake are attached to the angle iron. Sufficient power can be applied to the wheels to hold a car on the hump grade if necessary, but ordinarily each operator in succession notes the weight and speed of the car and checks its progress according to his judgment, knowing the destination in the yard for each car by means of a cut list in front of him. An auxiliary device, also controlled from the towers, is a pneumatic apparatus for placing a sliding skid or shoe on the rail to check a car that may be moving too fast or too far along a classification track.

Engine Terminals.—Line and division terminals at which engines end their runs must have facilities for handling the locomotives. Engine terminals are provided usually adjacent to freight yards and to large

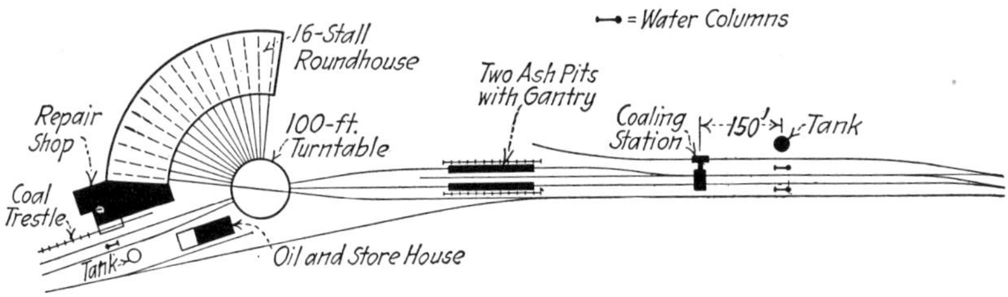

FIG. 77.—Engine terminal; Pere Marquette R. R.

passenger stations, but if passenger engines have to move any considerable distance from the station to the engine facilities, there should be an extra track to keep engine movements clear of main tracks. A locomotive entering the engine terminal stops first at a concrete washing platform and then at an inspection pit (both usually covered) and then proceeds to the ashpit to have its fire cleaned or drawn. After taking coal (or oil fuel), sand, and water, it proceeds to the turntable, where it is usually reversed before being put into the enginehouse. In the house the engine is inspected (if there is no outside inspection pit), the boiler is washed out and refilled, light repairs are made, and the engine is then put in condition for its next trip. Water columns must be provided near the turntable and on the lead track, to serve engines entering and leaving the terminal; also at the outlet from parking tracks, at the coaling station, and at special points for the convenience of switch engines. An engine terminal of the Pere Marquette R. R. is shown in Fig. 77.

In track arrangement, an engine terminal may be single or double end, having one or both ends connected with the main tracks. From the lead track at least two parallel tracks should extend to the enginehouse so that engines may move directly to and from the house without being blocked by engines at the inspection pits, ashpits, or coaling station, while engines needing to take coal or water promptly (as in the case of switch engines) may not have to wait their turn. Several short leads may be preferable, but if the number is restricted, there should be frequent crossovers to permit of these independent movements.

Enginehouses are usually segmental in plan, with radial tracks served by the turntable, and it is often advisable to have at least one of these tracks extended through the rear of the building to a yard track. Rectangular enginehouses may have parallel through tracks served by a ladder at each end; a turntable may then be omitted, engines being turned on a Y-track. Where engines have to be changed quickly, for through trains, a track for the new engine should be arranged close to the changing point at the passenger station or freight yard. Engines which are ready for service or which do not need to go to the enginehouse, but simply lay over at the engine terminal for the return trip, may be placed on outdoor parking tracks. These may be radial tracks served by the turntable, or parallel stub tracks served by a transfer table (perhaps opposite a repair shop served by such a table), or short parallel tracks connected by a ladder at each end.

Passenger Terminals and Stations.—Convenience and economy of operation are main considerations in planning passenger terminals. A compact arrangement of all facilities is usually desirable, but local conditions may prevent this, as by making it necessary to place engine terminals and coach yards at a distance from the station, thus increasing the switching movement and operating cost. In such cases an ample number of approach tracks should be provided, so that the movements of light engines and empty trains will not interfere with regular trains.

The ratio of approach tracks to station tracks depends upon density of traffic and ranges usually from 2.5 to 3 or even 4. Thus the South Station at Boston has 8 approach tracks for 28 station tracks, a ratio of 3.5. The Chicago Union Station has 10 stub station tracks at the north end and 14 at the south end, served by 6-track approaches. The St. Paul Union Station, with 8 stub and 14 through tracks, has a 7-track east approach serving 22 station tracks and a 4-track west approach for 14 station tracks. At the Indianapolis Union Station, 12 through tracks are served by 4-track approaches. The connections should be so arranged that any station track can be connected with any approach track, although this involves a complicated track layout, with ladder tracks, switches, and slip switches.

At stations which are terminal points for a large number of trains, either in main line, branch, or local traffic, these can best be accommodated by stub tracks, thus keeping such trains clear of the main or through tracks. Special equipment for passenger terminals includes the following: (1) mail, baggage, and express car tracks; (2) coach-storage and coach-cleaning yards, with water supply and other facilities; (3) coach-inspection and light-repair tracks; (4) sleeping-car tracks for cars which are picked up by late trains or set out for early trains; these tracks must have steam pipes for car heating and portable sewer connections for the toilets; (5) facilities for charging gas tanks or storage batteries for car lighting; (6) engine terminal.

Tracks at large stations are usually arranged in pairs, separated by platforms which are used both for passengers and baggage handling. Thus at Indianapolis the tracks are alternately 12½ and 26 ft. 9 in. c. to c., with platforms 16 ft. 9 in. wide; at St. Paul the spacing is 12½ and 29½ ft., with 19-ft. platforms. Another plan is to have single tracks, with alternate wide platforms for passengers and narrower platforms for baggage handling. The South Station at Boston has tracks spaced 17 and 23 ft. c. to c. with platforms 8 and 14 ft. wide. In the Chicago Union Station the track spacing is 21½ and 23¼ ft. alternately, with platforms 10 ft. 9 in. and 13 ft. 9½ in. wide; the narrower baggage platforms are 20 in. above top of rails, while the height of the others is 7¾ in. Extended details of the facilities at numerous large stations are given in the A. R. E. A. "Proceedings" for 1922 to 1925.

Design of St. Paul Union Station.—In the design of large terminals, the planning of the track layout for convenient handling of trains and passengers is an important feature. For the new union station at St. Paul, the number of station tracks to be provided was determined after careful consideration of past and possible future traffic developments. The number of approach tracks was determined largely by the number of leads of tenant railroads adjacent to the station property. Of the five principal entrances, two are used as through tracks exclusively, and it was considered necessary to provide double-track movement through the yard to only the other three entrances. To provide against possible interference to traffic by derailments, there are crossovers, pocket tracks, and auxiliary leads so arranged as to furnish alternative routes and thus prevent blocking of the yard. After the general plan had been completed, a definite time study was made on the basis of the train schedules or time tables then in operation. Track occupancy diagrams were then prepared and a careful study was made of the series of switching movements required with each train. The general plan was drawn on a scale of 50 ft. to the inch, and this was afterwards enlarged to 20 ft. to the inch for reference and for the field work (see "Switch Work").

Loops and Rapid-transit Terminals.—At the terminals of busy suburban and rapid-transit lines, special arrangements are required to insure prompt movements. Trains usually occupy one track for both unloading and loading. If the incoming locomotive has to back out before another train can be admitted, convenient sidings must be provided to hold the engine until it can be backed onto an outgoing train. In some cases, however, the inbound engine is not detached, but may back its train out to the car yard. Even with the best track arrangement there will be delays, but this condition is eliminated by the use of self-contained electric trains which can be operated from either end. With the greater facility of handling and the more rapid acceleration of electric trains as compared with steam trains, a larger number of the former can be operated in a given time on the same number of stub tracks.

A loop is convenient where space is available, as it provides for continuous forward movement. Some terminals of the Chicago Rapid Transit Ry. are kite shaped, with platforms at each side and a connecting loop of 100 ft. radius. In the suburban section of the Grand Central Station at New York, several parallel tracks and platforms are arranged on one side, with all tracks leading to a double-track loop of 137 ft. radius by which the trains proceed to the outbound tracks. For suburban traffic a loop of 8 to 12 deg. is often useful at coach yards; empty trains backing out of the terminal go round the loop to a group of parallel tracks and after being inspected and cleaned can continue around the loop to back into the station for another trip.

Freighthouses.—City freighthouses are usually long and narrow, accommodating a number of cars and wagons; convenient widths are 30 to 40 ft. for outbound and 50 to 75 ft. for inbound houses, the latter requiring greater width as the freight has to be held until called for. A very general plan is to have the two houses parallel, with tracks between them and a driveway outside each house. With not more than the four tracks they may be spaced 12 or 12½ ft. c. to c., but for a larger number alternate pairs may be 28 ft. c. to c. with 16½ ft. trucking platforms in these wider spaces. Where tracks are in the building, special designs must be made. In most cases the plan and dimensions are governed by local conditions and the size and shape of site.

For city terminals, it is a problem to combine the needed capacity with efficient and economical operation in the space available and to avoid excessive curvature or track complications. In large cities, high value of land and the tendency to separate street and railway grades is increasing the use of double-deck freighthouses having the driveways at the street level and the tracks above or below, necessitating the use of elevators. Upper warehouse floors are often provided in addition. A further development may be to place tracks on two decks, the cars being handled on inclines or elevators. Trackage on shipping piers with freight sheds on each side often consists of two to four depressed tracks in the center, putting the car floors level with the shed floors. A surface track along each side enables freight to be handled directly between car and ship.

Coal Mines and Piers.—Track layouts at coal mines usually include storage tracks on which the railroad places empty cars and which have a grade of $1\frac{1}{4}$ to $1\frac{1}{2}$ per cent extending along a ladder track so that the mine crew can let the cars run down to the tipple or loading point; a similar grade and ladder carries the loaded cars to storage tracks from which the railroad takes the cars. Coal and ore-shipping piers are usually high enough to deliver the materials to ships by gravity from pockets or bins under the track deck. Where the land is low, trains or cars may be pushed up an incline by a locomotive or handled by a cable and barney car. A light grade facilitates the movement of the car along the deck, at the end of which is a short steep upgrade with automatic switch, reversing the movement of the empty car and diverting it to an inclined track which returns it to the yard. Arrangements of this kind are shown in Fig. 78. Machines for tilting and dumping coal cars are used extensively.

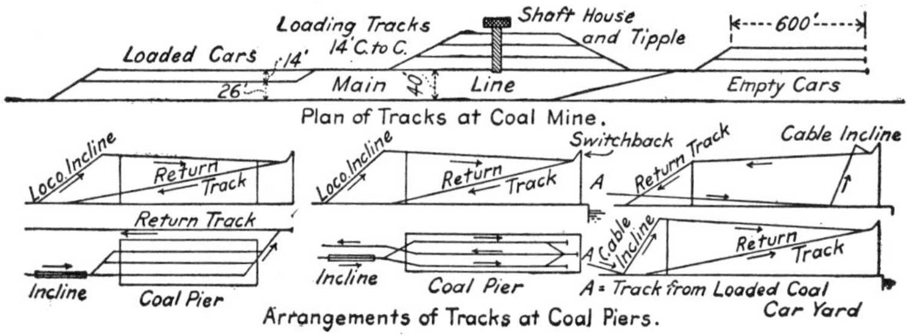

FIG. 78.—Tracks at coal mines and coal piers.

In transferring cars to and from harbor floats or barges, a hinged connecting bridge provides for varying water level and height of barge, the outer end of the bridge being suspended by cables or hanger rods which provide for adjustment. At car transfers or ferries on rivers, a common arrangement consists of a double-track trestle of about 3 per cent grade, extending below water and carrying a wheeled cradle about 180 ft. long, with its deck inclined upward about 3 per cent towards the river. This cradle is moved up and down the trestle by a cable and hoist, so as to keep the end of its deck level with the deck of the ferry or transfer boat. Feathered or tapered rails about 12 ft. long connect the tracks on the cradle with those on the trestle, while the outer end of the cradle may be widened for turnouts connecting with three or four tracks on the boat.

Sidetracks and Station Layouts.—Sidings or sidetracks are of three classes: (1) those for train service, commonly known as passing sidings or relief tracks; (2) those for placing or storing cars at yards and stations; (3) those for the private use of industries. Such tracks may be of relatively light construction, except that tracks on which trains continue running, instead of waiting, should be built and maintained as part of the main track. Furthermore, all turnouts from running tracks should be of main track material as far as the clearance point. Sidetracks on which trains wait or cars are set out may be slightly lower than the main track, so that cars and trains will not drift forward and foul the latter. Stop blocks or derails may be placed for the same purpose. Where several spurs or sidings occur, they should open from a lead track, so that only one main track connection will be required.

At small stations, the principal siding or house track may be on the rear side of a combination passenger and freight station, or between the main track and a separate freighthouse. A standard station layout for branch lines on the Union Pacific Ry. comprises a 3,000-ft. passing track and a 2,000-ft. warehouse track spaced 43 ft. c. to c., the latter having $6\frac{1}{2}$ deg. curves to the ladder track. Main track and siding turnouts have No. 10 and No. 9 frogs, respectively. The combination

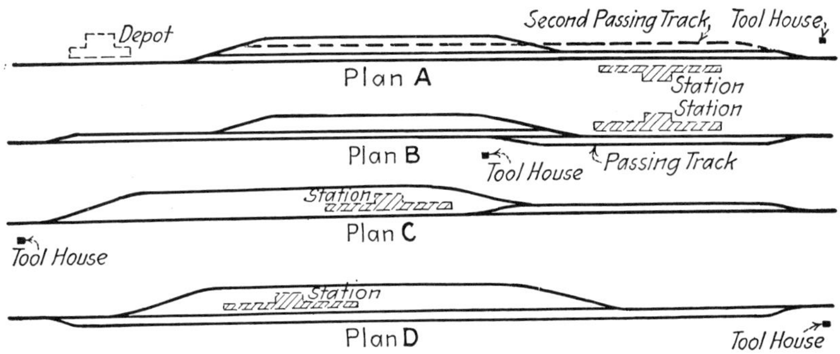

FIG. 79.—Station layouts and passing tracks at stations; Rock Island Lines.

passenger and freight station and agent's house has a 700-ft. platform with water column at one end. Other small layouts, as used on the Rock Island Lines, are shown in Fig. 79.

Industry Tracks.—For sidetracks to industries or manufacturing plants, the portion beyond the right-of-way line belongs to the industry, but the railway reserves the right to enter upon it for maintenance and operation. In some cases the industry undertakes the maintenance, which must be satisfactory to the railway. An agreement by which the railway does the work and charges it to the owner of the sidetrack is generally preferable to having the owner let the work to a contractor. The local officers of the railway should keep informed as to the condition of the track and warn the owner if the maintenance is neglected. In emergency the section foreman may be authorized to arrange with the local representative of the owner and proceed to do the necessary repair work. The length of track shoud be adequate for the business, to avoid delays and excessive switching. Where an industry requires a system of trackage, it is advisable to have the proposed layout examined by the railway engineers, to provide against operating difficulties or insufficient clearance. As a rule, the curvature of such trackage should not exceed 20 deg., but built-up districts or old plants may require curves of 40 deg.

When application is made for a new industry track, the railroad makes a survey for the proposed position and a suitable connection with the main track or an existing lead track, using preferably a turnout with No. 10 frog for the former or No. 8 for the latter. On double-track main line, the connection should be arranged with a trailing switch. In this survey all obstructions are noted, and also such features as street crossings or bridges; separate estimates are made for cost of work on the right-of-way and on private property. In the design of industrial plants, the track plan is often considered as of very secondary importance, whereas it has an important relation to the efficiency and cost of operation of the plant. Where buildings and other structures have been located or built in advance, the engineer called upon to provide tracks serving the several facilities may find it impracticable to obtain a logical or really efficient layout.

Passing Sidings.—These are provided to enable opposing trains to pass on single-track roads, and also to enable freight or inferior trains to keep out of the way of faster or superior trains moving in the same direction. Where the tracks are long enough for an inferior train to continue running (instead of waiting until the superior train has passed), they are known as relief tracks. On double track, the sidings may be placed between or outside the main tracks. The best location for a passing siding is at a summit of moderate grades. If it is on a ruling grade, this grade should be so reduced for the length of siding and a train length beyond each end that a train of maximum load can be started readily on the grade. Passing and main tracks should be at least 14 ft. c. to c. On a new line of the Santa Fe Ry., passing sidings 5,000 ft. long, for 100-car trains, are provided at intervals of about 7 mi. The turnouts have $16\frac{1}{2}$-ft. switch rails and No. 10 frogs.

Passing sidings with telegraph offices should be spaced systematically and not simply placed at stations for convenience. They should have a fairly uniform spacing (with due allowance for grades and other conditions) to insure a steady movement, for if the spacing varies considerably, the flow of traffic will be governed largely by movements on the longer runs. Each run between telegraph offices should have at least one intermediate siding to provide for heavy traffic. Delays incident to taking sidings may be reduced considerably by having the switches operated from the telegraph office. At the entering end of lapped sidings, a simple mechanical interlocking plant can be installed. At the farther end there may be a remote-control switch-operating device handled from the telegraph office, or a spring switch with automatic electric signals.

These switch-operating devices are desirable both for safety and to facilitate traffic with heavy trains. Otherwise, the train must stop to allow a brakeman to go ahead and open the switch, and must then start to pull into the siding. This means delay and expense, with possible blocking of other trains and liability of engine failures and pulling out drawbars in starting. Where a train takes an outlying siding, a bell circuit will enable the conductor to notify the tower man when this train is clear of the main track and when the superior train has passed; or an automatic device may give these indications.

Passing Track Layouts.—In the standard layout for center passing tracks on double-track lines of the Union Pacific R. R. (Fig. 80), the eastbound track swings out to increase the track spacing from 13 ft. to

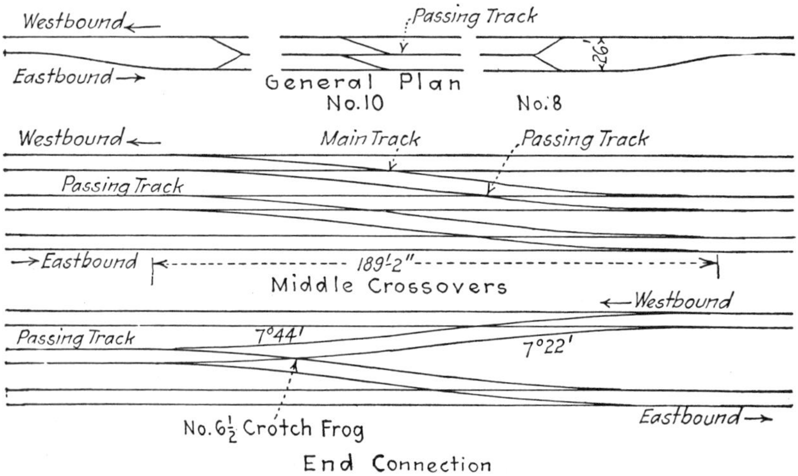

FIG. 80.—Center passing tracks: Union Pacific R. R.

26 ft., the curve ending about 10 ft. ahead of the switch for the passing track. The clear length of passing track is determined by traffic requirements, but at the middle there are crossovers connecting with the two main tracks, as shown. These crossovers have No. 10 spring-rail frogs in the main tracks and No. 9 rigid frogs in the passing track, with a curve of 3 deg. 54 min. between the heels of the frogs. At the end turnouts, No. 10 spring-rail frogs are used in the main tracks and a special No. 6½ crotch frog in the connection to the passing track. All switches have switch rails 16 ft. 8 in. long. Typical track arrangements at small stations on the Rock Island Lines, as shown in Fig. 79, have No. 10 frogs in the main track connections and No. 8 frogs elsewhere. Each layout includes a warehouse track as well as a passing track. Plan *A* provides for a second passing track, shown by a dotted line. The track spacing

is 14 ft. c. to c., except 28 ft. for the warehouse track in plan *B* and a minimum of 58 ft. in plans *C* and *D*.

Lap sidings are so arranged that opposing trains can pass without either one having to stop. In Fig. 81, plan No. 1 is a middle siding with the pulling-out switches at the tower; westbound trains enter at *A* and pull out at *B*, while eastbound trains enter at *D* and pull out at *E*. The sidings are connected across the lap at *B-E* and have the end switches *C* and *F*, so that the entire length can be used for trains in one direction if necessary. In No. 2, the pulling-in switches are at the tower, trains entering at *B* and *E* and pulling out at *C* and *F*, respectively. At these latter points, electric starting signals may be placed and interlocked with the switches, so that trains cannot pull out on the main track without the

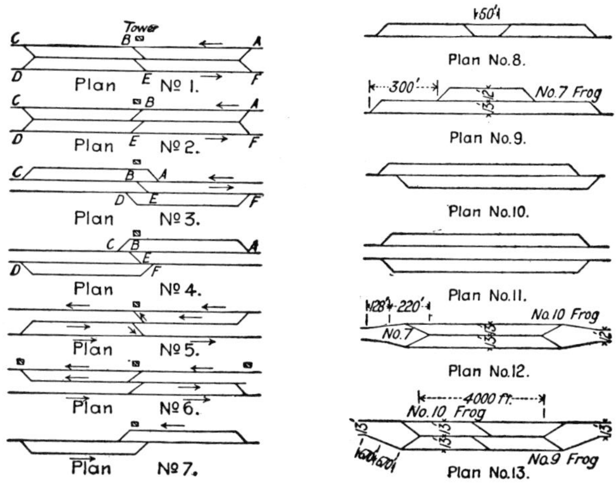

FIG. 81.—Different layouts of passing tracks.

knowledge of the tower man. Outside lap sidings, in Nos. 3 and 4, corresponding with Nos. 2 and 1, respectively, are used on long, straight lines to avoid the necessity of spreading the main tracks for a middle track; but each siding can be used for trains in one direction only, except by crossing over the main track.

Plan No. 5, in Fig. 81, is used where the sidings hold only one or two trains and are served by one tower. Where the sidings extend over two block sections, with their towers, the arrangement may be as in No. 6. For single track, plan No. 7 greatly facilitates the handling of heavy traffic, since two opposing trains waiting on the sidings can proceed as soon as the main-line train has passed. If the siding is long enough for two or more trains, the first pulls out as soon as it gets a clear signal (manual or automatic), and the next moves up to the clearance point ready to follow. Where the siding is long enough for two trains, two crossovers may be placed, as in No. 8.

A station layout with sidetrack and house track is shown in No. 9. Double sidings for single and double main lines are shown in Nos. 10 and 11. In relief track layout, No. 12, the main tracks are spread 26 ft., with No. 10 and No. 7 frogs in the main and relief tracks, respectively. In the center-passing track layout, No. 13, the main track on the station side is left straight in order to give a good view for the tower man or telegraph operator; the swing in the other track is made with 1-deg. curves 670 ft. long, so that trains can run at full speed.

CHAPTER XIII

TRACK TOOLS, STORES, AND LABOR-SAVING APPLIANCES

Tools and appliances for track work represent such a large aggregate investment that careful check should be kept on their supply and use. Each section gang must have a complete outfit, with a reasonable supply of spares or extras for use when others are sent for repair, but new tools should not be issued unless the worn or broken tools are returned or accounted for, or good reason is given for additional supply. Tool records are kept by the engineer and roadmaster (and sometimes the storekeeper). Section foremen must see that they have a sufficient supply, that tools are properly used, are not left on the rails or thrown aside, and are collected and checked at the end of the day's work. As good design and make are important in hand tools, to avoid excess weight and to facilitate their use, standard designs and specifications have been adopted by the A. R. E. A. Plants for reclaiming scrap may produce many such tools as bars and picks; they may also undertake repairs to tools and appliances. A few roads have maintenance-of-way shops for repair work, but such heavy equipment as locomotive cranes, ditchers and spreader cars are repaired more effectively at the shops of the mechanical department.

Typical tool lists are given in Table XII, but the equipment varies according to track construction, ballast, number of men, traffic, and other conditions. Thus, stone ballast is tamped mainly with picks, and gravel with bars; and where rock cuts are numerous, the section gang may have special appliances to break or remove fallen rocks. Tools used only occasionally, such as rail benders, heavy jacks, post-hole diggers, or ditching shovels, may be kept at the division roadmaster's headquarters and sent out to the gangs requiring them. For each division, also, there will be such heavy equipment as a ditching machine, spreader car, locomotive crane, or steam-derrick car, piledriver and snowplow. At the stores department there must be a sufficient supply of ordinary tools to equip extra gangs or for the use of emergency forces in case of washout, landslide, snow, or train wreck. In Table XII, the Baltimore & Ohio list is for a 3-mi. section on double track or a 5-mi. section on single track, with stone, slag, and gravel ballast. The list of the Cleveland, Cincinnati, Chicago & St. Louis Ry. is for a gang of six men. A number of track tools and appliances are shown in Figs. 82 to 97.

TABLE XII.—TOOL EQUIPMENT FOR SECTION GANGS

	Baltimore & Ohio	Cleveland, Cincinnati, Chicago & St. Louis	Louisville & Nashville	Great Northern	New York, New Haven & Hartford
Spiking mauls	6	4	6	3	6
Striking hammers	...	...	2	1	2
Ballast hammers	...	...	...	...	2
Trackwalkers' hammers	...	...	...	1	1
Nail hammers	...	...	...	1	1
Sledges	1	1	...	...	2
Tamping bars	2	2	...	6	10
Tamping bars, end	...	6			
Lining bars	8	6	4	6	10
Pinch bars	...	1	...	...	1
Raising bars	...	...	...	...	1
Claw bars	4	3	3	2	5
Spike pullers	1	1	1	1	1
Picks, clay	8	10	...	4	8
Tamping picks	12	2 stone	12	6	20
Scoops	3	6	...	4	3
Track shovels	12	7	12	10	10
Snow shovels	0	...	...	...	15
Long-handled shovels	1	6	...	...	2
Ballast forks	8*	6*	...	...	10
Axes	1	1	1	1	2
Adzes	3	2	2	2	4
Hatchets	1	1	1	1	2
Track wrenches	6	8	6	4	8
Frog wrenches	2	1			
Monkey wrenches	2	1	1	1	2
Trackwalkers' wrenches	0	...	...	...	1
Rail cutters	6				
Rail punch	1	1	...	1	1
Rail chisels	6	8	...	6	24
Track drill	1	1	1	..	1
Brace	0	..	...	..	1
Files	2	1	...	1	2
Hack saws	...	1			
Hand saw	1	1	1	1	1
Cross-cut saw	1	1	1	1	1
Scythes and snaths	4 to 8	6	8	4	8
Brush hooks	2	...	..	2	2
Scuffle hoes	...	6			
Grub hoes or mattocks	8	...	2	...	2
Wood rakes	0	...	...	...	1
Iron rakes	4	...	...	...	1
Timber hooks	2				
Hand car	1	1	1	...	1
Motor section car	1	...	1	1	1
Push car	1	1	1	1	1
Pony car	1				
Track jacks	2	2	4 or 6	2	1
Track gage	2	2	1	1	3
Track level	1	1	1	1	1
Sighting boards and blocks	few	...	few	1	1
Sighting bobs	...	...	...	1	2
Spirit levels	1	...	...	...	1
Tie square	...	...	...	...	1
Rail tongs	4	3	2	2	5
Tie tongs	4	3	4	2	6

TABLE XII.—TOOL EQUIPMENT FOR SECTION GANGS—*Continued.*

	Baltimore & Ohio	Cleveland, Cincinnati, Chicago & St. Louis	Louisville & Nashville	Great Northern	New York, New Haven & Hartford
Rail bender	1	...	...	1	1
Tape line, 50-ft	1	1	1	1	1
Ditching line, 100-ft	1	1	1	1	1
Wire stretcher for fencing	...	...	...	1	1
Switch lock	...	2	...	..	1
Padlock and chain for car	1	2	...	2	1
Oil cans	2	2	1	1	1
Oiler	1	1	1	1	1
Funnel, oiling	..	1			
Brooms	6	7	...	4	18
Grindstones	1	...	...	1	
Tool grinder	1	1	1	1	1
Whetstones	6	6	...	2	8
Wheelbarrows	1 to 4	1	...	2	1
Water pail and dipper	1	...	1	...	1
Water kegs, 15 gal	...	1	1	1	
Whitewash brushes	...	...	6	...	6
Tool boxes	1*	...	1		
Flags, red	6	2	2	2	4
Flags, yellow	2	2	3	2	4
Flags, green	0	2	3	2	2
Flags, white	0	..	...	0	0
Lanterns, red	3	2	2	2	3
Lanterns, yellow	...	2			
Lanterns, green	...	2	2	2	
Lanterns, white	2	2	2	2	5
Torpedoes	200	9	*A*	48	24
Fusees, caution signal	...	2			
Shims	0	...	*A*	as needed	2,000
Spike-hole plugs	1,000	...	*A*	1,000	500
Track bolts	1 keg	...	*A*	1 keg	3 kegs
Nut locks	$\frac{1}{2}$ keg	...	*A*	50	$\frac{1}{4}$ keg
Spikes	6 kegs	...	*A*	1 keg	5 kegs
Angle bars	20 pairs	...	*A*	...	20

* Baltimore & Ohio. Ballast forks for stone only; hand or motor car, not both; grindstone or tool grinders, not both; tool boxes for extra gangs only.

A, Louisville & Nashville. At discretion of supervisor.

COMMON TRACK TOOLS

Bars (Figs. 82 and 84).—For tamping ballast (mainly gravel and loose material, but sometimes for stone also) a $\frac{7}{8}$-in. steel bar 5 or $5\frac{1}{2}$ ft. long, weighing 13 to 15 lb., has a $\frac{1}{2}$-in. piece about 4 in. wide and 6 in. long, welded to one end at an angle of 24 deg. so as to strike under the tie. For stone, this head may be of diamond shape. The upper end of the bar is round or flattened to a $\frac{1}{2}$-in. chisel edge. A special bar for tamping under the ends of ties has the tamping head in line with the bar, the other end of which is bent up at a right angle and terminated in a loop handle. A lining bar for lining and throwing track is 5 to $5\frac{1}{2}$ ft. long, weighing 18 to 26 lb.; the lower part is $1\frac{1}{4}$ or $1\frac{1}{2}$ in. square with either a wedge or diamond point, while the upper part is $\frac{7}{8}$ or $\frac{3}{4}$ in. in diameter.

A pinch bar for heavy lifting is similar, but with a wedge-shaped edge on the square end.

Claw Bars and Spike Pullers (Figs. 82 and 84).—For pulling spikes, claw bars of various patterns are used, 5 to 6 ft. long, weighing about 30 lb. At the bottom the face is curved forward and widened, with a notch to engage the spike head. A block or bait may form the fulcrum,

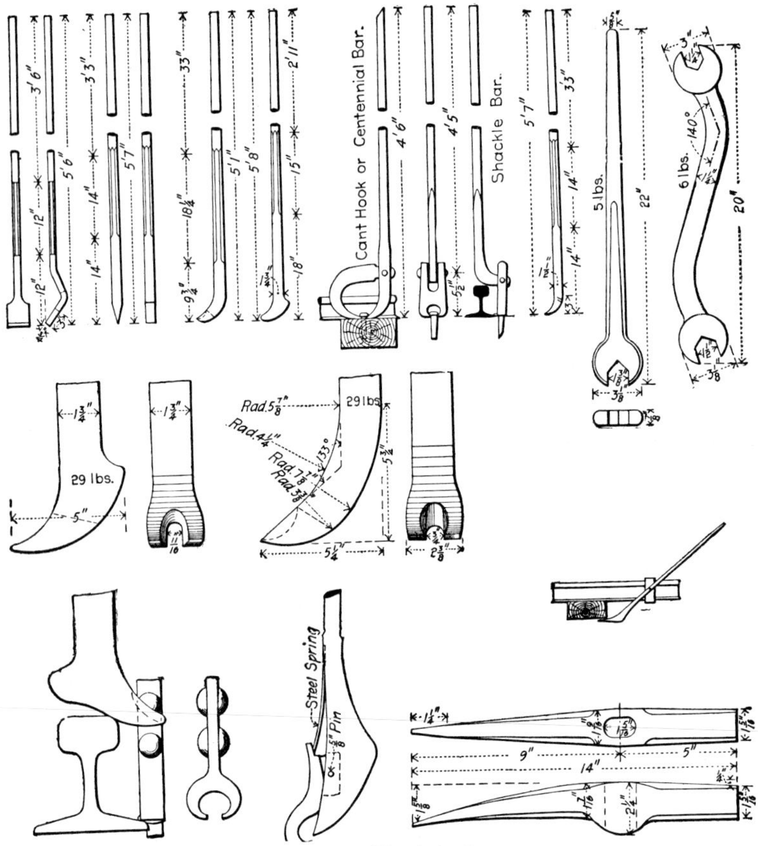

FIG. 82.—Track tools.

Top row: Tamping bar; lining bar; claw bars and spike pullers; track wrenches. Middle row: heads of claw bars; tie hold-up bar. Bottom row: spike pullers; track walker's hammer.

but many bars have the heel widened for this purpose. Of special designs to give greater leverage in pulling the spike, one bar has the lower end bent forward on a radius of 5 to 6 in. for a height of about 6 in., giving a distance of about 6 in. from point of claw to back of straight bar; another is bent backward and then forward in a reverse or gooseneck curve. To prevent the bar from slipping off the spike, an arm may be pivoted to it to hook over the rail; this is particularly for bridge work, where a slip of the bar might give the holder a bad fall. Special devices are required where the ordinary claw bar cannot be used, as inside the guard rails of elevated railroads and on bridges, at switches and frogs, and at station platforms. One of these is a bar with a hinged spring clamp or a curved canthook on the face of the bar, the spike being gripped between the clamp or hook and the claw, and the bar being pulled parallel with the rail. Another device has a shackle or bar with its lower end shaped to fit the neck of the spike (under the head) while the top is engaged by the claw of a bar having a projecting heel which rests on the rail head.

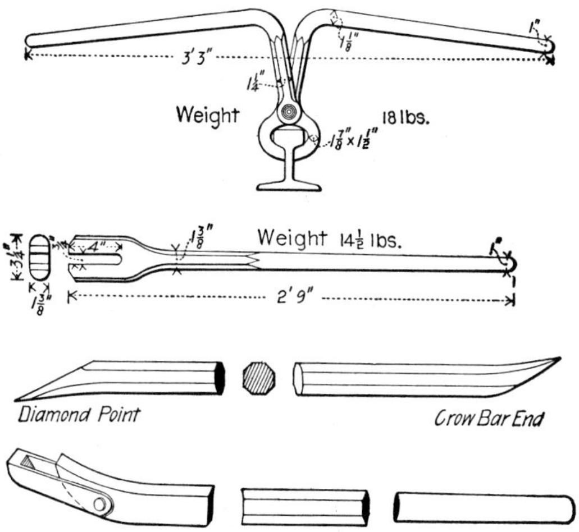

FIG. 83.—Track and bridge tools.
Rail tongs. Rail fork. Bridge inspector's sounding bar. Drift-bolt puller.

Hold-up Bar.—In driving spikes, the tie is usually held up to the rail by a crowbar (or two bars) underneath, but the bars are likely to give or to sink in the ballast. A device to hold the tie firmly against the rail (Fig. 82) has an L-shaped casting pivoted to the end of a long bar or to a short lever in which a pipe handle may be fitted, while a block on the side of the bar or lever rests on the rail head to form a fulcrum. With the horizontal leg of the casting placed under the tie, weight applied to the bar holds the tie up to the rail.

Hammers.—A maul for driving spikes has an 8- to 10-lb. head 13 or 15 in. long, tapering to $1\frac{3}{8}$ in. in diameter at the ends. A ballast or napping hammer for breaking stone ballast is 7 in. long and about $1\frac{3}{8}$ in. in diameter, weighing 4 to 6 lb. A sledge for knocking out ties and striking track chisels has an octagonal head 7 to 8 in. long, about $2\frac{1}{2}$ or 3 in. in diameter at the ends and weighing 8 to 15 lb. All these hammer heads are on straight wooden handles about 3 ft. long. A trackwalker's hammer, 14 in. long and 14 lb. weight, has one end straight like a hammer and the other curved like a pick, with either a point or a chisel edge; this is fitted to a short handle.

Picks, Shovels, Ballast Forks.—Ordinary picks, 25 to 31 in. long and weighing 6 to 11 lb., may have both ends pointed or one made with a chisel edge. Tamping picks (Fig. 84) have at one end a flat tamping head, usually 3 in. wide for stone or 4 in. for gravel ballast; they are 22 to 24 in. long, weighing 7 to 9 lb. A tamping puddle (Fig. 84) for gravel, cinder, and sand ballast resembles half a tamping pick and weighs about 5 lb. Tamping shovels about $9\frac{1}{2}$ by $11\frac{1}{2}$ to 10 by 12 in. with 26- to 30-in. handles weigh about 7 lb. In sand and gravel ballast the men will often use the shovel handle for tamping, and for this purpose a special iron handle may be fitted. Scoops for handling coal or snow are 11 by 15

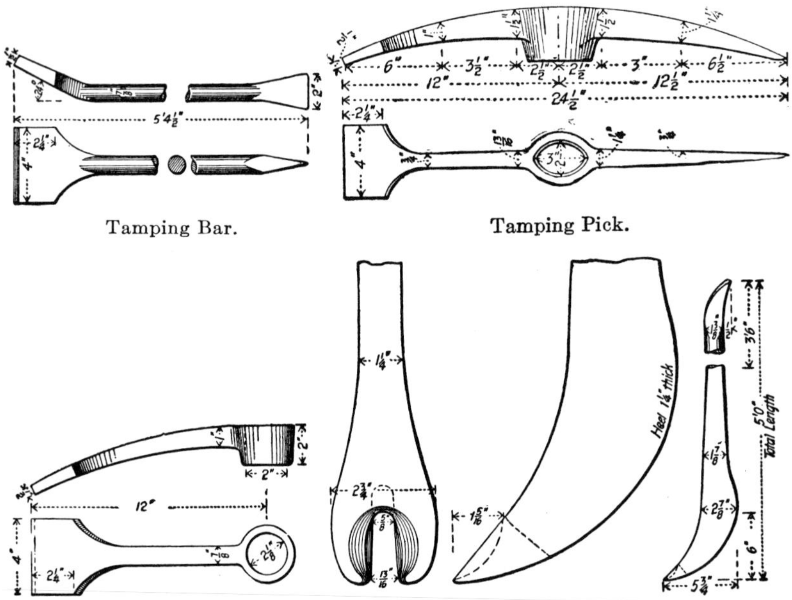

FIG. 84.—Track tools.

to 15 by 19 in. Long-handled shovels of special shapes are used for post holes. Spades are required for ditching, with special spades to dig trenches for tile drainage. In handling stone and slag ballast, forks like large stable forks are used, having 8 to 12 tines or prongs spaced to suit the size of ballast.

Post-hole Diggers.—For quick work in digging holes for fence posts and telegraph poles, post-hole diggers and augers operated by hand are used, the first having chopping blades to cut holes 6 to 8 in. in diameter while the second has a spiral blade 4 to 10 in. in diameter for posts or 10 to 14 in. in diameter for poles. For extensive work a power-driven machine with revolving cutter blades may be used, either mounted on wheels or suspended from the boom of a locomotive crane or derrick car.

In signal or electric railway work, a hole 24 in. in diameter and 6 ft. deep for the concrete footing of a steel post may be drilled in about 6 min.

Scythes, Hoes, Adzes.—A railway scythe for cutting grass and weeds is heavier than the common grass scythe; the weed and bramble scythes are heavier and shorter and the bush scythe is still shorter with a broad blade. Grass hooks and sickles are also used. The brush hook is a short blade in line with a straight axe handle, but having a curved or hooked end to cut overhead boughs. For ditching and heavy grubbing, the grub hoe has a single broad blade like an adze. The mattock resembles a pick, but has one end horizontal like an adze and the other end vertical like an axe; in the pick mattock, the latter part is shaped like a pick. Long-handled hoes are more convenient than shovels in cutting weeds on the roadbed. Adzes are required for dressing the rail seats of ties and for other wood work.

Rail Forks and Tongs; Tie Tongs (Fig. 83).—A fork for handling and carrying rails resembles a wrench about 33 in. long, 10 to 14 lb., but with a $\frac{3}{4}$-in. slot 4 to $5\frac{1}{2}$ in. deep in the jaw to receive the rail web or flange. In handling rails with a crane or derrick, two short forks may be attached to the ends of a cable sling; with the rail held at its ends by the forks it will not swing as when held at the middle by tongs. Rail tongs are shaped to grip the rail head and have two straight handles about $16\frac{1}{2}$ in. long, held by two men on opposite sides of the rail. Tie tongs are similar, but with a jaw wide enough to engage a tie, the sharpened ends fitting under or into the side of the tie. One-man tie tongs have the two handles brought together vertically and finished with loops for the hands; long-handled one-man tongs are used for pulling ties in and out of the track. The weight is about 16 or 18 lb. for rail and tie tongs, or 10 lb. for the one-man pattern.

Track Chisel, Punch, Wrench.—Chisels, 9 to $10\frac{1}{2}$ in. long, weighing 5 to $6\frac{1}{2}$ lb., for cutting rails, are forged to shape, oil tempered, and ground. When dull, they may be reground, but if the temper is lost, they must go to the blacksmith. For test, the chisel set at 45 deg. on a steel rail must stand three blows of a 15-lb. sledge without injury. A straight handle about 18 in. long is used, but both the chisel holder and the hammer man should wear goggles to protect their eyes. A track punch about 5 lb. in weight and struck by a sledge is used to make holes in rails, but it is better to drill the holes. Small hydraulic and screw punches are also used for heavy rails, the latter operated by two or three men. The track wrench is usually 15 to 24 in. long, weighing 3 to 7 lb., with the jaw shaped to take square or hexagonal nuts and the handle finished with a chisel edge or tapered for use as a drift in lining up the bolt holes. A double-end wrench has jaws to fit two sizes of nuts. With long wrenches, care must be taken that the men do not exert such force as to strip the threads.

Rail Saw and Drill.—Rails can be cut more accurately and with less damage by saws instead of chisels, particularly in fitting up frog and switch work. Cutting with hand hack saws is slow, but hack or circular saws in frames clamped to the rail may be operated by hand gears or by power. For drilling bolt holes and bond holes, ratchet and geared drills are commonly used. The drill frame is preferably hooked to the flange (Fig. 85) and not over the head of the rail; and the drill carrier slides on the frame, so as to drill the four or six holes at one setting. Crank-handle drills have an upright frame beside the rail, but this is hinged so that it can be lowered when a train approaches, leaving the drill in place. Portable pneumatic and electric drills are also used. For a long stretch

FIG. 85.—Track drill.

of work, the power drills may be mounted on a car having a gasoline engine, or steam engine, or an air-compressor operated from the locomotive of a work train.

Rail Bender.—To bend rails for sharp curves and turnouts the ordinary jim-crow bender (Fig. 86) has a curved frame with hooked ends to fit over the rail head or base, while pressure is applied between these points by a bending bar with a screw which extends through the middle of the frame and is turned by a long-handled wrench on a nut, or by bars in a capstan-head screw. When the rail is bent to the required ordinate, the screw is slackened and the machine pushed along the rail to repeat the operation. These machines weigh 100 to 200 lb. A hydraulic ram may take the place of the screw. Another design has a plunger or bending bar operated by an eccentric or cam which is worked by a long lever or sweep moving in a vertical plane at right angles to the rail. For use on rails in the track, there is a cast-steel lever having one end hooked over the rail and a capstan-head screw in the other end.

The roller rail bender (Fig. 87) has grooved rollers at the ends of the frame to bear against the outer side of the rail head, while a middle roller on the bending bar fits against the inside. When the bar is set up by a screw to give the ordinate of the required curve, the middle roll is turned by a long, cross-handled wrench, causing the machine to travel along the rail and give it a uniform curve. At yards, a roller rail bender may be

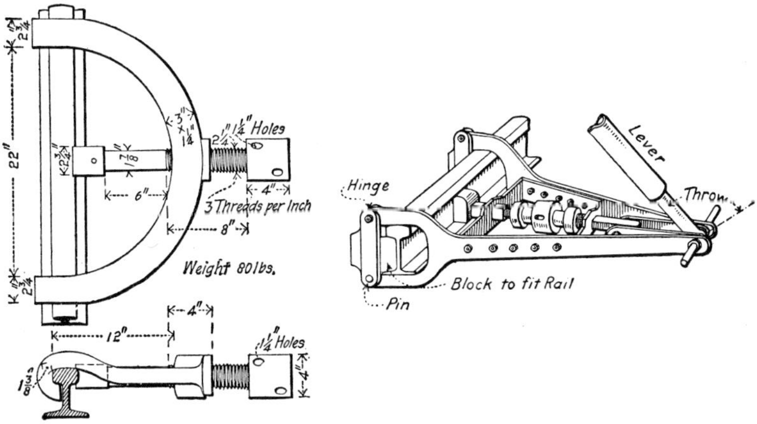

FIG. 86.—Rail benders.

mounted on a foundation and the rails pulled through by a cable and engine; or a portable plant may have the bender mounted on a flat car, with cars at opposite ends for the straight and curved rails, another car at the rear having a hoisting engine with cable for pulling rails through the

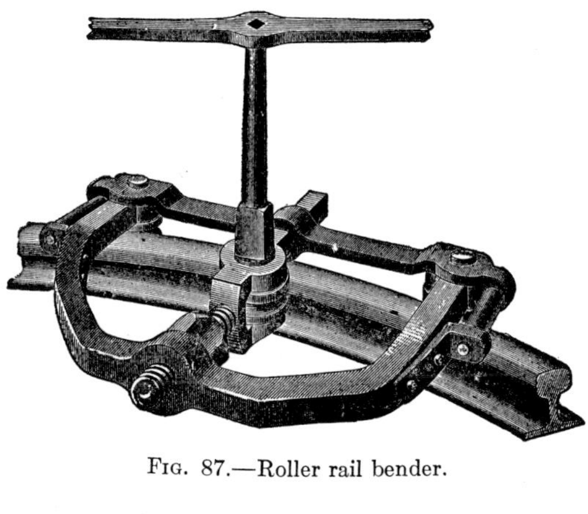

FIG. 87.—Roller rail bender.

bender. In another arrangement, the two outer rolls are driven by gearing and serve as feed rolls, thus dispensing with the cable.

Track Jacks.—In track raising and many other items of track work, jacks of various patterns are used. The jack stands beside the rail and has a lifting bar with a claw at the bottom to fit under the rail or tie;

this bar is formed with ratchet teeth engaging pawls or is a smooth bar passing through lifting rings, the pawls or rings being operated by a lever to which a long handle may be fitted. For tunnels, or in front of retaining walls or station platforms, the claw may be at the side of the jack, the lever then working parallel with the rail. Single-acting ratchet jacks raise the load a full notch at each down stroke, while double-acting jacks raise it half a notch at each up and down stroke. As the tool standing beside the rail may be a danger to trains, track jacks are made with a

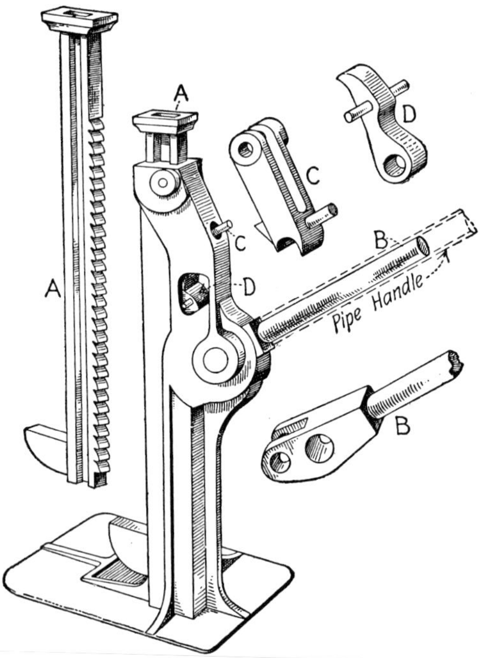

FIG. 88.—Track jack.

trip by which the bar can be released instantly to allow of removing the jack. In use, the jack should be always on the outside of the rail, so that the operator will not be in danger, and if the jack should be accidentally left in place, the train will knock it away. If the jack is left on the inside of the rail, the engine pilot might throw it across the rail and so cause an accident.

In the ratchet jack shown by Fig. 88 the rack bar *A* is operated by the lever *B* and pawl *D*, while the top catch *C* holds the bar in position at the height to which it is raised. The load can be lowered one tooth of the rack at each stroke or it can be tripped and dropped the full height by releasing the pawl *D*, without the use of an independent trip.

A jack of 10 or 12 tons capacity with a lift of 12 in., for ordinary work, will be about 24 in. high and weigh 60 lb.; with an 18-in. lift it will be 28 in. high and weigh 75 to 80 lb.; a 15-ton jack of 10- or 18-in. lift will be 15 to 28 in. high, weighing 50 to 75 lb. A light jack of 15 tons capacity but a lift of only 6 in. is about 16 in. high, weighing 30 to 35 lb., while a jack to be carried by a trackwalker weighs only about 5 lb. Lowering jacks (without the releasing trip) are preferred in bridge erection and wrecking work where a sudden release (perhaps accidental) might be dangerous. Both trip and lowering features are combined in some jacks. Hydraulic, screw, and geared ratchet jacks are used in heavy work of this kind. A 27-in., 25-ton, geared jack of 13 in. lift will weigh 200 lb.; a 36-in. 40-ton jack with 26-in. lift, for very heavy work, will weigh about 250 lb.

Track Gages and Levels.—To give the required distance between rail heads in tracklaying and to test the accuracy of gage of track, a convenient form of gage is an iron pipe or wooden bar having a lug at each end, these lugs being 4 ft. $8\frac{1}{2}$ in. apart. Both designs are shown in

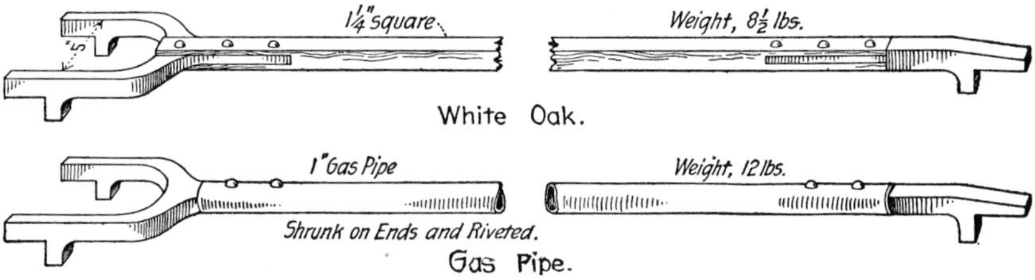

FIG. 89.—Track gages.

Fig. 89. One or both lugs may be $1\frac{3}{4}$ in. wide, to set guard rails with the proper flangeway width. In many designs the bar is a 1-in. gas pipe with cast-steel or malleable-iron end pieces riveted on; one casting is U shaped about 7 in. wide, with a lug on each arm, so that when both these lugs are set against one rail the bar will be at right angles with it, insuring the proper width to the opposite lug and rail. The weight is about 12 lb. Fiber bushings may insulate the ends. Track gages with special devices for measuring the widened gage on curves have been tried, but there is opportunity for making a wrong allowance, and such gages are little used.

To ascertain if the track on tangent is in a horizontal plane, a level board carrying a spirit level is placed across the rails; this is about 4 by $1\frac{1}{4}$ in., 5 ft. 3 in. long. For curves, where the outer rail is elevated, the proper elevation may be given or checked by a rod or plate graduated to $\frac{1}{4}$ in. and sliding in a vertical hole in one end of the level board (Fig. 90). This rod has its lower end shaped with a foot to rest on the rail head and held in any position by a thumbscrew. For a given curve, the rod is lowered the distance for the required superelevation and is seated on the low or inner rail while the other end of the board rests on the outer rail. If the elevation is correct, the spirit bubble will be at the middle of the tube; otherwise, the outer rail must be adjusted to bring the bubble into this position. In a combined track gage and level (Fig. 91), the board is about $6\frac{1}{2}$ ft. long, $1\frac{1}{4}$ by 7 in., with one end made in steps or offsets of $\frac{1}{2}$ in. so that the spirit bubble will be level when the bottom edge of the board is on the low rail and the offset corresponding to the proper elevation is on the high rail. To prevent wear, the level board and offsets are shod with a steel strap, but this is not continuous so that it will not affect track

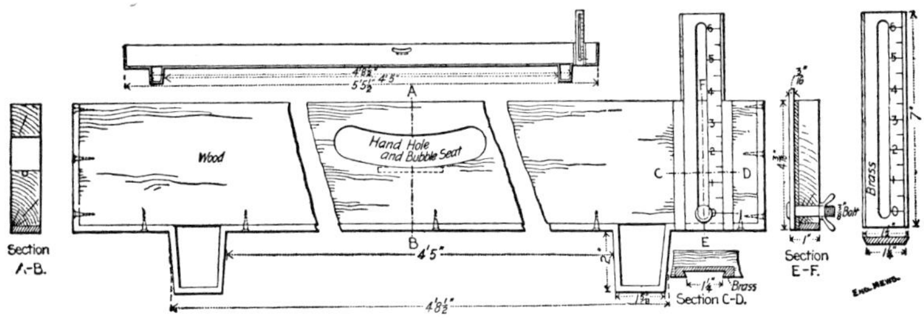

FIG. 90.—Adjustable track level and gage.

circuits when the board is placed on the rails. Spirit levels for such tools should indicate $\frac{1}{16}$ in. difference in elevation in a horizontal distance of 5 ft. For leveling one track from another, a 14-ft. board is placed with one end resting on the rail of the first track and the other on a support which is adjusted to bring the board level, as shown by the spirit bubble. The rail of the opposite track is then brought up level with the board.

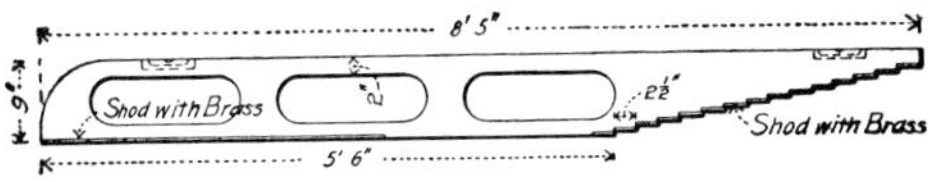

FIG. 91.—Track level and gage.

Surfacing or Spot Board.—In surfacing track or taking sags out of the grade line, the foreman uses a spot board and two blocks. The board, $1\frac{1}{4}$ by $8\frac{1}{2}$ in., is painted white, with a horizontal black stripe a little above the center line. The height of the blocks is equal to that from the bottom of the board to the top of its black stripe. With the board placed across the rails at a point where the track is at proper elevation, and with one block placed on a rail at a similar point, the other block is placed at intermediate points and the track is raised until the foreman sighting over the first block gets the second block in line with the black stripe on the board. For raising track, a similar spot board may be supported in brackets sliding on rods driven into the ground, the brackets being held in position by thumbscrews. If ballast stakes are set, only one rod and bracket is used, the other end of the board being set on the stake.

Tie-plate Gage.—To set tie-plates in correct position, the Buffalo, Rochester & Pittsburgh R. R. uses the Ware gage, consisting of a 1-in. pipe bar with a fixed head at one end, while the other end has a sliding head held in position by a thumbscrew clamp, this head being adjusted for plates of different sizes. The use of this gage (Fig. 92) is described in Chap. III while methods of applying tie-plates are noted in Chap. XIX. The bar A is of 1-in. gas pipe, with a fixed head B at one end and at the other end a sliding C having a lug D and held by a thumbscrew clamp E. This head is clamped in such position that tie-plates of the size being laid will be in proper position when placed with their ends butting against the faces F and G. The tool is set with its heads flat on the tie to test the level of the surface, and is then turned over with the faces F and G resting on the tie. After one plate is set in position against the edges F and B it is driven down and the other plate is gaged from it. Such gages are not generally used.

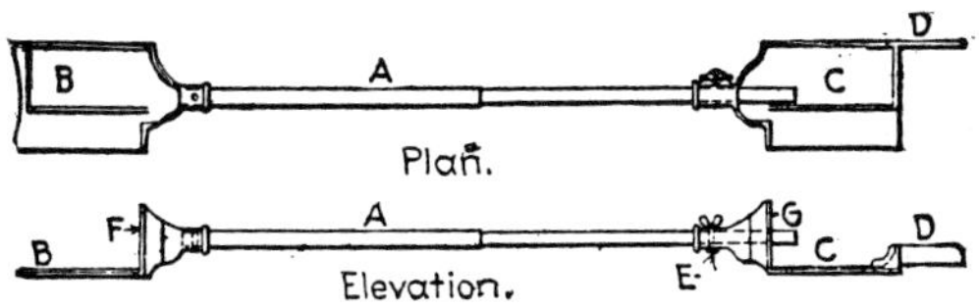

FIG. 92.—Gage for setting tie-plates.

Hand and Motor Section Cars.—Each section has a hand car or motor car for carrying the men to and from work, and a push car (which may be used as a trailer) for carrying material. As operation of the hand car consumes time and energy, there is a rapidly increasing use of motor cars to increase the working time and bring the men fresh to their work. A hand car with a platform 7 by $4\frac{1}{2}$ ft., on 20- or 24-in. wheels, will weigh 500 to 600 lb. On one axle is a pinion gearing with a spur wheel driven by a connecting rod from a walking beam, the gearing being usually $3\frac{1}{3}$ to 1. A cross-handle at each end of the beam enables four men to "pump" the car. Roller bearings and a loose wheel on the free axle assist the easy running of the car. A track velocipede with three or four wheels and weighing 150 to 200 lb. is propelled by one man by means of a rocking lever which drives the gear wheels. In three-wheel cars, there is usually a small wheel at the end of an arm or outrigger which can be folded back against the two-wheel portion for convenience of shipment in a baggage car. Some designs have wire spoke wheels and frames of bicycle tubing. These velocipedes are used by roadmasters, foremen, signal repairmen, and others, but it is now more general for roadmasters to have motor cars.

A four-wheel motor car with 4- to 7-hp. gasoline engine will weigh 500 to 1,000 lb. and can be run at any speed up to 20 or 25 mi. per hour, but some roads prefer cars that cannot exceed 20 mi. The load is usually so distributed that one man (for a 500-lb. car) or two men can put the car on and off the track, this being sometimes facilitated by extension levers or handles which can be pushed out beyond the end of the car. Belt, chain, or gear drive may be used. Wheels with wood or pressed-steel centers are 17 to 20 in. in diameter, with roller bearings and with a loose wheel on the undriven axle. There may be a band brake on the axle and a shoe brake (operated by treadle) on two or four wheels. To prevent interference with track circuits operating automatic signals, the wheels on one side may be insulated. These cars will carry 8 to 12 men and haul a trailer for ties or supplies. They are made to stand rough usage, but the foremen should see that they are not abused and that they are kept oiled and in proper repair. Inspection cars and roadmasters' cars may have spring seats. Push or trail cars have platforms about $5\frac{1}{2}$ ft. wide and 7 to 8 ft. long and weigh from 500 to 700 lb. To carry men instead of supplies, bench seats may be attached to the platform.

Wheelbarrows and Dolly Cars.—In ditching and other earthwork, iron or steel wheelbarrows are used, a special pattern having a wide-grooved wheel (or two tandem wheels) to ride on the rail. A dolly car has two small, grooved wheels in a frame which is held and propelled by an inclined lever on one side; this car, traveling on one rail, can carry a rail or tie or a small removable iron body for excavated material. Where ditching or slope dressing must be kept clear of the track, $\frac{1}{2}$- to $\frac{3}{4}$-yd. cars may be run on a track of 12- to 20-in. gage laid over the ditch.

Miscellaneous.—Heavy rakes are used for cleaning up brush, trimming station grounds, and handling hay where grass is cut on the right-of-way. Rattan or corn brooms serve to clean ties after adzing, to clear snow from switches and frogs, and to clean out the section house. A tool grinder clamped to the section car or to a bench in the tool house is useful for sharpening edge tools. To prevent theft, the shovel handles and other wooden parts should be marked; this may be done with a gasoline branding iron, which is used also for marking ties and poles.

Bridge and Building Tools.—Special equipment for bridge, building, and water-service gangs may include bars for sounding and moving timbers (Fig. 83), cant hooks, shackle bars, or pulling jacks for removing drift bolts (Fig. 83), mason and carpenter tools, wood-working outfits, concrete mixers, cement guns, heavy jacks, winches or stump pullers, light derricks or push cars, power pumps for foundation work, and spray-painting apparatus. Detailed lists are given in the "Proceedings" of the American Railway Bridge and Building Association (1922 and 1923).

LABOR-SAVING MACHINES AND APPLIANCES

An important step towards higher efficiency and economy in track work is the introduction of labor-saving machines, tools, and devices to secure a greater output of work at lower cost and with a smaller force. Thus in specific cases, ditching by machine has cost 2 to 6 cts. per cubic yard, as against 75 cts. to \$2 by hand labor; weed killing, \$19.50 instead of \$175 per mile, and rail renewals \$500 instead of \$1,400 per mile. Furthermore, the machines may do much of the rough and unpleasant work. While the introduction of machinery has been aided by the labor cost, shortage, and trouble of recent years, a much wider development may be looked for, partly in new machines but principally in a more general use of machines already available.

Motor section cars, already noted, conserve the energy and working time of the men otherwise expended in "pumping" the hand car, while the motor car can also haul trailers carrying men and materials, thus replacing work-train service to some extent. A ditching machine will do the work of a large gang of shovelers. In relaying rails, a light, portable hand or power derrick with 3 to 5 men does more work than a gang of 12 to 16 men, and eliminates the danger to men handling heavy rails. All such machines and appliances must be of simple design and substantial construction, as they are handled largely by unskilled men. In view of their high cost, such machines should be utilized as much as possible, and many of them can be applied or adapted to various purposes.

Pneumatic and Electric Tools.—An important development is the use of hand tools operated by compressed air or electricity, the power being

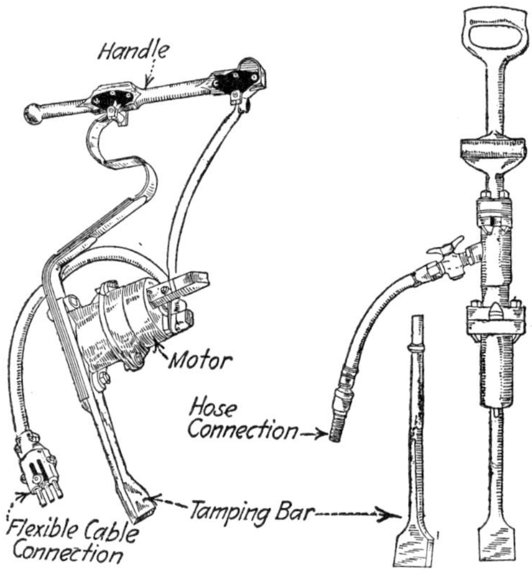

FIG. 93.—Electric and pneumatic ballast tampers.

furnished by self-propelled portable plants with gas-engine drive for compressor or generator. Flexible hose or conductor cords are connected to the tools. The principal tools used on track work are tie tampers, rail saws, rail drills for bonds and bolt holes, spiking hammers, screw spike drivers, tie- and wood-boring drills, grinders for dressing rail ends and switch points, and wrenches for running nuts on and off of track bolts. With these last, hand wrenches must be used for the final turns in tightening the joint and for the first turns in loosening a nut for removal; the remainder of the work is done rapidly by the machine.

Tamping ties is the operation most largely performed by power tools, and both the pneumatic and electric machines are shown in Fig. 93. The former has a long, straight barrel with hand holds and hose connections, the machine operating on the principle of a pneumatic riveting hammer. As the air is turned on, under control of the operator, a rapid succession of blows is delivered on the end of the 18-in. tamping bar which is inserted in the barrel, the bar having a tamping head suitable to the kind of ballast. In working order, the tool, about 40 in. long over all, weighs about 38 lb. Air at 65 to 75 lb. pressure is furnished by a compressor which may be of capacity to serve four to twelve tools. In the electric tamper, a small motor on the frame runs at 3,600 r.p.m. and has an unbalanced weight which sets up a vibration that is transmitted to the tamping bar and ballast. The portable generator plant serves four or more tampers, which are connected to it by 250-ft. cables, so that work can be done in a range of 500 ft. without shifting the generator plant.

Miscellaneous Machines.—To increase the speed and accuracy of adzing worn ties in dressing the seats for new rails or tie-plates, there are motor-driven machines with four circular saws to cut grooves on each side of the rails. The saw frame rides on the rails and thus all grooves are cut to the same depth below the rail head. With these grooves as a guide the adze men can readily dress the rail seat to an even surface at the required level. For cleaning tracks where large quantities of coal are handled, both the Pennsylvania and the Baltimore & Ohio have track sweepers. The latter machine consists of a car with 37-in., steel-wire, revolving broom sweeping the dirt into an apron conveyor which feeds another conveyor dragging the material along a horizontal trough on a boom extending over a gondola car. By means of traps in this trough, about 5 ft. apart, the material is distributed over the car. The machine is pushed along the track at a speed of 4 to 6 m. p. h. A track-shifting machine, used where there is continual shifting of track at dumps, is a four-wheel car with a central spud or vertical timber fitted with a rack geared to a gasoline engine. With the car held to the rails by clamps, the spud is set at an angle, with its shoe resting on the ballast, and is forced or racked downward; this raises the car and track, and the inclined position causes them to shift laterally.

Machinery and Appliances for Track Work.—An outline of the varied applications of power and other labor-saving devices in railway work is given below, and several of these devices are described in connection with the work for which they are used.

1. Electric ballast tampers and rail drills.
2. Pneumatic tampers and appliances for driving and pulling spikes, drilling rails, turning nuts on track bolts, cutting nuts off old bolts, grinding switch points, riveting and chipping on bridge work, boring ties and timbers, sand blasting, and spray painting.

3. Tie borers and trimmers; tie spacers and devices for pulling ties in and out of track.
4. Sod line cutters, ballast toe liners, ballast slopers; and machines for heavy tamping. These are mounted upon or attached to section cars.
5. Mowers and grass cutters for track and shoulders; attached to section cars.
6. Rail saws, and rail benders.
7. Track oilers, to prevent rusting of bolts and corrosion of steel parts.
8. Oxyacetylene outfits for cutting, drilling, and repairing rails and frogs, and building up worn joints.
9. Motor section cars carrying men and supplies.
10. Horse scrapers and narrow-gage dump cars for use in cleaning ditches and widening cuts.
11. Track dollies and rollers for moving ties, rails, and timbers.
12. Track-laying machines on new work.
13. Small rail-laying cranes, hand or power, for rail renewals.
14. Rail- and tie-handling devices for loading and unloading cars, distributing new rails and picking up old rails.
15. Locomotive wrecking cranes and derrick cars for handling and laying rails, ties, frogs, and crossings; handling earthwork and ballast with grab buckets; handling scrap with bucket or magnet, setting poles, driving piles, and placing bridge girders and deck slabs.
16. Ditching machines of blade type for clearing ditches and widening cuts.
17. Small rotating steam shovels, on cars, for ditching and light earthwork.
18. Spreader cars for leveling dumped material.
19. Ballast cleaning and screening devices; small for section gang or large for use with work train.
20. Ballasting cars and plows; power unloader plows for flat and gondola cars in ballasting or widening fills.
21. Rail-sawing and drilling outfits, stationary or on cars, for reclaiming damaged rails. Rail-bending outfits, stationary or on cars, for curving rails.
22. Weed killers; using chemicals, oil flame, or steam.
23. Track-oiling cars for spraying dusty ballast and road crossings.
24. Air-dump cars for use with ditchers and steam shovels.
25. Portable concreting outfits with power mixer on push car for small jobs; mixer or cement gun on flat car with work trains for more extensive work. Cement-gun outfits.
26. Earth-boring machine for post holes and footings; small pile driver for setting fence posts.
27. Wood-working outfits for bridge falsework and concrete form work.
28. Snowplows, flanger cars, and thawing outfits for clearing snow and ice from switches and in station yards.

Stores Supply System.—For efficient distribution of tools and supplies there is increasing use of supply trains or cars which go over each division monthly and are in charge of the stores department. Under the older system, individual requisitions are handled separately and supplies shipped by freight train, thus involving multiplicity of records and small accounts and much expenditure of time in wrapping, addressing, and handling the numerous shipments, besides involving the services of the freight department and freight agents. With the newer system, all requisitions are handled together and are carried on the train, which is loaded accordingly. On the day for the supply train, worn or broken tools for replacement are laid out at each section tool house, oil cans are made ready for filling, scrap is piled for loading, and other preparations are made. Inspection is made at the time of exchange, requisitions are checked, and any complaint or dispute as to loss or damage can be settled directly between the section foreman, roadmaster, and stores agent. Such an arrangement avoids much correspondence over what may be trivial matters, while it tends to eliminate friction and to promote cooperation and efficiency.

In detail, the system varies considerably; some roads run a completely equipped special train, including tank cars for oil and gasoline; others operate the cars in local freight trains. In either case the service includes section gangs; bridge, building, and extra gangs; signal maintainers, pumpers, watchmen, station agents, and camp outfits. On the New York Central R. R. the stores train delivers construction material along the line at points where it is needed, instead of unloading it at stations for further distribution. Such material includes spare or renewal rails, ties, joints, tie-plates, crossing planks, and fencing. The train has a gasoline crane with electric magnet hoist for picking up scrap and loading it into a car.

In the stores department records and accounts there should be included the material held at tool houses and tool cars or distributed along the line, as well as the material in store houses. These "line stocks" include rails, ties, and track material, bridge timbers, signal and telegraph supplies, building materials, culvert pipe, materials for coaling and water stations, and miscellaneous scrap. With the supply train service, these line stocks can be inspected and checked readily and periodically. It has been suggested that material yards may be established where the requirements of track, and bridge and building departments can be assembled, so as to be shipped to the job complete when required, thus reducing the chance of material being left exposed and unprotected.

CHAPTER XIV

ELECTRIC RAILWAY TRACK AND TRACK WORK

Electric railways are of two distinct classes: (1) interurban or cross-country lines, with track similar to that of steam railways; (2) street railways having track of innumerable types adapted to various forms of paving. On well-built, electric, interurban lines the maintenance should be lighter than on steam lines of comparable class, since the motor cars and electric locomotives are of less weight than steam locomotives and do not produce the pounding and hammer-blow effects incident to the counterbalancing and concentrated wheel loads of the latter. On city lines, maintenance is complicated by the surface wear due to street traffic and by the inaccessibility of track embedded in the paving; on the other hand, this track is protected from the disturbing influences of wind and weather. For all electric lines, efficient bonding of the rail joints is specially important, to provide ample capacity for flow of the return current, as any check may cause the current to seek easier paths through the earth and along water or gas mains, where it may cause electrolysis or electric corrosion.

Overhead Wire and Pole Lines.—In a great majority of cases, the current is distributed by an overhead wire and is taken off by means of a trolley pole or pantagraph on the car or locomotive. This wire, from 16 to 22 ft. above the rails, is suspended from bracket arms on poles beside or between the track, or from transverse span wires attached to pairs of poles. A pole spacing recommended by the American Electric Railway Association begins with 35 ft. for curves of 40 ft. radius and increases 5 ft. for each 10 ft. of radius up to 65 ft. for 100-ft. curves; then 70 and 75 ft. for 125 and 150 ft.; 80 ft. for radii of 200 to 500 ft., and 100 ft. for tangents and for curves having a radius of 750 ft. or more.

The trolley wire has considerable deflection or sag between its supports, but where long spans would make this sag excessive, the wire is sometimes kept practically level by suspending it from one or two catenary cables by means of hangers about 10 ft. apart and of such varying length as to compensate for the sag of the cables. These cables, with spans of 150 to 300 ft., may be carried by steel bridges or by cross-arms between pairs of side posts.

Wood poles for trolley lines are usually of cedar or chestnut, and if not creosoted, they should be painted with, or dipped in, tar or creosote to about a foot above the ground line. They are usually 5 to 7 in. in diameter at the top and 10 to 13 in. at the butt, and are sometimes dressed to a square or rectangular section. Steel poles of tubes or built-up shapes are used extensively, and concrete poles are used to a limited extent. In streets, the poles are placed usually at the curb line, but on open track they are $8\frac{1}{2}$ to 10 ft. clear from center of track. They are ordinarily 27 to 30 ft. long, or 40 to 50 ft. at road crossings and if they carry transmission lines.

Third-rail and Conduit Systems.—With the third-rail system, shoes on the car trucks slide upon a conductor rail laid at the side of the track and carried by insulating supports, every third or fifth tie being of extra length for this purpose. As a rule, this is a T-rail with the shoe riding upon the rail head, but double-headed and special sections are also used, sometimes hung from brackets so that the shoe rides against the bottom of the rail. These rails are usually of low-carbon steel on account of its greater conductivity. In some cases a protective covering is placed above and behind the third rail. This system is applicable only to open track on private right-of-way, or to elevated and subway lines in cities. Conduit systems, with the conductor in an underground conduit having a slotted opening in the pavement to admit a contact arm attached to the car, are used to a very limited extent and only in large cities, on account of the high cost.

INTERURBAN RAILWAYS

Track construction on electric interurban railways is similar to that of steam railways, having 75- to 100-lb. rails in 33-ft. lengths, sometimes 60-ft., with 17 to 18 ties per rail and 6 to 8 in. of ballast under the ties. Where these lines pass through small towns, the ordinary track construction is maintained, as a rule, but is covered with gravel or a concrete base for the paving. If there is a frequent service of heavy cars through the town, it may be economical to use concrete base and stiff rails. Curves on open line range up to 6 or 8 deg., but right-of-way conditions near terminals may necessitate curves as sharp as 12 to 16 deg.

Passing Sidings.—Double track is exceptional on lines of this class, but passing sidings 500 to 2,000 ft. long are provided at intervals of 2 to 5 mi. On some lines they are operated as stub sidings, the car backing out to proceed on the main track. Frogs of No. 8 or No. 10 are used for the turnouts. The proper spacing and length of passing sidings depend upon local operating conditions, the spacing being based upon scheduled speed and the headway between trains, but in many cases the construction of sidings is governed by financial rather than operating considerations. On double track and at sidings the tracks are usually spaced 12 ft. between centers.

Illinois Traction System.—The several lines in this extensive system are typical of interurban railways in rural districts and serving a number of towns. For open track, 70- to 80-lb. rails are used, with 100-lb. rails in paving, all 33 ft. long, with 20 ties 6 by 8 in., 8 ft. long, on 8 to 12 in. of ballast. Passing sidings 1,700 ft. long are spaced about 2 mi. apart and have No. 9 frogs. For superelevation on open line, the rule is 1 in. per degree up to a maximum of 6 in., but there is no rule for widening gage on curves. The sharpest curves are of 16 deg. Poles are spaced 100 to 130 ft. apart and $8\frac{1}{2}$ ft. clear from center of track. In paved streets, the track is as described, with creosoted ties, metal tie-plates, and not less than 8 in. of stone or coarse gravel ballast under the ties. Wood-block paving and brick, with asphalt filler, is laid on a sand cushion on 6 in. of concrete, but in small towns bituminous macadam or rock asphalt paving are used. Automatic block signals are used extensively.

An engineer of maintenance-of-way, reporting to the vice-president, has roadmasters in charge of divisions of about 100 mi. each, with track sections of 8 to 10 mi., averaging four men per section in summer and two men in winter, besides the section foreman. There is also a general foreman of bridges and buildings. Work trains are assigned through the cooperation of the offices of the engineer of maintenance-of-way and the general superintendent, while the work is directed by the roadmaster or the foreman of bridges and buildings. Maintenance of signals and trolley lines is under the electrical superintendent.

Chicago, Aurora & Elgin Ry.—This is a high-speed interurban line, operated on the third-rail system. It is largely double track, with cross-overs 4 mi. apart, while on single track there are sidings 700 to 5,000 ft. long at 4-mi. intervals. These turnouts have No. 10 spring-rail frogs. The track spacing is 13 ft. Rails are in 60-ft. lengths, with 28-in. splices, and with 32 oak or cedar ties to a rail length, with 6 to 9 in. of stone or gravel ballast under the ties. Besides the bonding of joints, the rails are cross-bonded every 500 ft. It is found that the 60-ft. rails are more difficult to keep in line than 33-ft. rails, but any difference in keeping surface is favorable to the 60-ft. rail. In lining track on the third-rail side, a chain is hooked over the flange of the track rail and passed under the third rail, so that men on the farther side can put their lining bars through a ring on the chain. The third rail is of 100-lb. A. S. C. E. section, in 33-ft. lengths, and has only 0.1 per cent carbon in order to give high conductivity. It is carried on insulating supports, every fifth tie being 9 ft. long; its head is $11\frac{3}{16}$ in. above the ties and $10\frac{1}{8}$ in. from the gage side of the track rail. At grade crossings the third rail is cut, the last rail being dropped 2 in. in its length and fitted with a 24-in. incline to engage the contact shoes on the cars. Current is carried across the road by cables. Ordinary curves are of 1 to 4 deg., with 16 deg. near terminals. A super-elevation of 1 in. per degree was given originally, but for high speeds it was found necessary to increase this to 3 in. for 2-deg. curves. This elevation is run out at 1 in. in 50 ft., and the ends of the curves are spiraled.

Chicago, North Shore & Milwaukee R. R.—This also is a double-track, high-speed line, but is operated on the overhead trolley system. Its standard track has 100-lb., 33-ft. rails, with 18 ties per rail and 6 in. of ballast under the ties. Turnouts for sidings or spurs have No. 10 frogs. For curves, the superelevation is determined by the formula: $0.00066 DS^2$ (see "Curves"), and gage is widened $\frac{1}{8}$ in. for each 2 deg. over 8 deg., to a maximum of 4 ft. $9\frac{1}{4}$ in. The sharpest curves are of 263-ft. radius on open track and 50-ft. on track in city paving. In paved streets of small towns, white oak ties 6 by 8 in. and 8 ft. long are laid on 6 in. of stone ballast and embedded in concrete, with a 1-in. sand cushion for brick paving with cement grout filler. Poles are spaced 180 and 100 ft. apart for steel and wood, respectively, and 10 and 9 ft. clear from center of track, carrying the trolley wire 22 ft. above the rails. Maintenance is in charge of an engineer of maintenance-of-way, with two roadmasters for 70 mi. of main line and 14 mi. of branch line, all double track. Track sections are 6 mi. long in open country and $4\frac{1}{2}$ to $5\frac{1}{4}$ mi. in and near towns; they have an average of seven to eight men per section in summer and five in winter. The average tool equipment of a section gang on this road is given in Table XIII.

TABLE XIII.—EQUIPMENT OF SECTION GANG ON ELECTRIC RAILWAY

2 Adzes	1 Grindstone	4 Snaths
1 Axe, chopping	4 Spike hammers	8 Track shovels
1 Hand axe	1 Nail hammer, claw	4 Scoop shovels
2 Claw bars	1 Hatchet	1 Long-handled shovel
6 Pinch bars	1 Track jack	1 Hack-saw frame
1 Raising bar	1 Switch key	12 Hack-saw blades
8 Tamping bars	2 White lanterns	1 Hand saw
2 Brooms (coarse)	2 Red lanterns	1 Spade
12 Switch brooms	2 Green lanterns	1 Tape line, 50-ft.
1 Hand car	1 Level board	4 Rail tongs
1 Push car	2 Switch locks	1 Tool box
2 Car chains	2 Mattocks	24 Torpedoes, with box
12 Track chisels	1 Oil can, 1 gal.	1 Spike puller
2 Cold chisels	1 Oil can, 2 gal.	1 Vise
1 Chalk line, 100-ft.	8 Picks	1 Water pail or jug
2 Cups or dippers	8 Tamping picks	6 Weed scuffles
3 Files	1 Hand punch	3 Wheelbarrows
4 Flags, red	1 Rake (garden)	4 Whetstones
2 Flags, green	1 Ratchet	3 Track wrenches
4 Ballast forks	6 Ratchet bits	1 Monkey wrench, 8-in.
1 Track gage	4 Grass scythes	

STREET RAILWAYS

Track in Paved Streets.—Track construction for electric railways in the larger cities follows no uniform type, but varies according to local ideas, paving systems, and municipal regulations. A firm and well-drained foundation is necessary, since the rails must conform to the level of the paving and there is no opportunity for the continual tamping and other maintenance work done on open track. For this reason the subgrade is well rolled and may have a trench for a pipe drain which is surrounded with broken stone and leads to sewer connections. Gratings in the track paving may divert storm water to this drain. A bed of stone or gravel rolled to a depth of 6 to 8 in. supports the ties, similar material being filled between them to carry a concrete base for brick, stone, wood block, asphalt, bituminous macadam, or other paving. This concrete may also form the paving surface. In small towns the track is generally of T-rail construction, with gravel or stone filling to carry the paving.

Where a concrete foundation is used, instead of stone or gravel, it may enclose the steel or wood ties, and sometimes also the base of the rail, while in certain designs it has a curb wall along each side to retain the ballast and to make a separation between the paving of the track and that of the sides of the street. Ties are omitted in one type of concrete construction, used to a limited extent, the rails being laid on the concrete base and secured by anchor bolts. In a modification of this plan, concrete blocks or a few widely spaced ties hold the rails, which are carefully lined with the aid of a transit and then braced in position while concrete is poured under and around them. Tie rods holding the track to gage are placed about 5 ft. apart on tangents and $2\frac{1}{2}$ to 3 ft. on curves.

Flangeways in the paving, where grooved rails are not used, may be formed by nose bricks or blocks beveled to fit under the rail head, or by various arrangements. With concrete paving, a wooden strip of the desired shape may be placed against the rail head; or the wet concrete may be shaped by a template or by a roller which rides on the rails and has ribs or flanges of shape corresponding to the groove to be formed. The groove or flangeway is usually about $1\frac{1}{2}$ in. deep, or to the bottom of the rail head, and slopes up to the surface in a width of about 5 in., a broad flaring groove being usually better than a narrow groove. Car wheels on street railways have smaller flanges than those of interurban lines, but where the latter cars run over city tracks, all flangeways must be designed to accommodate the larger flanges in order to prevent damage to the paving. On the outside of the rail, the paving is about $\frac{1}{4}$ in. below the top of the rail, so that with worn wheels and rails the wheel treads will not ride on the paving. As a rule, the track space is level and the sides of the street slope down to the curb or gutter.

Rails and Rail Joints.—Four different types of rail sections are shown in Fig. 94: (1) ordinary T-rails of 70- to 100-lb. sections; (2) T-girder rails 7 to 9 in. high, to give greater stiffness; (3) grooved girder rails 6 to $10\frac{1}{2}$ in. high, weighing 75 to 150 lb. per yard and having grooved heads for the wheel flanges; (4) tram girder rails of about the same size and weight, but having a step-shaped head. High, stiff rail sections are commonly used for heavy traffic. With T-rails, flangeway grooves must be formed in the paving, or the paving between the rails may be lower than that of the main part of the street, this latter plan being used also with tram or step rails. Grooved rails have usually a groove about $1\frac{1}{8}$ in. deep and flaring to a top width of $1\frac{5}{8}$ in., with the guard or outer side of the groove about $\frac{1}{4}$ in. lower than the rail head. This guard may be a narrow bar (heavier for curves) or it may be extended as a flange wide enough to carry wheels and prevent them from cutting ruts in the paving along the rail head. With T-rails in asphalt paving, this action is prevented usually by lines of stone or brick blocks laid against the rails. Various forms of compound track have been tried, with renewable rails of special section attached to supporting chairs which form a permanent part of the track, but such construction is now obsolete.

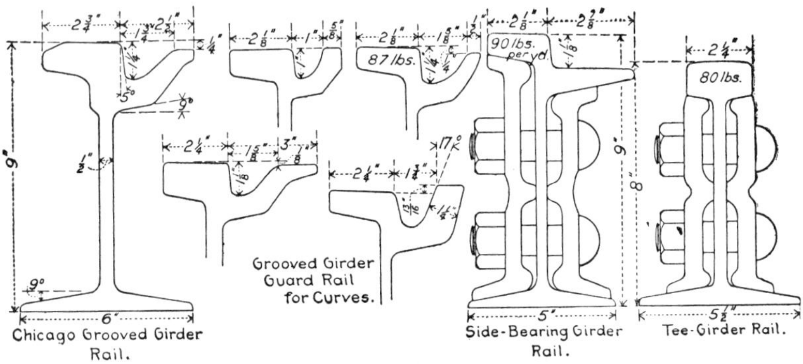

FIG. 94.—Rails for street railway track.

Corrugation of rail heads is a peculiar form of wear (see "Rails") that is often experienced on street railways, both with soft or flexible roadbed and with rigid concrete base. One cause appears to be a tendency to lay track to a tight gage, about $\frac{1}{8}$ in. less than standard, causing the wheels to rise and fall as the flanges are "pinched." This section gives a variable load on the rails and sets up considerable vibration. If the wheels can force the rails outward, the gage may be widened before corrugation results, but where the rails are held rigidly, being perhaps partly embedded in concrete, the tight gage is likely to cause trouble. In busy streets, the rails are subject to severe wear by the continual passage of heavy cars and iron-tired wagons, and by the grinding of dirt and grit under the wheels.

Rail-joint design and maintenance demand special attention for street railways, since low joints make a rough-riding track but cannot be continually tamped or repaired as on open track. Rails 60 and 66 ft. long are used frequently to reduce the number of joints, and various methods have been designed for making the joints rigid and permanent. Welding is used extensively, mainly by the electric and autogenous processes, but also by the cast-weld process of pouring molten iron in a mold around the rail ends. With electric welding, the rails may be actually welded together or splice bars may be welded to them, while bolts are sometimes spot welded to prevent slackening. Welded joints are sometimes set a little high, or cambered, and then ground down to normal level. Riveted splices are also used, made with 32-in. bars and 10 rivets; the contact surfaces are cleaned by sand blasting, and molten spelter is poured between the rails and splice bars after the joint has been riveted. Bolted splice joints for very high rails have two rows of bolts, and the splice bars may be stiffened laterally by a horizontal rib at midheight, bearing against the web of the rail.

Ties and Special Work.—Wood ties are usually sawed, 6 by 8 in., and 7 to 8 ft. long, and there is increasing use of treated ties. The spacing is usually 18 to 36 in., or as much as 5 to 10 ft. if the ties serve mainly to hold the rails while a concrete base is being poured. A graded spacing used with creosoted ties and 66-ft. rails at New Orleans, with concrete base, provides 18 in. at joints, 26 and 32 in. for the two shoulder ties on each side and 36 to 38 in. for intermediate ties. With slag or stone base, this is reduced to 18-in. joint, 24-in. shoulder, and 26-in. intermediate spacing. Tie-plates are commonly used, with combined tie-plates and rail braces for high rails not connected by tie rods. Both screw and drive spikes are used, the former holding the rail directly or by heavy clips. Steel ties of I-beam section or built of angles and other shapes are used to some extent, usually embedded in concrete. Such steel and concrete construction permits the use of shallow ties and a thin bed of concrete, with reduced thickness of track. Steel ties are sometimes welded to the rails to help in carrying the return current.

Switches in paved streets have usually a single pivoted tongue moving in the grooved surface of a base casting, the opposite side of the turnout having a fixed point or mate like a long frog. Frogs, switches, turnouts, and track crossings or intersections, which constitute what is termed special work, are largely of girder-rail type, with solid or rail-bound castings of specially hard steel for parts subjected to severe wear. Manganese steel is used extensively, and also chrome-nickel steel on account of its combined hardness and strength or toughness and the fact that it is more readily welded than manganese steel. All such complicated special work needs to be designed, built, and assembled with great accuracy, large plans on a scale of 5 to 10 ft. to the inch being used frequently. It is estimated that in reconstructing special work in streets, 50 per cent of the cost is for removing old work, excavating, installing, and paving; 35 per cent for switches, mates, and frogs, and 15 per cent for rails, joints, and fastenings.

Gage, Curves, Grades, and Sidings.—Sharp curves of 35 to 75 ft. radius, required in turning street corners and at terminal loops, are sometimes spiraled for easy riding. The track spacing is usually 10 ft., but it may be necessary to increase this on sharp curves to enable long cars to pass. Probably the maximum grades operated by electric cars without assistance are 14.6 and 15.5 per cent, at San Francisco, but an electric line with light cars and cable counterweight system at the same city has grades of 24 and 25 per cent. Cable lines, which are now almost obsolete except for occasional steep-grade lines, operate on grades of 20 per cent. Standard gage is commonly used, but various lines have gages from $3\frac{1}{2}$ to $5\frac{1}{2}$ ft. Passing sidings on single track may be arranged in three ways: (1) with the main track a continuous straight line and the siding on one side; (2) with both tracks diverted an equal distance from the center line; (3) with main track offset to overlap for the length of the siding, so that cars in both directions continue on a straight line through the facing switch.

Examples of Track Construction.—At Indianapolis, the standard construction (Fig. 95) consists of 7-in. T-girder rails with thermit or feralite welded joints and with white oak ties 6 by 8 in., 8 ft. long, spaced 2 ft. c. to c. On a concrete base 6 to 9 in. thick is a 2-in. layer of dry concrete (1 cement to 10 sand and gravel) which is tamped under the ties by pneumatic tampers. A wet concrete (1 to 7) is then filled flush with the ties as a base for brick, granite block, or asphalt paving. To reduce the depth of track construction at Kansas City, shallow steel ties were substituted for wood ties, as shown also in Fig. 95. Each tie, 3 ft. wide, consists of two 3-in. channels connected by two 36-in. top plates supporting the rails. These ties are spaced 6 ft. c. to c. and those at the joints are welded to the rails. The main body of 1:3:6 concrete is finished with a 2-in. wearing surface of 1:2:3 concrete, which is kept $\frac{1}{4}$ in. below the rail head on the outside of the track. A design used at Toledo has the concrete formed as a stringer under each rail, $8\frac{1}{2}$ in. deep below the steel ties and 12 in. wide at the base, the concrete slab between the stringers being only $1\frac{1}{2}$ in. below the ties.

At Milwaukee, the standard track consists of 102-lb., $7\frac{1}{4}$-in. T-girder rails on shoulder tie plates, with pine ties 6 by 8 in., 7 ft. long, spaced 24 in. c. to c. Screw spikes hold the rails, while gage is maintained by tie rods 6 ft. apart. Crushed stone ballast rolled and tamped to a depth of 6 in. is covered with 2 in. of finer stone on which the ties are laid, after which a 6-in. concrete bed is placed as a base for paving. If the concrete is to form the paving, it is 7 in. thick over the ties and 10 in. between them, wooden templates being used to form flaring flangeways 5 in. wide and $1\frac{1}{2}$ in. deep along the rail heads. For the Detroit municipal railways, the track is assembled on concrete blocks placed on a 2-in. slag bed, and concrete is then filled in, extending over the rail base. Crushed granite of 2- to 3-in. size is spread and rolled and then grouted with cement and again rolled, an 8-ton roller being used for all this work. A dense concrete is obtained by rolling, while the large-sized hard stone gives a good wearing surface. Such a surface may be obtained also by spreading coarse, hard, broken stone on the wet concrete and driving it in with vibrating or tamping machines. In some cases moist concrete is tamped under the ties like stone ballast, and traffic is turned onto the track at once, instead of leaving the concrete several days for setting. At Baltimore, 7-in. grooved-girder rails with electrically welded joints are laid on wood ties 24 in. c. to c., with stone ballast 6 in. deep tamped under the ties by pneumatic tampers. With asphalt paving, two rows of granite stretcher

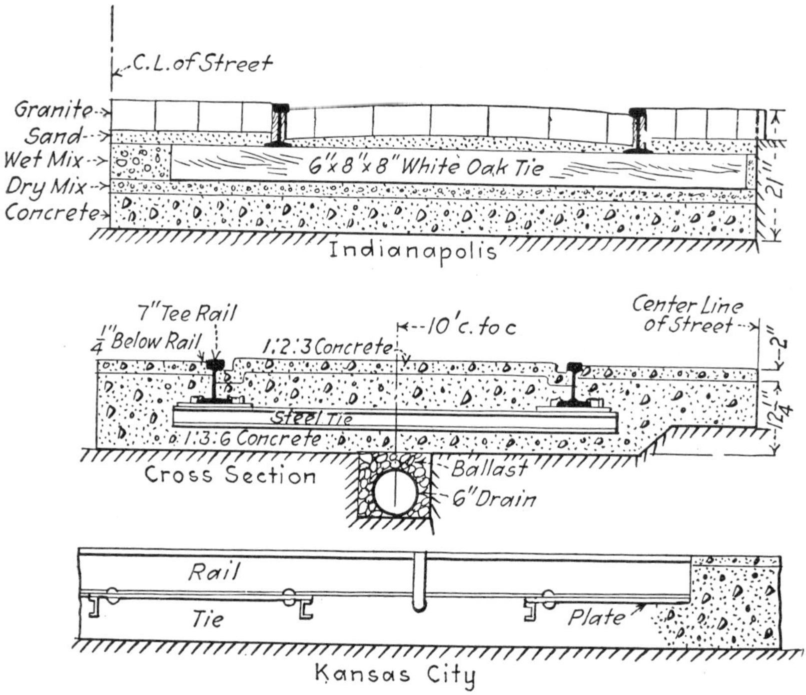

FIG. 95.—Types of street railway track.

blocks are laid on each side of the rail, but with granite paving the blocks are laid against the rails as headers in the ordinary way.

Maintenance and Reconstruction.—For small repairs in busy streets the paving can be removed in patches and the work done without interrupting traffic. Extensive repair or reconstruction usually requires tearing up the whole street. Rapid but thorough work is necessary in such cases, and also in the repair of intersections, as it obstructs street traffic and interferes with business on the street. Closing the street and detouring or rerouting cars during such work is undesirable, and a general plan is to lay a temporary track along one side of the street, all cars being handled on this one track under protection of flags or signals and the track work being thus clear of all traffic. Such work is done sometimes under traffic; on double track, traffic may be handled on one track while the work is in progress on the other, but this causes more or less interference. For concreting, street traffic may be excluded from a stretch of one or two blocks for about 10 days, during which time all work is completed; this is likely to be less troublesome to the shopkeepers and the construction force than if the street is kept open and the work is continued for a longer period owing to the interference of traffic.

Track reconstruction work is aided greatly by the use of machinery, which not only hastens progress but may reduce the labor force by 30 to 50 per cent. Paving may be cut and torn up, concrete base broken, track ripped out or post holes excavated by special machines; power shovels remove debris and grade new roadbed, loading the material into trucks or dump cars. Old rails may be cut by torches and loaded on flat cars, which also bring in new rails. Material for concrete may be hauled in by trucks or service trains and placed in stock piles, to be measured and loaded mechanically for delivery to concrete mixers which travel on the rails or beside the track and either discharge the concrete into place or deliver it to dump carts. Ballast may be tamped, ties bored, and rails drilled by electric or pneumatic machines, which may also drive screw spikes. Rails, frogs, and special work may be handled by cranes or crane cars, while material for maintenance and repair can be distributed by motor trucks and cars. Whether such service can be handled better by motor truck or car equipment will depend largely upon local conditions.

Organization on Street Railways.—In the organization and equipment for maintenance-of-way and reconstruction on street railways, there is greater diversity of practice than on steam or main-line railways, since the size, character, and local conditions vary so widely with individual lines. The close subdivision into track sections is not followed, and there is more extensive use of gangs assigned to special kinds of work.

Baltimore Electric Railways.—The system of the United Railways & Electric Co., at Baltimore, Md., consists of 250 mi. of track (on a single-track basis) in paved streets, with girder rails, and a total of 410 mi. including suburban lines which have T-rails. Four city and four suburban sections range from 37 to 58 mi. There is a chief engineer, who has a principal assistant engineer, a bridge engineer, a roadmaster, and a superintendent of work cars.

Maintenance work is under the roadmaster, who has jurisdiction over the entire system. Each of the four city divisions is headed by a division foreman, and each suburban division is headed by a section foreman. Each foreman has on an average about three gang foremen, and the number of men varies with the time of year and amount of work on hand. The divisions and sections are not further subdivided, but the work is assigned to the gang foremen by the division and section foremen. In addition there is a curve-cleaning force of about 35 men under the superintendent of curve cleaning, who reports to the roadmaster. There is also a special gang, in charge of a foreman reporting to the roadmaster, which has charge of such work as rail welding, grinding, and renewing hard-center plates in track crossings.

Reconstruction work is handled by a separate organization under the jurisdiction of the principal assistant engineer. This force has a supervising foreman and gang foreman, the number of men varying with the amount of work on hand. The skeleton of this organization is carried through the winter months and may at times be employed on maintenance work, although the climate generally permits a limited amount of reconstruction work even in the winter. Bridge carpenters and masons, consisting of about 25 men, report to the bridge engineer. Work-car motormen and conductors are in charge of the superintendent of work cars, who also has charge of the handling of material from the material yard to the job where it is to be used.

In planning the work, a general outline of maintenance for the entire year is based on an annual inspection made as early in January as weather conditions will permit. The detail work is then planned at weekly conferences, and is still further detailed by daily meetings of the division and section foremen with the roadmaster. Track reconstruction work is also planned on the basis of an annual inspection, but is generally changed considerably by the city's paving program. Maintenance-of-way work is done on the budget system, the budget being approved by the president from month to month. Reconstruction work is done by special authorization. Equipment for all this work includes crane cars, dump cars, flat cars, trucks, electric shovels, gasoline road roller, pavement plow, portable air compressors for operating tie tampers, concrete breakers, electric welders, grinder, concrete mixers, and miscellaneous tools.

Philadelphia Rapid Transit Co.—In the organization of the maintenance-of-way department, the entire surface line system of 631 mi. is under an engineer of way, and is in six divisions, each with a superintendent in direct charge of all track and paving work. Five of these divisions have approximately 120 mi. of track each, the other division being in a suburban district. In addition there are 26 mi. in a suburban district, and 39 mi. of track on elevated road; each of these has its separate departments. For the city surface lines, the work for each month is outlined or scheduled by the engineer in frequent consultation with his superintendents; written and verbal orders are issued daily as required. Each division has men assigned to the cleaning and oiling of curves and switches outside the regular repair work. Each division also has its welding force, under a foreman directly responsible to the superintendent for all welding and bonding work within the division. Record of all work done is forwarded from the division offices to the general way office, where various reports are compiled. Charges against the individual jobs are forwarded direct by the division timekeepers to the accounting department on individual time cards, and from these are compiled the costs of the various items of work.

Other City Railways.—At Milwaukee there is a superintendent of way and structures, with a general roadmaster, three division roadmasters, and six track foremen. The Kansas City Railways Co., with 317 mi., of which 230 mi. are in paved streets, has a chief engineer with an assistant engineer, office engineer, assistant superintendent of way and structures, supervisor of work trains, and three roadmasters. Each roadmaster has about six foremen with floating gangs, different gangs being assigned for particular kinds of service, such as repairs to straight track or renewal of special work. The maintenance department has in all about 500 men. All power-operated equipment is in charge of a separate division, which furnishes the necessary service to the roadmaster.

PART II

MAINTENANCE WORK AND ECONOMICS

CHAPTER XV

ENGINEERING AND MAINTENANCE-OF-WAY ORGANIZATION

Division and Department Systems.—In the organization of the railway service, there are two distinct systems; the divisional and the departmental. In the former, each operating division is practically an independent unit in charge of the division superintendent, who has under him a division engineer, division master mechanic, and trainmasters. The head or system organization, then, may comprise a general manager, a chief engineer, an engineer of maintenance-of-way, a mechanical superintendent, and a general superintendent. In the departmental system, division officers report directly to the heads of their departments; division engineers to the chief engineer or engineer of maintenance-of-way, master mechanics to the mechanical superintendent, and division superintendents to the general superintendent. Besides the usual branches of the engineering department, such as bridge engineer and valuation engineer, special branches are featured on some roads to meet special conditions; thus there may be an engineer of track elevation, a drainage engineer (on lines in wet or swampy country), a sanitary engineer (for malaria-mosquito control and sanitation of camp trains, stations, and water supply), a landscape engineer, a locating engineer, an engineer of construction, or an engineer accountant.

Maintenance-of-way is organized as a separate department, as a section of the engineering department, or as a section of the operating department. Logically, it belongs under engineering, and on large roads, where the chief engineer has wide responsibilities, there may be also a chief engineer of maintenance-of-way, both engineers reporting directly to executive officers. In such cases, the chief engineer has jurisdiction over new construction, improvements, bridges, drainage, contracts, standards, valuation, right-of-way records, etc. On smaller roads, the organization is simpler and is more likely to be departmental. In general, it is better to have maintenance-of-way directly under the engineering (or maintenance) department rather than under the division superintendents, since the latter are more particularly interested in, and concerned with, matters of operation and traffic. The division engineer, however, may report to the superintendent as to work done and planned and at the same time be responsible to the engineer of maintenance-of-way (or similar official) for proper methods of carrying on the work.

Organization systems vary widely, as will be seen from specific examples given at the end of the chapter. Particulars of organization on electric railways are given in Chap. XIV.

Engineers of Maintenance-of-way.—In general, each operating division has an engineer responsible for maintenance of track and structures. On the Chicago & Northwestern Ry., the roadmasters are under the division engineers, who report on maintenance matters to the engineer of maintenance-of-way. On the Pennsylvania R. R. System, the division engineer, reporting to the superintendent, has charge of the maintenance of all property under the latter. This includes the routine maintenance of the telegraph and telephone facilities, which is handled directly by the supervisor of telegraph and signals. He is in charge of the operation of water stations and does a limited amount of construction work, depending upon local conditions. On different roads, various titles and positions are included in maintenance-of-way service, according to local organization or conditions; in some cases there is an engineer of track and roadway.

On the Cleveland, Cincinnati, Chicago & St. Louis Ry., maintenance-of-way is under the chief engineer, who has on each division an engineer of maintenance-of-way in charge of construction, renewal, and maintenance of roadway, track, bridges, buildings, water supply, coaling stations, track scales, signals, and other structures and appliances. He is responsible for the safe, efficient, and economical maintenance of the property and for the construction of such improvements as may be assigned to him, but heavy construction work is assigned to a separate construction department. The engineer sees that standard and special plans are furnished, and that they are followed by those in charge of the work; he makes such inspections as are necessary to enable him to direct the work on the division, prepares reports as required, keeps detailed accounts of maintenance and other work, reports derailments, and, if practicable, goes to wrecks and washouts where serious damage has occurred to track and structure. Under him is an assistant engineer of maintenance-of-way, who is responsible for the work of the engineer corps of the division and sees that lines and levels required are given promptly and accurately; he keeps complete files of standard plans and specifications and property plans, making such periodical corrections as may be necessary, and he also keeps records of tracks and structures and other engineering information. He is in responsible charge during the absence of the engineer of maintenance-of-way.

A monthly allowance is assigned for the engineer of maintenance-of-way, from which he is required to meet ordinary expenses for the operation of the division and for the construction of new work under his jurisdiction. He is required also to be familiar with all industrial tracks and crossings and to know that they are covered by agreements with their owners. He has charge of assistant engineers, track supervisors, and other subordinate officers, and one of his duties is to confer with representatives of labor organizations regarding grievances or promotions. While his office prepares such regular and special reports as may be required by the railway, state laws, or the Interstate Commerce Commission, the only report issued personally by the engineer of maintenance of-way is a weekly letter to the chief engineer concerning progress in maintenance and construction work under his charge.

Engineers in Track Work.—Modern maintenance-of-way requires a high degree of technical knowledge, but the engineer in charge should have also practical experience with track work and track forces, though he need not be an expert trackman. A good plan is to select men from the engineer corps to fill a certain percentage of the positions of supervisors of track, water service, signals, or bridges and buildings, on an equal footing with other applicants. This plan should be handled on a definite system, including office experience, so that the young man becomes familiar with standard plans, business methods, and economic aspects, as well as with the rules and methods of track work and the handling of men. It should be recognized, however, that a man with such training is a valuable asset to the railroad, and if he undertakes this training (beginning perhaps in the rough work on the track), it is only fair that the railroad should give assurance of permanent employment, promotion, reasonably good pay, and allowance for the value of his experience.

On some roads, the young engineer begins as roadmaster's clerk and then works up as foreman, general foreman, or assistant roadmaster, with the maintenance-of-way department as his objective; or engineering students may be worked in the section gangs in the summer, those who prove satisfactory being offered employment after graduation, although such opportunities should by no means be limited to graduate engineers.

To the engineer belongs the important duty of organizing and superintending the track work, seeing that it is done according to instructions and with regard to economy and efficiency. His subordinates are responsible to him for the details, but he must have some personal knowledge of such details. It is for him to study the economic relations of track to rolling equipment and traffic and to maintain good and safe track even under adverse conditions. In view of the large proportion of operating revenue absorbed in operating expenses and of the great aggregate expenditure on maintenance-of-way and structures, it is important that the officer in charge should be qualified to consider the combination of efficiency and economy. From his training and experience, the engineer should be specially qualified for such work, as well as for the equally important work of managing men, although too many engineers limit themselves to the technical side of their work, to the neglect of its executive or economic aspects. In other lines of railway work, such as construction, improvements, and bridge renewals, the engineer has to organize forces, plan and schedule the work, control and record the expenditures, direct large bodies of men, and provide for carrying on the work with rapidity and a minimum of interference with traffic. In the maintenance-of-way department, he may have the same exercise of his ability, but although there is growing recognition of the value of technically trained men, comparatively few railroads have a definite policy or practice of offering opportunity or encouragement to young men to make a specialty of this department.

ROADMASTERS

The roadmaster (called the track supervisor on some roads) is the direct head of track work on his road or division, and as the intermediary between the engineering and operating officers and the working force, he should be not only competent as to the work, but of such character that his men will feel he is friendly and just in his dealings with them. As to whether he should be an engineer or what is termed a "practical" man, promoted from section foreman, both may be equally good; but the former has the advantage of technical training, which is specially important under modern conditions, and should have had practical experience, as already noted. The engineer is likely to be more familiar with the principles of curve and switch work, yard design, the problems of tie preservation, and the relation of rail sections to wear of wheels and rails. He is familiar also with instruments and mathematical work, but needs to realize the limits of precision in track work so that he will not figure a turnout lead too closely, when a little variation may save the cutting of a rail and will make no difference in the turnout. In the same way, he must understand the limitations which traffic and operating conditions impose upon superelevation and the widening of gage on curves. The more general practice, however, is to promote roadmasters from among the section foremen, care being taken to select or train men of particular ambition and capability.

In regard to the work of the roadmaster, he is responsible for the care, maintenance, and safe and neat condition of the roadbed, tracks, and track appliances, right-of-way, crossings, fences, cattleguards, track signs, yards, and station grounds; also for the tools, tool houses, section boarding houses, camp cars, and work trains. He employs section foremen, road watchmen, and foremen of extra gangs and work trains, and is responsible for them and the laborers under them. On smaller roads, his work may include also stations, bridges, trestles, culverts, turntables, coal and water stations, and minor structures.

Much of his time is spent in going over his division on trains and going over every section periodically on foot or on his motor car. If any section is not maintained properly, he must walk over it with the foreman and plan its early improvement. Prompt report is made of any defective conditions likely to delay or endanger trains, and of any evidence of damage to track by worn wheels or other defects in rolling stock. He records the dates of his trips, with method of traveling, and notes such matters as progress of important work. Proper time must be given also to his office duties in checking time books, payrolls, and requisitions, making reports, and attending to correspondence. He has usually a clerk as office assistant. He informs the division engineer and superintendent daily as to his probable movements.

The roadmaster must be thoroughly informed as to the physical condition of the line and the daily disposition of his forces. During his trips, he discusses the details of work with the foremen, calling attention to necessary or defective work, instructing (but not "nagging") and seeing that his orders are understood and properly executed. He compares their watches with his, which should be inspected every 6 months. He also inquires as to their plans for future work and notes if they have too few or too many men. He does not issue orders to the laborers, as the foreman is the head of the gang. The roadmaster also instructs new foremen as to work and rules, and sees that they are supplied with the rule book, time books, and forms for reports. When a foreman leaves, the roadmaster sees that his time books are correct and that tools and other company property are accounted for. A report is also made, giving the cause of leaving. Changes in foremen should be made on the first of the month. Record is kept and monthly report made of all foremen, noting their employ, discharge, resignation, promotion, transfer, or injury, and also their character or ability; a card index is useful for this purpose.

The roadmaster sees that all section and other foremen under him are supplied each month with the necessary time books and blank forms for reports. During his trips, he should examine these to see that they are being filled in correctly. These returns are sent to him on the last day of the month (or on specified dates where laws require more frequent payments of wages), and after examination, he forwards them to the proper officers. He accompanies the pay car to identify the foremen and settle disputes. He also accompanies the supply train if requisitions for tools and supplies are filled in this way. He personally receives and inspects rails, ties, ballast, and other materials delivered on his division and attends to the unloading and storing. Requests for section men to assist fence or bridge gangs or to clear yards and station grounds should be made to the roadmaster and not to the foremen.

In inspecting the track, he notes if it is in proper condition as to line and surface, gage and superelevation on curves, frogs and switches, policing, and other details; when on a train, he notes any points that do not ride easily. Lamps and tool houses are inspected; also camp cars and boarding trains. He plans and supervises new work ordered, such as ballasting, renewing rails, putting in switch or interlocking work, or laying additional tracks. He has full charge of work trains, lays out their work, and makes orders for running them. If unserviceable locomotives are assigned, this must be reported, as these trains are expensive and need good engines for economical operation.

In track-circuit territory, the roadmaster notifies the signal supervisor before doing work which may interfere with the signal system. Work being done by contractors must be watched to see that it does not interfere with the safety of track or trains. He notes in a journal the location and dates of beginning and completing all such work as rail and tie renewals, line revision, ballasting, or laying sidings. Ties marked for renewal by the section foreman are inspected by the roadmaster as a basis for his requisitions for new ties. He may be directed to investigate the wear of rails or the performance of special or experimental appliances.

A special inspection of all waterways is made before the winter or rainy season, any necessary repairs being made and obstructions removed. In storm, flood, or heavy snow, the roadmaster keeps the superintendent advised of the condition of the road and sees to the distribution of trackmen to help the snow or repair gangs. Accidents attributable to track or structures must be investigated and reported on. Broken rails especially must be studied carefully, report made, and the rail set aside for examination. On receiving notice of a wreck, he proceeds to it at once, taking charge of clearing the track or building a runaround or shoofly track with the least delay. Broken rails, switches, axles, or other material which may be useful in determining the cause of an accident must be preserved. This outline of his duties and responsibilities may seem somewhat overwhelming, but as a matter of fact much of his work is of a routine or supervisory character, which is more easy to do than to describe.

ORGANIZATION FOR TRACK WORK

Organization of the track forces is practically uniform, the line being divided into sections, each having a foreman and laborers constituting the section gang. For special or heavy work there are supplemental floating or extra gangs which move from place to place. These section gangs and extra gangs are in charge of roadmasters (or supervisors) having 50 to 200 mi. of line, or 30 to 60 mi. of four-track. This length may include the full or partial length of an operating division; in the latter case, there may be division roadmasters, above whom, on comparatively few roads, are general roadmasters. The length of track section varies with the number of tracks, character of track construction, amount of traffic, and number of frogs and switches; it averages 4 to 8 mi. on single-track, 3 to 5 mi. on double-track, and 2 to 4 mi. on four-track lines. All work does not vary directly with the number of tracks; ditching, clearing right-of-way, and cutting grass and weeds are about the same in any case.

The number of men in a section gang varies according to the season and the conditions of track and traffic. In the spring, there may be 5 to 10 or 12 men; in winter from 2 to 4, and where traffic is very light, there may be only one laborer, or the foreman may be left alone, but given authority to employ men in emergency. In the warmer parts of the country, there is less seasonal variation, while there is a growing tendency to reduce the variation even in colder parts in order to secure a better distribution of the work. For the working season, the average is $1\frac{1}{4}$ to 2 men per mile of single main track. A large yard or terminal may have its own section gang. At the Indianapolis Union Station, where there is heavy traffic at all seasons, a foreman with 15 to 22 men maintains $14\frac{1}{2}$ mi. of track having 86 switches, 26 slip crossings, and 23 solid and movable point crossings.

Practically the only variation from the system described is where floating gangs do nearly all the regular track work, with small section gangs for trackwalking, tightening bolts and spikes, general lining and surfacing, inspecting, and policing. This plan is very little used. The weight of opinion is decidedly in favor of having larger section gangs to do as much regular work as possible, putting responsibility on the section foremen. In Table XIV is shown the distribution of track forces on different railroads.

TABLE XIV.—ORGANIZATION OF TRACK FORCES

Railroad	Roadmaster's divisions, miles	Track sections, miles	Section gangs	
			Spring, men	Winter, men
Chicago, Burlington and Quincy	100 to 300	D.T.; 3 to 6	5 to 10	2 to 3
Chicago, Burl. & Quincy	100 to 300	S.T.; 6 to 12	3 to 6	1 to 3
Southern Pacific	D.T. 60	4 to 5	....	....
Southern Pacific	S.T. 183	7 to 10*	....	....
Baltimore & Ohio	25 to 130	D.T., 2 to 4	....	....
Baltimore & Ohio	25 to 130	S.T.; 5 to 8	....	....
Pennsylvania	25 to 75	D.T. & 4-T.; $2\frac{1}{4}$	5 to 9	....
Michigan Central	Ave. 100	D.T.; 4	8	6
Michigan Central	Ave. 100	S.T.; 5	4	3
Erie	50	D.T.; 4	8 to 12	4
Erie	50	S.T.; 8 to 12	8 to 12	4

* Southern Pacific Ry.; branch line sections, 12 to 14 miles.

SECTION FOREMEN

The section foreman in charge of the working forces on the track needs to be competent and trustworthy, able to rule his men and have their respect, and also able to read and write and to handle the necessary reports and accounts. Procuring new foremen who meet these requirements is a difficult problem on many roads, owing to the low rate of wages in track service. With the cooperation of the foreman and the roadmaster, a laborer who shows ambition and executive ability may be trained in various lines of work, taught to think and to plan, and eventually made assistant foreman with a slight increase in pay, although some roads have found that variations of wages in the gang cause dissatisfaction.

For training trackmen, both to get good laborers and to develop a supply of section foremen, the Union Pacific, the Pennsylvania, and other roads have organized bureaus with correspondence courses in general education and railway operation, giving instructions in track work, with simple questions and problems that will cause the men to think. Foremen having such training in addition to their practical knowledge are likely to be specially efficient and capable. Annual track inspections, with prizes, are helpful; also periodical meetings of foremen and laborers with the roadmaster and other officials for free discussion on railway affairs and track work. It is a good plan to establish a graded rate of pay, giving higher wages to foremen who remain in the service for 3 to 5 years. Track foremen, foremen of extra gangs, and supervisors of water service, signals, bridges, and buildings should be trained to plan their work in advance.

The section foreman walks over his section at least once a week, or daily if no trackwalker is employed. This time will depend upon the season, length of section, work in hand, condition of track, and amount of traffic. He also makes a monthly inspection and report on culverts, trestles, bridges, and tunnels. He should be subject only to the orders of the roadmaster. Train dispatchers, station agents, bridge foremen, and others should have no authority to give him orders or to call upon him to supply men, except in case of emergency. Where two or more section gangs work together, as in case of accident, the foreman of the section on which the work is being done is in charge. With a gang of four or five men and such simple work as cutting grass or cleaning up, the foreman may work with the gang and at the same time watch the progress. But it is his main duty to direct and supervise the operations. On a busy line and at such dangerous places as tunnels, he sees that the men are warned of approaching trains and that they get clear of the track and have all tools clear before the train is dangerously near. When the work makes it necessary for trains to reduce speed, he sets out the necessary "slow" signs and flagmen. If the men do not live at a section house, he must know their addresses and arrange for one man to call others in case of emergency at night.

The section foreman is responsible for the inspection and safety of track and structures. It should be his aim not only to keep the track in safe running condition, but steadily to improve its condition and the appearance of the entire section. His routine duties are much the same as those of the roadmaster, but on a smaller scale. He sees that ditches are clear, weeds and grass cut, rails spiked and bolted, ties tamped, track in line, and surface and switches in proper working condition; he sees that the gang has the proper equipment of tools and supplies, and that these are properly used and stored. In case of flood or heavy storm, he patrols the section with enough men to warn trains by signals and torpedoes in case of danger.

Entries of work and materials are made in his time book at the end of each day, with the work done and time for each man. He has also to make up reports and requisitions. He supplies boarding masters with blanks for "board due," which they fill out and have signed by the men. These blanks are returned to the foreman on the last day of the month, and from these he enters on the time book the amount due by each man. The time books and material reports are sent to the roadmaster by a night train on the last of the month or by the first train next day, as noted above. All requisitions for material and supplies are sent to the roadmaster. On some roads the foreman keeps duplicates of his reports.

Trackwalkers and Watchmen.—As a rule, the entire length of each section is inspected at least once a day (except Sundays) by a trackwalker who is one of the section men. He may walk along one rail in going and the other in returning; on double tracks he may go and return over different tracks, or he may make a round trip over each track. His routine and duties vary considerably; on some roads he is mainly an inspector and performs only emergency work, while on others he performs all kinds of track repairs. The importance of trackwalking depends largely upon the density of traffic, the character of the track, and local conditions such as the existence of forests or rock cuts or the prevalence of storms. For lines of moderate and light traffic, an inspection on alternate days should suffice. Instead of a trackwalker, one man of the gang may be assigned to walk over that part of the section not covered by the foreman during the day's work. Still another practice is for the foreman or one of his men to go over the section daily on a motor car to see that it is safe, leaving all repair work to the section gang. Where there are many switches or signals, the lamp man may act as trackwalker. But with the general use of motor section cars and the consequent facility for frequent inspection, trackwalking is less important and is being abandoned in some cases.

In general, the day trackwalker looks out for defective or broken rails and joints and burned ties, removes rocks or other material fallen on or near the track, tightens loose bolts and spikes, examines switches and frogs, notes that switchstands are properly connected and that the lamps are burning, sees that cars on sidetracks are clear of the fouling point, notes broken fences and open farm gates, looks for signs of fire on bridge and notes water and ice conditions at bridges and culverts. If there are only a few switches, he may attend to the lamps. The night trackwalker does no repair work except where there is an unsafe condition, and then his first duty is to set out signals. On the Union Pacific Ry., the day trackwalker on single track carries a trackwalker's hammer, track wrench, two red flags (three on double track) and twelve torpedoes; at night he carries only a red lamp (two on double track), a white lamp, twelve torpedoes and three fusees. A shovel, pick, spike maul, and a few spikes, bolts, and nuts are sometimes carried.

Watchmen may be stationed at tunnels, bridges, rock cuts, or sliding cuts, and at bridges and trestles in time of flood. Where they are stationed permanently, cabins are provided, and the men are furnished with tools for track work and with lamps, flags, and torpedoes for warning trains. A flag or lamp signal should be given to each train if the line is safe. Tunnel watchmen make frequent trips, noting the condition of rails, track, and drainage, and looking for fallen rocks and signs of bulging of walls. Bridge watchmen keep water barrels or sand boxes full, for extinguishing fires, follow trains to look for cinders or signs of fire, remove combustible material from around the structure, keep the bridge seats and copings clean, and watch for any defects in the bridge or masonry. They should prevent persons from walking over the bridges. Drawbridge tenders must look after the locking and other gear, signals, and gates. All watchmen may be required to do incidental work when not engaged in their special duties. Where trackwalkers or watchmen find defective conditions, they set out stop signals as required and notify the section foreman and other officials according to instructions. At dangerous points telephones may be provided. On rocky portions of the line, there is often a permanent rock gang for removing loose rocks which may be dangerous.

Separate track sections for maintenance and inspection are the practice on light lines of the Paris & Orleans Ry., in France. The inspection sections are 7½ to 10 mi. long, each in charge of a trackwalker who attends to switch lamps, dresses ballast, cleans ditches, and does all such repair work as can be done by one man. On each maintenance section, 12½ to 20 mi. long, the gang works systematically, doing all repair work simultaneously on successive lengths of about 330 ft. (except for emergency repair at any point), the entire section being thus covered in 3 or 4 years.

FORCE AND LABOR

Skilled Labor on Track.—An unfortunate but general policy to consider the cheapest and least-experienced grades of casual labor as suitable for track work has resulted in much poor but expensive work done by inefficient forces with heavy labor turnover. Intermittent employment with alternate hiring and firing and the continual training of green men is still common practice; employment of gangs of foreigners who have to be handled through interpreters is bad for the railroads and the track work. These conditions are improving to some extent, since track work as skilled labor and as an economic factor is being more widely recognized, especially with the heavy track construction and high class of maintenance required on main lines. More attention is being given to the stabilizing of labor and to the education and training of the section forces, as noted under "Section Foremen," but much of this training is wasted if there is no policy for steady employment and good pay. In fact, a fundamental difficulty in the railway labor question is the uncertainty of employment. Much time and thought are given to means of securing good men. But at the close of a season, numbers of men are discharged (often because this is the custom), while at any time during the season the forces may be cut down arbitrarily by the management because of a temporary slackening of business.

It takes time and experience to train good track men. Green men will require so much of the foreman's attention that he has little time for thinking and planning, and they use more time and material in poor work than experienced men will use in doing good work. The responsibilities of the trackmen are apt to be overlooked, but if these men fail in their duty not all the skill of the engineer, the faithfulness of enginemen and train crews, or the high quality of rolling stock can give satisfactory train service or insure against accident.

The not uncommon practice of doing extra or unfinished work on Sunday is detrimental to discipline and good work. If men are worked 13 consecutive days, they will not do a fair day's work during the last 6 days; and if 5 or 6 days are spent in preparing for a large job to be done on Sunday, there is likely to be loafing and dissatisfaction, with little real progress. In emergency or accident, of course, the men may be called. But they are entitled to rest and freedom on Sunday. Men (whether laborers or foremen) who get fair treatment are the best to be depended on.

Permanent Forces and Winter Work.—Track work has been regarded as necessarily seasonal, resulting in the almost universal practice of recruiting the force every spring and reducing it in the fall, just when the men have become trained and efficient. It is not unusual, also, to have the force cut down arbitrarily at short notice when there happens to be a reduction in traffic, thus upsetting the program of engineers and foremen.

The money thus saved in wages may be more than offset by the loss in work done inefficiently and material wasted or damaged by fresh inexperienced labor. Thus every season the foreman has to spend time and energy in training new men. Experienced trackmen are valuable employees, and means should be found to retain their services. Apart from reasonable pay, an obvious method is to spread the track over the entire year and thus provide permanent employment. Careful study has shown that many kinds of work can be done successfully throughout a cold winter, except during severe storms or intense cold. But even in the warmer southern states the practice of concentrating the work in a few months is largely followed, although some roads assign in January the number of men for the gangs for the entire year.

The permanent force should be reasonably large; if it is kept at a minimum, the advantage is lost, as only light work could be done and it would still be necessary to employ a large number of green men for each summer season, while the best men probably would not care to remain on the small winter force. Four men throughout the year would cost no more than two men for 6 months or six men for 2 months, and would probably keep the track in more uniform condition. Where eight men are required in summer and three or four in winter, it might be better to have a permanent force of five or six men. Conditions will vary on different roads and in different sections of the country, but even where snow is fairly heavy, some roads carry on regular winter work. Such work includes distributing and renewing rails, laying ties in sidetracks, repairing and renewing switches and frogs, tightening bolts and placing new splice bars, sorting rails, reclaiming scrap material, widening fills and shoulders, repairing track in tunnels, and doing fence and ditch work. In this way the full working season is left for tie renewals, ballasting, and surfacing.

For the heavier work, two or three section gangs may be worked together, but the Norfolk & Western R. R. and Chicago & Alton R. R. keep permanent extra gangs at work all winter in distributing and laying rails, as there is enough other work for the section forces. On the former road, section gangs and bridge and building gangs are kept almost at full strength during the winter, the latter working mainly on bridges in the summer and on repair and interior painting of stations and other buildings in winter; buildings and other structures may also be erected on foundations built in advance. It is found that by thus providing continuous employment a better class of both laborers and foremen is retained. Rail renewal carried on with a large force throughout the winter, even in the snow, is a feature of the organization for permanent work on the Lehigh Valley R. R.

Extra Gangs.—Floating or extra gangs, which are moved from place to place and may be housed in camp trains, are generally employed for such work as raising track, ballasting, rail and tie renewing, widening fills or cuts, fencing, ditching, switch work, or other work which is too heavy or extensive for the section gangs to handle. These floating gangs under the direction of the roadmaster consist of 10 to 25 men, or 20 to 50 on large railroads, with a foreman, assistant foreman, and sometimes a timekeeper. With rail-handling machines and other mechanical devices, smaller gangs may now do as much work as the large gangs which were formerly necessary. An alternative method for handling such extensive work is to group adjacent section gangs under the senior foreman to make a force of 15 to 30 men. Opinion is divided about equally as to which plan is the better; on one hand, it is said that section foremen will be more interested in thorough work, while the extra-gang foreman wants to finish up and move on. On the other hand, the extra gang and its foreman are likely to be better trained in their special lines and the foremen are experienced in handling large forces. Section foremen have a greater variety of duties to perform, and as their men change more frequently, they are less experienced. Furthermore, section gangs grouped temporarily are likely to be less efficient than a regular extra gang, while with such grouping on special work there may be more or less neglect of routine work on the sections.

Bridge and Building Department.—The bridge and building department is connected either with the engineering, the maintenance-of-way, or the operating department, under different systems of organization, but, more properly, it is an adjunct of the first named. It has charge of construction, repair, maintenance and renewals of bridges, buildings, and roadway structures in general, and one of its special duties is to plan for repair or rebuilding of wrecked or damaged structures. Bridges, trestles, culverts, foundations, turntables, track and stock scales, ashpits, water stations, coaling stations, mail cranes, stations, and shop buildings are usually in charge of this department, except that on large roads there may be separate engineers of bridges, buildings, and water service.

The department is generally in charge of a superintendent or engineer of bridges and buildings who reports to the division superintendent or division engineer, the chief engineer, and the bridge engineer. He must have not only complete records of all structures, but also intimate knowledge of them and their condition, making a personal inspection of each one at least once a year. Under him are the supervisors of bridges and buildings, who report to him and to the division engineers and who have charge of the bridge foremen and master carpenters. Supervisors of bridges and buildings are recruited both from assistant engineers and from the foremen; the former are generally the better, if properly trained, but the latter often have the advantage of having grown up with the railroad and being familiar with its work and structures.

Logically, the bridge and building work should be under the engineers. On the Duluth, Missabe & Northern Ry., the bridge and building department is in charge of an engineer of bridges and buildings, who reports directly to the chief engineer. This department handles the construction and maintenance of all bridges, buildings, docks, wharves, and other structures; its engineering staff makes surveys of bridges, docks, wharves, water stations, and other structures for construction and renewal, and prepares plans and estimates of cost for submission to the chief engineer. Construction details are recorded by assistant engineers or inspectors, while construction costs are obtained by the distribution of labor and material by timekeepers, material clerks, and others. The coal and water supply department is under the jurisdiction of a foreman, who reports to the engineer of bridges and buildings. On the Soo Line, this work is included in the engineering department; bridge and building foremen are under the division engineers, who report to resident engineers, who report in turn to the chief engineer. The duties of supervisors of bridges and buildings are handled by the division engineers, which positions are filled by advancing engineers from minor positions. Material for repairs or other regular work is distributed as far as possible by regular trains; work trains are used only where such equipment as pile drivers or derrick cars is required.

Water Service Department.—On large railroads, the water supply for locomotives, shops, stations, and other railway purposes is under a separate organization which is logically a part of the engineering department, but must cooperate with the mechanical and operating departments. This organization is headed by an engineer or superintendent of water service, reporting to the chief engineer and engineer of maintenance-of-way, and being more or less in charge of construction and operation of pumping and treating plants. He has a chemist who assists in investigating the quality of water and its suitability for boiler and other purposes and supervises the operation of water-treatment plants.

This department may also maintain a continuous study of the effects of treated water on boiler repairs and engine performance. Water testers and inspectors in the field make periodical analyses of the raw and treated waters at the different plants. On each operating division, there is a supervisor of water service or master carpenter who has charge of the pump men and repair gangs, and reports to the division engineer or division superintendent. On smaller roads, the water service is usually under the bridge and building department, except for large pumping plants in charge of the engineering department. On the Great Northern Ry., the engineer of water service reports to the chief engineer and is charged with general supervision of all matters pertaining to water service. Under him are assistant engineers who prepare plans for pumping machinery and make requisitions for material. The drilling of new wells and the repairing of old wells are done by his crews. On each district, there is a general water inspector who sees that treated water is in proper condition; he reports jointly to the general superintendent and the engineer of water service. There are several water inspectors on each division who report jointly to the general water inspector and master carpenter, and who make at least weekly tests of the treated waters on their districts. The master carpenter is charged with the proper maintenance of the pumping plants, and has water-service foremen to do this work.

Signal Department.—Two distinct systems place the signal department under the engineering or the operating department, the former being the more general. Construction and maintenance are associated with engineering problems, but as signal operation and the planning of new work is closely related to operation, the signal engineer must be in co-operation with the operating department. With the introduction of automatic train control, he is also brought into touch with the motive-power department, but this is so new that his jurisdiction and system of reporting on such matters have not been settled definitely. He prepares plans, specifications, and estimates for signal equipment, consulting with the division forces as to the plans. Maintenance and operation are usually in charge of the division superintendents, since they are responsible for the trains and the traffic operation on their divisions. A division supervisor of signals, in charge of maintenance, reports to and receives his instructions from the division engineer, the division engineer of maintenance-of-way, or the division superintendent. He also reports to the signal engineer in matters involving standard methods of maintenance.

On the Chicago & Northwestern Ry., with divisional organization, the signal department is a part of the engineering department, and the signal engineer reports to the chief engineer, the engineer of maintenance, and the general superintendent in regard to construction, operation, and maintenance, respectively. The division engineers have direct charge of all signal matters and report to the signal engineer; signal supervisors report to the division engineers. In automatic signal territory, there is a maintainer to each 20 mi. on double track, with an assistant or a helper, and additional assistants if interlocking plants are included. Outlying divisions have one or more repairmen, with the necessary assistants. Train control apparatus is under the signal engineer. On the Michigan Central R. R., with departmental organization, the signal engineer reports to the chief engineer and attends to all interdepartmental correspondence. His force designs, constructs, and inspects the installations and has charge of the division supervisors who handle maintenance and construction. On the Baltimore & Ohio R. R., the signal engineer reports to the vice-president in charge of operating and maintenance.

TYPICAL ORGANIZATION SYSTEMS

Chicago, Burlington & Quincy R. R.—Maintenance-of-way is under the jurisdiction of the division superintendents, who report on maintenance matters to the general superintendents, and the latter report to the general managers of the two grand divisions. District engineers of maintenance-of-way on each district are on the staff of the general superintendent, reporting to him and acting in an advisory capacity on maintenance matters. They also have more or less authority over roadmasters. There is a general inspector of permanent way and structures on the staff of the operating vice-president, acting in an advisory capacity and devoting his full time to maintenance-of-way matters, study of labor-saving devices, and track and track-tool experiments. The district engineers of maintenance-of-way work in close contact with the general inspector of permanent way and structures and are frequently at headquarters for consultation on matters of general practice. Particular features of track maintenance are covered by standard practice circulars, prepared by the general inspector of permanent way and structures and issued over the signature of the general managers and the chief engineer.

The engineering department proper is not in charge of maintenance work, but (as is customary on many railroads) it is in charge of standards and handles such work as the staking out of track work, bridge work, building work, the rerunning of curves, and the placing of ballast stakes. In this department, the division engineers report to the district engineers, and the latter report to the two assistant chief engineers, one for lines east and one for lines west of the Missouri. The assistant chief engineers, the signal engineer, the bridge engineer, and the engineer of buildings all report to the chief engineer.

Roadmasters, reporting to the division superintendents, are recruited mainly from among the section foremen. Work-train conductors are under the direction of roadmasters or master carpenters so far as the operation of the trains is concerned, but they report to the trainmaster and division superintendent in regard to train movements, discipline, and observation of rules of the operating department. Roadmasters' divisions cover from 100 to 300 mi., depending on the importance of the line. Double-track sections average 3 to 6 mi. long, with five to ten men in summer and two to three men in winter. Single-track sections average 6 to 12 mi., with three to six men in summer and one to three men in winter. The number of men in spring and summer depends upon the importance of the line, the extent of work, and the availability of men. In winter, there may be only the foreman, or he may have from one to three men, but the number may be increased in an open winter when outdoor work can be carried on. Extra gangs are employed for rail laying, surfacing, ballasting, bridge construction, or any extensive work of maintenance or construction. On some divisions these are permanent gangs, while on others the gangs may be disbanded for the winter.

Southern Pacific Ry. (Pacific System).—The chief engineer is in direct charge of the construction of new lines and second-track construction and is also in direct supervision of all maintenance matters, besides having charge of technical studies. He reviews and reports upon any proposed changes of operating standards relative to maintenance and is a member of all boards of inquiry. On each construction project, there is an assistant engineer in charge, reporting direct to the chief engineer, and under him are assistant engineers detailed to each residency, the length of residency depending upon the class of construction work. Assistant engineers in charge of preliminary and location survey parties report directly to the chief engineer. Track laying and ballasting on new construction work are under the supervision of a superintendent of construction, who reports directly to the chief engineer.

Maintenance-of-way comes under the operating department, which is in charge of a general manager reporting to the president. The Pacific System is divided into three districts, each under an assistant general manager, reporting to the general manager; each one has responsible charge of all matters in his territory and cooperates with the engineer of maintenance. The superintendents report to the assistant general managers. Under the general manager is an engineer of maintenance-of-way and structures, in responsible charge of maintenance on the entire system. Division superintendents receive instructions from, and report directly to, him on all maintenance and engineering matters on their divisions. The chief engineer reports to the president, but there is close cooperation between the chief engineer, the general manager, and the engineer of maintenance. Construction of new railroad lines is directly under the chief engineer.

On the staff of the engineer of maintenance are the assistant engineer of maintenance-of-way and structures, the bridge engineer, the signal engineer, the general fire inspector, the general scale inspector, the architect, the chief draftsman (engineers), the track inspector, and the rail-welding inspector. Under the bridge engineer are the general bridge inspectors, who work with the division inspectors in inspecting structures. Maintenance as well as the construction of all new work directly connecting to existing lines is under the engineer of maintenance; this includes such work as siding extensions, second-track extensions, terminal track facilities, buildings, and roundhouses. Under each superintendent is a division engineer in responsible charge of all maintenance matters and having his own engineering organization. All plans, estimates, and requests for authority pass through the superintendent, to assistant general managers, to assistant engineer of maintenance-of-way and structures, to engineer of maintenance, and thence to the chief engineer and the general manager.

Reporting to the division engineers are subordinate officers such as roadmasters, bridge and building supervisors, signal supervisors, and water and fuel supervisors. All trackmen are under their respective roadmasters; bridge and building foremen and division bridge inspectors are under the bridge and building supervisor; signal maintainers are under the signal supervisor; water-service foremen and pumpers are under the water and fuel supervisor. The length of a roadmaster's district varies from about 60 mi. on double-track mountain territory to 183 mi. on single track. Track sections on double track are 4 to 5 mi. long; on single track they are 7 to 10 mi. long for main line or 12 to 24 mi. for branch lines. The number of men in the section gangs during summer and winter varies from 0.5 to 1.5 per mile. As the lines of the Pacific System pass through eight states, various climatic conditions prevail, so that throughout the year work is being prosecuted vigorously at some points on the system. In Nevada and Utah, where track freezes up, the winter forces are cut down to three men per section.

Baltimore & Ohio R. R.—The organization is on the divisional system, and the maintenance-of-way department is separate and distinct from the engineering department. A chief engineer of maintenance has supervision over the maintenance-of-way department for the entire system and reports to the vice-president in charge of operation and maintenance. An engineer of maintenance-of-way, reporting to the general manager, has charge of each general manager's district. A division engineer, reporting to the superintendent, has charge of maintenance-of-way on each division. General supervisors of track, track supervisors, master carpenters, and signal supervisors report to the division engineer. Foremen of extra gangs report to track supervisors; work-train conductors report to the superintendent. Track supervisors are mainly promoted from section or extra-gang foremen, although some are young engineers specially trained for the position. The signal engineer reports to the vice-president in charge of operation and maintenance. The engineer of bridges and the engineer of buildings are in the engineering department and report to the chief engineer. Operating divisions are 285 to 650 mi. in length, while track supervisors have 25 to 130 mi.

Erie R. R.—In 1920, this railroad adopted a regional system of organization, but within each region the organization is on the division system, the entire arrangement being very similar to that of the Pennsylvania R. R. The railroad is divided into three regions, under the jurisdiction of three regional vice-presidents reporting to the operating vice-president, who has on his staff an engineering assistant to handle maintenance-of-way and construction matters. Each regional vice-president has a regional engineer on his staff, with similar duties.

The chief engineer, reporting to the president, has the title of chief engineer and assistant to the president. He establishes and revises all standards, has charge of all surveys and construction on important projects, and passes on the plans of minor projects carried out by the regional forces. He maintains, also, an independent inspection of bridges, buildings, water tanks, and other structures where the element of safety is concerned; thus he calls the attention of the regional organization to any matter of maintenance which, in his opinion, is not being handled with safety, although maintenance responsibility rests with the regional organization. The engineer of bridges, the engineer of construction, and the engineer of grade crossings all report to the chief engineer. The signal engineer reports to the operating vice-president.

For each region, the regional engineer, who takes the place of the former engineer of maintenance-of-way, has charge of all matters pertaining to maintenance of track, bridges, buildings, signals, water supply, and similar matters; he also handles such construction work as may be assigned to him by the chief engineer. He reports to the regional vice-president and not to an operating officer. Instead of a general superintendent, there is a superintendent of transportation, who also reports to the vice-president.

Division engineers report to the division superintendents, but are held responsible to the regional engineer for standards and for proper methods of carrying on the work. The division engineer has an engineering corps reporting to him; also track supervisors, master carpenter, and signal supervisor. The master carpenters correspond to what are known on some other roads as superintendents of bridges and buildings. The bridge engineer, reporting to the chief engineer, prepares plans and supervises the erection of important bridges, and also inspects the plans of minor structures as prepared by the division engineers. Track foremen report to the track supervisor. Carpenters, masons, and water-supply forces report to the master carpenter. Signal maintainers, electric maintainers, and motor-car maintainers report to the signal supervisor; they also take care of electric lighting, electric pumps, and similar apparatus.

There are no night trackwalkers, and day trackwalkers are employed only on territory not equipped with automatic signals, where it is necessary to cover the line to attend to switch and signal lamps. On automatic-signal territory, the signal department takes care of all switch and signal lights, except that in yards having a large number of lights a special lamp man is assigned from the track department. All sections are equipped with motor cars, and as frequent trips are made by the section foremen, this kind of inspection is found more dependable than trackwalking. All repair work, tightening bolts, and driving down spikes is done by the section gang. The number of permanent watchmen has been reduced considerably by the replacement of timber trestles with permanent structures, and by the clearing and scaling of rock cuts. Watchmen are assigned permanently, however, to portions of line having heavy rock cuts. They are supplemented by a rock gang composed mainly of quarrymen, who are experienced in climbing and in the detection and removal of loose rock.

Track supervisors have divisions of about 50 mi.; double-track sections are 4 mi. long, with eight to twelve men in summer and four in winter. Single-track sections are 8 to 12 mi. long, with the same force.

Michigan Central R. R.—The chief engineer has charge of all construction and also of the maintenance of track, bridges, buildings, water supply, interlocking, and signals. To him report the assistant chief engineer, the principal assistant engineers, the bridge engineer, the signal engineer, and the superintendent of buildings and water service. Roadmasters and assistant engineers report to the division engineers, who in turn report to the assistant chief engineer. Assistant bridge engineers report to the bridge engineer. Foremen of repairs report to assistant roadmasters. Carpenters, painters, masons, and building carpenters report to division foremen of buildings and water service; these foremen, as well as division foremen of buildings, report to the superintendent of bridges and water service. Signal maintainers report to inspectors, inspectors to supervisors, and signal supervisors to the signal engineer.

Pennsylvania R. R. System.—The chief engineer of maintenance-of-way reports to the general manager, and is one of the four "cabinet officers" assisting the latter (chief engineer of maintenance-of-way, general superintendent of motive power, general superintendent of transportation, and superintendent of telegraph and signals. He has an assistant chief engineer of maintenance-of-way, and is represented on some of the grand divisions by an engineer of maintenance-of-way as the engineering advisor of the general superintendents. This engineer is represented on each operating division by a division engineer, to whom the supervisors or roadmasters report. Each supervisor is in charge of from 25 to 75 mi. of line, and even more in some cases, having an assistant for the harder divisions. Subordinate to them are the track foremen, whose sections average $2\frac{1}{4}$ mi. on double-track and four-track lines. On each division, there is a master carpenter, who looks after general repairs to bridges and buildings; he reports to the division engineer. The maintenance-of-way department has general charge of the track, bridges, buildings, turntables, water and coaling stations and signals. It is also charged with watching the developments or improvements in track construction.

The chief engineer is the officer having general responsibility for all engineering work, including construction, maintenance of standards, preparation of plans, estimates and specifications for new lines, bridges, buildings, signals, and general supervision of engineering of all kinds. He also keeps records of the cost of all construction work and has charge of the distribution of rails for both construction and renewals. He reports to the vice-president in charge of operation. The division engineer, reporting to the superintendent, has charge of the maintenance of all property under the latter, including the telegraph and telephone facilities, which are handled directly for him by the supervisor of telegraph and signals. He is in charge also of the operation of water stations, and does a limited amount of construction work, depending upon local conditions.

Nashville, Chattanooga & St. Louis Ry.—The chief engineer, reporting to the general manager, is head of the engineering department, which is divided into five sub-departments: construction, track, bridges and buildings, signals, and telegraph and telephone. The assistant chief engineer, in general charge of the office and its affairs, gives special attention to rails and track appliances, the checking of expenditures against budgets, the preparation and execution of contracts, and the preparation of statistical data. The senior assistant engineer has charge of construction, handling work too extensive for the division forces; he also assigns roadway machinery for the various pieces of work. The assistant engineer of design is in charge of the drafting room and the filing of plans. The assistant engineer of bridges is in charge of bridge design and keeps all bridge records; he makes an annual inspection, in company with the bridge inspectors and division officers. Another assistant engineer has charge of design, maintenance and operation of water and coaling stations (except for the machinery), timber treatment plants and similar installations.

The general roadmaster is occupied mainly in inspection of track and roadway, and the instruction of roadmasters in maintenance work. He contracts for ballast, supervises its distribution, and approves profiles for raising track, these being submitted to him by the division engineers. The signal engineer has a designer and a general inspector; there is a maintainer on each division, and a signal construction gang handles work which is too much for the division forces. The superintendent of telegraph reports to the chief engineer and general manager. A chief clerk is in charge of correspondence files, records and clerical work of the engineering department. On each of the five operating divisions there is a division engineer in charge of maintenance of way and general engineering work, reporting through the superintendent to the chief engineer. The division engineer has a bridge and building supervisor, track supervisors (or roadmasters), signal and telephone maintainers, an assistant engineer, instrument man, rod man and chief clerk.

CHAPTER XVI

TRACKLAYING AND BALLASTING

TRACKLAYING

Staking Out Track.—Preparatory to track laying on a new line, the engineer in charge runs in the alinement on the completed roadbed. He has the location field notes but frequently has to shift the P.C. backward or forward to bring the P.T. on the given tangent. Pieces of rail 4 ft. long are often used to monument the curves. In general, center stakes are driven 100 or 200 ft. apart on tangents and 50 ft. on curves (25 ft. on transition or spiral curves); a stake marks each end of each curve. For double track, the stakes may be set between the tracks, the tracklayers having measuring boards to give the distance from center mark to gage side of line rail.

At sidetracks, the switches and turnouts must be laid out, positions of head block and frog indicated, and center stakes set for the sidetrack. On bridges and trestles, the center line is marked by tacks about 20 ft. apart. Offset measurements indicate the position of the inside edge of the rail base, and a chalk line struck between these offsets shows the position of the line rail; when this is spiked, the opposite rail is set in position with a track gage. Daily reports made by the engineer are either letters or on printed forms, and show the progress by 100-ft. stations. A form recommended by the A. R. E. A. is a loose-leaf sheet $9\frac{1}{2}$ by 11 in., showing the length of main and side track laid, the location by mile posts, weather conditions, number of men, and causes of delay. This report is made in triplicate, the engineer in charge retaining one copy and forwarding the others to the chief engineer and division engineer.

Tracklaying Records.—In the days when long lines of railways were being pushed across new and open country, with light track construction, daily records of 3 to 4 mi. of track completed were numerous; 6 and even 8 mi. are reported on the first transcontinental line of the Central Pacific and Union Pacific in 1868 and 1869. With slightly heavier material, the Canadian Pacific Ry. in crossing the prairies made records of 4 and 6 mi. per day in 1882 and 1883, respectively, while the progress averaged $2\frac{1}{4}$ to $3\frac{1}{2}$ mi. This was accomplished by gangs of about 180 men without tracklaying machines, and speed was a primary object. In 1888, the Atchison, Topeka & Santa Fe Ry. laid track at the rate of 2 mi. per day with a gang of 164 men. Since the completion of a great network of trunk lines, this spectacular feature of railway construction has disappeared, as the filling in of the network does not call for the feverish rush and haste of the old days of railway pioneering. Furthermore, rails of 85 to 100 lb. per yard cannot be handled with such speed and facility as rails of 50 to 65 lb.

Modern Tracklaying.—For the tracklaying on a Kentucky line of the Illinois Central R. R. in 1924, two material trains were used, one being at the head of the track while the other was being loaded at the supply yard by a gang of 40 men. The working train consisted of a pioneer car, two cars of ties, two cars of rails, and a car with spikes, bolts, tie-plates, and splices; then the locomotive, caboose, and three cars of ties. At the head end of the pioneer car was a hand-power derrick, and on each side was a chute or trough with rollers. Similar troughs were attached to the cars, along which men pushed the rails and ties. With a force of 162 men making an advance of 4,000 ft. of completed track in an 8-hr. day, the organization was as follows:

Tracklaying gang, 60 men; 1 foreman, 1 derrick operator, 2 unloading material from pioneer car, 3 throwing ties into left-hand chute, 2 throwing rails into right-hand chute, 6 pushing ties and rails along the chutes, 18 carrying and placing ties (3 men to a tie), 5 carrying bridle rods ahead, 2 lining ties, 3 unloading tie-plates and spikes, 2 distributing splice bars, bolts, and spikes, 6 placing rails, 2 putting on splices, 1 setting track gage and placing expansion shims, 4 unloading ties from rear cars, 2 at air compressors (one ahead and the other in rear of train). Back gang, 102 men; foreman with 50 men jacking up track, putting ties under, and spacing them; foreman with 50 men tie-plating, full bolting, and spiking.

Ties were delivered 16½ ft. ahead of the pioneer car by the chute, and six ties were laid to each rail length, their ends being lined against a cord and the spacing being governed by a measuring stick. As each 90-lb., 33-ft. rail was delivered 16½ ft. ahead of the car by the chute, it was picked up by the derrick and laid with the aid of six men; the line rail was first set to a chalk mark on the ties and then the other rail was set in position with a track gage; the two rails were secured by bridle rods punched to receive a spike on the inner side of the rail. No rail spikes were driven ahead of the pioneer car. Meanwhile, the heel joints were spliced with angle bars and two bolts. As the curves were light, no bent rails were carried. Behind the tracklaying force was a work train with material for turnouts and passing tracks, which were installed by a gang of 40 men. This train also carried materials for the fence gangs.

Ballasting followed at a distance of several miles behind the track laying, so as not to interfere with the train loading material or the train serving the construction of passing tracks. Gravel ballast hauled in open-end cars and plowed off by a rapid unloader was applied in two lifts; a first lift of 9 in. and a final lift of 3 in., with lining and surfacing. For this work the track was raised with jacks by the ballasting gang.

Tracklaying without Machinery.—Where no tracklaying machine is used, the first step after the arrival of the material train at the end of the track is to unload the ties from the cars onto wagons for distribution alongside the roadbed. The tie gang places and spaces the ties, locates the joints with a 15-ft. pole, selects the larger ties for joints (if required), and lines one end to a cord stretched between stakes set at a half-tie length from the center stakes. On curves, the middle and quarter ordinates are measured from the cord, and stakes are set by which the cord is adjusted. If the ties are laid too far ahead, there may be troublesome shifting at the joints when the rails are laid. At least half the full number of ties should be laid in advance of the material train to prevent kinking of the rails when the train moves ahead.

A push car or horse car takes rails to the head of track, where the rail gang takes off two rails, half bolts them at the heel joint, sets the head ends in position with a grooved track gage, and secures them by a few spikes or by bridle rods which are removed by the following spikers. Rails bent for curves are kept separate, as well as short rails, $29\frac{1}{2}$ ft. long (distinguished by painted ends), for use on the inside of curves to keep the joints in relative position. The foreman has a list of curves and the number of short rails required for each. The joint or splice gang completes the bolting and is followed by the spiking gang; the latter spikes the line rail and then the other, which is adjusted to position by a track gage. In some cases a few rails and ties were thrown from the material train, which then moved back while they were loaded on cars drawn by horses and hauled to the head of track. When this supply was laid, the material train moved up and unloaded another lot. About 75 to 80 men would be required to lay a mile of track per day by this method.

Tracklaying Machines.—In modern work, the distribution of ties and rails is simplified by the use of so-called tracklaying machines, which effect a great saving in force and cost and avoid the cutting up of roadbed by teams distributing the ties. At the head of the material train is a pioneer car with appliances for delivering the rails and ties ahead of the car and with an elevated platform from which a flagman signals the engineman to move ahead as each rail length of track is laid. Runways at the sides of the material cars may carry rails and ties to the head of the train, and the rails may be swung out and placed by a derrick boom or by trolleys riding on an overhead cantilever frame, thus dispensing with the large gangs required to handle heavy rails by hand.

On work done by company force, a locomotive crane is sometimes used as the pioneer car. In branch-line construction on the Union Pacific Ry. in 1921, the crane handled the 90-lb. rails, there being 3 men on the rail car and 12 on the ground to place, bolt, and spike the rails. Ties were distributed by a force of 8 teamsters and 13 laborers. A gang of 65 men followed the train to do the backspiking, lining, and bolting. An average of 4,400 ft. of track were laid by a force of about 100 men in a 9-hr. day. At the material yard, a gang of 25 men loaded a second train, ready for taking to the end of the track.

Roberts Tracklayer.—In this system, there is along each side of the material train a horizontal tramway or runway carried by brackets in the stake pockets of the cars and having live rollers driven by gearing from longitudinal shafts with flexible couplings. Power is supplied by an engine on the pioneer car, taking steam from the locomotive. One tramway has corrugated rollers for the ties, and in the other the rollers are double, so that the two lines of rails can be moved independently. Plain or dead rollers are placed between the live rollers. An extension runway held by guys from a gallows frame on the pioneer car delivers the ties 60 ft. ahead of the car and clear of the rails; the other runway extends 6 ft. ahead of the car, so that the heel joints are 8 or 10 ft. in front of the car, giving plenty of room for splicing. A hand derrick on the car, to place the rails by a cable and tongs, reduces the number of men formerly required to handle the rails. Splice bars, bolts, and spikes are carried on the front end of the pioneer car, while behind its engine are tool boxes, track tools, and a few short rails.

For laying a mile of track, the material train is made up with the pioneer car, seven cars of ties, the locomotive, and four cars of rails, a rail runway being slung alongside the locomotive. In some cases, the supply is for only $\frac{3}{4}$ mi., but in older days it was usually 1 or $1\frac{1}{4}$ mi. The track is half bolted ahead of the train, but the rails are held by bridle rods instead of spikes. All spiking, full bolting, tie spacing, and lining of track is done behind the train. For laying 2 mi. of track per day, using two material trains with supply for 1 mi. each, the force would be about 126, distributed as follows:

With the Tracklaying Train (42)

1 Foreman.
1 Engineman on pioneer car.
7 Throwing ties into runway.
2 Rolling rails into runway.
4 Placing angle bars and bolting.
1 Carrying angle bars to front of car.
2 Distributing spikes and bolts.
14 Receiving and placing ties (2 men per tie).
1 Assistant foreman with rail gang.
4 Receiving and placing each rail (with hand derrick).
1 Placing bridle rods.
1 Nipper.
2 Spacing ties.
1 Oiler.

Behind the Tracklaying Train (84)

2 Assistant foremen.
40 Spikers.
20 Nippers, holding up ties.
6 Bolting splices.
6 Spacing and lining ties.
10 Lining track.

Clyde Tracklayer.—This system, first developed for logging railway work, also has power-operated devices, but of a different type. Upon the pioneer car is a steel frame which carries an overhead runway 90 ft.

long, extending about 26 ft. beyond each end of the car. On the frame also is an overhead platform for an engine which operates the hoisting and traveling cables of a 2-ton trolley hoist on the runway. The trolley carries rails by tongs and bundles of ties in slings, running them forward and lowering them upon the subgrade. Steam is taken from the locomotive. For small pieces of work, rails for $\frac{1}{4}$ or $\frac{1}{2}$ mi. of track are carried on the pioneer car, and ties are on a car behind. An average force then consists of a foreman, engineman, two men on the car to attach and guide the rails and bundles of ties, one spike peddler, four spikers, two nippers, and one bolter. For heavier work with 100-lb rails on the Florida Western & Northern Ry., as many as four cars of rails and ties were placed between the tracklayer and the pioneer car, while on part of the work the ties were distributed by teams, leaving the tracklayer to handle only the rails. With a crew of about 24 men and laying only half the full number of ties ahead of the machine, the average advance was 3,400 ft. of track per day; or 6,600 ft. when ties were distributed by teams. The highest daily record was 7,936 ft.

Earlier Tracklayers.—At the time when many long lines of railway were being built, a great mileage of track was laid with the Holman and Harris tracklaying machines, both now obsolete. In the Holman tracklayer, the cars of the material train were fitted with an inclined trough or roller runway along each side, carried by brackets in the stake pockets, the troughs being connected between the cars and having a slight inclination towards the head of the train. Rails and ties were thrown from the cars into the troughs and moved down to the head of the train, the tie runway having an extension guyed from a frame on the head or pioneer car so as to deliver ties about 35 ft. ahead of the car and thus keep the tie gang clear of the rail gang. Each rail was taken from the runway by the rail gang, and heeled into place against the previous rail, an expansion shim having been placed by the joint gang. In a modified form, to reduce the labor force, a cantilever frame was fitted, having trolleys by which the rails were run out and lowered into place. On the Northern Pacific R.R., 11,200 ft. of track were laid in 10 hr. with 63 men; the Chicago, Burlington & Quincy R.R. laid $1\frac{1}{2}$ mi. with 85 men, using two material trains daily, each with four flat cars of 65-lb. rails ahead of the locomotive and four cars of ties in the rear.

In the Harris tracklaying system, timbers laid across the floors of the flat cars carried plank gangways on each side and a 2-ft. track in the middle. This track carried a tie tram, and on the rail cars iron rollers were fitted between its rails. The rails were piled on each side of the tram track. This track, with its rollers, extended along the pioneer car and beyond it on stringers about 20 ft. long which were supported by guys from a gallows frame on the car. A cross-frame on the stringers carried rollers for the rails. The tie tram had a sliding deck or platform mounted on rollers in the frame. Men on the rail cars threw rails onto the central rollers and pulled them forward to the pioneer car, where splice bars and two bolts were placed. Men on the tie cars loaded ties onto the tram, which was then run forward; as it struck a stop block on the end of the pioneer car, the sliding deck shifted forward and tilted, dumping the ties on the roadbed. When the tram was run back, two pairs of rails were run forward over the rollers on the extension and on portable dollies or frames 3½ ft. high, standing on the ties. From these dollies, the two line rails were taken, placed, spiked, and bolted to the previous rail; the two gage rails were then placed and the train moved ahead two rail lengths. More than 2 mi. of track were laid in 9 hr. of steady work.

Later devices included the Hurley and Westcott machines, both used to a limited extent. The former had a self-propelling pioneer car hauling the material train and having a cantilever truss with trolley hoists for rails on the bottom chords and a conveyor for ties, on the top chords. The rails were bolted together on the cars and were moved forward by live or driven rolls, thus forming a conveyor for the ties, which were in this way delivered to the conveyor on the truss. The Westcott machine also had cantilever trusses, carrying hoisting trolleys with cables and tongs to handle the rails and a conveyor to handle the ties. Conveyors on the material cars carried rails and ties to the pioneer car.

BALLASTING

The ballasting of track has to be done both in construction and maintenance, being required in the latter case to make up for loss by settlement, to increase the quantity, or to provide better material. It is economical to keep a good depth of ballast and to increase it gradually when necessary, as this will reduce maintenance work; but the ballast should not be used for filling sags or for other work where cheaper material would suffice.

Ballasting New Track.—For a new line, the ballast train runs on the track already laid on the subgrade (see "Tracklaying"). The material is discharged, and the track is then jacked up while the ballast is shoveled and tamped under the ties. As a rule, it is applied in two lifts or courses, the first one being the heavier. After this first lift, the track may be left for a few days so that the passage of work trains will compact the material, the second lift being then placed and the track lined and surfaced. A certain amount of ballast per mile is specified, any excess being classified as filling. The ballasting is usually kept several miles behind the tracklaying, so as not to interfere with the material trains for the track and switch work; but as little traffic as possible should be allowed on a half-tied and unballasted track as the rails and joints are likely to be bent. In English practice, a light temporary track is used in ballasting, after which the permanent track is laid. Thus the British term of "permanent way" for track is used to distinguish the track proper from the temporary construction track, and is not intended to indicate (as is sometimes assumed in this country) that the track itself is permanent.

Grade or ballast stakes for the level of the top of the rail are set on one or both sides of the track, so located that they will not be displaced in dumping the ballast. They are placed 100 ft. apart on tangent or 50 ft. on curves; also at all changes of grade and at the ends of curves and spirals. On curves, the stakes are set usually for the lower rail, which is kept at grade, all superelevation being given in the outer rail. Where vertical curves are used at grade intersections the stakes are set for the proper profile. If the stakes indicate insufficient or excessive depth of ballast, the roadbed should be trimmed or filled accordingly, and all earth above the bottoms of the ties should be removed.

Reballasting Track.—The ballasting of operated lines is done mainly by work trains and extra gangs, either under traffic or with a stretch of 5 to 10 mi. of track (on a double-track line) given over to the work. The latter method has advantages similar to those in rail renewals, already noted, and may be handled in much the same way. With the former method, trains must be run at slow speed as the stability of the track is reduced temporarily by the removal of the old ballast. If the roadbed needs to be widened at any point, this should be done well in advance.

As the first step, the track is skeletonized by removing all ballast between the ties for the full width of roadbed, down to or a little lower than the bottom of the ties. If the track is to be raised and this material is suitable, the track may be jacked up and the old ballast spread under and between them to form a bed of sub-ballast. Otherwise, the material may be used to widen or level the roadbed shoulders, being thrown over the fill or loaded by work trains in cuts. Any cross-drains needed are placed to carry water from the space between the tracks, and any decayed or damaged ties are renewed. Grade stakes are then set by an engineering party in preparation for the ballasting. In yards, where the material may have to be carried or wheeled across several tracks, and also in tunnels, the men may simply clear out the ballast from between the ties, shoveling it to the outside of the track, where it is picked up by a ditcher which also excavates the material outside the ties and loads it all into a car coupled to the ditcher.

Ballast is delivered usually in two lifts or courses of 6 to 8 in. and 6 to 4 in., respectively, for a 12-in. depth of ballast. For greater depths, two of the heavier lifts may be made. With ballast deposited for the first lift, the track is raised to the required height by jacks, which are placed in pairs, with usually two to four pair to each rail; they should be at least two tie spaces from the ends of the rail to prevent bending at the joints. A long runoff is also given so that trains will ride easily over the change in elevation. Both rails are raised together, the track level being used to keep them in the same plane. Then the ballast is thrown under the ties with shovels or forks and is shovel tamped to a uniform surface and bearing. After this the jacks are removed, and traffic is allowed to compact the ballast, a few men going over it occasionally to raise and tamp low spots where settlement is irregular.

After a few days, the track is lined, and the second lift of ballast is placed. If this is the finishing lift, the track will be jacked up to the exact height indicated by the ballast stakes, and after the material has been placed under the ties, it will be thoroughly tamped with picks, bars, or tamping machines. The foreman uses the spot board and track level frequently in finishing the track to a good surface. After 2 or 3 days of traffic, a gang goes over the work, lining the track to the center stakes and raising and tamping any low spots. A following gang completes the the work by dressing the ballast to the standard cross-section, clearing the roadbed and ditches, and leaving the right-of-way clear of material that would interfere with the use of scythes or mowing machines for cutting grass and weeds. For more than one lift with stone ballast, bar tamping may be allowed for the first lift, but only picks or pneumatic tampers for the others. With gravel, shovel tamping may be allowed for the first lift.

Ballasting with gravel may cost from 72 cts. to $1.30 per cubic yard in place, depending largely on first cost and length of haul. An actual itemized cost is as follows: cost at pit, 20 cts.; hauling in work train, 51 cts.; unloading, 13 cts.; skeletonizing track, 39 cts.; lifting track on gravel, 30 cts.; lining track, 11 cts.; dressing track, 37 cts.; supervision, 6 cts.; total, $2.07 per cubic yard in place.

Ballasting Gang.—A typical force of 77 men for raising and ballasting track already skeletonized is suggested by the A. R. E. A. as follows: foreman, 4 assistant foremen, 2 flagmen, 2 men ahead digging holes for jacks; 13 with head jacks (8 with jacks, 4 spaders, and 1 with track level); 11 men with the raising jacks (5 with jacks, 4 tampers, and 2 with forks); 34 in tamping gang (28 tampers and 6 with forks); and 8 in the lining gang. There is also a water carrier, a tool man, and a time and material clerk. One assistant foreman is with the head jacking gang, another with the raising gang, 200 ft. behind, a third with the tamping gang 100 ft. behind and covering 100 ft. of track, and the fourth with the lining gang. The flagmen keep about 4,000 ft. in front and rear of the work. A small emergency gang of 24 men with 20 laborers would include: foreman, 2 flagmen, 2 men digging jack holes and setting the spot board, 6 men raising with jacks and handling the level board, 12 tampers and a water carrier. With stone ballast there would be 4 jack men and 14 tampers. The jack-hole diggers keep about 60 ft. ahead of the jack men, who in turn are 10 ft. ahead of the tampers, these latter being spread over about 30 ft. of track. In ballasting by contract, the final surfacing and lining is done by the contractor after the track has been settled under traffic.

Distributing Ballast.—Ballast is deposited usually by shoveling or plowing it from gondolas or flat cars, or by dumping it from drop-bottom gondolas or special hopper cars, with the train at rest or moving at such speed as to distribute the ballast in desired quantities. With material dumped between the rails, the flow being regulated by the opening of the hopper doors, the rear car of the train may have a plow and flangers to level the ballast, throw excess material outside the rails, and leave proper flangeways. Snowplows and flanging cars are utilized for the same purpose. Another method is to place a tie or heavy timber across the rails in front of the rear truck of each car; for the last run, in dressing the ballast to shape, this timber has angle irons on the bottom to clear the required depth below the rail head. The timber should be braced from the truck frame or car so that it is not in contact with the wheels, except where the wheels ride on shoes so that they do not revolve. Center dumping requires less shoveling, but ballast for new sidetracks or additional main tracks may be side dumped from cars on the adjacent track and then spread and leveled by the use of a spreader car to the width of 15 to 20 ft. on one or both sides.

Special cars of hopper type for dumping ballast have a longitudinal hopper with the sides formed by doors hinged at the top and opened by means of levers to control the flow of the material, which is then spread in the way already noted. In another design, special door arrangements provide for dumping between or outside the rails as may be required.

Small quantities of ballast may be deposited along the track by one or two cars hauled in local freight trains; the train may halt at the specified place while the crew dump the cars or while section men shovel off the load. If this will cause too much delay, the cars may be set out on a siding to be taken on by a work-train engine and then picked up by another freight train. On the Delaware & Hudson R. R., which is ballasted mainly with hard-coal cinders from the engine ashpits, both freight and work trains are employed. In winter, the cinders for ballasting, raising sags, or widening fills are loaded into air-operated, side-dump cars which are handled largely by local freight trains, being dumped at designated points either by the section gang or by a man riding on the train. In summer, cinders for ballasting are loaded generally in self-cleaning, hopper-bottom coal cars and handled by work trains, the air-dump cars being then used only for widening banks. It is considered preferable to close one track to traffic when making heavy ballast lifts, but this depends upon grades and local conditions. Ballasting gangs range from section gangs with 10 or 12 men to extra gangs with 50 men, according to the lift and the extent of the work.

Plowing Ballast from Cars.—Gondola cars for unloading by shoveling or plowing have removable ends and the sides composed of panels hinged to a top rail and latched at the bottom. Such cars are of about 30 cu. yd. or 50 tons nominal capacity. Flat cars with 10 to 12 cu. yd. (or about 15 yd. if 12-in. sides are used) have low-hinged sides or loose boards supported by short stakes in the stake pockets. With the boards removed, these stakes and the posts between the doors of gondolas guide the plow along the cars. Steel aprons are laid between the cars to prevent material from falling on the rails. On the rear car of the train is a heavily weighted, wedge-shaped plow extending the full width of the floor and shaped to throw off the material on one or both sides. To the plow is attached a steel cable laid along the loaded cars; this is led through sheaves or snatch blocks or chains hitched to the sides of the cars if the unloading is to be done on a curve. If the plowing is done by the locomotive, the car brakes are set and the cable is attached to the locomotive, which is then uncoupled and moves slowly ahead, pulling the plow along the cars.

A more efficient method is to have the cable wound on the drum of a steam hoist or "rapid unloader" mounted on a car next to the locomotive and taking steam from the boiler. With this arrangement, the material can be dumped at one or several places, or distributed in any desired quantity by moving the train ahead while the plow is throwing off the ballast. Near the point where ballasting is to be done, the train stops while the end of the cable on the head or unloader car is hooked to a boom or gallows frame beside the track; the cable then unwinds from the drum as the train moves slowly ahead, until at the rear car the cable is hooked to the plow. Men run ahead to unlatch the doors, and a man riding on the plow signals the unloader operator or locomotive engineman by flag or signs. Behind the plow come men to clean out the cars and latch the doors. When the plow is on the last ballast car, the cable is detached and the car is switched out, ready to be coupled to the rear of the next train. The unloader car is not cut out, on account of the pipe connections to the locomotive.

Cleaning Ballast.—In time, the bed of ballast becomes choked with dirt and cinders so that it does not drain properly, but hard ballast such as stone and hard slag may then be restored to condition by screening out the fine stuff. This may be required at intervals of 1 to 3 years at terminals, 3 to 5 years on heavy-traffic main lines, and 5 to 8 years on lines with light traffic. From 15 to 25 per cent of new material will be required to make up for lost ballast and to maintain the standard section. If the track has become centerbound on hard ballast, the only remedy is to raise the track, clear out the old material, and replace with new ballast.

For cleaning, a simple method is to shovel out the ballast between the ties (level with bottom of tie) and about 6 in. below them along the shoulders, and then to throw it back with forks; or it may first be cleaned by throwing it on inclined wire screens placed at the side of the roadbed. Machine methods are cheaper and quicker. In one method, a locomotive crane with grab bucket picks up the ballast between the tracks (where the ballast from between the ties is shoveled), and dumps it on an inclined screen mounted on an adjacent gondola car, so that the dirt will fall through into the car. The clean stone collecting at the bottom of the screen flows down an inclined chute hung to the side of the car and is thus returned to the roadbed, ready to be thrown in between the ties. A lighter equipment for the same purpose consists of a rotary screen mounted on a small push car or motor car and driven by a gasoline engine. With a gang of about 20 men, ballast is shoveled from the track into a long hopper or trough, at the bottom of which is a belt conveyor discharging into the screen. The cleaned stone is returned to the track by a chute, while the dirt falls through the screen to a lateral belt conveyor which discharges it clear of the track.

For the cribs between the ties, handling with forks may be as efficient as machine work, for after the first few cribs are cleared, the ballast forked from the successive cribs may be shaken and thrown back with the preceding cribs, although in this method the dirt remains on the track. In any case, the ballast under the ties is left undisturbed and that between and outside them is removed by shoveling, except that a grab bucket may excavate the shoulders to about 12 in. below the ties. But in a pneumatic process tried on the Pennsylvania R. R., a vacuum cleaner mounted on a car has three 8-in. pipes hanging between and outside the rails through which the ballast and dirt are drawn by suction into a horizontal cylindrical tank. The pipes are telescopic for adjustment in length and have a lateral swinging motion, one moving between the rails and the others on the outside. An inclined screen across the tank retains the ballast, while cinders and dirt pass through. With the expansion of the air in the tank, its velocity is reduced and the dirt settles to the bottom, baffles preventing it from escaping with the air exhaust. A suction pipe at the rear end provides for removing the collected dirt and discharging it into a car coupled behind the machine. The cleaned ballast settles into small hoppers which have revolving valves through which the ballast falls into chutes which return it to the track, the valves forming air locks to prevent loss of air or pressure from the tank. A steam turbine at the rear of the cleaner car is connected to the locomotive at the front end and operates an exhaust blower. This ballast cleaner is moved along the track at about 10 or 12 ft. per minute.

CHAPTER XVII

RAIL RENEWAL

On all existing track the rails have to be renewed from time to time, either on account of wear or to provide heavier rails for increased traffic, since maintenance-of-way becomes expensive and unsatisfactory when the rails are too light for the wheel loads or traffic which they carry. This work of renewal is more or less troublesome since it interferes with traffic, and, therefore, thorough cooperation between the engineering (or maintenance) and operating departments is essential to insure safety, reasonable progress, and a minimum of delay to trains. To some extent rail renewal may be done by two or three section gangs grouped together, but for any extensive length of track it is handled by an extra gang. It is done usually during the winter or at seasons when traffic is light and the climate is not favorable for other kinds of track work. It is notable, however, that the regular practice of the Lehigh Valley R. R. is to carry on its rail renewals throughout the winter, special steam-jet appliances being used to clear snow and ice when necessary, and the ballast and road-bed being so well consolidated that there is little or no heaving to require shimming of the new rails.

Single-track Operation.—On single track and on very busy sections of double-track lines, the work is done necessarily in the intervals between trains, but on double track it is usually better to close a stretch of one track for a day or half a day. In this latter case, all traffic is handled on the other track, under the protection of flags and signals or a pilot. To direct the traffic, a trainmaster should be on hand, and telegraph operators may be installed in cabins at the ends of the single-track section, having telephone communication with each other and with the dispatcher. If the permanent crossovers are not located conveniently for this diversion of traffic a temporary crossover with facing switch may be installed to form the connection at each end of the section; an alternative method is to disconnect the track, throw it over, and put in a switch, like an ordinary turnout. Under this plan, as soon as the track is closed to traffic, the main work will be started with a large force that has been organized carefully, so as to complete the work within the allotted time.

Rail renewal can be done effectively, and generally with less delay to traffic by the above method than if it is done between trains. With many trains and allowing ½ hr. for each interruption in connecting up the track, waiting for a train to pass, and starting work again, the loss in wages for non-productive time of a large gang would be considerable, in addition to the amount involved in the delay, stopping, and starting of trains (especially heavy freight trains). With intensive traffic, however, as near large cities, it may be impracticable to handle it even temporarily on one track.

Preliminary Work.—New rails are distributed along the track some time in advance, and the curves and spirals may be staked out to insure accuracy. As a rule, all the work is concentrated in the main rail-renewal force, but in some cases a gang is sent over the selected length of track in advance of the actual rail laying, and before traffic is diverted, in order to do some preliminary work. This includes placing new rails upright on the ballast shoulder, so as to be ready for the tong men, lining the track, pulling the extra spikes on curves, scoring the ties to indicate where adzing is required, and also oiling the bolts of joints at switches and crossings and at intervals of 10 to 25 rail lengths on curves, to facilitate opening these joints when the new rails are being laid. If the rails are to be bent on curves, the gang may do this also, the curve stakes being then marked with the degree of curve for this purpose. Distribution of splice bars, bolts, spikes, tie plugs, tie-plates, and rail anchors should be done only shortly before work is started, or some of the material will be scattered and lost.

Method of Rail Renewal.—In the ordinary method of work, rails are renewed consecutively, each old rail being removed in turn and the new one placed, spliced, and spiked. The old rails are usually shifted inward, but are sometimes thrown out on the roadbed shoulder. After the new rails have been laid, the old rails and fastenings are picked up and loaded for removal. In disconnecting the old rails, the nuts should not be knocked off with hammers, but unscrewed with wrenches (after oiling, if necessary), so that the material will be fit for further use.

Where the track is of correct gage, only two lines of spikes need be drawn; if the gage must be corrected or the new rails have a different width of base, it may be necessary to pull three rows, while if new tie-plates are to be laid, all spikes will be drawn. On the Pennsylvania R.R., it is required that one rail be laid from the outside and the other from the inside, in order to maintain the gage, retaining the north row of spikes for each rail. Where shoulder tie-plates are used, however, both lines of rails are laid from the inside. At the end of each day, the track must be left in completed condition, full spiked and bolted.

In following this method on the electrified zone of the Chicago, Milwaukee & St. Paul Ry., requiring much heavier joint bonding than for signal circuits, special care is taken that the bonding crew is adequate to keep up with the relaying and to have all joints bonded when work stops for the day. Care must be taken also that all rails are secured by an ample number of spikes, before a train is allowed to pass, and all trains should be signaled and required to pass over such work at reduced speed. Neglect of these precautions has resulted in serious accidents.

Temporary Connections.—For temporarily connecting up the new and old rails to permit the passage of trains over work done under traffic, most railroads permit the use of a switch rail of the same section as the new rail. This rail, 10 or 12 ft. long, is placed on riser plates and faces traffic so as to prevent worn or false-flanged wheels from spreading the rail and being derailed. At the end of the day, however, the connection must be closed by cutting and drilling the old rail for a standard splice joint, full bolted and spiked. Some roads require a splice joint in any case, and this will be a compromise joint if the rails are of different sections. Two 12- or 15-ft. rails may be kept on hand for these closures in order to avoid the use of very short pieces. A combination of short rails may be used in connecting with a turnout to avoid cutting the lead rails. For cutting new rails, a saw is preferable to a chisel, unless its slower action would delay traffic. Where rails are bonded for track circuits, bond wires must not be broken or rails removed (except in emergency) unless a signal repairman is present with facilities to keep the signal apparatus in working order and to bond the new rails.

Rail Joints and Expansion Spacing.—Care must be taken to provide the proper expansion spacing at the joints by shims of proper thickness (see "Rail Joints"), the foremen being provided with thermometers which are placed on the rail base in the shade, and left long enough to show the correct temperature. These shims are left in place for a sufficient time to insure against disturbance of the joint spacing by the laying of adjacent rails. Rail anchors should be placed the day the rails are laid (or not later than next day) to insure against creeping. All joints should be bolted up before the spiking is done. They should be kept clear of road crossings, cattle guards, and water stations, and at least 6 ft. away from the ends of bridges, whether on the embankment or the bridge deck. Rails of different lengths, but not less than 14 ft., may be used for this purpose, although some roads provide 60-ft. rails for use at road crossings. Rail joints will require tightening the next day after the rails are laid and again within a week or 10 days. Insulated joints require special care to prevent injury to the fiber parts, and one man may be assigned to assemble and fit up all these joints.

In position, the joint in one rail should not vary more than 6 to 12 in. from the middle of the opposite rail, although the Pennsylvania R. R. allows a range of 18 in. on curves. The inner rail on a curve will gain upon the outer rail at the rate of 1 in. (or 1.03 in.) per 100 ft. for each degree of curvature, so that on a 3-deg. curve 1,000 ft. long the inner rail will be the shorter by $2\frac{1}{2}$ ft. Occasional short rails are inserted to keep the joints in proper relation, but no rail shorter than 14 ft. should be permitted in main track.

Laying Rails in Strings.—An alternative method of rail renewal is to connect the new rails in strings on the ends of the ties for 1,000 ft. or up up to a mile or more. Each string is bolted and electrically bonded, ready to be thrown or shifted into place with bars as the old rails are thrown inward, leaving the string of old rails to be disconnected later. The length of the string of rails is such as can be thrown into place in the time available. It is estimated that this practice costs twice as much as laying rails singly, but it is followed to some extent, principally on main lines having electric traction and heavy traffic, where the application of large copper bonds in the track would take so much time as to delay trains. Some railroads prohibit this practice; the Pennsylvania R. R. requires that, if it is necessary on account of traffic, not more than 30 rails must be bolted together.

With long strings of rails handled in this way, it is difficult to maintain uniform expansion spacing on curves, since the rails in the outside string, on the ends of the ties, will close together as the string is thrown into position, while those of the inside string will draw apart. For this variation allowance must be made at the rate of about $\frac{1}{4}$ in. per 100 ft. excess spacing on the outside and less than the normal spacing on the inside. The bolts may be left somewhat slack, the expansion shims being removed, spacing adjusted, and bolts tightened as soon as the rails are thrown into position. Two men can remove the shims, one raising the joint with a bar to open it sufficiently to release the shim. The joints at ends of curves should also be left slack. In throwing the rails into position with bars, care must be taken to avoid bending the splice bars laterally. Two gangs, 300 to 400 ft. apart to avoid interference, may relay both lines of rails simultaneously; otherwise, the second line should be relaid as soon as possible.

Machinery in Rail Laying.—To speed up the work and reduce the labor force, machinery has been introduced in renewing rails as well as in tracklaying. Of particular importance are small, portable, rail-layer cranes or derricks, with which 4 or 5 men can do the work of a gang of 16 to 20 handling heavy rails with tongs. While the time per rail is about the same with a machine and a fresh gang, the former can maintain its pace all day, while the tong men necessarily slow down. Locomotive cranes are also used for placing rails. Other machines used are pneumatic wrenches to screw nuts on the splice bolts and pneumatic drills for bond holes and bolt holes, these tools being connected by hose with the portable compressors which serve pneumatic tampers. Several self-contained machines having their own power and propelling mechanism are used on the Chicago, Burlington & Quincy R. R., including rail-laying derrick cars, bond-hole drilling cars, and tie-scoring cars with circular saws to make cuts across the ties indicating the depth for adzing the ties to fit the new rails.

Rail-laying cranes are usually small hand or power derricks on motor cars which travel on the track, with boom projecting in front to pick up new rails at the side of the roadbed and place them in position on the track. A different design has a steel frame mounted on grooved wheels of $3\frac{1}{2}$ ft. gage, two wheels riding on the track rail and two on a short portable rail laid on the ties and inside the line of rails that is being relaid. A fixed, inclined boom transverse to the track extends beyond the ties and has a trolley hoist with rail tongs to pick up rails laid along the ballast shoulder or between the tracks. A hand crab or winch operates the cable, first raising the rail and then bringing it into position by traversing the trolley backwards along the inclined boom.

A force of 19 men with such a machine has laid rails on $\frac{1}{2}$ mi. of track in a day, closing up four times to allow trains to pass. For laying 3 mi. of rails in 10 hr., completely bolted and spiked, there may be two rail-layer machines with 80 men each and a third gang of 15 men putting bonds and rail anchors on both lines of rail, making a total of 175 men. In each 80-man gang there would be 30 ahead of the machine to pull spikes, throw out old rail, plug and adze ties, and shift tie-plates; 8 men (exclusive of operator) with the machine pick up, place, line, bolt, and spike the rail; about 40 men follow the machine to do the full bolting and spiking. The old rails are sometimes disconnected and loaded on a work train by a separate gang during the same time.

EXAMPLES OF RAIL-RENEWAL PRACTICE

Lehigh Valley R. R.—A typical rail-renewal force on the Lehigh Valley R. R., where extensive use is made of locomotive cranes, is given herewith, listing the men in order of work from the head end to the rear end. Such a total force of 140 men with a locomotive crane has not infrequently laid 3 mi. of track with 39-ft., 136-lb. rails in a working day of 8 hr., using pneumatic nut tighteners and pneumatic bonding drills. The rail is laid usually on one side for the entire day, in order to avoid loss of time in backing up the crane and interrupting or disorganizing the force. But by duplicating the forces, both rails can be laid at the same time, each gang having a locomotive crane.

In replacing 100-lb. with 136-lb. rails, it is necessary to strip the track of all fastenings and materials down to the ties, replacing them with the larger tie-plates and other material for the heavier rail. With extended experience in work of this kind, the forces have increased in efficiency, so that the number of men now required is somewhat less than in former years. Much of the rail-renewal work on this road is done throughout the winter, and this plan has proved very satisfactory.

RAIL-RENEWAL FORCE: LEHIGH VALLEY, R. R.

A. MEN AHEAD OF CRANE

Removing Old Rail and Fastenings	
Cutting bolts at approximately every tenth joint	2
Removing old rail anchors from rail	2
Pulling spikes (claw bars, winter season)	29
Throwing out old rail	8
Removing old tie-plates	2
Digging out road crossings	2
Preparing Ties for New Rail and Fastenings	
Putting creosoted tie plugs in old spike holes	2
Driving down tie plugs and old spike stubs	1
Adzing ties for new and larger tie-plates	8
Putting on new tie-plates	5
Water boy	1
Laying Rail and Fastenings	
Swinging rail in place with crane (39-ft. rail)	3
Holding rail in line with bars	2
Spiking new rail down at ends and center (on curves)	3
Locomotive crane operator	1
Locomotive crane fireman	1
Total men ahead of crane	72

B. MEN BEHIND CRANE

Applying Splices, Bolts, Spikes, and Bond Wires	
Gaging	6
Putting on insulated joints and restoring road crossings	2
Spiking	20
Putting on splices	8
Operating pneumatic nut tighteners	4
Driving up splices	1
Operating portable compressor	1
Final tightening of bolts by hand (about three-quarter turn)	8
Digging out and applying rail anchors	4
Water boy	1
Operating pneumatic bonding drills	4
Operating portable compressor	1
Distributing bond wires and supervising bonding work in general (usual signal construction gang foreman)	1
Applying bond wires	3
Changing wires at signal connections	2
Handling push truck equipped with grinder for sharpening tools	2
Total men in rear of crane	68
Grand total	140

Baltimore & Ohio R. R.—A gang of 60 men with a light rail-laying crane has relaid 100-lb., 39-ft. rail at the rate if $2\frac{1}{4}$ mi. per day, but without changing tie-plates. One track was given up to the rail-renewal force for the entire day. Rails and material were distributed on both sides of the track, with the proper number and lengths of short rails to fit the curves of 2 to 6 deg. To expedite its work, the rail-laying machine was filled with long rollers under the front end, thus dispensing with the short portable rail (see "Machinery for Rail Laying") and giving a more stable support.

Spikes were drawn on the inside of the line of rails to be changed and occasional joints were disconnected, the rail then being thrown in far enough to support the rollers of the rail layer. The new rail was then swung in, spiked, and spliced, and the old rail was thrown out. A following gang tamped loose ties and did any final dressing, so that the track was left in finished condition at the end of each day. As section gangs and extra gangs were combined to build up the rail-renewal force, there were 7 foremen included; but they were utilized to advantage by dividing up the force of 51 men (including a water boy) as follows:

Foreman with 11 men pulling spikes and disconnecting rails. Foreman with 3 men moving old rail in and 1 pulling outer spikes of old joints.

Foreman and 5 men with rail-layer crane.

Foreman with 2 men spiking new rail and 2 men distributing new joint material.

Foreman with 11 men splicing and full bolting.

Foreman with 11 men full spiking.

Foreman with 4 men throwing out old rail.

Behind this force came a foreman with 10 men for surfacing, and a foreman with 6 men applying rail anchors and drilling the rails for bond wires.

Rock Island Lines.—For rail renewal under traffic on single track with train speed restricted to 20 m. p. h., a practice followed by the Rock Island Lines is to divide the rail-renewal force into two gangs, each in charge of an assistant foreman, as follows: First gang—3 men pulling spikes and connecting up track for passage of trains, 7 to 12 throwing out old rails and adzing ties, 2 men placing tie plugs, 5 men with rail layer putting in new rails, 2 men flagging, 1 man noting temperatures and applying expansion shims. Second gang—4 men splicing and bolting joints, 4 distributing tie-plates and track fastenings, 12 spiking, gaging, and applying tie-plates, 4 shimming on old ties and applying rail anchors. A water boy serves both gangs. A surfacing gang of 50 to 75 men follows the work to adjust and tamp ties, remove shims, complete the spiking, and dress the ballast.

New York, New Haven & Hartford R. R.—A gang of about 146 men is required for each line of rail, two gangs some distance apart being worked if the track is closed to traffic. For each gang there is a steam rail-laying crane, two portable air compressors (for tampers) serving the pneumatic wrenches, another compressor serving drills for signal-circuit bonds, two compressed-air rail-loading machines, and a work train. The distribution of the 146 men is as follows: 12 pulling spikes, 7 throwing out old rail, 12 removing tie-plates, adzing, and plugging ties, 4 placing new tie-plates, 10 with hoist to set new rail and spike it sufficiently for machines to pass, 9 placing splice bars and bolts and starting the nuts, 4 with pneumatic wrenches tightening the bolts, 45 men full spiking, 10 tightening bolts with hand wrenches, 5 putting on rail anchors, 12 starting nuts off old bolts, 4 taking off old nuts with pneumatic wrenches, 12 loading old rails and material on work train and clearing up the track.

Other Railways.—On the Santa Fe Ry., four men with a motor rail-laying car have laid 100 rails in an hour, keeping about 100 men busy spiking, gaging, and bolting behind the machine. On busy track, 300 rails have been laid in the morning, the afternoon being then occupied in full bolting and spiking, dressing the track, and disconnecting the old rails. On the Norfolk & Western R. R., 36 men with a rail-laying crane can lay rails on about 1,500 ft. of track per day, or 1,000 to 1,200 ft. where the heavy traffic averages 40 to 50 trains daily. In this force, 14 men would be pulling spikes and throwing out old rails, 2 putting on splice bars and 20 with and behind the machine to set in new rail and do the spiking, bolting, and gaging. Rapid single-track work on the Canadian Pacific Ry. has reached 8 to 10 mi. daily with a specially organized force of about 260 men. In this case, the 33-ft. rails were bolted up in pairs or threes along the ends of the ties, each unit being handled by a large number of men with tongs. After the rails had been laid, the force was divided into gangs of 30 to 50 men to insert tie-plates and complete the spiking.

Renewing Rails on Elevated Lines.—With the dense traffic on elevated rapid-transit lines, the renewal of rails is a difficult proposition. On the Chicago Elevated Rys., much of this work is done under traffic, although in some cases, especially on the downtown loop, the work is done on Saturday night with all traffic handled on one track.

DISTRIBUTING AND HANDLING RAILS

Care in Handling Rails.—Since the rails, which are the most expensive part of the track, are flexible when not supported uniformly, it is important that care be taken in handling them to prevent bending, fracture, or injury. Surface-bent rails make rough-riding track and are difficult to straighten. New rails should never be dropped or thrown from cars but unloaded by means of skids or other devices; relayer or second-hand rails are best distributed in the same way, but if thrown from flat cars they should be dropped on soft level ground, clear of boulders or other rails, and not left so unevenly supported as to become bent. They should never be dropped from gondola cars.

Rails for renewal should be distributed in proper position for laying, but not so as to obstruct road crossings or to be dangerous to trainmen. At yards and station grounds, the rails should not be distributed along the track if they are not to be used at once, as trainmen and yardmen are likely to fall over them; the new rails should be stacked by means of skids or unloader devices at places in the yard convenient for distributing them just before laying.

Cars for Rails.—Rails 30 and 33 ft. long can be carried ordinarily on flat or gondola cars, with end planks on the former to prevent the rails from shifting; but gondola cars are more difficult to load and unload. Box cars and stock cars are also used. Rails 39 and 45 ft. long can be loaded in two groups on three flat cars or open-end gondolas, the middle car supporting the projecting ends of both groups.

Rails 60 ft. long may be carried on two cars, being supported mainly on bolsters just below the outer trucks, but having intermediate bolsters over the inner trucks to prevent undue sagging and swaying of the load; an alternative plan is to load them on alternate cars, with the ends supported on the intermediate cars, giving three loads on seven cars.

Unloading Rails on Track.—For unloading rails of moderate weight from flat cars by hand, two gangs working from opposite ends of the train may throw two rails from each car; when the gangs meet, the train is moved ahead a train length, and the groups then work back to the end cars. To avoid dropping the rails from full height, the car may have from two to four hangers hooked to the stake pockets and each carrying a horizontal roller. The hangers are of varying length so that the rollers form an incline down which the rails slide and drop only from the last hanger, unless men receive the end and lower it.

In stacking rails for later distribution, they may be handled by a locomotive crane or slid down skids formed of rails or of a pair of timbers about 3 by 4 in. and 6 to 10 ft. long, with a sheave on the underside at the upper end. Each skid has a rope passing up over the sheave and fitted with a hook to engage the rail, these ropes being handled by men on the ground.

In unloading from the end of a flat car, men with tongs on the car slide the rail over the sill; men on the track carry it out and lay the end on the ground, then returning to lower the other end. The men should not be permitted to drop the rail, but there is liability of their dropping it accidentally and being injured. A simpler plan is to pull the rails off by ropes, the men lowering the end next the car, but the number of men may be reduced materially by using an inclined apron or pair of skids attached to the end sill of the car and riding on the track; as the rails are hauled off, their ends slide down the apron.

For gondola cars with fixed ends, a second and shorter apron may be hooked to the top, with its lower end resting on the longer apron. Unloading from the rear of flat or stock cars may be done by anchoring the ends of the rail and by moving the train ahead. For this purpose two ropes or chains 30 to 50 ft. long are used, having at one end an L-hook to engage a bolt hole in the rail. The other end of the chain has a claw hook which is fitted over the rear side of a tie; this hook is attached to a bar, so that the man does not have to stoop in attaching and detaching it.

An apron may be used, as described above, but another arrangement is to couple a push car about 10 or 15 ft. behind the rail car, with a gable roof of steel plate on two or three rails bent to shape, the apex of the incline being level with the floor of the rail car. A man hooks the ropes to the rails in the car and guides their ends to one side or other of the inclined plate so that the rails will be thrown to opposite sides of the track.

Rail-handling Machines.—In general, it is quicker, cheaper, and safer to unload rails by mechanical devices, with a smaller number of men and less chance of injuring the men or the rails. Bridge derrick cars and locomotive cranes are sometimes used, but require regular crews and usually are not adapted for quick work in unloading rails along the track, so that there are advantages in unloaders which can be operated by the foreman or an experienced laborer.

These machines are of two classes: (1) portable unloaders to be placed on the cars; and (2) unloader machines which are coupled in the train. A machine of the latter class is a flat or box car having an A-frame and mast at each end, with a detachable 22-ft. boom which is guyed from the head of the mast. The hoisting cable passes around a sheave on the piston rod of a horizontal or inclined air cylinder and leads back to an attachment on the rear of the mast. Hoisting drums may be used instead of the direct-pull cylinders. Such a machine coupled between two rail cars can do quick work with a crew of 10 men; one at each hoist and four with each car.

Portable derricks with hand or gasoline hoists or compressed-air cylinders may be mounted on the cars or travel along them, or the small rail-layer cranes already described may be placed on the cars. Portable hoists or derricks of this kind consist of a wheeled platform with mast or gallows frame and boom, the mast sometimes guyed by cables hooked to the car sill. The hoisting line, with tongs or a sling for two rail tongs, is operated by an air cylinder on the mast or boom, air being supplied by a connection from the train pipe of the brake system to the hoisting cylinder or to a reservoir on the derrick platform.

An unloading gang with such a derrick may consist of six or eight men: foreman, machine operator, 2 or 3 on the car to attach tongs and guide the rails, and 2 or 3 on the ground to place the rail and detach the tongs. A gang of 16 men with two such machines has unloaded four to five rails per minute as the train moves along. A ditching machine or small locomotive crane mounted on a flat car between two rail cars has unloaded six cars of 75-lb. rails in 8 hr., including the time consumed in clearing trains and running to and from work.

Loading Old Rails.—In removing old rails from the track, the loading on cars is done to best advantage by the derricks and other machines used in unloading and distributing new rails. With the ditcher arrangement noted above, a work train was loaded by six men: a machine operator, one man on the ground hooking tongs to rails, two on the ground guiding the rails, and two on the car arranging the rails and detaching the tongs. With the machine between two flat cars, the front car was loaded from one side of the track and the second from the opposite side; then the loaded cars were switched out and replaced with two empty cars.

Where it is necessary to load rails by hand from the side of the track, there should be an ample number of men for the weight of rail and one man should be assigned to call all directions. No attempt should be made to load in this way on narrow fills or other places where the man cannot get away readily if the rail should fall back. A safer plan is to slide the rails in at the end of the car.

CHAPTER XVIII

DRAINAGE AND DITCHING; HOLDING LOOSE SAND

Drainage.—A dry and firm roadbed greatly reduces the labor and cost of track maintenance, so that one of the most important factors in securing good track and reducing the expense of maintenance is effective drainage, which includes keeping water away from cuts and fills as much as possible and also providing for the removal of water from the subgrade. In cuts, side ditches are provided for draining the roadbed and carrying off surface water, as described under "Roadbed," while tile subdrains may supplement the ditches in wet ground or at soft spots.

Where the ground is inclined towards a cut, surface drainage may be intercepted by a ditch far enough from the top of the slope to prevent water seeping through to the cut, and with grade sufficient to carry off the water readily and not allow it to accumulate. The excavated material is thrown out on the side adjacent to the cut. Wet cuts and long cuts on easy grades should have wide and deep side ditches; the slopes should be flat enough to prevent slipping, and intercepting ditches should be built outside the cut, as noted above. If sufficient width for flat slopes and wide track ditches cannot be obtained, as on cuts through valuable property, retaining walls of masonry or cribbing may be built outside the ditches and the slopes commenced from the top of the walls.

For fills on wet ground, ditches near the toes of the slopes will catch the drainage flow towards the fill and will also tend to drain water from under the fill and thus give a more stable foundation. This latter drainage may be facilitated by lateral pipes or stone-filled trenches extending from under the fill to the intercepting ditch, especially where springs are under the fill. On sidehill lines, ample provision must be made for carrying water across the roadbed and under the fills by means of culverts, ditches, and drains. For the drainage of yards, pipes may be laid between the tracks, with cross drains where necessary, but in fine or silty soil the drains may be trenches filled with coarse, broken stone and having pipe laterals to a main ditch or natural watercourse. Lateral drainage is sometimes provided by stepping down the tracks in succession, the higher side of the yard being along the main track. At stations, a pipe drain may be laid outside the edge of the platform or between the tracks and connected with suitable outlets.

Culverts and Waterways.—All culverts and waterway openings must be kept clear. Where freshets or floods occur, the banks of the stream and the slopes of the embankment (especially at trestles) may be protected against scour or erosion by cribbing, rough stone riprap, or trees anchored and weighted down. To prevent silting beyond the outlet of a culvert, a concrete flume of V-section may be provided. If freshets bring down much drift, a catcher may be built upstream, consisting of old rails set vertically in a concrete footing or one or more V-shaped dams formed of trees having their butts lapped on a heavy post at the apex, with their ends resting on the banks. No fences or wires should be allowed across drainage openings, as they may collect drift and block the channel until the water bursts through and perhaps causes a washout.

Openings that may be inadequate for flood flow should be kept in mind by the engineer, the roadmaster, and the section foreman, and examined carefully after a storm to note any signs of scour or failure. Small pile trestles on flashy streams need particular attention of this kind. High-water flood levels should be marked by the foremen for the information of the engineer in investigating the necessary capacity of drainage openings or waterways. Much has been done in replacing trestles at waterways with solid fills having box, arch, or pipe culverts. In shallow fills, a double-barreled concrete box culvert may be placed, with the ballast laid directly upon the deck. Pipe culverts are of concrete, vitrified clay, corrugated sheet iron, and cast iron; this last including ordinary bell-and-socket pipe and special plain or corrugated pipe in short lengths with interlocking joints to prevent the pipes from pulling apart.

Building New Culverts.—In providing additional waterways, pipes have been forced through earth fills with jacks, a 50-ton jack being used on 30-in. pipe while a man inside removes the earth. Gas and oil pipes of 3- or 4-in. diameter have been put through with an earth auger or water jet. In high fills, tunneling may be employed for the central portion, working from both ends and perhaps from a middle shaft in a very wide fill. Where a sheeted trench is necessary, the track may be supported by timber or I-beam stringers placed under the ends of the ties and resting on heavy sills beyond the trench. For narrow trenches, a rail may be placed under each end of the ties, with another rail on the ties, the two rails being held together by heavy clamps; or two short rails with end bearings may be placed on either side of each track rail, the latter resting on U-shaped hangers hooked over the side rails. In replacing old box culverts with pipe, the pipe may be hauled in on skids; the middle length is laid first and backfilling rammed tightly around it, after which the other lengths are placed and backfilled successively.

Track Ditches and Ditching.—Routine work of the section gangs includes the enlargement and clearing of side ditches, especially where there is frequent heavy rainfall, since maintenance work and expense is reduced materially with a firm and well-drained roadbed. In general, these ditches must conform to the standard plans, but may be enlarged where necessary for prompt removal of storm water. Earth and ballast slips into the ditches and obstructs them, while soft ground will gradually close. The section foreman, therefore, must have his ditches cleaned in the spring, as soon as frost is out of the ground, and again in the autumn to put them in condition for winter. In wet cuts, the ditches should be at a good distance from the track, so that water in them will not seep into the roadbed, and they must be deep enough to drain water from the roadbed. Ditching is commenced at the low point, so that the work will be drained as it proceeds.

Ditches should be kept parallel with the track and not led around stumps or boulders, as such bends catch floating drift, check the flow, and cause silting. The ditches must be large enough to carry off heavy rainfalls without flooding the roadbed, deep enough to drain the ballast and roadbed, and with as steep a grade as practicable. The grade must be uniform, to avoid pools. Pipe or box drains of ample capacity carry the ditches under road crossings, and special attention should be given to the drainage of these crossings. Heaving or wet spots must be investigated, and special drainage provided if necessary. In soft material that washes readily, the ditch may be lined with old ties or dry stone walls on each side. Concrete ditches or guttters are used in some cases, particularly where the slopes are turfed or sodded.

At the ends of cuts, the ditches must be diverted from the roadbed so that the water will not wash or gully the adjacent fill but will flow safely to a stream or culvert. If natural waterways are not available as outlets, the ditches must not be connected to private waterways without special permission; nor must farmers or others be allowed to lead their drains or ditches into the railroad ditches. On curves it may be desirable to dam one ditch at different points and divert the water to the opposite side by cross-drains in the roadbed; similar drains on double track may lead the water from a central ditch to the side ditches. In wet sidehill cuts, cross-drains may relieve the ditch on the upper side, the discharge on the downhill side having an apron or gutter on the slope to prevent gullying the fill; or if an iron pipe drain is used, it may project far enough to discharge the water clear of the fill.

Cleaning Track Ditches.—In cleaning the side ditches, the material must not be thrown on the shoulder or the slope of the cut. If it cannot be cast outside of the cut, it must be hauled out and dumped over the fill or at some convenient place. This may be done by wheelbarrows, by pony cars riding on one of the track rails, by push cars (preferably with dump boxes mounted on them to avoid shoveling), or by small cars on a narrow-gage track over the ditch if there is room to clear trains. Where material has to be carried across the tracks and it is undesirable to lay planks for wheelbarrows, the men can carry it in boxes having two straight handles at each end. Wheeling planks must not be laid close to the rail, as they may be tilted or displaced so as to engage the wheel flanges of trains and perhaps cause accidents. Wheelbarrows having grooved wheels to run on the rail are sometimes used. Where a push car is used on the running tracks the work must be protected by flagmen.

The ditch should be set out with a ditching line, and grade may be given by sighting with a spirit level over stakes 100 ft. apart, measuring the required depth at each stake. In yards, the foreman should arrange with the yardmaster and do the work when the sidetrack next the ditch can be kept clear of cars.

Water Pockets.—As railroad embankments are formed by the dumping of loose materials, usually of varying character, soft spots may develop where water is trapped in pervious material. Ballast also is gradually driven down so that the roadbed surface becomes concave, while the shoulders may be built up from time to time by adding impervious material. As a result, water collects in the trough or enclosed pockets and forms soft spots, necessitating frequent filling and surfacing, and thus increasing the expense and work of maintenance. The concave roadbed or natural subsoil conditions may cause soft spots also in cuts. Such troubles often increase with heavier engines and traffic, so that it is economical to find the cause and provide a permanent remedy rather than to carry on continual filling, surfacing, and track work. Rolling new roadbed with a 10-ton roller would compact the material and reduce such troubles materially, but such rolling is rarely practiced.

The remedy is usually drainage in some form, such as deeper side ditches, tile subdrains, or lateral drains of pipe or stone-filled trenches extending from the water pocket to the side drains or to the outside of the fill. Other methods that may prove advisable are: (1) digging out the soft material and refilling with clean cinders or gravel, or perhaps with loose rock covered with a bed of cinders; (2) cutting down the shoulders and dressing the roadbed to a convex section, and then refilling with good material or sub-ballast for proper grade; (3) widening the roadbed to give better drainage in cuts and better distribution of the load on fills. On the Kansas City Southern Ry., stable roadbed in water-bearing soil has been obtained by digging 3-ft. trenches along the sides of the roadbed, 6 ft. wide and 8 ft. apart, and filling these with loose rock to provide both drainage and side support for the roadbed material between the trenches. Roadbed in soft material may be improved by the application of a bed of sub-ballast to support the main ballast and distribute the load.

Wet Cuts.—In cuts through loose and wet material, there is constant trouble from the gullying and washing of the slopes by rain or drainage water For such cases, the intercepting ditch at the top may be lined with concrete. Diagonal drains of bundled saplings, broken stone, or semicircular tile are sometimes benched in the face of the cut to intercept water flowing down the slope. Encouraging the growth of grass and sod will aid in reducing slips and gullying. Where springs or drainage water break through the face of the slope, drain pipes may be inserted and a concrete gutter or an apron of stone riprap laid from this outlet down to the track ditch. If there is heavy seepage at various points, French drains of broken stone in trenches may be built up the slope from the side ditch.

Subdrainage.—In wet ground and soft spots and where trouble is caused by heaving, the drying of the roadbed by subdrainage will effect a material reduction in the labor and cost of track maintenance, especially where the ground has become saturated through side ditches and intercepting ditches being inadequate to carry off the surface water. For such work a pipe drain is laid usually 8 to 10 ft. from the center of the track under one or both of the side ditches, with inclined laterals extending across the roadbed in the former case or nearly to the center line in the latter. The drains should be below frost line and at least 8 or 12 in. below the bottom of soft material or they will soon be displaced. In very soft soil, the pipes must be laid in a V-shaped trough made of two 1 by 6 in. planks. The drains must be laid carefully to the grade set out by levels and stakes, with as much fall as can be given.

Bell-and-socket pipe is preferable for subdrains, as the bells hold the pipe in position, while plain-end pipe with butt joints is likely to be displaced. Both clay and concrete pipe are used, but the latter may be affected by some subsoil waters; second-class vitrified sewer pipe is utilized to a large extent, as its strength is not impaired by slight distortions and hair cracks which render it unsuitable for sanitary purposes. The joints are not calked but left slightly open to admit water, being covered with straw, hay, or strips of sod or tar paper to exclude dirt. For the main drains, 8- to 12-in. pipe is used, with 6- to 8-in. laterals. Manholes are made of large pipe set vertically, with the drains entering through the side. Cinders, clean gravel, or broken stone are filled in at least 12 in. deep over the pipe and then the trench is backfilled. Laterals have the upper ends closed to exclude dirt; both ends of the main drain are open, for air circulation, but should have rods or gratings to exclude small animals. The outlet must be kept clear and may have a head wall to hold the pipe in place and a concrete apron to prevent wash by the discharge. To prevent overtaxing the tile subdrains, the track ditches should be made of ample capacity and kept clean.

Sliding Ground.—Hillside ground which is unstable and has a continual movement downhill necessitates constant work in maintaining line and surface and keeping the track in safe condition. Drainage will generally prove a remedy but may involve heavy expense to be effective, and each case must be investigated thoroughly before adopting any plan. Where wet soil slides on a smooth stratum of shale or soapstone, light charges of explosive may break up the smooth surface, but the danger of starting a big slip by the shock and vibration must be considered. In sliding ground along the Columbia River on the Union Pacific Ry. (Oregon-Washington line), test shafts were driven to explore the subsoil conditions and then drifts about 4 by 6 ft. were driven to tap the wet spots, starting from points where water flowed from the ground and following the flow to its source. Similar work has been done on the Southern Pacific Ry.

Sliding ground at Raccoon Mountain, on the Nashville, Chattanooga & St. Louis Ry., was checked by surface trenches cut along the lines of marked surface flow and filled with loose rock, forming French drains on a large scale. Another effective method on this road was the consolidation of loose ground by grouting with cement. A sliding hillside at the Boone viaduct of the Chicago & Northwestern Ry. was checked by rock-fill drains. Trenches 250 to 350 ft. long were constructed in the direction of the movement, 5 ft. wide at bottom and increasing from 4 ft. depth at the lower end to 10 ft. at the top; a 3-ft. bed of one-man riprap stone was covered with 2 ft. of willow brush and then backfilling. Wooden box drains about 10 to 15 ft. apart were put between the ties and extended up the hillside, with occasional lateral drains leading into them from wet spots.

Piling or heavy retaining walls are sometimes required to hold the cut through sliding ground, the piles or walls being extended into the underlying firm stratum and ample means being provided for the escape of water by brush filling around the piles and by drains behind and through the walls. In a cut 40 ft. deep on the Cleveland, Cincinnati, Chicago & St. Louis Ry., with wet yellow clay sliding on the smooth inclined surface of a stratum of shale, a road along the top was held by two rows of piles 4 ft. apart and 4 ft. spacing, with brush packed between them; two similar rows, but without brush, were driven near the foot of the slope, all piles pentrating the shale. Trenches 3 ft. wide and 4 ft. deep were cut in the slope and filled with loose rock. The track ditch was enlarged and a 2-ft. riprap wall built to protect the roadbed, 18-in. pipe drains carrying the water under the roadbed and discharging on the downhill side. An 8-in. tile drain between the tracks also helped to keep the roadbed dry. To eliminate trouble from slides in sidehill cuts at tunnel approaches, the Southern Pacific Ry. has built flat-roof concrete structures like snowsheds, the falling material passing over the roof. Similar protection provided by the Great Western Ry. (England) consists of a double-track gallery like a tunnel along the base of a cliff where the line is bombarded by loose rock and falling material.

Machine Ditching.—Improving track ditches and widening cuts is a permanent class of work on a large proportion of railway mileage, especially where the original cuts are narrow and have steep slopes in wet or loose ground. Such work is too extensive to be handled by the section gangs, with their various other duties, so that work trains with ditching machines or large extra gangs are kept in service. The excavated material is utilized for widening banks, filling old bridges replaced by culverts, or filling low ground, the earth being leveled with a spreader car when necessary. In plowing or dumping the material from cars, it should be thrown clear of the roadbed to avoid getting the dirt on the ballast and the expense of shoveling it off the track. Digging by hand and shoveling onto flat cars is slow, tedious, and expensive, particularly in wet and heavy ground. Machine ditching is economical and practicable under almost any conditions, irrespective of the amount of traffic and depth of cut.

Steam-shovel Ditchers.—Ditching machines are of three classes—steam-shovel, wing, and scoop. The first consists of a small revolving steam shovel with $\frac{1}{2}$- to $\frac{3}{4}$-yd. bucket, mounted on the work train to excavate the ditch and load the material on the cars. With a train of flat or gondola cars to be unloaded by shoveling or plowing, the machine travels along the cars, loading them in succession. This method is adapted for work in long, deep cuts where traffic is light and there is a long haul to the dumping point. For more rapid work, where the train cannot occupy the track for any length of time and a short haul is practicable, the flat car with machine may be coupled between two large air-dump cars operated from the locomotive. After loading one car, the ditcher moves to the opposite end of its car to load the second dump car. In some cases, however, the ditcher is complete in itself, built on a steel frame with railroad trucks. Besides the train crew, there will be the engineman and fireman on the ditcher and a few men to trim the slope. Several roads use a work train made up of two flat cars with ditchers (or two self-contained ditchers), four 20-yd. dump cars, tank car for water, caboose and locomotive.

Wing Ditchers.—The wing type of ditcher consists of a 40- to 50-ton spreader car having its wings fitted with templates to cut to the standard section of ditch and shoulder. This wing, operated by compressed air and braced to the car when swung out to its working position, cuts away the earth and pushes it ahead as the ditcher is propelled along the track by the locomotive until it reaches the adjacent fill or level ground, where the material goes over the slope or is spread out and leveled by the wing. About 10 cu. yd. of earth can be moved in this way without fouling the shoulder or ballast, but in long cuts the material may be pushed into piles to be loaded onto cars by a steam-shovel ditcher, the wing machine making a final run to finish the work to standard section.

A supplementary wing at the outer end of the main wing will widen the cut and trim it to the required slope. In heavy ground and for deep ditches, the machine may make several runs, the wing being adjusted to cut deeper each time. The speed is usually 3 to 8 mi. an hour, depending upon the depth of cut and character of material; but it may be as high as 10 mi. an hour in making the finishing cut, cleaning a good ditch, shaping a ditch excavated in some other way, or dressing the shoulders on fills. On single track, both ditches can be cleaned at the same time, two men being on the ditcher to regulate the position of the wings by the compressed-air cylinders.

Scoop and Bucket Ditchers.—In the scoop type of ditcher, which is not used extensively, a derrick traveling along the train of flat cars handles a 1- or $1\frac{1}{2}$-yd. scoop riding on a curved guide. With the scoop lowered into the ditch, it is loaded by moving the train forward and is then swung up and dumped. A modified machine has two large scoops on each side, and when these are filled by pushing the machine forward, they are raised while the machine is run to the end of the cut for dumping, thus dispensing with the cars. The frequent moving and stopping of the train in the first case is not a convenient or economical method of handling, while the second case involves much running to and fro.

Where traffic is light and long haul makes it necessary to load a number of cars at one operation, an ordinary steam shovel may be placed on one track to load a train of flat or gondola cars on the other track. A locomotive crane with grab bucket may be used in the same way, but will require one or two men on the ground to spot the bucket in position. On parts of the Pennsylvania R. R. where heavy traffic makes work trains undesirable, a derrick traveling along the top of the cut and handling a narrow $\frac{1}{2}$-yd. grab bucket has been used with success; its boom is 18 ft. and 28 ft. long for light and deep cuts, respectively. This machine weighs about 12 tons, and where the right-of-way between cuts is too rough for travel, a wrecking crane picked up the derrick bodily and carried it to the next piece of work. A ditching line marked the track side of the ditch, and the grade was kept by measuring from top of rail. The crew consisted of an engineman, a fireman, a helper, a teamster hauling coal and tank wagon, one man spotting bucket, two men with mattocks and shovels dressing the bank slope to a template, and one man leveling the spoil bank along the top of the cut.

WORK IN SANDY COUNTRY

Control of Drifting Sand.—In sandy districts along the coasts, and in the interior arid regions, trouble from sand drifting over the tracks may be reduced by sprinkling oil, spreading cinders and gravel, erecting barriers, or promoting the growth of vegetation. Cuts should be wide, with very flat slopes. Sand barriers are planned like snow fences, their size and position depending upon wind and other conditions. In some cases they are parallel with the track and 8 to 10 ft. high. Some railways in Africa and Australia, having special difficulties with sand drifts and dunes, have tried numerous experimental methods, of which the most satisfactory is the spreading of a heavy layer of gravel along the crests and slopes.

Oiling the right of way in desert country to check drifting sand is practiced in some cases, but with varying results owing to varying local conditions. On the Oregon-Washington Ry., heated crude petroleum is sprayed by means of hose and compressed air, so as to form a crust which holds the sand beneath it, while moving sand will pass over it without forming drifts. A train of tank cars, with a heater in a freight car, serves a crew of about 12 men with four or five hose lines. The amount of oil and frequency of application vary with the conditions, but ordinarily 500 to 600 gal. per acre is applied every year or on alternate years. On the Los Angeles & Salt Lake R. R., the experience has been that the results of similar work did not warrant the expense. On an 8-mi. stretch in light loamy sand, continually shifting and disturbed periodically by high winds causing sand storms, the oil did form a crust over which sand could pass freely, but owing to the obstruction caused by the rails, there was enough loose sand to make heavy dust when trains passed. Another 8-mi. stretch through fine, clean, drifting sand was oiled for a width of 400 to 500 ft. in order to check sand dunes which were moving toward the railroad. The benefit to trains was slight, and the effect upon the dunes was doubtful. It was concluded that, as the sandy area is so extensive and the winds come from all directions, it would not be possible to curtail sand movement by methods applied along the railroad.

Oiling dusty ballast, which was practiced extensively at one time, has been largely discontinued, partly due to the increasing use of washed gravel and partly to the objectionable oily dust raised by fast trains, while in districts having an average rainfall, the results were often unsatisfactory. But oiling may be of advantage at road crossings, yards, and yard or station driveways.

Grass in Sandy Cuts.—Planting grasses in sandy soil has been tried with varying success, but grass which grows in bunches with deep roots is not always satisfactory. Better results have been obtained with vines which creep over the slopes of cuts and fills, rooting in the soil and forming a thick mat which holds the sand even during severe storm of wind and rain. For this purpose, the wild cranberry vine is used extensively on the Cape Cod line of the New York, New Haven & Hartford R. R., Marram, bermuda, and johnson grasses are among the many varieties used near the coast and in the desert regions, but different kinds are needed for sunny and shady exposures and for different locations. On the lines of the Chicago, Burlington & Quincy R. R. in Wyoming and Nebraska, where sand fills the ditches, chokes the ballast, and drifts along the rails, it has been found difficult to get local vegetation to grow, owing to the sandy soil and the aridity. Better success has been obtained with sweet clover. The bare sand is first covered with manure mixed with straw, so as to prevent the sand from shifting before the vegetation gets a start, and this mat or layer is then seeded with the clover.

CHAPTER XIX

TRACK WORK FOR MAINTENANCE

Track Maintenance.—Maintenance-of-way includes the various kinds of work required to keep the track in safe and proper condition for the traffic which it carries. It includes, also, improvements designed to raise the standard of construction and thus reduce the work and cost of maintenance. Details of this work vary with the climatic conditions, the character of the track, and the amount of traffic. Light track carrying heavy traffic is an unfavorable condition for operation and economy. Favorable conditions attend a combination of heavy and stiff rails, sufficient ties to distribute the load on the ballast and sufficient ballast to distribute the load over the roadbed. The labor item is from 55 to 65 per cent of the cost of maintenance, with 45 to 35 per cent for material.

Maintenance work on a new line, during its early period of operation, or approximately 5 years, will be materially higher than after the road bed has become compacted or seasoned. This excess may be charged properly to the construction account. If maintenance is neglected or deferred and the property deteriorates, additional or excess expense, beyond that of normal maintenance, will be incurred in restoring the property to its normal condition.

Defects in Track.—It may seem unnecessary to remark upon the necessity of good work and safe track, but even on leading railroads and on main tracks carrying fast and heavy traffic, there may be found decayed, rail-cut, loose or ill-tamped ties, low joints, loose or missing bolts and spikes, battered frogs and switches, uneven surface and line, and irregular gage and curve elevation. Such conditions are revealed sometimes by investigations following accidents. For such unsafe conditions, the responsibility may be traced, perhaps, to the railroad management, the direct cause in many cases being a lowering of morale and discipline resulting from continual reduction in force, restricted expenditures, neglect of requisitions for material, or the disregard of warnings.

Under such circumstances, the foremen and the roadmasters are likely to get slack in their work and inspection and fail to do their best. More frequent personal inspection by division engineers or engineers of maintenance would often prevent the men from getting into such an attitude or habit. Unfavorable financial condition is no excuse for such defects, since no railroad is so poor that it cannot provide the men and material required to keep track safe for operation, even at reduced speed if necessary.

YARDS, BRANCHES, TUNNELS, AND BRIDGES

Work on Branch Lines.—On branch lines, the section forces are usually smaller, depending upon the traffic, but there should be a recognized standard condition of track to be maintained. The work may be reduced materially by providing substantial track support for the rails, with sufficient ballast, treated ties, tie-plates (light or second-hand), and other equipment. As the rails are frequently light (though it is economy to use heavy rails taken from main track), while the loaded cars may be as heavy as these on main lines, this substantial construction will give increased life and decreased maintenance expense. With small gangs, it may help the work if two adjacent gangs are combined, spending alternate weeks on each section. For long sections it is desirable to equip the small section gangs with motor cars.

Whatever the character of track, it should be kept in good and safe condition, properly lined and surfaced, bolts and spikes tight, ties well tamped, switches in good working order, ditches clean, and ballast dressed and weeded. Track construction may range from practically main-line standards to good 80- to 85-lb. relayer rail with 18 ties per rail and 8 in. of ballast, or down to 60-lb. rail on soft ties with little ballast. Track of this last class is rarely economical, as an excessive amount of work is required to keep it in reasonable condition, and derailments are likely to occur. Cinder ballast will give good drainage and keep down grass and weeds. Ditching may usually be done more economically by machine.

Yard Maintenance.—Track maintenance in large yards is done under different conditions from that on open track, owing to the continual movements of cars and the large number of switches. Large yards may have a general foreman with assistant foremen, or the yard may be divided into sections, each in charge of a foreman. Good drainage and ballast and substantial track construction are aids to economical maintenance and good yard service. The roadmaster and the yard section foreman should plan the season's work, and there should be close cooperation with the yardmaster, especially for such work as rail and tie renewals or drainage which will necessitate taking a track out of service for a time. After all such work, care should be taken to have all old material removed and scrap picked up promptly.

For track repairs in yards, a designated track may be put out of service for a day or half a day, and a car with the necessary supplies set out on this track. The old material is loaded on the car, and the track is left in finished condition. If the force is sufficient, the work may be started at both ends of the track.

Work on Tunnel Tracks.—In tunnels where traffic is heavy, the work of surfacing and rail and tie renewals is not only difficult and dangerous but slow and expensive, so that there may be practical and economic advantages in some special type of track construction which will require less attention (see "Roadway"). Even with ordinary conditions, the men cannot see to work to advantage, ties and rails are difficult to handle, tools and small materials are easily misplaced or lost, and it is difficult for the foreman to sight for line and surface. The difficulties are most serious in long tunnels where smoke and gas remain for some time after a train has passed. In damp tunnels, there is likely to be rapid corrosion of the rails and metal parts, so that oiling is advisable.

For work in tunnels, special attention should be given to the effective lighting of a length of 150 to 200 ft. of track, by means of large lamps and torch or flame lights, or by portable electric-lighting equipment, this last being the safest and giving generally a more diffused light. For track work and general repairs in tunnels, the Southern Pacific Ry. uses portable gasoline-generator outfits of about 5-kilowatt capacity; these are placed at the portals, and the wires that are led to the work have lamp sockets at 50-ft. intervals. Flood lights with 200-watt lamps serve for general illumination, while 50-watt guarded lamps are used at special parts of the work.

Tunnels should have refuges for the men. Special care must be taken to have proper signals set out, and the men should be warned in ample time to get clear of trains. A portable telephone outfit may be installed, or a visual or audible signal operated by the flagman or by connection with the block-signal system. If smoke and gas conditions in a tunnel become severe, men may lie down at the side of the track, covering their faces to prevent inhaling the gases, but except in long tunnels, it is customary for men to come out when trains are due and stay out until the smoke clears away. Gas masks have been tried experimentally but men do not like them and cannot do much work when wearing them. There appears to be little danger of serious "gassing," as the men soon become accustomed to working in the tunnel air and learn how to protect themselves under bad conditions.

Work on Bridges and Trestles.—Track maintenance on bridges and trestles with open floors is of a special character, and if the roadmaster finds the track out of line or surface, he must investigate the trouble and notify the bridge department. Temporary repairs may then be made and "Slow" signals put up if necessary. On ballasted floors there is no such trouble, all work being done by the section gangs, but the work is hampered more or less by traffic and the restricted working space. Refuge platforms and gangways for walks should be provided to enable the men to get out of the way of trains, and all work of this kind should be protected by flagmen and "Slow" signs. Special difficulties attend work on elevated and subway rapid-transit lines, where much of the work must be done at night on account of the density of traffic.

PROTECTION OF TRACK WORK

Safety for Track and Trains.—All work must be done with due regard to the safety of trains and workmen. If the track will be made impassable for trains, the Union Pacific Ry. requires a flagman to be sent out in each direction to place a torpedo on the rail $\frac{1}{4}$ mi. from the work, and two torpedoes $\frac{1}{2}$ mi. from the work, the man then remaining at the $\frac{1}{4}$-mi. point until relieved or recalled. Where curves are numerous or grades heavy, and also in foggy weather, he will go farther, placing two torpedoes at each $\frac{1}{4}$ mi. If a train approaches, he places one torpedo on the rail and displays a flag by day or a red fusee by night and in foggy weather.

Where the work necessitates only reduced speed of trains, a yellow flag or light is placed $\frac{1}{4}$ mi. in advance and a green flag or lamp beyond the work, but the Cleveland, Cincinnati, Chicago & St. Louis Ry. requires a yellow "Slow" target and light to be placed not less than 5,000 ft. in advance, and a green "Resume speed" target and signal 100 ft. beyond the work. Such work should not be undertaken in stormy or foggy weather, except in emergency.

Where track or a bridge is found unsafe, a red flag or light is placed between the rails, and torpedoes are set out as noted above. If a broken rail is found in main track, all trains must be stopped at least 200 ft. distant, and then allowed to proceed slowly if this is considered safe. If a suitable rail for replacement is not immediately available, the broken rail may be drilled for a bolted splice; but defective, damaged, and broken rails should be removed as soon as possible. Report is made of all such cases.

Operating Track Cars.—Section or track cars must be run slowly through towns and across grade crossings; they should not be used in foggy weather if the men's work is less than a mile from the section house. If used at night or in tunnels, a white light must be carried in front and a red light at the rear. Loaded push cars that cannot be removed quickly from the track must be operated under the protection of flagmen, and if run through switches, the foreman or the employee in charge of the car must open and close the switch. Hand cars and unloaded push cars must be lifted from one track to another at switches. Rails and ties should not be carried on the motor car or hand car, except in emergency, but should be placed on a trailer or push car. To insure rapid removal of the car and to avoid confusion in an emergency, it is well to assign two men who will always lift the front end off the track, while the others handle the rear end and then push the car clear of the track.

Protection of Track Gangs.—When two gangs work close together, the foremen must arrange between them as to the proper signals, both men being held equally responsible for the protection of the work. A repair gang must not work between another gang and the latter's signals. If such a position is necessary, as for a section gang working inside the flag of an extra gang, the section gang must put a flag in the middle of the track 100 ft. ahead, to warn enginemen that a second gang is at work. The foreman must not only protect his gang by flagmen, but see that they get clear of the track and have the track in safe condition before the time of regular trains. Extra trains may be expected at any time, and he must look for signals carried by the trains. It is advisable for him to have a loud whistle to warn the men to get clear. With a portable telephone outfit, the foreman can keep informed as to train movements and so save much time.

Train Signals.—Two green flags (or yellow on some roads) or unlighted marker lamps by day indicate the rear of the train. At night, the markers are tail lamps showing a green (or yellow) light forward and at side and red in the rear, except that when the train is on a sidetrack or passing track all three lights are green (or yellow); there is also a red light on the rear platform of a passenger train and on the cupola of a freight train caboose. Two green flags, or lamps at night, on the head end of the engine indicate that another section of the train is following on the same schedule, so that the section foreman must look out for these extra trains, each of which carries two white flags (or lamps) on the engine, except that the last section carries no such markers. A white light on the rear of a passenger car, freight car, or tender indicates a backward movement, the engine then carrying green flags or red tail lights on the bumper beam. A yard engine carries two white lights on the rear at night, except when it has a headlight at each end.

In signaling to trainmen, a lamp or flag swung horizontally across the tracks means "Stop"; raised and lowered vertically, "Proceed"; swung in a short circle across the track when train is at rest, "Back up"; swung at arm's length across the back when train is running, "Train parted." held horizontally at arm's length while train is moving, "Reduce speed." One torpedo on the rail means "Stop"; two torpedoes 100 to 200 ft. apart, "Reduce speed and look out for obstructions or stop signals." Burning fusees are dropped from a train which is losing time and may be overtaken. A blue flag on a car, engine, or train indicates that men are working on or under it and that it must not be moved on any account. The workmen display the signal, and they alone are allowed to remove it. A blue target on a rod attached to a portable derail is placed by car-repair men to protect them while at work.

SEASONAL TRACK WORK

For the ordinary work on the track sections, no set program can be given as the requirements vary with the climate and weather and under different conditions of track and traffic. Work trains and extra gangs can be worked from early spring to late fall on general improvement work, tile drainage, heavy ditching, widening cuts, rail renewal, reballasting, yard extensions and bridge and building work. These operations must be planned with relation to the regular work, so that, for example, a ballasting gang will not disturb the track just after the section gang has finished surfacing. For this reason the roadmaster should discuss the season's plans with the section foremen and inform them as to work which is to be done by other forces. In general, the year may be divided into four seasons, with work as outlined below.

Spring Work.—After the clearing of snow, ice, and debris from ditches, culverts, and waterways, and after the gradual removal of shims as the track settles with frost coming out of the ground, the section is gone over quickly for attention to the worst low spots and swings as well as to adjust bad gage. Spikes, bolts, and rail anchors are tightened, joints oiled, worn rails replaced, switches and frogs examined, road crossings and cattleguards repaired, fences patched, portable snow fences removed, sign posts straightened, and the right-of-way cleared and burned over. Sidetracks, yards, and station grounds may be overhauled and scrap picked up. Tool requisitions for the summer work must be prepared early in the season, foremen are examined and selected for the extra gangs, and orders are given for tools, camp cars, and other equipment for these gangs. Work trains distribute rails and ties, if this has not been done during the winter. With the section gang increased to full number, tie renewals should be undertaken and finished as soon as possible, new rails may be laid, and on warm days, the expansion spacing may be adjusted and bolts tightened. Ballasting is sometimes done at this season, with consequent lining and surfacing. Drainage and ballasting of soft spots that have required shimming should be attended to.

Summer Work.—About a week is spent in quickly skirmishing over the section to adjust bad spots in surface and line. Switch ties are renewed and fire guards dug or plowed where necessary. Unless extensive ballasting or rail renewal is to be done, the general surfacing and lining is commenced at one end of the section and carried on continuously; this is followed by mowing, clearing, and burning over the right-of-way, the time for which may be set by state laws. Grade crossings, road crossings, fences, and track signs are also overhauled. Drainage, dressing slopes, and work not interfering with the track can best be done during the heavy-traffic summer season. Badly worn rails should be noted and reported to the roadmaster, so that he can arrange to have them renewed in the winter, when the work will cost less and cause less interference with traffic.

Autumn Work.—This is the time to finish the general surfacing (if not already finished) and attend to swings and low spots that may have developed. About a month before the winter or rainy season, the track should be surfaced, lined, and gaged from one end of the section to the other, thorough work at this time reducing the amount of shimming during the winter. Switches, frogs, guard rails, road crossings, and sidetracks are repaired, loose spikes driven down, joints again oiled, and bolts and rail anchors tightened. Trenches are cut in the ballast at switch rods and interlocking work to prevent water and snow from collecting and freezing up the mechanism. After cutting and burning weeds on the right-of-way and burning old ties, all ditches, drains, culverts, and other waterways are cleared and adjacent streams examined for obstructions, riprap or other protection against erosion being placed where necessary. Flanger signs and portable snow fences are set up, and sidetracks and yards cleared. Where heavy snow is expected, planks may be removed from farm crossings, and spare splice bars and other track material placed on platforms or piles of ties. The tools and supplies for winter work may be checked and estimates made for new ties, crossing planks, and other materials.

Winter Work.—With section gangs reduced at this season, as is usual, they are busy inspecting track and switches, placing shims where the track heaves, doing light repair work, erecting snow fences, repairing and renewing right-of-way fences, clearing up yards, and cutting brush that is too heavy for scythes. Ties, rails, bridge timber, and fence material may be distributed and rail seats of ties adzed for the new rails or to straighten rails that are tilted. On some large roads, with fair-sized section gangs, fence building and repair is done during the winter. Heavy-rail renewal may also be carried on as noted under "Organization." In time of snow, the men clear the track, switches, crossings, and yards, open flangeways along the rails, form channels for escaping water, and keep the tracks clear of ice at water tanks and water columns. Inspection work includes looking out for heaving track, embankments heaving at bridges, spreading of shimmed track, broken rails and splices, and obstructed culverts and waterways. For heavier work, such as distributing ties, two adjacent gangs may be doubled up, as the work could not be handled by two or three men. But, as noted under "Organization" there are good economic reasons for maintaining a practically uniform track force and performing much work in the winter which is ordinarily left for busier seasons. In a moderate climate the "winter" may be a rainy season, requiring special attention to ditching and drainage.

ROUTINE TRACK WORK

Summary of Track Work.—In the routine work of maintenance, the largest item is that of keeping the track in proper line and surface, both these features being subject to continual disturbance by traffic. This item includes tamping of ties and bolting and spiking of rails. Switches and turnouts require considerable attention, and ditching is another important item; besides these, there is the mowing and clearing of right-of-way, weeding roadbed, dressing ballast, and general policing and inspection. The routine work is supplemented by such special operations as raising track out of face, renewing rails and ties, ballasting, putting in new turnouts and sidings, and building tile drains. Where any important work will affect or interrupt traffic, the operating department must be notified in advance. The regular work on the section should be done systematically and not at scattered points, and standards should be followed as closely as practicable. How much time and labor may be expended in maintaining or improving the appearance of the track will depend largely upon financial conditions. Hand-dressed ballast and turfed slopes may be expected only on comparatively wealthy roads, but neatness should be seen in every case, as it involves no extra expense but rather tends to economy.

To insure easy and smooth riding of trains, it is essential to have the track in good line and surface, tangents and curves free from kinks, the grade in even longitudinal planes without low spots and humps, and the rails in the same horizontal transverse plane except for the proper superelevation on curves. To maintain track in this condition requires periodical surfacing, which usually includes lining and gaging.

Raising Track.—About once in 3 to 5 years, the track will require to be raised out of face or brought up to its original grade. This may be done by extra gangs, but it is more economical for the section foremen to raise about a third or a fourth of their sections each year. Grade stakes should be set and ballast distributed, as described under "Ballasting." No single lift should exceed 6 in., any greater raise being made in two separate lifts. The track may be raised ½ or ¾ in. above grade and well tamped to allow for settlement, but the finished track must never be left above grade. In raising, the joint and shoulder ties may be tamped first, then the center tie, the quarters and the intermediates in succession, finishing off by tamping the joints again. The joint and quarter ties are sometimes tamped at the same time to avoid placing too much strain on the former after the jacks are removed. An incline or runoff should connect the old and new levels to prevent bending the rails and to allow trains to ride easily. A day or two after raising, a few men may give a final surfacing and lining, and also dress the ballast to standard section.

Surfacing.—In the general surfacing, such as is done once or twice a year, the track is raised only sufficient for proper tamping, to bring the low points to a uniform surface. In both surfacing and raising, the track must not be raised above grade in tunnels, at water stations, or under structures with a headway of less than 22 ft. In stone, slag, or washed gravel, thorough tamping can hardly be done without raising the rails about 1 in., but in loose material the lift may be only ½ in. Care must be taken that the men do not raise it too much, with the idea of letting the traffic settle it. With a raise of 2 in. or more, a runoff or incline should be formed at each end so that trains will ride easily. Surfacing should be done after the renewals in the spring and again before winter; also immediately after the laying of new rails, the track being then tamped to give an even bearing. The cost is said to increase about 1 per cent with each 12 deg. of curvature.

Regular surfacing, as distinct from attention to low spots that may develop, should be commenced at one end of the section and continued toward the other end, leaving the track in finished condition each day. Two or four men go ahead to tighten all spikes, so that the ties will come up when the rails are raised; one man in each pair holds the tie up to the rail, while the other drives in the spikes. Bolts also should be tightened. After the foreman has tested the track with the level and determined which side he will take for the high rail, any low spots on this side are raised, and the opposite side is raised to the same elevation as shown by the track level. If the raise is heavy, the level is set at the joints first, then the centers, and then the quarters of the rail length; but for a light raise, the joints and thirds or "long quarters" may be successively tested and raised. On curves, the inner or low rail is taken as the grade rail. The length of track raised should be such that the men can finish it between regular trains. After the track has been put in good surface, it is thoroughly tamped; then it is lined, and finally the ballast is dressed to proper cross-section.

Frequent use should be made of the track level and the spot sighting board (see "Tools"); this board, which is set across the track at a point beyond the raise, has a black or white stripe 5 in. from the lower edge. A block 5 in. high is placed in advance of the raise, and with an intermediate block at the low spot, the track is raised until the top of this second block sights in with the first block and the spot board. If each of the blocks (or targets) is mounted on the middle of a track level, sighting will give the longitudinal surface, and the spirit level will give the transverse surface.

With earth and other inferior ballast which will not give a uniform support, there is continual requirement for surfacing low points, and on a long section with a small gang the traffic must be made to help the tamping. For this purpose the low spots are raised about $\frac{3}{4}$ in. above grade for small lifts or $1\frac{1}{2}$ in. for lifts of 4 to 5 in., and the track is put in good line. One or two trains will compact the ballast, after which the track is sighted for a final light surfacing if necessary, and the ballast is dressed to shape. Surfacing on earth-ballasted track should not be begun until the soil is dry, as disturbing wet ground may make matters worse. At soft spots, cinders may be dumped to raise and surface the track, as the earth slowly dries out.

Lowering Tracks.—Lowering high spots is more troublesome than raising track to surface. Ballast is shoveled out between alternate ties, which are then shifted into these spaces while their original beds are cut down as required. These ties are then put back in place, and the other alternate ties are shifted into the open spaces while their beds are cut down, after which they are replaced and all ties are tamped. Track lowered in this way should be tested for surface after the passage of several trains.

Lining.—Putting the track in good line should be done ordinarily after the surfacing and then finished up each day to leave the track in good condition, but if the track is badly out of line it may be corrected before surfacing. All kinks in rails should be taken out with the rail bender. Where line stakes are set, as for ballasting or new tracks, the track gage is placed on the rails at each stake, and the track is thrown or shifted by bars to bring the center mark over the center point on the stake. After this has been done at four or five stakes, the men go back and throw the intermediate points, the foreman lining the rail by eye.

Short sights, as on bent rails, are made with the eye close to the gage side of the rail; but for longer sights and where no stakes are set, the observer stands up at some distance beyond the work so as to avoid putting a swing in the track. For a long stretch of track, a rod or target may be fitted to a track gage, with its center line directly over the inside of the rail head or at the center of the gage. Three sighting boards are used, each having lugs or brackets to fit against the gage side of rail heads and at its center a vertical flat bar with its edge on the center line. Two boards are set on the straight track at either end of the swing. The third is shifted along the rails, and the track is thrown until this middle sight is in line with the two end sights. The third or far sight may have a small target to make it conspicuous at a distance. All sighting is done along one side, the opposite rail being adjusted in the subsequent operation of gaging. After proceeding ahead for some distance, a backsight should be taken to check the accuracy of the work. For double track, a special gage may be used to line the inner rail of the second track from the track already lined.

Track is ordinarily thrown to line by six or ten men with lining bars, half the men at each rail, with bars stuck firmly into the ballast at an angle and bearing against the rail. At the word, the men give a short, steady heave, this being repeated until the desired movement is completed. If this is done after surfacing, care must be taken that the track is not lifted but only thrown laterally. As this work is severe, it is often slighted when the section gangs are small, being left until the force is increased by new men or adjacent gangs. In some such cases, it is necessary to send an extra gang. To eliminate this labor difficulty, lining jacks and levers have been devised with which three or four men can shift track more easily then eight or ten men with bars. When tangents or curves are badly out of line and the foreman cannot correct it, the engineer is called on to run a line with transit and stakes.

Lining Curves.—With the lateral forces exerted by trains and the continual lining or neglect of lining, curves often become irregular and ride badly. Without waiting for test with a transit, the foreman can check a curve by holding a cord with its ends touching the gage side of the rail head (or the outside if there is much rail wear) and then measuring the ordinate to the rail at the center of this cord (or chord). On a regular curve, these ordinates would be of uniform length. The length of the measured ordinate in inches, divided by the middle ordinate for a given length of chord on a 1-deg. curve, gives the degree of the curve. With a 62-ft. chord, each inch of middle ordinate represents 1 deg. of curve; a 30-ft. chord, $\frac{1}{4}$ in.; 44-ft. chord, $\frac{1}{2}$ in., 50-ft. chord, $\frac{5}{8}$ in., and 100-ft. chord, $2\frac{5}{8}$ in. These figures may be used also in bending rails for curves (see "Bending Rails").

In lining a curve, a part which seems to be true is selected, and chords are taken for measuring the middle ordinates. If it is a 5-deg. curve, the ordinate of a 62-ft. chord will be 5 in. Then beginning at the point of curve (or point of circular curve if there is an easement), successive chords are taken, measured from the rail at each middle ordinate, thus getting ordinates 31 ft. apart. These ordinates are recorded successively. If most of them are uniform, the track is lined to make the others of the same length; but if there is great irregularity, the ordinate may be averaged, and the curve lined to make all the ordinates of this average length. The ballast should be cleared from the end of the ties to facilitate shifting the track. This method of lining is not suitable for transition curves, which should be staked out and monumented by the engineers.

Gaging.—Under the combined effect of the thrust of wheels, wear of rail heads, tie cutting, and spike loosening, there is a gradual widening of the gage. Correction must be made periodically, but is not necessary where the widening is not more than $\frac{1}{4}$ in. on tangents or $\frac{1}{2}$ in. on curves and is uniform for a considerable distance The gaging is done usually with, or immediately after, lining. All measurements are made from the line rail on one side (already lined); a track gage is set at various points, the spikes of the opposite or gage rail are withdrawn, the rail is shifted with bars to bring it against the lug of the gage, and the spikes are then driven. With allowance for both spreading and wear of rails and the extra widening on curves (see "Curves"), the gage must never exceed 4 ft. $9\frac{1}{2}$ in. At frogs on the inside of curves, the gage must be 4 ft. $8\frac{1}{2}$ in.

Tamping.—In tamping stone, slag, and clean gravel ballast, it is important to have this done very thoroughly from the ends of the ties to 10 or 15 in. inside the rails, leaving the ballast loose at the middle of the tie and thus giving a firm support to the parts which receive and distribute the load. Some roads prefer to tamp about 22 in. inside, but practice depends largely on the kind of ballast. With earth ballast, only about 12 or 15 in. at the middle of the track need be left loose or shovel tamped.

The ends outside the rail may be tamped first, and two or three trains allowed to pass before tamping inside the rails, but it is generally considered better practice to tamp each tie thoroughly at one operation.

If the ballast is tamped hard at the middle of the ties, the track is likely to become centerbound, so that with the less firmly supported ends the ties will rock under the load. Such track will not ride easily and will require excessive maintenance work, while if neglected it may result in a dangerous condition. As it is difficult to loosen again the ridge of hard ballast, a remedy is to give the track a slight raise and to insure that the new ballast is tamped mainly under the ends of the ties. One advantage of certain substitute ties is that they give a large bearing surface at each end, with little bearing in the middle.

Tamping machines may reduce the cost of hand work by 60 to 75 per cent, 6 or 8 men with machines doing more work than a gang of 16 to 20 men with bars and picks. While these machines are used mainly in stone ballast, they are used also in gravel, and there is a field for their wider development in the looser materials. Tamping picks and bars are used with stone and hard ballast, but bars principally in gravel and inferior material. Shovel handles make good substitutes for bars in light work and loose ballast, but shovel blades should be used only for the first lift of gravel or cinder ballast, as the shovel has not force or weight sufficient to pack ballast hard enough to carry heavy loads. Joint and shoulder ties should be tamped somewhat harder than the others on account of the heavy pounding of wheels at the rail joints. Machine tamping is more uniform and durable as well as more economical, as men with picks do not tamp with uniform force or precision.

Distributing Ties.—New ties are distributed usually by work trains during the winter, under instructions as to the places for unloading or the number between each pair of telegraph poles, the roadmaster and the foreman being on the train to check the counting and distribution. If the ties are thrown off at random, there will be some waste and rehandling. All ties for each cut should be placed at the end of the cut. The tie train can be run at 5 or 6 mi. an hour, or slower on high fills, care being taken that the ties do not go down the bank, but the section gang should follow the train to move any ties that fall too near the rail or into the track ditch. For small lots and during a dull season, the ties may be distributed from local freight trains by the section gang. In some cases, the ties are unloaded in lots at specified points, such as near stations or section houses, or on sidetracks, being then neatly piled, ready to be distributed along the line on track cars by the section gangs.

Renewing Ties.—Tie renewals should be commenced as early in the spring as possible and continued steadily until completed, so as to have the track in condition for the heavy traffic. An alternative method is to carry them on in conjunction with other work, but tie renewal continued throughout the season may keep the track more or less disturbed and increase the maintenance cost. All ties to be removed should be plainly marked in advance by paint or notching. A method to avoid disturbing the bed of ballast under the ties is to remove the spikes from the old tie, loosen these in the two adjacent ties, and then raise the rails slightly by jacks, so that the old tie can be pulled out and the new one put in, the rails being then lowered. In the absence of jacks, the rails may be raised with bars, spikes being then slipped under them on the adjacent ties to hold them clear. Otherwise, the ballast between the ties is removed to a sufficient depth below the ties to enable the old tie to be withdrawn, the bed being then leveled to receive the new tie.

After spikes have been removed from the old ties on a certain length of track, the men work in pairs, far enough apart to avoid interference. Each pair takes a tie, loosens the ballast, pulls out the old tie, puts in the new one, and tamps it to a firm bearing. Where traffic is heavy, the outside spikes may be driven. Some roads specify that not more than two consecutive ties in main track, or three in other tracks, may be renewed without permission of the roadmaster. It is unsafe to allow trains to pass when more than two or three ties in a rail length are unspiked, unless "slow" signs are put out, and this is not always practicable for ordinary track work. When a number of ties have been renewed, the gang turns back to spike the ties and tamp and dress the ballast, leaving the track in good condition. At the end of the day, also, old ties should be piled for inspection and disposal, those suitable for use in side-tracks being kept separate. About a week later, the track should be gone over for tamping all new ties and any low spots that may have developed. In burning old ties that have been inspected, this is done most safely in small piles of 6 to 12 ties each; these piles should be at least 15 ft. from the track to prevent damage to the paint and varnish of cars that may stand near, and the fire should be watched to prevent its spreading to fences or fields.

Spotting in Ties.—It is usual practice to renew ties by "spotting;" that is, removing each bad tie on the section, as described. This may be commenced at the section house or at one end of the section and continued steadily; but if the ballast has not thoroughly dried, the work will be done first, probably, at the driest parts. In a few cases, the practice has been tried of renewing all ties out of face; that is, each year a certain length of track on each section has all the ties removed and new ones placed. The objection is that since ties (even treated ties) have a varying life, a number of good ties will be removed, while on the remainder of the section a number of bad ties will be left in place. At switches and turnouts, however, when needed renewals represent a large percentage of the ties, it is generally advisable to dig out all ballast between the ties and replace the entire set of switch ties or timbers. New ties should be handled with tongs and not moved by striking picks into them. Although this rule is important it is actually enforced on comparatively few roads.

Adzing Ties.—When ties have been cut deeply by the rails, the rail seats must be spotted or leveled with an adze to give a proper bearing for the rails or tie-plates. This must be done with care to insure an even surface and a uniform level on all ties, avoiding insufficient or excessive cutting. To facilitate the work, portable machines which score or groove each tie to the required depth have been used, these grooves forming guides for the adze men. Four circular saws, cutting a groove on each side of each rail, are mounted in a frame on a small car with power plant, and the depth of cut (which is adjustable) is gaged from the head of the rail, so that the depth from head of rail to bottom of grooves will be the same in all ties. The frame can be raised to allow the saws to clear the rails at turnouts. Cutting of treated ties should be avoided where possible; when it is necessary, the men should be supplied with preservative to be mopped on the exposed surface. If the adzing of ties is to be done in advance of rail renewal, a sufficient force should be employed to keep ahead of the rail-laying force.

Setting Tie-plates and Rail Anchors.—In preparing to place tie-plates in old track, as many spikes as permissible are withdrawn and adzing is done outside the rails. When traffic permits, the remaining spikes are drawn, rails thrown out, spike holes plugged, and the rail seats adzed smooth and brushed clear of chips and dirt before placing the plates. Then the rails are replaced, the line rail being lined and the tie-plates adjusted to it, with shoulders tight against the outside of rail base before the plates are spiked. The gage rail is then adjusted and its tie-plates are shifted to proper position and spiked. If the track is being reballasted, the tie-plates should be applied a few weeks in advance so that they may become firmly bedded on the tie before the final surfacing is done. Tie-plates should be laid "out-of-face;" that is, on all ties and not "spotted in," or placed here and there, where they may seem desirable. This latter method is likely to make a rough track, and may lead to rail breakages, but if new ties are "spotted in" on tie-plated track, the new ties should have tie-plates.

Before applying rail anchors or anti-creepers, the track must be put in good condition, with all bolts and spikes tightened. Both rails must be anchored to the same tie. Two pair of anchors for each rail length is the minimum, placed at the joints. If others are needed, they will be applied first at the quarters and then at intermediate points. When laid with new rails, they should be applied promptly after the rails are in place. Trackwalkers should inspect them and test by tapping to see if they are tight.

Bending Rails.—Practice and opinion vary widely as to the precurving or bending of rails for curves, but the tendency is to limit this to very sharp curves. It is a question whether the cost of bending with a tool or machine is any less than the labor cost of bending a straight rail in track by means of lining bars and spiking; and also whether any better results in maintenance are obtained by the former method. In general, it is considered that precurving is not necessary for such curves as occur ordinarily in main tracks, although many roads formerly required this for all curves over 2 or 4 deg. Rails for curves of even 200 ft. radius have been laid without precurving and have caused no trouble. In machine bending, special care must be taken with the ends of the rails, as usually the curve will begin and end about 2 ft. from the rail end.

Crooked or bent rails must be straightened before laying, especially those with bent ends, and such rails in track must be replaced or straightened in a rail bender. In Table XV are given the middle ordinates for rails of different lengths and different degrees of curve; the side or quarter ordinates are three-fourths of the middle ordinate. This table also enables the degree of curve to be determined for any measured ordinate (see "Lining Curves"). For the middle ordinate of a 30-ft. rail, multiply the degree of curve by 0.02; for the quarter ordinates, multiply the middle ordinate by 0.75. For the middle ordinate (N) of a rail of any other length, (L): $N = M\ x\ (L^2 \div 30)$.

TABLE XV.—MIDDLE ORDINATES FOR BENDING RAILS

Degree of curve	Length of rail, in feet							
	12	16	20	24	26	28	30	33
	Middle ordinates, in inches							
1	. .	. .	. .	$\frac{1}{8}$	$\frac{1}{8}$	$\frac{1}{8}$	$\frac{1}{4}$	$\frac{1}{4}$
2	. .	$\frac{1}{8}$	$\frac{1}{8}$	$\frac{1}{4}$	$\frac{3}{8}$	$\frac{3}{8}$	$\frac{3}{8}$	$\frac{5}{8}$
3	$\frac{1}{8}$	$\frac{1}{8}$	$\frac{1}{4}$	$\frac{1}{2}$	$\frac{1}{2}$	$\frac{5}{8}$	$\frac{5}{8}$	$\frac{7}{8}$
4	$\frac{1}{8}$	$\frac{1}{4}$	$\frac{3}{8}$	$\frac{5}{8}$	$\frac{3}{4}$	$\frac{7}{8}$	1	$1\frac{1}{8}$
5	$\frac{1}{8}$	$\frac{1}{4}$	$\frac{1}{2}$	$\frac{3}{4}$	$\frac{7}{8}$	1	$1\frac{1}{8}$	$1\frac{3}{8}$
6	$\frac{1}{4}$	$\frac{3}{8}$	$\frac{5}{8}$	$\frac{7}{8}$	$1\frac{1}{8}$	$1\frac{1}{4}$	$1\frac{3}{8}$	$1\frac{3}{4}$
7	$\frac{1}{4}$	$\frac{1}{2}$	$\frac{3}{4}$	1	$1\frac{1}{4}$	$1\frac{1}{2}$	$1\frac{5}{8}$	2
8	$\frac{1}{4}$	$\frac{1}{2}$	$\frac{7}{8}$	$1\frac{1}{4}$	$1\frac{3}{8}$	$1\frac{5}{8}$	$1\frac{7}{8}$	$2\frac{1}{4}$
9	$\frac{3}{8}$	$\frac{5}{8}$	$\frac{7}{8}$	$1\frac{3}{8}$	$1\frac{5}{8}$	$1\frac{7}{8}$	$2\frac{1}{8}$	$2\frac{1}{2}$
10	$\frac{3}{8}$	$\frac{5}{8}$	1	$1\frac{1}{2}$	$1\frac{3}{4}$	2	$2\frac{3}{8}$	$2\frac{7}{8}$
11	$\frac{3}{8}$	$\frac{3}{4}$	$1\frac{1}{8}$	$1\frac{5}{8}$	2	$2\frac{1}{4}$	$2\frac{5}{8}$	$3\frac{1}{8}$
12	$\frac{1}{2}$	$\frac{3}{4}$	$1\frac{1}{4}$	$1\frac{7}{8}$	$2\frac{1}{8}$	$2\frac{1}{2}$	$2\frac{7}{8}$	$3\frac{3}{8}$
13	$\frac{1}{2}$	$\frac{7}{8}$	$1\frac{3}{8}$	2	$2\frac{1}{4}$	$2\frac{5}{8}$	3	$3\frac{3}{4}$
14	$\frac{1}{2}$	1	$1\frac{1}{2}$	$2\frac{1}{8}$	$2\frac{1}{2}$	$2\frac{7}{8}$	$3\frac{1}{4}$	4
15	$\frac{5}{8}$	1	$1\frac{5}{8}$	$2\frac{1}{4}$	$2\frac{5}{8}$	$3\frac{1}{8}$	$3\frac{1}{2}$	$4\frac{1}{4}$
16	$\frac{5}{8}$	$1\frac{1}{8}$	$1\frac{5}{8}$	$2\frac{3}{8}$	$2\frac{7}{8}$	$3\frac{1}{4}$	$3\frac{3}{4}$	$4\frac{5}{8}$

For 30-ft. rail: 17 deg., 4 in.; 18 deg., $4\frac{1}{4}$ in.; 19 deg., $4\frac{1}{2}$ in.; 20 deg., $4\frac{9}{16}$ in.

Cutting Rails.—For cutting rails on the track, as for turnouts, it is best to use a portable power saw or a hack saw, the rail being carefully measured and a pencil or chalk line drawn around it as a guide. Where a chisel must be used, a cut should be made around the rail, which is laid on the ties or on blocks, and with a cutting block near the cut. The cut is made first on top of the base and then across the head; then with the rail turned on its sides a $\frac{1}{4}$-in. nick is made in each edge of the base and side of rail head; finally it is turned over while a $\frac{1}{8}$-in. cut is made across the base. With the rail laid on its side and held down firmly on each side of the cut by a lining bar placed under the track rail, a heavy cut is made on each side of the web, causing a crack to start at the base. Then the rail is set in normal position and held by lining bars while another cut is made on the head, causing the crack to extend through the rail and making a clean fracture. The chisel should be sharp and should be struck with a sledge, not a spiking maul. Both the chisel man and hammer man should wear goggles, and the men with the lining bars should face away from the cutting. The edge of the cut should be filed to remove burrs which might interfere with the fit of the splice bars.

Spiking.—Main tracks should have at least four spikes in every tie, with the outer spikes near one edge and the inner spikes near the other edge, so as to hold the tie square to the track, but with the spikes at least $2\frac{1}{2}$ in. from the edge of the tie. On curves of more than 3 deg., there should be two spikes outside each rail to resist the thrust of the wheels, except that if tie-plates are used, the extra spikes may be on the inside, where they will not only resist spreading but also hold the rail against any overturning tendency. In spiking, the tie should be supported by bars, but not raised above its bed. Each spike should be set vertical and driven straight down, with its shank touching the edge of the rail base. It should not be slanted to or from the rail to tighten or widen the gage, but the rail should be adjusted to position with a bar and then spiked. Care must be taken not to strike the rail. In Continuous joints, which grip the rail base, the spike heads may be turned away from the rail. When pulling old spikes, a tap on the head will often loose a tight spike; if the spikes are good, they can be used again, but spikes worn or necked by the rail base should not be redriven in main track. All spike holes should be filled promptly with wood plugs.

At joints, the spikes are usually driven in the slots of the splice bars, except that with new rails the bars are not flange spiked until the final tightening of the bolts, about 10 days after laying; after that time, the spikes are pulled and then driven in the slots, the old holes being plugged. On open-deck bridges and trestles, the spikes are not driven in the slots, as longitudinal movement of the rails must not be checked. There is an increasing tendency also to eliminate all slot spiking and to rely on other methods for holding the rails against longitudinal movement or creeping.

Although section men and trackwalkers attend to the spikes, it is sometimes the practice to send two men over the section twice a year to drive down the spikes and replace those which are broken, worn, or misplaced.

Bolting.—It is becoming general practice to oil the bolts and the inner surfaces of splice bars, before placing, to avoid rusting and to leave the rails free to slide as their length varies with temperature changes. As the nuts are tightened on the bolts, the bars may be tapped lightly to drive them home; screw clamps have been used also for this purpose. In laying new rails, the nuts are usually tightened the next day and again in a week or 10 days. On old track, the bolts are usually slackened and retightened in the spring and autumn (before hot and cold weather) to insure that the rail is not held too tightly but is free to move in the splice. This tends to reduce battering of rail ends and thus to increase the life of rails. Each joint should have its full number of bolts, nuts, and washers or nut locks, and broken or damaged bolts should be replaced promptly. In removing bolts, the nuts may be oiled first to loosen them, and while too much time should not be spent in getting off rusty nuts, the men must not get in the habit of knocking off the bolt ends with a sledge. If the bolt must be cut, this should be done with a chisel and hammer, care being taken in any case not to strike the rail.

Shimming.—Inferior ballast may freeze in winter and cannot be tamped when the track becomes uneven by the heaving of the ballast. To surface the track, therefore, it is necessary to put shims or wooden plates under the rails on the low ties. The high ties should not be adzed to lower the rail, unless this is absolutely necessary. Shimming is required also with ballast which is so soft after heavy rains that it cannot be tamped. In an emergency, blocking may be placed under the ties, but it must be removed when the shimming is taken out or the ballast has become firm. As the frost comes out of the ground and the track settles, thinner shims are substituted for the others to prevent surface bending of the rails. All shims must be removed when the frost is entirely out of the ground, the extra spike holes being then plugged and the track given a light surfacing if necessary.

Shims may be cut by the section men, but it is better to have them made at lumber mills or car shops. They are about 6 by 18 in., $\frac{1}{8}$ to 2 in. thick, with two spike holes in a diagonal line. Where the shimming exceeds $1\frac{1}{2}$ or 2 in., the rail must be held by four long spikes, and a piece of 1-in. to 2-in. plank may be placed along the full length of the tie if the lift is uniform, or half the length if it is irregular, the rail being secured, also by a regular rail brace or one improvised from a piece of plank or a tie-plate, well secured by long spikes. This bracing is specially necessary on curves. Where tie-plates are used, they must not be removed, but with high shimming a second tie-plate may be placed on top.

Shimming is but a makeshift way of providing a practicable and safe track under conditions that should not be allowed to continue on main line. The proper remedies are drainage, the replacing of mud or bad material with cinders, slag, or gravel, and the application of a bed of good ballast. Where the trouble occurs frequently, the ties will become so damaged and spike killed as to need renewal, thus increasing the maintenance cost. Although shimming is temporary work, it must be done with great care and long runoffs provided to adjacent good track, so that trains will ride easily. Further, the shimmed track must be watched carefully and frequently tested as to line, surface, and gage, as the heaving and subsequent settlement are irregular and the shimming must be adjusted accordingly.

Shifting Rails.—It is common practice to transpose the inner and outer rails on curves when the outer rail becomes worn, replacing it with the less-worn inner rail and making the unworn outer side of each rail head the gage side. To insure good results, this change must be made before the rails are so badly worn as to be rough, and with sharp curves there is sometimes liability to fracture due to the reversal of stress, the force being now applied to the opposite side of the rail.

SPECIAL KINDS OF TRACK WORK

Lining Track for Ballasting.—In the following method of lining track preparatory to ballasting, as developed by R. A. Rutledge on the Santa Fe Ry., the object is to get tangents as nearly straight as possible, but to leave easy swings where exact lining would require an excessive amount of work. The transit being set near a curve point, the line of track is obtained, and then produced to the next intersection by the following method: Two points are located 500 or 1,000 ft. apart, according to the clearness of the atmosphere and the accuracy of the object glass. The transit is next set over the second point and a backsight taken on the first, after which a third point is located the same distance ahead by reversing the transit. The line is thus continued forward to a point near the P.C. of the next curve, and at each point located, the distance to the right-hand rail is measured and recorded. Inspection then shows whether this is the line that best fits the track. If not, a simple calculation will show at which point to swing the line, the idea being to get a tangent that will fit the track as it is on the ground and save as much relining as possible. When the line is decided on, a calculation is made as to how much each point, already set, has to be moved to get it on this line. The points are then measured in, using a tape graduated to hundredths.

The transit is set over alternate points on the selected line and run each way by foresights, in this way getting as near a straight line as can be run, and at the same time reducing the track work to a minimum. On curves, the intersection points are obtained either by running the tangents to their intersection, or by running in long chords and calculating the curve point if the intersection is difficult of access. After establishing the curve points, the curve is run in and temporary points are established at each station. The distance from each temporary point to the outside rail is measured and recorded, after which (by calculating the external for the curve at the given degree, and for another curve either 1 min. more or less) it can be seen whether the curve can be improved by flattening or sharpening it. In all the work, the reduction of track work is kept in mind. This method makes work for the instrument man, but the work of a party of three engineers may save that of a track gang of 50 to 100 men.

General Realinement.—After track centers have been destroyed and traffic has shifted the track more or less, swings will gradually develop in the line and irregularities on curves which are not rectified in the routine work of lining. It may then be desirable to give a thorough realinement of an entire section or a long stretch of road, and to set up permanent monuments from which to check alinement in the future. The transit should be used and track centers marked by tacks in stakes, as for new track. Short, sharp swings or bends in tangents should be corrected, but long easy swings are not serious. It is not necessary to bring the track to mathematically straight tangents or exact curves; in fact, such changes might involve an amount and cost of work out of proportion to the practical benefits resulting, especially with long, easy curves. Both curves and tangents may be shifted and adjusted, the aim being to obtain practically straight tangents and curves of such regularity that trains will ride easily.

With sights of 1 to 5 mi., as on long tangents, the center line may be sighted upon a red and white target 36 by 18 in., placed over the track at a water tank or signal bridge. Center stakes may then be put about 750 ft. apart, or opposite every fifth telegraph pole. The transit is then placed over the gage side of the line rail and a foresight taken on a rod also set on the gage side and attached to a track gage, the track being shifted until the center line of the gage corresponds with that of the stake. Intermediate sighting is then done on a small target attached to a second track gage which is moved along about 50 ft. at a time. A lining gang for this work may consist of three men ahead of the moving target and five men behind it. A useful backsight to expedite such transit work is a triangular sheet-iron target about 3 in. wide and $4\frac{1}{2}$ in. high, with a point which is driven into the stake or into the tie behind it. These targets give a well-defined point and are so low that trains will not knock them out.

Oiling Track.—Brine dripping from refrigerator cars causes serious corrosion of the metal parts of railway track, and as car builders have not been successful in attempts to prevent such dripping, many railroads have adopted the practice of oiling the track to prevent corrosion. This is done, also, in tunnels and at places (as along the seacoast) where damp, salt air or other atmospheric influences cause corrosion. The oil is applied to the web and base of the rails and covers spikes, rail anchors, tie-plates, bolts, nuts, splice bars, frogs, and the working parts of switches or turnouts. It is done usually twice a year and only when the track material is thoroughly dry. At switches and in stations, it may be applied with a brush or small mop, but for long stretches of track it is more economical to use a car with spraying device, and most roads use for this purpose the Neafie machine developed on the Delaware, Lackawanna & Western R. R.

This machine consists of a car with adjustable spraying apparatus, the oil supply and position of apparatus being controlled by an operator at the head end. A forked pipe over each rail has nozzles discharging a spray against the rail web and along the base. A shield prevents the oil from reaching the rail head, and, to remove any spray that may be blown on the head, there is behind the shield a wiper or pad and then a sand pipe. A 2-in. pipe delivers the heated oil and a ½-in. pipe delivers air at 40-lb. pressure to the nozzles to spray the oil; this pipe is led also to the domes of the tank cars to force the oil to the spraying car. The train, with one or more tank cars, is pushed by a locomotive at 15 to 30 m. p. h. One marked result is the great reduction in track bolts damaged by removal in maintenance work. This oiling is distinct from the oiling of splice bars and bolts before application (see "Bolting"). A third method of track oiling is the spraying of oil to prevent dust on poor ballast, road crossings, and freight yards, and also to check drifting sand (see Chapter XVIII).

Moving Track.—In minor revisions of alinement and in building additional track, it may be desirable or necessary to shift the track laterally to a new position, which may be done by: (1) tearing up the track and relaying it; (2) shifting it bodily on skids; or (3) shifting it with lining bars or a machine. In some four-track stretches on the New York, New Haven & Hartford R. R., the two old tracks were skidded 20 or 30 ft. to a new roadbed about 8 ft. higher, the old roadbed being then filled to the new grade. Joints were unbolted at every five-rail lengths, planks were spiked upon the ties to hold them in position, and six skids of old rails on 6 by 8-in. timbers were placed for each section. A gang of 35 men with six jacks raised each section ready for the skids; a gang of 150 men then moved the sections by block and tackle (about 3 min.), and ballasted and roughly surfaced them; a third gang of 75 men connected up the sections, and lined and surfaced the track. Where a section had to be moved longitudinally, sometimes as much as 3 ft., the 150 men put 75 bars under the rails and moved the section as required. Initial cuts were made where the raise was only 2 ft., and the first sections were shifted to the new alinement with lining bars. In track elevation and grade revision, sections of track on the old level may be tilted up on the ends of their ties and then pulled up to the new roadbed by derrick cars and cables.

For a short move by lining bars, center stakes for the new alinement should be driven down to clear the rails, the ballast should be leveled (and inclined to give superelevation on curves), and the new and old lengths of track measured carefully so that rails may be cut to fit if necessary. Two or three gangs following each other along the track then throw it successively not more than 12 in. each time to avoid bending rails or splice bars. As soon as the first end of the track is in its new position, two or four men with sledges tap any misplaced ties into position in advance of the lining and surfacing gang. This work may be done at once if traffic is light, or may be done gradually under traffic, easy curve connections being maintained at each end and all trains being flagged to run slowly.

In removing second track or sidetrack for use elsewhere, one road has loaded the track in three-rail lengths on three flat cars placed between two wrecking cranes or locomotive cranes. A long truss or equalizer was placed upon the track length and hitched to the rails by chains, the two cranes then lifting the track and placing it on the cars. About ten lengths were loaded on each set of cars. The spikes were pulled from the joint ties at the end of each length, these spikes and ties being loaded by hand. Complete slip-switch installations in busy main tracks have been renewed bodily by lifting out the old one with a pair of wrecking cranes and replacing it with a new one already built up and delivered on cars. Switches, frogs, and track crossings can be shifted on skids by men with bars or by a locomotive with cable; derrick cars can also be used in such work.

Track-throwing Machines.—The Bierd machine is a self-propelled derrick car having a lifting boom which swings vertically, and a 28-ft. shifting boom which swings laterally. A chain sling on the hoisting cable is hooked to both rails and lifts the track clear of the ballast as the boom is raised by the topping lift; the shifting cable is then hooked to one rail and pulls the track over as the boom is swung. Two moves are made in each rail length, and a throw of 4 ft. can be made without injury to the track. For a greater throw, the machine makes another run, using the shifting boom only. In the Creese machine, a stout 30-ft. pole, braced against one corner of a steel-frame car and held by a cable, has a wheel which runs against the web of the rail on the opposite side; this can throw the track 6 to 36 in. with the ballast loosened along the ties. A smaller machine for use on dumps is noted in Chap. XIII.

Fencing.—On new lines, the fences are often built by contract, the railroad distributing the material and its engineers staking the right-of-way line. Fence building and reconstruction on existing lines may be done by extra gangs with work trains, the section gangs doing only repairs or small jobs. Several roads have on each division from one to four fence gangs of about four men each to work exclusively on the building and repair of fences of all kinds. Material may be distributed by work trains, or by freight trains for minor jobs.

For an extensive program of fence construction, the Southern Ry. had gangs of ten men, each provided with camp and material cars, motor track car, and push or trail cars to distribute material. These gangs were spaced about 5 to 6 mi. apart. Each had a foreman, an assistant foreman, seven laborers, and a cook. Steel posts $16\frac{1}{2}$ ft. apart were driven with a cap and hammer; the fence was of 47-in. woven wire, with a barbed wire 3 in. above the top. Progress averaged 6 mi. per gang per month.

WEEDING AND POLICING TRACK

Killing Weeds.—Grass and weeds in the ballast and along the roadbed are cut and cleared usually twice a year, or oftener if necessary. In the growing season, they may be cut only to clear the rails, owing to the amount of work required for complete clearing of the track. In the autumn, the grass is cut to sod lines or grass lines, all growth between these lines being dug and removed and kept down until the next season. A cord and stakes are used to set out the grass lines, usually 7 or 8 ft. from the rail. On stone ballast, the New York Central R. R. uses a rake or scarifier device attached to a car handled in a work train, the rake being raised or lowered by compressed air under control of an operator on the car. A hinged wing on each side of the car is fitted with vertical bars or teeth 7 in. apart, this spacing being sufficient to allow the stone to pass through and not be pushed ahead. In operating position, the teeth extend about 6 in. below the ties, cutting and tearing weeds and leaving the ballast dressed to shape. This device can be adapted also for cutting sod lines.

Weed-killing Machines.—This clearing of vegetation from the ballast is necessary for maintaining good track, giving drainage to the ballast, and preventing fires, as well as for appearance. If done by hand, with hoes and shovels, it is slow, laborious, and costly work, especially where the ballast is poor and the growth is heavy. In order to reduce the time, labor, and money expended in weeding track by hand, especially where the section gangs are small and the growth is heavy or rapid, various devices have been introduced to destroy grass and weeds by spraying them with poison or brine or burning them with flame or steam. A modification of the former plan is to scatter coarse salt from a car, but where track circuits are used for automatic signaling, it is necessary to keep salt or brine away from the rails.

Chemical Process.—For the chemical or poisoning process, as used extensively by steam and electric railways, the usual material is a solution of arsenic and caustic soda, varying in strength according to conditions. This mixture does no damage to rails or track. The spraying may be stopped near highway crossings to avoid injury to cattle which may browse on the vegetation, and signs may be posted to give notice that poison has been spread on the right-of-way. At the head of the weed-killing train is a car with spray pipes across the track for a width of 12 to 18 ft., the flow being controlled by a man operating valves mounted on the end platform; this car also has pumps, an air compressor, and quarters for the crew. A tank car carries the concentrated solution, which is pumped to other tank cars of water, the mixture being agitated by compressed air and piped to the head car. A locomotive pushes the train at a speed of 12 to 20 m. p. h.

The treatment and results vary with different kinds of vegetation. Some grasses are not killed by one application, and others are practically immune, so that in some cases the destruction is completed with a weed burner or by hand weeding. For small pieces of work where it would not pay to use the train, the Illinois Central R. R. has solution tanks and sprinkling pipes mounted on a trailer attached to a motor section car which is equipped with an air compressor.

Steam and Hot-blast Weed Killers.—With the steam process, a locomotive boiler, with superheater, oil and fuel tanks and an operator's cab, is mounted on a steel car frame, and the steam is piped to coils of ¾-in. perforated pipe under a flat hood or tray which is placed between the trucks and carried by 10-in. wheels. This hood, 10 by 10 ft., is in three parts, the two sections outside the rails folding up when not in use. A heavy canvas curtain confines the steam, and the feed to each part of the hood is controlled separately. A locomotive pushes the machine at 2 to 4 m. p. h. In heavy tropical growth, the cost has been 16 per cent that of hand work, besides being quicker and more effective. In a modified plan, the locomotive supplies the steam to an oil-heated superheater on a flat car which also carries the hood or pan, the steam being discharged from pipes under a hood 20 ft. long at 75-lb. pressure and a temperature of 700 to 900°F. From two to four trips are made each year.

A hot-blast weed burner used on the Texas & Pacific Ry. has a 60-ft. car with forced-draft oil furnace from which the gases at 1400 to 1500°F. are forced by a blower into a conduit extending under the car and having openings into a hood 38 ft. long and extending about 10 ft. beyond the rails. When the blast is shut off at bridges or road crossings, it is diverted to a large stack by automatic dampers. This machine, with a crew of three men, is moved at about 4 m.p.h. Still another type of weed burner using an oil flame has been used on the Illinois Central R. R. Oil, which has been found more satisfactory than gasoline, is vaporized by steam in tubes placed under hoods between and outside the rail, making a fierce flame which can be regulated according to the character of the vegetation. With two or three trips per year, the cost is 5 to 8 per cent that of hand work.

The weed-burner train consists of the burner, camp car, sleeping car, and oil-tank car. Good ties do not catch fire, and the crew looks out for any fires in old or damaged ties or on bridges, the machine having water tanks under pressure for fire protection if necessary. It is the practice also for the foreman on each track section to follow the machine on a motor car with a few men to watch for and extinguish fires. A lighter and smaller plant consists of a four-wheel motor car with blower fan, a four-wheel car in front with fuel tank and driver's seat, and a four-wheel trailer with hood 10 ft. wide extending close to the rails and having a refractory lining. Only two men are required; one to drive and one to attend to the oil feed, which is shut off when crossing roads and bridges.

Clearing Right-of-way.—Brush, grass, and weeds on the right-of-way must be cut by the section gangs at least once a year, and preferably twice, the last cutting being in the autumn before the seeding time of the plants. Brush and stumps should be cut close to the ground. On level ground where heavy brush has not been allowed to grow, a farmer may be hired to do the clearing with a mowing machine. The section gang should previously cut with shovels for about 3 ft. around the telegraph poles, and a man ahead of the machine locates wire, boulders, old ties, holes, and other obstructions which might cause damage to the team or the machine.

There is an increasing use of mowing machines, consisting of motor section cars fitted with blades for mowing 6 to 8 ft. wide on each side of the track or with shafts carrying disk cutters that turn over the sod, a scraper or loose heavy chain at the rear of the car smoothing the slope behind the cutters.

After the cutting and mowing, the material is raked into heaps and burned as soon as it is dry. The right-of-way is also cleared by burning in the dry season, to avoid risk of fires started by the locomotives. In all such burning, which should not be done in windy weather, care must be taken to keep the fire under control, so that it will not damage fences, telegraph poles, and buildings, or spread into adjacent fields. Water buckets or a barrel should be on hand, with sacks to be wetted and used to beat out fire if necessary. The foreman must see that all fire is extinguished before the end of working hours. Loose stone may be piled around telegraph poles and sign posts to protect them from fire and to kill vegetation. The clearing of right-of-way is governed somewhat by laws and local conditions, while in forest lands it must comply with fire regulations. Reports of locomotives which throw sparks badly, and of fires started by sparks, should be made by the section foremen and roadmasters.

Policing.—To keep the line in neat and proper condition, the section gang should give a few hours each week to clearing up and putting things in order around the tool houses, station grounds, yards, sidings, crossings, and right-of-way in general. Weeds must be cut, ditches cleared, loose sign posts straightened, rubbish picked up and burned, and scrap and spare material properly placed. A track sweeper used by the Pennsylvania R. R., where heavy traffic causes an accumulation of dirt and coal dust, consists of a flat car with rotary steel broom mounted between the trucks and revolved at 100 r. p. m. by a gasoline engine and chain drive. The broom, 3 ft. in diameter and 7 ft. long, placed diagonally to the track, sweeps the dirt over a steel apron to a 7 ft. belt conveyor feeding a hopper; from this, another conveyor belt on an inclined boom delivers the material to a gondola car in the rear. Such sweeper cars are noted in Chap. XIV.

No material must be piled within 8 ft. of the rail or in such a way as to obstruct the view at crossings. Grass, weeds, brush, and refuse must be kept cleared away from bridges, trestles, culverts, buildings, passenger and freight platforms, road crossings, and piles of cross ties and crossing plank. Ties and brush obstructing the view of crossings or signs, obscuring the view of enginemen, leaning toward the track or likely to touch telegraph or signal wires must be removed. If the obstructions are on private property, permission must be obtained from the owner, or report made of his refusal. Dirt and gravel must be removed from bridge seats, and ballast prevented from working over onto bridge abutments or falling into streets below.

The track also must be in good condition and appearance, ballast trimmed and sloped, roadbed shoulders of proper width, rails full bolted and spiked, switches clean and oiled, road and farm crossings in good condition, fences in repair, and wing fences at cattleguards kept whitewashed. It is much easier to keep the line neat by a little continual attention than to give it a cleaning at long intervals.

More attention should be given to keeping people from walking along the track or using the railway as a public path. It is the duty of section foremen and crossing watchmen to order trespassers off the property, while fences or other protection should be placed where necessary.

Station Grounds and Gardens.—At stations, the grounds and yards should not only be kept neat and free from rubbish, but attention should be given to providing a definitely attractive appearance, particularly at junctions and important stations and in suburban districts. Many roads making a feature of lawns and flower beds at stations have a landscape gardener with a special force of men to grade, plant, and seed the grounds, remove weeds, mow the grass, and trim the shrubbery. This force may be included in the roadway department. On the Philadelphia & Reading

R. R., a greenhouse and nursery supply flowers and shrubbery for station grounds and ornamental shrubbery along the right-of-way. Intricate patterns for flower beds are prepared in the office of the landscape gardener, the plans indicating the kinds of flowers to be used.

Care of the grounds at the smaller stations is sometimes delegated to the station agent, and annual prizes are awarded for the most attractive grounds. But, as a rule, this is part of the section gang's work; in such case, it should be under the direction of the roadmaster, the station agent having no authority to employ the men on this work. The agent and section foreman, however, may well cooperate, the former seeing that the platform and grounds are kept clear of old paper and other rubbish. The approaches and surroundings on the town side of the station should be included in this policy of good appearance. Cinders or gravel should be spread neatly over the spaces between tracks, and all refuse swept from the station should be placed in a can or bin to be emptied periodically into the car of a work train. At large passenger terminals, one man may be kept busy clearing up papers and rubbish. Station inspectors should see that stations, toilets, boarding houses, and other places are kept in proper and sanitary condition. Cleanliness and neatness may be required in every case, but the standard of appearance will depend upon local and financial conditions.

Scrap and Reclamation.—In all renewal work and the periodical policing of track, old and scrap material should be collected and carefully saved. Such material as well as that dropped from trains or resulting from wrecks should be piled at the section house, the smaller parts in boxes or kegs and the larger parts on a platform outside. Sorting, cleaning small serviceable material, and removing nuts from broken bolts can be done at the section house in stormy weather. Old rails may be sorted for relaying on main or side track or for scrap; all splice bars and bolts should be removed, the good bars being sorted in pairs and good nuts put on good bolts. Old bolts may be oiled to loosen them, and the nuts then removed with a wrench, but if the nut is rusted or wedged tight, the bolt may be cut with a chisel and sledge. Bent spikes may be straightened or turned in as scrap. Brass and copper scrap should be kept in a locked box. Old ties have not much value, but if sold or used for cribbing, all tie-plates and unbroken spikes should be pulled, and when ties are burned, the ashes should be raked to recover the spikes.

Section foremen should see that the men do not throw away bolts, spikes, or tools; even if not usable, they have a value as scrap. Inspection of the scrap by the roadmaster may be a check upon men who are inclined to throw away worn but serviceable material and make requisition for new supplies. This is checked also by a proper system of handling requisitions and supplies (see "Tools"). All this material is collected monthly by a work train in charge of the store department. Many large railroads have reclamation plants for working over the scrap; bent spikes may be straightened, broken bolts cut to shorter lengths, stub ends of bolts reheaded, nuts retapped, and various supplies manufactured. Such a plant may have shears, riveters, a bulldozer, a bolt-heading machine, and small rolls for finishing rods to standard sizes for bolts. The equipment and extent of operations will depend upon the economy of the work as compared with the purchase of new material. In determining this economy, and the cost of the reclaimed or manufactured material, full account must be taken of the investment cost for machinery and apparatus, the interest and other fixed charges, operating costs, maintenance costs, and all overhead expenses involved in the railway plant for such reclamation.

Salvaging Abandoned Track.—In salvaging track material on abandoned lines, a work train may be run from the far end, picking up the material as it goes. Ahead of the train, a gang removes the outer bolts of each joint and all but a few spikes. Behind the train, the remaining bolts and spikes are removed, and the rails picked up by a rail-handling derrick or more usually loaded by hand onto a flat car at the rear of the train, a dolly or a roller on the end of the car assisting this work. Serviceable ties may be loaded onto cars ahead of the rail car.

CHAPTER XX

GAGE, GRADES, AND CURVES

Railway Gage.—The gage of track is the transverse distance between the inner sides of the rail heads, usually measured at $\frac{1}{2}$ or $\frac{5}{8}$ in. below the top in order to clear worn corners and to allow for rail heads having sloping sides. The gage of 4 ft. $8\frac{1}{2}$ in. is now universal or standard in this country. In the early days of railway construction, there was little uniformity as to gage, each line being regarded as an individual and isolated enterprise, with no idea of the linking up of various lines into connected systems. In the northern states, many railways were built to the English gage of 4 ft. $8\frac{1}{2}$ in.; the Camden & Amboy Ry. was 4 ft. 8 in.; the Ohio gage was 4 ft. 10 in., and various lines were of 5-, $5\frac{1}{2}$-, and 6-ft. gage. In the southern states, the usual gage was 5 ft., as recommended by Horatio Allen for the South Carolina Ry. As the principal later lines were of a gage of 4 ft. $8\frac{1}{2}$ in., a change was necessary to provide for through traffic; on some of the broad-gage lines, this was effected temporarily by laying a third rail, while in 1885–1886, several thousand miles of 5-ft. gage on southern railways were changed to standard gage. The 4 ft. $8\frac{1}{2}$-in. gage originated in England, but its basis is indefinite. It seems probable that the old colliery wagons had a gage of 5 ft. over the wheels; flat and L-shaped rails were laid to carry these wagons, and when edge rails and flange wheels were introduced, the width of 4 ft. $8\frac{1}{2}$ in. between rails was adopted to fit existing cars. A gage of 7 ft. was used by some of the earlier English railways, but the growth of the railway system resulted in the adoption of 4 ft. $8\frac{1}{2}$ in. as standard. This is the standard gage for the greater part of Europe, but in Russia it is 5 ft.; in Ireland it is 5 ft. 3 in., and in Spain, 5 ft. 6 in. Other countries have various standards, and have usually secondary lines of narrower gages, from 24 to 42 in. Thus in South America there are important lines of 5 ft. 6 in., 5 ft. 3 in., 4 ft. $8\frac{1}{2}$ in., and 3 ft. $3\frac{3}{8}$in.; South Africa has 3 ft. 6 in. as its standard; India, 5 ft. 6 in. and 3 ft. $3\frac{3}{8}$ in.; Japan, 3 ft. 6 in.; China, 4 ft. $8\frac{1}{2}$ in.; Australia has 3 ft. 6 in., 4 ft. $8\frac{1}{2}$ in., and 5 ft. 3 in. Both India and Australia have serious difficulties due to multiplicity of gages, and the necessity of transferring freight and passengers at interchange points.

Narrow Gage.—At one time there were several important railroads of 3-ft. gage, but their isolation among standard-gage systems led to these lines being changed, and it has been conclusively shown that there is practically no economy in construction due to any but an extremely narrow gage. There are now no railroads of more than 4-ft. $8\frac{1}{2}$-in. gage in the United States; but there are some minor local lines of 2-ft., 3-ft., and $3\frac{1}{2}$-ft. gage. For logging roads, gages of 36 and 42 in. have been used extensively, but in most cases standard gage is practicable and has the advantage that the use of standard-gage equipment eliminates the rehandling of shipments. In grading, a steam shovel usually requires a subgrade width sufficient for standard gage, but with long sidehill cuts in steep mountain sides, the excavation quantities may be considerably greater than for a narrow-gage roadbed. As a rule, the curves are such as can be transversed by standard-gage engines. In general, a narrow-gage track is justified only by special conditions, and these apply mainly to the grading, while in operation the permanent additional expense due to transfer of loads is likely to wipe out any saving in first cost.

RAILWAY GRADES

Grades and Maintenance.—Steep grades increase the maintenance work and expenses materially, especially with heavy traffic, owing to the wear of rails from the use of sand by ascending trains and the application of brakes in descending, as well as the general displacement and disturbance of track and creeping of rails. These conditions are aggravated usually by numerous and sharp curves and the use of heavy locomotives with long wheelbase on lines of steep grades. On a mountain division with grades of 2 to 3 per cent (or even 4 per cent), curves of 8 to 12 deg. and heavy traffic, engines, and train loads, the maintenance work will be very severe.

Besides maintaining good track, care must be taken that the grade line is not broken up in surfacing and ballasting so as materially to increase the actual grade in places, as such changes may have a serious effect in reducing the hauling capacity of locomotives. For this reason, the engineer or roadmaster should occasionally run a line of levels over the division or over parts having heavy grades.

In view of the higher cost of operation and maintenance on steep grades, it is economy to keep the grades as low as possible in the construction of a line for heavy traffic. Steep grades may be used for light traffic or for temporary use to avoid the immediate construction of costly permanent works, and also where train loads may be reduced on standard trains assisted by pusher engines. On many railways, the traffic has outgrown the original grade conditions to such an extent that it has been wise economy to spend large amounts in grade revision (see "Permanent Improvements"). The ruling grade is that which limits the maximum weight of train; it is not necessarily the maximum grade, as heavier engines or additional pusher engines may be used for maximum trains on maximum grades. A pusher grade is not necessarily steep, except in relation to the characteristics of the division. Thus, where grades were reduced to 0.5 per cent on the Louisville & Nashville R. R. to enable the maximum train load to be increased from 3,500 to 5,300 tons, one grade of 1.25 could not be reduced below 0.65 per cent except at great expense and the 0.65 per cent grade was operated, therefore, as a pusher grade for maximum trains.

Virtual or Momentum Grades.—Under certain conditions, the momentum or energy stored in a moving train may be utilized to assist it in ascending a grade. Momentum grades are rarely warranted in reducing cost of construction, as the new track and roadbed would not be in condition to permit the speed assumed, so that the intended momentum grade would be actually the ruling grade. The system is used more generally in grade revision; as in a case where the desired reduction of grade from 1.25 to 0.7 per cent was found to be financially impracticable, while momentum grades of 1 per cent enabled trains to be operated with the full tonnage rating for a grade of 0.7 per cent. Since about 1900, it has been the general practice of the Canadian Pacific Ry. to use momentum grades on all new construction, and the operating results have been entirely satisfactory. In planning to convert the Ontario & Quebec division to a new ruling grade of 0.6 per cent, it was proposed to introduce momentum grades, which would have reduced the amount of reconstruction from 30 to 10 per cent of the length of the line. In place of this reconstruction, however, a new line for through traffic was built, having a ruling grade of 1 per cent.

Stopping points, severe curves, and grade crossings must not be included within the distance for acquiring and utilizing the momentum, as the train might be checked at such points and thus become stalled on the grade. A momentum grade should not exceed the rate of grade over which a locomotive loaded for the ruling grade can handle its load by doubling, in case the locomotive is stalled in the sag. Its rate should be such that the speed of freight trains in the sag will not exceed the speed limit of the division, and such that the speed at the top of a momentum grade running into a regular ascending grade will be not less than 10 m.p.h. If the top of the momentum grade is at a summit, a lower speed limit may be allowed.

Compensating Grades for Curvature.—In order to equalize train resistance, it is usual to reduce the normal grade on curves, so that the combined grade and curve resistance will not exceed the grade resistance on tangent. A general rate of reduction is 0.04 per cent per degree of curve, a normal grade of 2 per cent being thus reduced to 1.76 per cent on a 6-deg. curve. Other rates are 0.03 to 0.05, while in tunnels on the Canadian Pacific Ry. it has been made 0.06 on account of damp rails. This reduction of grade will increase the distance required to reach a given elevation with the normal grade. The elevation thus lost is equal to the rate of compensation multiplied by the central angle of the curve, and this amount divided by the length of grade will give the rate by which the maximum grade on tangent must be increased to introduce the compensation without a final loss in elevation. Compensation is advisable even upon easy grades, especially those which are nearly the rate of ruling grade, in order to provide for future increase in train loads or reductions in grades. The reduced grade may begin at the nearest even station and not necessarily at the point of curve, and its upper end usually runs out on the tangent.

Physical and operating conditions should be taken into account in this adjustment of curves and grades. In a few cases, the rate is increased on sharp curves, but Wellington's "Economic Theory of Railway Location" shows that this is based upon the erroneous assumption that curve resistance increases with the degree of curve. On a curve immediately below a stopping place, the compensation may be increased to allow for trains that have not acquired speed; thus the Pennsylvania R. R. specifies 0.05 per cent where trains frequently stop, but 0.04 per cent for general use. On a curve immediately below a stopping place, the compensation may be reduced about 0.01, but Webbs' "Railway Construction" states that this is not necessary, since the curve resistance will correspondingly reduce braking required to stop a train. In this as in other matters affected by train resistance, modern practice is governed by arbitrary rules based upon experience and upon old tests made with relatively light engines and trains. Few investigations or tests have been made to determine the accuracy of existing methods or the extent to which mathematical analysis needs to be modified to meet conditions of modern equipment and operation.

It has been suggested that the different rates of compensation might be used as follows: 0.03 when the length of the curve is less than half that of the longest train, or where the curve is within the first 20 ft. of a rising grade; 0.035 per cent when the length of the curve is 1.75 that of the longest train and is between 20 to 40 ft. from the bottom of the grade; 0.04 per cent where the curve is operated at low speed, or where the length is more than 75 per cent of that of the longest train; 0.05 per cent wherever the loss of elevation involved can be spared.

Vertical Curves on Grades.—Train operation is facilitated by rounding off grade intersections with vertical curves, where the break in grade or algebraic difference of grades is 0.2 or 0.3. The advantages are greater at summits than at sags. Both parabolic and circular curves are used, the former being generally employed on main-line work and based usually on a rate of change of 0.10 per station at summits and 0.05 at sags, but on the Santa Fe and some other railroads, the change is 0.15 and 0.10, respectively. On minor roads, it may be 0.20 and 0.10, respectively. With the rate selected, the length of curve in 100-ft. stations is found by dividing the algebraic difference in grades by the rate of change per station. This gives the points of curve, the intersection point of tangents to a parabola being at equal distances from the tangent points. The length of offsets from grade lines to the curve is half the rate of change multiplied by the square of the distance between the offset and the point of curve. The Pennsylvania R. R. rule is as follows:

A vertical curve shall consist of as many equal chords as there are whole tenths of a foot in the algebraic difference of the two gradients. The length of the chords should be 100 ft. except where it is necessary to use a shorter vertical curve. The change from one gradient to another should be made by a change in rate for adjacent chords, equal to the algebraic difference between the two gradients divided by the number of whole tenths in the algebraic difference; except that the change from main gradient to the first chord and from the last chord to the second main gradient shall be half the rate of change between grades on curve. In sags, a vertical curve twice the length of the curve above specified should be used. Grade stakes should be used to insure uniform surface.

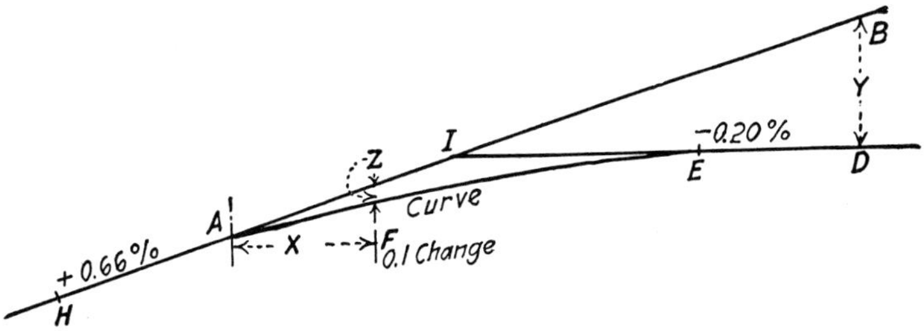

FIG. 96.—Diagram of vertical curve.

As to the maximum grade limits at which vertical curves should be used, probably a minimum variation of 0.3 per cent might be adopted for summits and 0.2 per cent for sags, although it might be more convenient to establish a standard minimum length of curve. Instructions on the Chicago, Milwaukee & St. Paul Ry. provide that to locate the point of intersection, I (Fig. 96), from any point D whose elevation is known, another point whose elevation is known is to be selected on the other grade, as at H, and a vertical drawn from D to the other grade line produced, at B. Designating this vertical distance as Y, and the algebraic difference of the intersecting grades as C, then the horizontal distance D–I is $Y \div C$. Taking A as the rate of change of grade per station, the length of curve, L, in 100-ft. stations will be $C \div A$. Thus, if C is 0.86 and A is 0.10, then L is 8.6 stations or 860 ft. Taking X as a given distance A–F in 100-ft. stations, the correction Z for any point F is:

$$X^2\left(\frac{A}{2}\right)$$

Thus, if $X = 3.3$ stations; $Z = 3.3^2 \times \frac{0.10}{2}$, or $10.89 \times 0.05 = 0.54$

From this formula, it will be found that the grade between any two consecutive stations may be calculated from the preceding grade by deducting the rate of change, A. It is to be noted that the first grade on the vertical curve varies from the approach grade by half A instead of A. This method has the advantage that a correct final result at the end of the vertical curve makes a check on all intermediate results. While the curve in this case is parabolic, it approximates very nearly to a circular curve.

Vertical curves are necessary at the sharp grade intersections of gravity switching hump and coal-pier inclines. Here they are usually circular curves 20 to 35 ft. long. The radius may be 85 to 110 or 200 ft. at the bottom of the cable inclines of 10 to 12 per cent, and 200 to 300 ft. at the top of such inclines. The curves must not be so sharp as to cause displacement of the king-pins or center plates of the trucks or to cause the cable to foul the brake-rigging, hopper or other parts. It is advisable to lay out the curves to a large scale and then plot the equipment upon it to insure against such interference. For switching humps, the curves may be of 750 to 5,000 ft. radius.

RAILWAY CURVES

Curve Degree and Angle.—Curvature is reckoned usually by the number of degrees of central angle subtended by a chord of 100 ft., except that sharp curves are designated by their radius. For a 1-deg. curve with 100-ft. chord, the radius is 5,730 ft., and the radius on the center line or the degree of any curve may be ascertained by dividing 5,730 by the degree or radius, respectively. In Fig. 97, the "central angle," A, is the angle contained within radial lines to the ends of the curve; that is, to the P.C., point of curve, and P.T., point of tangent. The "degree of curve," B, is that portion of the central angle contained within radial lines to a 100-ft. chord on the curve. The "angle of intersection," C, is the exterior angle at the intersection of the two tangents produced, and is equal to the central angle, A. The "angle of deflection," D, is contained within a tangent produced and a 100-ft. chord on the curve. Taking X as the length of curve in feet, Y as the degree of curve, and Z as the central angle, the relations may be obtained by the following formulas:

$$X = 100\frac{Z}{Y};\ Y = 100\frac{Z}{X};\ Z = \frac{YX}{100}$$

A handy means of ascertaining the degree of a curve in track is afforded by Table XVI, with the method shown by the small diagram in Fig. 97. A sight is taken from a point on the outer rail across the gage side of the inner rail to another point on the outer rail. From the length of measured chord AC, or arc ABC, the degree of curve is found in the table. For methods of ascertaining the lengths of inner and outer rails on curves, see "Maintenance."

Setting out Curves with Tape.—For spurs and industrial branches or temporary work, it is often convenient to lay out a curve without waiting for the use of an instrument, and the middle diagram in Fig. 97 shows how to set out a circular curve. The tangent approach is carefully lined, and a stake, *B*, set at the P.T., and another, *A*, 100 ft. back on the tangent (both on center line of track). The tangent is then lined in for 100 ft.

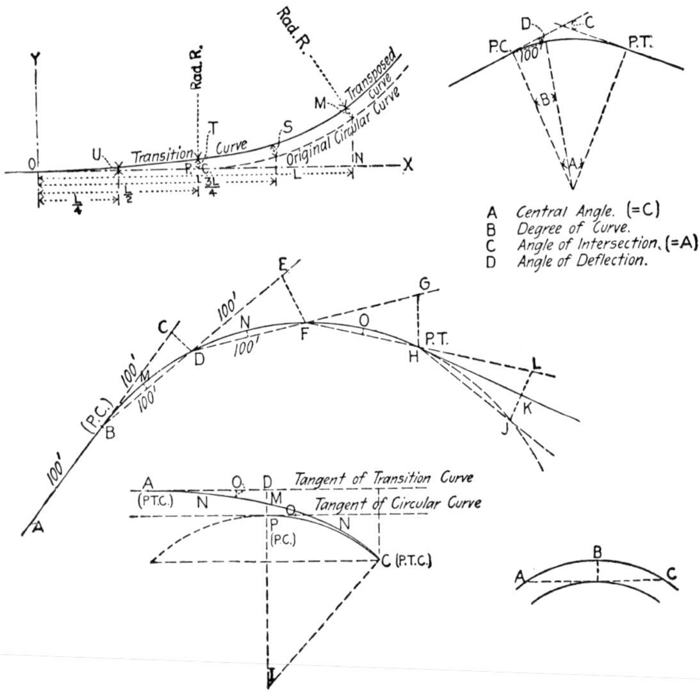

FIG. 97.—Curve diagrams.

beyond the P.C. at *A*, giving point *C*. The 100-ft. tape or cord is then held by one end at *A*, while its other end is moved inward from *C* for the distance given in Table XVII as the tangent deflection on a chord of 100 ft. for a curve of the required degree. This gives point *D* on the curve, and a stake is set. The chord *AD* is then extended to *E*, 100 ft. The tape being held at *D*, its free end is moved inward from *E* for the distance given in the table as the curve deflection on a chord of 100 ft. for a curve of the required degree. This gives another point *F* on the curve. The curve deflection, *EF*, *GH*, is always twice the tangent deflection *CD*, *LK*. The points *H* and *J* on the curve are set out in the same way from *F* and *L*. If the curve ends at *H*, as shown, then from the point *L* the tape will be swung in for the distance previously used for the tangent deflection *CD*,

TABLE XVI.—CURVE FUNCTIONS

Degree of curve	Radius of center line, feet	Number of 33-ft. rails in *ABC*	Number of 30-ft. rails in *ABC*	Length of		Central angle	
				Arc *ABC*, feet	Chord *AC*, feet	Degrees	Minutes
1	5,730	14.0	15.5	463.5	463.4	4	38
2	2,865	10.0	11	328.6	328.4	6	34
3	1,910	8.0	9	268.1	267.9	8	02
4	1,433	7.0	8	232.5	232.2	9	17
5	1,146	6.3	7	208.0	207.7	10	24
6	955.4	5.7	6.3	190.0	189.7	11	22
7	819.0	5.3	5.8	175.8	175.5	12	16
8	716.8	5.0	5.5	164.8	164.5	13	08
9	637.3	4.7	5.2	155.2	154.8	13	54
10	573.7	4.5	4.9	147.5	147.1	14	40
11	521.7	4.2	4.6	140.6	140.1	15	22
12	478.3	4.0	4.5	134.8	134.3	16	04
13	441.7	3.9	4.3	129.5	129.0	16	42
14	410.3	3.7	4.1	124.8	124.3	17	20
15	383.1	3.6	4.0	120.6	120.1	17	56
16	359.3	3.5	3.9	116.8	116.3	18	30
17	338.3	3.4	3.8	113.3	112.8	19	04
18	319.6	3.3	3.7	110.3	109.8	19	38
19	302.9	3.2	3.6	107.4	106.8	20	10
20	287.9	3.1	3.5	104.7	104.1	20	40

TABLE XVII.—CURVE OFFSETS AND ORDINATES

Degree of curve		Tangent offset*	Middle ordinate		Degree of curve		Tangent offset*	Middle ordinate	
Degrees	Minutes	Feet	Feet	Inches	Degrees	Minutes	Feet	Feet	Inches
1	0	0.873	0.218	2½	7	0	6.105	1.528	18¼
1	30	1.309	0.327	...	7	30	6.540	1.637	...
2	0	1.745	0.436	5¼	8	0	6.976	1.746	21
2	30	2.181	0.545	...	8	30	7.411	1.855	...
3	0	2.618	0.654	8	9	0	7.846	1.965	23½
3	30	3.054	0.763	...	9	30	8.281	2.074	...
4	0	3.490	0.872	10½	10	0	8.716	2.183	26¼
4	30	3.926	0.982	...	10	30	9.150	2.293	...
5	0	4.362	1.091	13	11	0	9.453	2.402	29
5	30	4.798	1.200	...	12	0	10.585	2.620	31½
6	0	5.234	1.309	15¾	15	0	13.053	3.277	39¼
6	30	5.669	1.418	...	20	0	17.365	4.374	52½

* The offsets around the curve are double the length of the first and last or tangent offsets.

and this will give a point K on the tangent, 100 ft. from the P.T. at H. Intermediate points on the curve may be set out by measuring the middle ordinate for each 100-ft. chord, as given by the table, thus marking the points M, N, R. The table gives the tangent offset CD (Fig. 97) for a tangent BC and chord BD, both 100 ft. long, and also the middle ordinate of a chord 100 ft. long, so that the table may be used in setting out curves by offsets or ordinates. In the former case, it must be remembered that the curve offsets, EF, GH, etc., are double the first or tangent offset.

Divergence of Curve.—In track laying, as well as in location, it is often desirable to ascertain the divergence of a curve from its tangent, and this is given by a formula prepared by Mr. Muenscher: X is the

$$X = 0.875N^2D. \quad \text{Example: } 0.875 \times 5^2 \times 6 = 131.25 \text{ ft.}$$

offset, or distance of one end of a curve from a tangent passing through the other end; N is the length of the curve in chords of 100 ft.; and D is the degree of curvature. Thus the divergence of a curve of 6 deg. from its tangent in a length of 500 ft. will be as shown. By making D equal to the difference of the degree of curvature of two curves of different radius but having a common origin, X will be their divergence from each other at the end of N stations. In tracklaying, if X is the gage of track, and D the degree of curvature of a turnout track (or corresponding to a given number of frog), N will be the lead of the main track. If X is half the gage, N will be the lead of the crotch frog, and if X is the throw of a stub switch, N will be the free length of the switch rail. The formula is sufficiently accurate for practical purposes, and may be of great use in field work, with the aid of a table of actual tangents for a 1-deg. curve. For example, suppose a 5-deg. curve to the right, eight stations long, has been located, and its extremity falls 28 ft. too far to the right to throw the tangent on the best ground. Taking X as 28 would give $D = \frac{1}{2}$ deg., showing that a $4\frac{1}{2}$-deg. curve starting from the same origin would pass through the required spot. Suppose that in the same case the new curve is to commence 200 ft. back of the first one; then the required divergence from the tangent will be $0.875 \times 8^2 \times 5 - 28 = 252$. Substituting this value for X and taking N as 5×2, gives $D = 2.88 = 2$ deg. 53 min.

Curve Economics.—Curves constitute a considerable proportion of railway mileage, and have an important relation to easy riding of trains, to train loading, to track maintenance, and to the wear of both track and rolling equipment. Wear and displacement of track are increased on curves owing to the lateral pressures exerted by the wheels, and it is commonly assumed that both rail-wear and track-maintenance expense increase directly with the central angle of curve, although investigations noted in the 1919 *Report* of the A. R. E. A. indicate that they increase as the square of the curve. Formulas and rules employed in problems where train speed and resistance are factors are based mainly upon tests and investigations made so long ago that it is doubtful how far they apply to modern conditions. As a result, such matters as superelevation, gage widening, and grade compensation on curves are governed by opinion, custom, and experience rather than by scientific analysis (see *Engineering News-Record*, Mar. 16 and 23, 1922, pp. 428, 446 and 489).

Furthermore, the relation of the locomotive to curve problems is rarely taken into account, although the heavy modern engine with long driving wheelbase is certainly a factor in the adjustment of curves to secure a safe and easy-riding track which shows no excessive wear or maintenance expense (see "Widening Gage"). It is true that every curve is operated under varying conditions of speed and weight of trains, and that grades and other local conditions are factors in individual problems, but this was true when earlier investigations were made, and it is very desirable that tests should be made to determine present conditions of train resistance due to curvature. This resistance is increased at each end of a curve, as the trucks have to be shifted to and from an approximately radial position. The increase will be higher with cars that are heavily loaded or badly balanced so that the trucks cannot swivel easily. This increased resistance occurs even where transition or easement curves are used, so that in improving alinement it is important to reduce the number as well as the sharpness of the curves.

Under prewar conditions, it was estimated that with ordinary traffic the operating cost due to curvature was $1 per degree per daily train per annum, increasing with heavy traffic, high speeds, and curves of over 5 deg. Cost of maintenance is considered to increase 1 per cent with each 12 deg. of curvature per mile. Not only are the rails worn but the lateral pressures tend to force rails outward by a combination of sliding and overturning and also to shift the ties in the ballast, with consequent general wear and displacement that needs continual attention. Sharp curves are also likely to increase derailments, especially on grades and where speed limits are set, as such limits are often exceeded. Many railways have made heavy expenditures in taking out or flattening curves to eliminate sharp curves and reduce the total curvature, especially where there is heavy and fast traffic. Curves of 4 to 6 deg. should be the maximum for main lines in ordinary country, though 8- and 10-deg. and even 12-deg. curves are used extensively on mountain divisions and on branch lines.

Wheel Action on Curves.—In a train passing around a curve, the wheels tend to travel in a straight line but are forced to follow the curved track by the outer rail, which guides the wheels by their flanges. The outer wheel on each axle has to travel a slightly greater distance than the inner wheel in the same period of time, but as both wheels are rigidly secured to the axle and must revolve together, the outer wheel tends to slide ahead bodily, swinging from the inner wheel as a center and with the axle as a radius. With two axles in a truck frame, the truck tends to swing horizontally around the inner rear wheel as a center. From experiments on the Canadian Pacific Ry. it was concluded by J. G. Sullivan that all skidding or sliding to make up for difference in length of travel takes place in the inner rear wheel.

In a car truck, the inner wheels have a lateral sliding motion across the rail head, due to the combined effect of the difference in length of travel, the variation of the axles from a truly radial position, and the tendency of the outer wheels to run against the rail. In rounding a curve, the flange of the outer leading wheel bears against the outer rail and tends to grind or cut it, or to climb over it when the pressure is excessive. At the same time, the flange of the inner rear wheel bears against the inside rail, but does not cut it, as the edge of the flange is running away from the rail; on the other hand, the pressure and lateral slipping on the inside rail tends to crush and deform the head, or to cause the metal to flow and form a lip along the side of the top of the rail head. The crushing and flow may be caused also in the outer rail. These conditions are aggravated by the failure of trucks of heavily loaded or badly balanced cars to swing promptly to the radial position in entering curves. With a locomotive, the driving axles cannot shift to an approximately radial position, and the wheels are adjusted on the curve by slight variations in width back to back of wheels on successive axles and by lateral play of the axles, while in some cases certain of the driving wheels have blind or flangeless tires (see "Widening Gage"). In spite of these provisions for adjustment, locomotives with long driving wheel-base have considerable destructive effects upon the rails. Both the front truck wheel and the first driving wheel are found to run against the outer rail, and with this, at the same time, there is pressure against the inner rail by the middle wheel, which is near the center of rotation of the engine frame, while the rear driving wheel exerts pressure against the outer rail. These conditions are largely theoretical or empirical, for careful examination indicates that, in general, the actual positions of the wheel flanges in relation to the rails do not conform to assumed positions. The actual positions are influenced to an indeterminate extent by various factors, such as the wheel gage, wheel spacing, flange wear, and lateral play of axles.

Lateral thrust exerted against the rail heads on curves tends to shift the rail outward horizontally and also to overturn it by tilting it on the outer edge of the base, the latter effect causing this edge of the rail base to cut into the tie, unless tie-plates are used. In both cases, the resistance of the rail is increased by the weight imposed upon it. At the same time, the centrifugal force tends to increase the weight on the outer rail. The various forces entering into curve mechanics are not static but dynamic, and there is a large amount of shock, impact, and oscillation in the movement of a train around a curve. Furthermore, the parts which receive the stresses are not rigid, since the rail may slide on the tie or tie-plate (forcing the spikes outward), and the ties may shift outward in the ballast. Tests of lateral thrust made by G. L. Fowler with special apparatus and an 87-ton 2-8-0 locomotive showed 13,430 lb. for the truck wheel and 11,450, 13,000, 12,215 and 11,170 lb. for driving wheels; this was at a speed of 30 m. p. h. on a curve of $4\frac{1}{2}$ deg. having a superelevation of $3\frac{7}{8}$ in. Tests with cars showed that the front truck exerts about 60 per cent of the total pressure. Later and more comprehensive tests are recorded in the "Proceedings" of the A. S. C. E. and the A. R. E. A. by the joint committee on "Stresses in the Railroad Track."

TRANSITION CURVES

Where circular curves spring directly from tangents, there is trouble in making the track ride easily, especially at high speeds. The cars will pass as smoothly around a well-laid curve as along straight track, but at each end of the curve there is a jar or shock due to the change of direction and to the shifting of the trucks. This jarring and lurching is not only unpleasant in passenger cars, but causes increased wear of rails and disturbance of track, indicating the lateral force that has to be resisted in guiding the car on and off the curve. It is now general practice to eliminate this trouble by an easement or transition curve which will eliminate the abrupt change. Where such curves are not laid out by the engineers, they will be introduced roughly by the section foreman, who will try to reduce the track trouble by throwing the ends of the curve inward so as to extend practically about 100 or 200 ft. on the original tangent. While the curve is flattened or eased at this point, it will be sharpened a little farther along beyond the P.C., where the true curve begins, but there the motion is less severe, as the car has already begun to change its direction. If this modification is not done by the section gang in order to make an easy-riding track, it will be introduced by the trains shifting the track to the position giving the easiest passage, which is an objectionable means of obtaining a desirable end. Transitioning of the curves results in easier riding of trains, economy of maintenance, reduced rail and flange wear, and reduced drawbar pull.

Transition curves for scientifically easing the change from tangent to circular curves were developed in England in 1828 and actually used in 1842. Their first application in this country was in 1880 when Holbrook applied his "spiral" on the Pennsylvania R. R. Numerous other methods followed, but the curves did not come into general use here until about 1900 (see *Engineering News-Record,* May 8, 1919). A majority of railways now use them in view of their very practical advantages in forming an easy riding track with reduced maintenance and less wear of wheels and rails. As with superelevation, the theoretical condition for any particular velocity is modified to a compromise suitable to the prevailing or governing velocity. In applying the curve to new or existing track, the computation and field work are little more difficult than for a circular curve, and theoretical niceties are rendered unnecessary by the fact that the curve must be a compromise to suit varying conditions. Different railways use them for all main-line curves over 1 deg. in some cases and 2 or 3 deg. in others or where an elevation of 2 in. or more is required for the highest permissible speeds. Special systems of transition curves are used also by elevated and underground rapid-transit lines and street railways.

As explained in Wellington's "Economic Theory of Railway Location," the purpose of the transition curve is: (1) to ease the curve by a rapidly changing radius for a short distance at the ends; and (2) to leave the body of the curve of uniform radius. This may be effected in three ways, of which the first is in general the best: (1) by a spiral curve; (2) by end curves of greater radius than the main curve; (3) by compounding short circular arcs of gradually increasing radius until the radius of the main curve is reached. The Holbrook spiral is of the first and the Searles of the second type.

The ten-chord spiral is recommended in the *Manual* of the A. R. E. A., which mentions also the Searles' spiral, cubic parabola, Holbrook, Crandall, and Talbot easements, but not the recent development of the lemniscate for transition curves, which has had little practical application. The *Manual* states that any system is satisfactory (1) in which the rate of increase in degree of curve can be changed readily to suit particular curves, so that the length of easement will be the same as the runoff of superelevation; and (2) which can be run in by deflection or offset with chords of any desired length. The minimum length of transition for 6 deg. and over is 240 ft.; for less than 6 deg., the speed for 8-in. elevation is multiplied by 5.33; for easy curves, the length in feet is thirty times the elevation in inches for ultimate speed or not less than two-thirds the ultimate speed in miles per hour multiplied by the elevation. A longer transition may be of advantage in some cases, provided that it does not alter the curve too much, but any great increase in length is unnecessary and may increase the cost unnecessarily.

In both the cubic parabola and the lemniscate, the curve starts with an infinite radius and at any point its degree is proportional to its length, until at the point where it connects with, and becomes tangent to, the main curve it is of the same degree as the latter. Conclusions presented by Mr. Wellington in regard to the cubic parabola were as follows: (1) The transition curve will deflect exteriorly from the main curve, because the rate of curvature becomes continuously less. (2) It will terminate in a tangent parallel to the given tangent, because the same central angle is consumed. (3) The offset D-P (or O) in the bottom diagram of Fig. 97, and the transition curve A-C, bisect each other at M. (4) At any intermediate points O,O' at equal distances from the ends or from the middle point, the offsets to the transition curve from the tangent or main curve produced will be equal. The curve may be set out by deflection angles or by offsets, the latter method being the more convenient and less expensive as a rule, while it is quite close enough for all useful purpose. The offsets quarter points are so small, only $\frac{1}{16}$ of the total offset, that for grading purposes the actual length of curve is little more than half its theoretical length.

A transition curve used on the Virginian Ry. and based on a plan devised by J. G. Sullivan, forms a spiral approximating a parabolic arc, but the method of laying it out is simpler. It has twice the length of the circular arc which it replaces, though the angle subtended remains the same. The spiral curve is divided into as many stations as there are degrees in the main curve, and the curvature increases 1 deg. for each station. By varying the length of station, the spiral may be made as sharp or as easy as the tangent conditions allow. This method leaves the tangent in its original position, while the center of the main or circular curve is drawn in towards the intersection angle. This transition is used for all curves of 2 deg. and over on some parts of the line, while on other parts the limit is 3 deg.

Easements on Existing Track.—The use of transition curves has increased greatly in recent years, and simple methods have been developed for applying them to existing track without excessive staking out or shifting of track. In most cases, the length of transition is twice the length of the simple curve which it replaces, while its angle is the same as that of the latter. In relocating a circular curve to apply transition curves at the ends, the center line of the main part of the curve will be thrown inside the original line, which may be objectionable by requiring a considerable shift of track and causing the rail joints to be closed. To meet this condition, a method has been introduced by which the original length of curve is maintained. The center portion of the curve is slightly sharpened and moved a few inches outward. The ends are drawn slightly inward, so that the transitioned curve intersects the old circular curve near the ends of the latter. In this way, the improved curve can be placed on an existing roadbed where no great change in center line position is practicable. These revisions and refinements are set out easily, and the track changes also made without difficulty, while the practical advantages in an easy-riding track, less wear, and maintenance work are well worth the expense and trouble incurred.

In an offset method specially applicable for existing track, as given by David Molitor and used on the Illinois Central R. R., the circular curve is staked out in the usual way and stakes are set at the P.C. and on the curve and the tangent at distances $L \div 4$ and $L \div 2$ (see top diagram in Fig. 97); the distance L being the length of the transition curve as given in Table XVIII. From these five stakes, offsets are measured for stakes

TABLE XVIII.—OFFSETS FOR STAKING OUT TRANSITION CURVES

Degrees of curve		Radius, R, feet	L, feet	Offsets			
				$X = \frac{L}{4}$	$X = \frac{L}{2}$	$X = \frac{3L}{4}$	$X = L$
Degrees	Minutes			U, feet	T, feet	S, feet	M, feet
2	00	2,864.9	60.7	0.003	0.027	0.051	0.054
2	15	2,546.6	68.3	0.004	0.038	0.071	0.076
2	30	2,292.0	75.9	0.006	0.052	0.098	0.105
2	45	2,083.7	83.4	0.008	0.069	0.130	0.139
3	00	1,910.1	91.0	0.011	0.090	0.176	0.181
3	15	1,763.2	98.6	0.014	0.115	0.216	0.230
3	30	1,637.3	106.2	0.018	0.143	0.269	0.287
3	45	1,528.2	113.7	0.022	0.176	0.330	0.352
4	00	1,432.7	121.3	0.027	0.214	0.401	0.428
4	15	1,348.5	128.8	0.032	0.257	0.481	0.513
4	30	1,273.6	136.4	0.038	0.304	0.570	0.608
4	45	1,206.6	144.0	0.045	0.358	0.671	0.716
5	00	1,146.3	151.7	0.052	0.418	0.783	0.836
5	15	1,091.7	159.3	0.060	0.484	0.907	0.968
5	30	1,042.1	166.8	0.069	0.556	1.042	1.112
5	45	996.9	174.4	0.079	0.636	1.191	1.271
6	00	955.4	181.9	0.090	0.722	1.352	1.443

set as track centers. The position of the circular curve between its two transition curves is given by the constant offset M. In the equation for the curve, LR is a constant, 173,800, which is found to give good results without excessive values for M. The length of transition curve L is then as in the first formula below. The other formulas give the ordinate U for points between O and the p.c., and the offset S for points between the P.C. and N.

$$L = \frac{173{,}800}{P},\ U = \frac{X^2}{6LR} = \frac{X^3}{1{,}042{,}000},\ S = \frac{X^3}{6LR} - \frac{(X - L)^2}{2R}$$

Parabolic Curves.—Although parabolic curves have been used to a limited extent with the idea that they ride more easily then circular curves, they have little effect on the transitioning. In the field book which the late A. M. Wellington left unfinished at the time of his death, it was pointed out that if the same tangents and tangent points are connected by parabolic instead of circular curves, track curves will be obtained which ordinarily are of only slightly longer radius at the P.C. and P.T. than the circular arc. Thus the objectionable features of the curve connections are only slightly alleviated, and there is the disadvantage that the curve is correspondingly sharpened in the center. Another unfavorable effect is that the degree of the curve is constantly changing, and as the change is very slow, the coefficient of friction of the center plates of the car trucks becomes a maximum and the flange pressures are increased materially. With the parabola, it is true, it is unnecessary to have the tangents equal, as with circular curves. If unequal tangents are used, the conditions become more favorable at one end of the curve and less favorable at the other end. It was concluded that, on the whole, the parabola is not a desirable railway curve, even if it were more easy, instead of more difficult, to lay out than the circular curve.

Compound Curves.—A combination of curves of different radius is used in some cases, the main or center part of the curve running into end curves of larger radius. This arrangement may be necessary or desirable at times, but it does not give the spiral effect of transition curves. In fact, transition curves might still be introduced with advantage to connect the end curves with the tangents.

Superelevation on Curves.—Since a moving train tends to travel always in a straight line, as explained, all its force would be exerted normally against the outer rail guiding the wheels around a curve. To offset this force, the outer rail may be elevated to such an extent that the rails will be in an inclined plane at right angles to the resultant of the gravity force and the centrifugal force acting upon the train, thus creating a centripetal force which tends to cause the train to travel in a curved path. The factors then remaining to be dealt with are the pressures and frictional resistances involved in deflecting the wheels from the straight path which they tend to follow, and compelling them to travel around the curve. The centrifugal force is about 58 lb. per ton on a 1-deg. curve for a speed of 50 m. p. h., while the centripetal force is about 35 lb. per ton per inch of elevation. The former force tends to increase the weight on the outer rail as well as the pressure against the rail.

Although the amount of superelevation for any given weight, curve, and speed can be calculated, no such exact relation can be provided in the track, as the important factors of speed and weight are variable, every curve carrying light and heavy trains at various speeds. Under such conditions a mean or compromise elevation must be adopted, taking into consideration the proportion of fast and slow trains and also local conditions which may influence train movements.

Thus, where passenger traffic is the more important, a high speed factor may be used, as insufficient superelevation may lead to a derailment. But at the same time the elevation must not be so great as to affect injuriously the slower freight trains, reducing the weight on the outer rail and thus making it easier for the wheel flange to mount the rail. High elevation also introduces the liability of getting heavy, slow trains stalled on the curve owing to the severe flange pressure and weight on the inner rail, especially if the grade is not compensated for curvature. In general, it is desirable to keep the superelevation low. As to location, a curve at the top of a grade should have less elevation than a curve of the same degree at the foot of the grade or on level track. On double-track curves, the ascending trains should have less elevation than the descending trains, since ordinarily the former will have a lower speed. Sharp curves near stations or grade crossings, where trains normally stop, will require less elevation than those on open track.

Length of engine wheelbase is not a factor in the calculations but has proved to be a troublesome practical factor in some cases. In the superelevation for any particular curve, therefore, judgment based upon observation and experience is fully as important as mathematical calculations which may support this judgment. Where curves are numerous, the engineer, the roadmaster, and the foreman should travel on the trains occasionally and note any curves where the trains do not ride easily. The superelevation is given by raising the outer rail, which is the line rail, keeping the inner rail at grade. In some cases, the inner and outer rails have been lowered and raised, respectively, by half the superelevation, in order to maintain the center of gravity of the train uniform on curve and tangent. The benefit is doubtful and the practice is objectionable in disturbing the roadway section.

Of numerous formulas for calculating the centrifugal force, C, of a moving train and the superelevation, E, to counteract the force, the following are given as examples, in which W is weight of train in tons, F, velocity in miles per hour, D, degree of curve, and G, distance c. to c. of rails.

$$C = W \times F \div 32.2 \times R$$
$$E = V^2D \div 1{,}450$$
$$E = 0.00066 \text{ (or } 0.0007) \times DV^2$$
$$E = GV^2 \div 32.2$$

Tables of superelevation for different curves and speeds are issued by individual railroads, but these serve only as guides, since the elevation is calculated usually for passenger or fast trains, and the elevation adopted in any instance depends partly on conditions of grades and character of train service. It is useless, of course, to carry out the tables to high figures, since the maximum elevation is limited to 6 or 8 in., and fast trains are usually required to reduce speed on very sharp curves. There is, however, always a chance of these speed limits being exceeded. The Canadian Pacific Ry., with a permissible maximum elevation of 6 in., prescribes a limit of 4½ in. for curves over 6 deg. on maximum grades, in order to avoid any tendency to derailment of slow trains.

For any particular curve, the actual elevation is a compromise more or less arbitrarily selected, and the maximum is usually confined to tracks used extensively by high-speed passenger trains, as it is too high for heavy freight trains at moderate speed. In some of these tables, the maximum safe speed is given for each elevation. Thus all 12-deg. curves are noted as safe for 30 m. p. h. and all the 1-deg. curves for 80 m. p. h.

Whatever the amount of elevation, it should be maintained uniformly around the curve to insure an easy-riding track, and foremen should test the condition frequently. Where the degree of curve is not known, the proper elevation may be ascertained by taking a string of given length, usually 62 ft. but sometimes 44 ft. for branch lines, holding the ends against the rail head and measuring the ordinate from center of the string to the head. Modifications sometimes specified to suit varying traffic conditions are as follows: (1) main passenger tracks, twice the middle ordinate of a 62-ft. string for curves under 2 deg., or middle ordinate plus 2 in. for curves of 2 deg. and over; (2) main freight tracks, $\frac{3}{4}$ of middle ordinate; (3) combination tracks, middle ordinate; (4) sidetracks, such elevation as may be specified to meet local conditions.

In these various tables of superelevation, there is considerable lack of uniformity. Thus, for a speed of 40 m. p. h. on a 5-deg. curve, there are several different elevations given, ranging from 4 in. on one road to $5\frac{3}{4}$ in. on another. Some carry the curves only to 8 or 10 deg. and others to 20 deg. or more; the speed range is from a minimum of 10 or 20 to a maximum of 50 to 80 m. p. h., and the maximum elevation allowed is from 6 in. to $9\frac{3}{4}$ in. This last is for 75 m. p. h. on a 3-deg. curve, but it is doubtful if such a great superelevation is used to any extent. As a rule, 6 in. is a practicable maximum, suitable for 25 m.p. h. on 15-deg. or 60 m. p. h. on 3- or 4-deg. curves. The Pennsylvania R. R. specifies 7 in. and the A. R. E. A. says 8 in. On sidetracks, a maximum of 2 in. is sometimes specified. As to sharp curves, the Boston Elevated Rys., with curves of 82 to 200 ft. radius, have a maximum elevation of 4 in., with runout on the transition curves; on the Chicago elevated lines, the elevation is $4\frac{1}{2}$ in. for curves up to 300 ft. radius, 3 in. up to 500 ft., $1\frac{7}{8}$ in. up to 700 ft., and 1 in. for curves of more than 700 ft. radius. On industry tracks, it is rarely advisable to use curve superelevation and then only about 2 or 3 in.

In the three condensed examples of tables of superelevation, given in Table XIX, one varies the elevation by increments of $\frac{1}{8}$ in., another by $\frac{1}{4}$ in., and a third by $\frac{1}{2}$ in. In view of the extreme range of speeds on any one curve, it is doubtful whether variations less than $\frac{1}{4}$ in. are worth the time and care involved in adjustment. The A. R. E. A. table is for 1 to 20 deg., with speeds increasing by 5-mi. increments from 10 to 70 m. p. h. Although it shows figures as high as $9\frac{1}{2}$ in., the instructions say that the superelevation should not exceed 8 in. The Union Pacific gives a range from 1 to 12 deg., with speeds increasing by 10 m. p. h. from 10 to 60 m. p. h. The Pennsylvania R. R. range is by half degrees from $\frac{1}{2}$ deg. to 8 deg., with speeds increasing from 10 to 70 m. p. h. by 10-mi. increments. Even allowing for differences in speeds, loads, grades, and operating conditions on different roads, the wide variations in the superelevation tables indicate divergent opinions rather than scientific analysis. They also suggest the need of careful investigation.

TABLE XIX.—SUPERELEVATION ON CURVES

American Railway Engineering Association

Degree of curve	Miles per hour						Degree of curve	Miles per hour	
	20	30	40	50	60	70		20	30
	in.	in.	in.	in.	in.	in.		in.	in.
1	$\frac{1}{4}$	$\frac{5}{8}$	$1\frac{1}{2}$	$1\frac{5}{8}$	$2\frac{3}{8}$	$3\frac{1}{4}$	11	$2\frac{7}{8}$	$6\frac{1}{2}$
2	$\frac{1}{2}$	$1\frac{1}{8}$	$2\frac{1}{8}$	$3\frac{1}{4}$	$4\frac{3}{4}$	$6\frac{1}{2}$	12	$3\frac{1}{8}$	$7\frac{1}{8}$
3	$\frac{3}{4}$	$1\frac{3}{4}$	$3\frac{1}{8}$	$4\frac{7}{8}$	$7\frac{1}{8}$	$9\frac{1}{4}$	13	$3\frac{3}{8}$	$7\frac{1}{4}$
4	1	$2\frac{3}{8}$	$4\frac{1}{4}$	$6\frac{5}{8}$	$9\frac{1}{2}$	..	14	$3\frac{5}{8}$	$8\frac{3}{8}$
5	$1\frac{1}{4}$	3	$5\frac{1}{4}$	$8\frac{1}{4}$	..	..	15	$3\frac{7}{8}$	$8\frac{7}{8}$
6	$1\frac{5}{8}$	$3\frac{1}{2}$	$6\frac{1}{4}$	..	..	..	16	$4\frac{1}{4}$	..
7	$1\frac{7}{8}$	$4\frac{1}{8}$	$7\frac{3}{8}$	..	..	..	17	$4\frac{1}{2}$	..
8	$2\frac{1}{8}$	$4\frac{3}{4}$	$8\frac{3}{8}$	..	..	..	18	$4\frac{3}{4}$	..
9	$2\frac{3}{8}$	$5\frac{3}{8}$	..	..	..	..	19	5	..
10	$2\frac{5}{8}$	$5\frac{7}{8}$	..	..	..	..	20	$5\frac{1}{4}$	..

	Union Pacific Ry.					Pennsylvania R. R.					
Degrees of curve	Miles per hour					Miles per hour					
	20	30	40	50	60	20	30	40	50	60	60
	in.	in.	in.	in.	in.	in.	in.	in.	in .	in.	in.
1	$\frac{1}{4}$	$\frac{3}{4}$	$1\frac{1}{4}$	$1\frac{3}{4}$	$2\frac{1}{2}$	$\frac{1}{2}$	$\frac{1}{2}$	1	$1\frac{1}{2}$	$2\frac{1}{2}$	3
2	$\frac{1}{2}$	$1\frac{1}{4}$	$2\frac{1}{4}$	$3\frac{1}{4}$	$4\frac{3}{4}$	$\frac{1}{2}$	1	2	$3\frac{1}{2}$	5	$6\frac{1}{2}$
3	$\frac{3}{4}$	$1\frac{3}{4}$	$3\frac{1}{4}$	5	7	1	2	3	5	7	
4	1	$2\frac{1}{2}$	$4\frac{1}{4}$	$6\frac{1}{2}$	..	1	$2\frac{1}{2}$	4	$6\frac{1}{2}$		
5	$1\frac{1}{2}$	3	$5\frac{1}{4}$	..	..	$1\frac{1}{2}$	3	$5\frac{1}{2}$			
6	$1\frac{3}{4}$	$3\frac{1}{2}$	$6\frac{1}{4}$	..	..	$1\frac{1}{2}$	$3\frac{1}{2}$	$6\frac{1}{2}$			
7	2	$4\frac{1}{4}$	$7\frac{1}{4}$	..	..	2	4				
8	$2\frac{1}{4}$	$4\frac{3}{4}$	..	..	..	2	5				
9	$2\frac{1}{2}$	$5\frac{1}{2}$									
10	$2\frac{3}{4}$	6									
11	3	$6\frac{1}{8}$									
12	3	7									

Runout of Superelevation.—In gradually raising and lowering the outer rail at the ends of the curve, the usual practice with circular curves is to form the runout on the tangents and to give the full superelevation for the entire length of the curve. By slightly raising the rail on the tangent, the wheels tend to run toward the high rail and are thus put in good position for taking the curve. In some cases, only 50 to 66 per cent of the runout is on the tangent. A general rate of inclination for the runout is $\frac{1}{2}$ in. per rail length, but in high-speed trains this rise may be perceptible or unpleasant, so that it is sometimes reduced to $\frac{1}{4}$ in. per rail length up to 3-in. elevation, while for greater elevation the length must not exceed 300 ft. Another rule is to give a rise of 1 in. in a distance in feet equal to the speed for which the curve is elevated, multiplied by 1.75. Where transition curves are used, the runouts should coincide with them; the Pennsylvania R. R. requires that transitions should be of a length that will give an inclination not exceeding ½ in. per 33-ft. rail. In some cases, however, this inclination is too steep to give easy-riding track, and the runout is commenced, therefore, at two rail lengths along the tangent, giving 1-in. superelevation at point of transition curve and increasing to full elevation at the end of the transition. To insure proper maintenance of such curves, a diagram of the elevation should be given to the section foreman.

On the Pennsylvania R. R., it is provided that main-track curves with 2-in. or greater superelevation should have transition curves of such length as to give a rate of runoff not exceeding ½ in. per 33-ft. rail. Where the branches of compound curves have a difference in superelevation of 2 in. or more, these branches should be connected by an easement curve giving the above rate of runoff. On the Union Pacific, the same runoff is used for circular curves and is entirely on the tangent; if the tangent between two curves is too short for this, the runout is divided into two parts in proportion to the degrees of the two curves and is made level for one rail length at this point, beyond which the runoff extends at the above rate as far along each curve as is necessary to reach the full elevation. On the Chicago Elevated Rys., the elevation of 1 to 5 in. is run out at $\frac{3}{32}$ in. for each 18-in. tie spacing.

Superelevation on Bridges.—Where curves occur on bridges of the open-floor type, the usual methods of putting in the superelevation are to use taper ties, to put shims or blocks under the outer rail, or to put

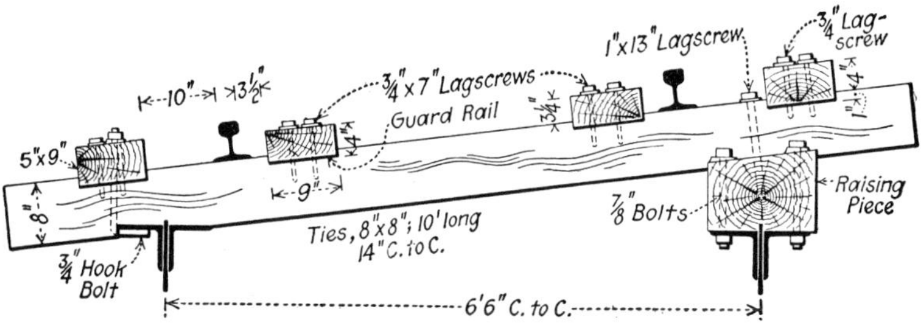

FIG. 98.—Curve elevation on bridge.

wood blocks under the outer ends of the ties. Tapered ties are used sometimes up to a maximum thickness of 14 in. at the outer rail or 16 in. at the butt, with the standard thickness under the inner rail. Beyond this, a cushion block may be placed on the outer girder, the bottom of the tie being cut so as to have a horizontal bearing on the inner girder, but the thickness under the inner rail must not be less than that of the standard tie. The block is the full width of the tie, 2 in. thick at the smaller end and about 4 ft. long. Its inner end is spiked to the tie, and the outer end has a bolt through the tie, while the guard-rail bolt passes through both the tie and the block. With ballasted floors, the ties are inclined in the ballast in the usual way, as on ordinary roadbed.

In the plan shown by Fig. 98, the girders have no top cover plates and the web extends $\frac{3}{4}$ in. above the chord angles. A timber of the proper depth and full width over chord angles is laid on the outer girder and bolted to it at intervals, being grooved to fit the projecting web plate, as shown. Alternate ties are fastened to the timber by long lag screws. Over the inside girders, the ties are boxed or dapped to give a level bearing on the chord. This plan is for use where tapered ties would have to be more than 14 in. thick under the outer rail.

On timber trestles, the elevation may be given by blocks between the caps and the outer stringers, or more generally by blocks under the outer rail. Tapered ties are not often used. Cushion ties are tapered timbers about 3 in. thick at the smaller end, with the full width and length of the ties, and spiked on top of the latter. Cushioned caps are similar but larger timbers placed on the caps. In both cases, water is likely to work between the contact surfaces of the two pieces and cause decay. Neither of these methods is used to any extent. A solid floor without ballast may have the floor system built to give the superelevation, but if the floor is ballasted, the ties will be inclined in the usual way.

WIDENED GAGE

Wear and displacement of the rails by trains causes a gradual widening of gage, but where this is slight and uniform for long distances, it need not be rectified, as the regaging would cause unnecessary work and expense, together with damage to the ties by shifting and respiking the rails (see "Gaging"). But sharp kinks in gage, like low spots in surface, must be rectified promptly.

Widening Gage on Curves.—It is usual practice to widen the gage on sharp curves in order to allow locomotives to pass without undue friction and pressure, wear of rails, and wheel flanges, or a dangerous tendency of the wheels to climb the outer rail. There is great diversity of practice, the widening beginning with 3 deg. on some roads and 8 to 10 or 12 deg. on others, but the maximum permissible gage, including rail wear and curve widening, is usually 4 ft. 9$\frac{1}{2}$ in.

Several railways follow the A. R. E. A.'s recommendation of standard gage up to, and including, curves of 8 deg. and then widening $\frac{1}{8}$ in. for each 2 deg. or fraction up to a maximum gage of 4 ft. 9$\frac{1}{4}$ in. On the Union Pacific R. R., the limit is 5 deg. for standard gage, while on the Pennsylvania R. R., it is 10$\frac{1}{2}$ deg., both with the same rate of widening and maximum gage as above. The Pennsylvania R. R. also provides that where a transition curve is used, the distance from its tangent end to the point at which the gage begins to widen is found by dividing the length of transition curve by the degree of circular curve and multiplying by eight. In widening, the low or inside rail is shifted, the high or outside rail being the line rail; this widening is made at the rate of $\frac{1}{8}$ in. to a rail length, and the full widening is carried around the entire length of curve. For gage of turnouts see "Switch Work."

An apparent reason for the diverse opinions in widening gage is the use of locomotives of different classes, with varying lengths of rigid wheelbase and width of wheel gage, but so many railroads now operate engines of long wheelbase that this factor does not explain the variations in practice. With four or five driving axles, the wheelbase may be 16 to 21 ft.; and although blind or flangeless tires of extra width have been used on one or more pairs of wheels, the more general practice now is to use flanged tires on all wheels. Thus, in a 2-10-2 engine, with all wheels having flanged tires $5\frac{1}{2}$ in. wide, the distance back to back of flanges is $53\frac{3}{8}$ in. for both trucks and for the second and fourth driving axles, while for the other axles it is $53\frac{1}{8}$ in. and the first driving axle has a lateral-motion device which allows 1-in. play on each side of the engine. In a 4-8-2 engine, the distance is $53\frac{3}{8}$ in. for the trucks and $53\frac{1}{4}$ in. for all drivers, while the lateral-motion device in this case gives $\frac{3}{4}$ in. play one ach side. The same distances are used in a 4-6-2 engine, but this has no lateral motion device.

The wheel gage, inside of flanges, ordinarily allows $\frac{5}{8}$ to $\frac{7}{8}$ in. clearance from the track gage, or $\frac{1}{4}$ to $\frac{3}{8}$ in. on each side, and there is no necessity for widening the track gage as long as the clearance between the wheel flanges and the rails allows the engine to travel around the curve without binding or exerting excessive pressure. A factor to be considered in relation to widening gage and width of guard-rail flangeway is the length of wheel flange which extends below the top of rail head. For a flange $1\frac{1}{8}$ in. deep, this distance will be 12 in. for 30- to 34-in. wheels, 13 in. for 40-in., 14 in. for 46-in., 15 in. for 53-in., 16 in. for 60-in., 17 in. for 68-in., and 18 in. for wheels up to 76-in. diameter.

Track on Curves.—Rails on curves may be double spiked, or for curves of 4 deg. and over they may be reinforced by braces on the outside to resist the lateral pressures mentioned above. Metal tie-plates prevent the cutting of the tie by the rail and also cause the inner and outer spikes to act together, but they reduce the friction between the rail and tie unless provided with a shoulder to bear against the rail base. On heavy curves, the gage may be maintained by the use of bridle rods or tie bars fitted to the base of the rail like switch rods, there being from two to four bars to a rail length. These bars are not used extensively. For sharp curves on double track, the standard track spacing must be increased in order to give the normal clearance between cars of passing trains (see "Roadway").

Specially heavy rails are sometimes used on curves, having deeper and wider heads and thicker webs than the rails on tangents, the height remaining the same. Rails of specially hard and tough steel have been used (see "Rails") and also an outer rail having a narrow, deep head, with the idea of reducing friction. As used by the Southern Pacific Ry. on curves of 19 deg., and by the Boston Elevated Ry. on curves of 9 to 57 deg., the 90-lb. outer rail was $6\frac{1}{8}$ in. high with a head $1\frac{15}{16}$ in. deep and only $1\frac{3}{4}$ in. wide. The advantages, however, have not been such as to lead to any extensive use of this special rail section. To facilitate the passage of curves and reduce wear of rails and wheel flanges, several railroads have used flange oilers on the engines, applying to the driving-wheel flanges a thick asphaltic oil which will stick to the wheel flange and side of rail head without spreading over the top. On 8- to 10-deg. curves, this practice increased the life of rails by 50 to 200 per cent.

Sharp Curves.—Many railroads have curves of 8 to 10 and even 15 deg. in main track, especially on branches and on mountain divisions, but 4 to 6 deg. is the desirable maximum for main lines carrying heavy and fast traffic. Sharp curves introduced for economy in construction where light traffic is in prospect may become objectionable with increased traffic, so that much work has been done during recent years in reducing such curvature to improve operating conditions even at very high cost. The advantage is greater with the sharp curves, and the expenditure warranted depends largely upon the number of trains daily. Exceptionally sharp main-line curves were 30 deg. on the Lake Erie & Western Ry. at Lafayette, Ind., and 22 deg. on the Canadian Pacific Ry., but the first was reduced to $10\frac{3}{4}$ deg. and the latter eliminated by a tunnel and an 8-deg. curve. In the Grand Central Station at New York, there is a loop of 140-ft. radius for multiple-unit electric suburban trains, while the Pennsylvania R. R. terminal on Long Island has a loop of 600-ft. radius for trains drawn by large electric locomotives. For rapid-transit lines, sharp curves are necessary and practicable; the Boston Elevated Ry. has 82 ft. as its minimum radius, with numerous curves of 100 and 200 ft.

Curves on sidetracks are limited by the Pennsylvania R. R. to 16 deg., except that in special cases a radius as short as 176 ft. (33 deg.) may be used. Curves of 20 to 35 deg. are required occasionally at yards and industries, and for spur tracks into warehouses or shops, but it may be necessary to handle cars singly or to separate them by bars between the couplers to prevent the corners striking. The length and construction of cars is a factor in this respect, and even on curves of 200-ft. radius, there has been trouble from the couplers swinging so far from the center line as to knock off the angle cocks of the air-brake pipe. Single cars have been handled on 40-ft. curves, while at Cincinnati the Pennsylvania R. R. has right-angle curves of 50-ft. radius entering an industrial plant and operated by four-wheel switching engines. Locomotives and cars can be operated on curves of little over 100-ft. radius. But although such curves are necessary at times to meet special conditions, it is generally desirable to make the limiting radius about 250 ft., on account of difficulties in operation, high cost of maintenance, and liability of trouble from derailments.

Four-wheel switch engines are rarely used in railroad service but six-wheel engines of 10- and 11-ft. wheelbase operate on curves of 75- and 90-ft. radius. Switch engines of the 0-10-0 class, with 19-ft. wheelbase have operated freight trains on curves of 260- and 360-ft. radius. Although very sharp curves may be necessary to meet governing conditions, they are more or less troublesome in operation and are expensive in causing excessive wear of rails and wheel flanges, this wear being often aggravated by poor condition of track. On all sharp curves, whether in main line or yards, the track should be well maintained, with the curvature uniform, gage properly widened, ties firmly tamped, and rails not only securely bolted and spiked, but reinforced by rail braces or tie rods.

Guard Rails on Curves.—On sharp curves, the pressure of the wheel flanges against the outer rails causes severe wear and may cause the wheel to climb the rail, especially with heavy trains and steep grades. To prevent this, a guard rail or check rail may be placed inside the inner rail to hold the wheel flanges away from the outer rail. As a rule, the guard is intended only to prevent derailment and comes into action only if the outer wheels tend to climb the rail, but in some cases it is placed closer to the inner rail so that the backs of the inner wheels are guided and the outer rail thus relieved of some of the wear. The width of flangeway is sometimes only 2 in., plus the widening of gage for curve, but it may be $2\frac{1}{2}$ to $3\frac{1}{2}$ in., or even more if the rail is intended only to prevent derailment. The wheelbase of the engines has some influence on this, but is not often considered. To carry blind or flangeless driving wheels on very sharp curves, a guard rail is laid sometimes on the inside of the outer rail, with about $1\frac{3}{4}$ in. flangeway, and another rail or a heavy plank close against the outer side of the inner rail.

The degree of curve to which guard rails are applied is as low as 8 or 10 deg. on some roads, and as high as 20 deg. on others. With heavy coal traffic on 10-deg. curves of the Louisville & Nashville R. R., the outer rails had to be replaced in about 12 months on account of side wear of the rail head, while there was a continual spreading of track and widening of the gage, even when tie rods were used. With the application of guard rails, giving $2\frac{1}{4}$ in. of flangeway, it was found after about 6 months that there was no widening of gage, no spreading of rails, and little wear on the side of the rail head, while the guard rail showed only $\frac{1}{16}$ in. of wear. On similar curves without guards, but provided with tie rods, the gage had been widened $\frac{1}{2}$ in. by wear and spreading of rails, both rails were tilted outwards, and the rail head showed $\frac{1}{4}$ in. of wear on the side and $\frac{3}{32}$ in. on top. Such guard rails are needed more particularly where high-speed electric locomotives are used, especially where the motors are on the driving axles and tend to cause greater lateral thrust than that of a steam locomotive. On the sharp curves of elevated and subway lines, where the guard rails are in continual action and are kept well greased, the flangeway width is ordinarily $2\frac{1}{4}$ in. for curves up to 125 ft., $2\frac{1}{8}$ in. up to 175 ft., 2 in. to 200 or 300 ft., $1\frac{7}{8}$ in. to 500 or 700 ft., and $1\frac{3}{4}$ in. for curves of larger radius.

Monumenting Curves.—With repeated displacement of track under traffic and its rectification by the section forces, the actual curve will in time vary from its original or true location, but permanent monumenting or referencing of curves will assist the maintenance, reduce the liability of serious irregularity, and also assist the occasional adjustment or resurvey. Ordinary center stakes are likely to be lost, destroyed, or displaced in a short time, and the permanent marking consists in placing stakes or posts outside the track to indicate the beginning and end of each curve and the points of change from spiral or compound curve to the main curve; also sometimes at 500-ft. intervals for convenience in checking or rerunning the curve. Comparatively few railways follow this practice systematically, but it is sometimes introduced as lines are resurveyed.

Another method is to mark the center line securely at points of curve, spiral, and tangent and at intervals of 50 ft. on curve and 25 or 40 ft. on spiral. On double track, the center line between tracks may be marked. For this purpose, $1\frac{1}{2}$- or 2-in. iron pins 42 to 48 in. long may be used, or oak stakes 2 by 2 by 24 in. or stone posts 6 by 6 by 36 in., these last particularly in slag ballast. Chisel marks on pins and posts, and tacks in stakes, indicate the exact center points. For the guidance of trackmen, many railroads indicate the superelevation at each curve by posts or signs set opposite the points of curve, spiral, and tangent at $9\frac{1}{2}$ to 12 ft. from center of track. If there is no spiral, the posts may be set opposite the ends of runout of superelevation. When used as signs only, these posts are not located so exactly as the monuments. On each post is marked the elevation: 0 at the ends and the full figure at point of main curve; sometimes the degree of curve and inclination of runout are marked.

Switchbacks.—Where a railway is developed by a zigzag line on a hillside in order to maintain practicable grades and there is no room for connecting curves, the inclines may be brought together in a Y, with a tail track or switchback of train length extending beyond the switch on a rising grade. As the train enters the switchback, a brakeman drops off and opens the switch as soon as the entire train is on the tail track, so that the train can be reversed at once and backed onto the next incline, with little loss of time. On the Crown King branch of the Santa Fe Ry. in Arizona, there is a switchback section 9 mi. long with 10 switchbacks. The inclines are 1,500 to 4,000 ft. long, with maximum grades of $3\frac{1}{2}$ per cent (compensated) and 16-deg. curves having the gage widened $\frac{3}{4}$ in. Both grades and curves are easier on the back-up inclines. The average speed is about 15 m. p. h. This arrangement cannot be used for main lines with heavy traffic. In reducing grades from 4.5 to 2.2 per cent in the Kicking Horse Valley, on the Canadian Pacific Ry., spiral tunnels in the hillside were built on 10-deg. curves to connect the inclines.

Switchbacks have been introduced in several cases where railways crossing mountain ranges have been built at first with open lines through passes at high elevations, for economy in first cost and for promptness in opening the railway to traffic. The Northern Pacific Ry., had a 7-mile switchback line with grades of 5.6 and 5.2 per cent, reduced to 2 per cent on the tail tracks, 400 to 500 ft. long. There were two switchbacks on each side, with split switches set normally for the ascending track, so as to trap runaway trains. On the Great Northern Ry., a switchback line in the Cascade range had five switchbacks on the west side and three on the east side. Stub switches were used. The grades were $3\frac{1}{2}$ and 4 per cent, with tail tracks 1,000 ft. long and rising at 1 to 5 per cent from the headblocks. A special crew took the trains over this section, an engine being coupled at each end and movements being governed by whistle signals. The track is usually level between the frog and headblock, but it has been thought that operation would be more convenient with the ascending grade continued unbroken to the end of the tail track, though this would sacrifice distance and be more expensive.

On the transandine railway now being built from Salta, Argentina, to Antofagasta, Chile, some switchbacks are used in steep sidehill country. In one case, the upper and lower stretches of the line are only 147 ft. apart horizontally but the vertical distance is nearly 100 ft. and the track distance, including two switchbacks, is 6,022 ft. There must be an even number of switchbacks, so that when the direct line is again reached the locomotive will again be at the head of the train. (See also "Summit Tunnels" in Chapter XXIII.)

CHAPTER XXI

SWITCH WORK AND TRACK LAYOUTS

Turnouts.—Where a train is to be diverted from one track to another, a turnout is installed, which is essentially a curve connecting two parallel or diverging tracks. This curve, however, is composed of three principal parts: (1) a switch, consisting of two movable rails to direct the train onto one track or the other, as desired; (2) a frog to allow the wheel flanges to pass the intersection of the rails; and (3) rails, known as lead rails, connecting the frog with the switch rails. The construction of switches and frogs has been described in Chap VI. Practically every railway has its standard plans and tables of angles and dimensions for a wide range of turnouts.

Main-line turnouts usually have frogs of Nos. 9 and 10, giving curves of 6 to 8 deg. But at junctions and ends of double-track sections, where trains run at high speeds, frogs Nos. 15 to 20 or 24 are used, giving curves of $2\frac{1}{2}$ to 1 deg. Yard and sidetrack turnouts may have frogs Nos. 7 and 6, with curves of 12 to 16 deg., but No. 8 is generally the desirable low limit, except for special conditions. While sharp turnout curves and special arrangements may be necessary in freight yard and terminal yards, it is advisable to avoid them as far as possible, especially where locomotives of long wheel base will be operated, as these are likely to distort the lead and to widen the gage beyond the frog, in addition to causing severe wear.

A turnout is right hand or left hand according to whether it turns to the right or left of a man standing on the main track and looking towards the point of the frog. The switch angle of a turnout is that contained between the point and stock rails of a split switch or between the two positions of the moving rails of a stub switch. Any switch is a facing switch to trains approaching the points, and a trailing switch to trains approaching from the heel of frog. On double track, most of the turnouts can be arranged with trailing switches, so that there will be no danger of accident due to misplaced switches, as a train taking the sidetrack must trail through the switch, and then back through it at slow speed as a facing switch and onto the turnout. This arrangement is not always practicable and cannot be used at junctions, but there the switches are usually interlocked with signals and are under close supervision. Nor is it desirable for sidetrack connections under certain conditions, as, for instance, on a descending grade where heavy trains would have to start and back upgrade through the switch onto the turnout.

Turnout Leads and Curves.—In the turnout, the lead is that part of the main-track rail between the switch and the frog. The turnout curve, or curved lead, is that part of the turnout rail between the same points. For calculating the length of lead, the turnout curve may be assumed to extend from point of switch to point of frog; this is termed the theoretical lead. In practice, this distance is too long for frogs above No. 7 and results in sharpening the curve near the frog or flattening it at the switch. This is due to the fact that the switch rail cannot be placed in the theoretical line of the curve, since it is of measurable width and is also set necessarily to allow sufficient flangeway space at the heel. In general, therefore, the turnout curve is assumed to extend from the heel of the switch to the toe of the point of the frog, which gives what is termed the short lead. The length of theoretical lead is equal to twice the gage of track multiplied by the frog number. In two formulas for the lead given below, G is the gage of track; N, number of frog; T, throw of switch; L, length of switch rail; and F, length of frog from toe to point.

$$\text{Lead} = 2N(G - \sqrt{GT}) + L$$
$$\text{Lead} = 6N + (L + F)$$

Frogs of the same number but of different lengths will give slightly different curves; as will also split and stub switches with the same frogs. Increasing the lead does not reduce the curvature of the turnout unless a higher frog number is used, corresponding to the longer lead. On curves, the length of lead for any given frog number is the same as on tangents.

The actual length of lead is a matter of minor importance for ordinary work and simple turnouts, and it may be varied several inches, or even feet, to avoid cutting rails or for other reasons of convenience. The reason is that the rails are of measurable width and cannot be laid exactly to follow the theoretical center lines of a diagram, so that a variation in actual length makes practically no difference in the turnout. For turnouts with short leads, it is only necessary to place the frog point at the place selected, letting the point of switch come where it may. This in no way disturbs the alinement behind the frog.

To economize in cutting rails, the short lead may be shortened. Thus a turnout with No. 8 frog may have a 15-ft. switch rail, a 30-ft. rail, and a 15-ft. piece, which with $6\frac{1}{2}$ ft. to frog point gives a lead of 66 ft. 6 in. A 30-ft. rail can be cut into two pieces, 15 ft. 1 in. and 14 ft. 11 in. long, and with these laid in the curved and straight leads, respectively, the switch points will be kept square across the track. With spring-rail frogs, the turnout wing is usually longer than the other, so that the closure lengths are the same for the straight lead and the turnout curve. It is very desirable to utilize rails of such lengths as are generally available and to avoid cutting rails to special lengths which will cause waste. For this reason, several railways have standard lists or diagrams of the rail lengths to be used in making up closures for turnouts with different frogs, thus utilizing ordinary lengths and securing the maximum length of curved lead. Some plans for No. 10 turnouts allow a variation of 5 ft. in length of lead to avoid cutting rails to fit theoretical conditions.

Gage at Turnouts.—At turnouts from tangent track, the gage is usually 4 ft. 8¾ in. at point of switch; but is 4 ft. 8½ in. at the bend in stock rail, at the second switch rod, and on the straight track through the turnout. On the turnout, with frogs of No. 15 and over, the gage is standard at the heel of the switch and beyond; with frogs Nos. 8 to 10, inclusive, standard gage at heel of switch and toe of frog, widening to 4 ft. 9 in. between these points by lining out the inside rail; with frogs Nos. 4 to 7, inclusive, 4 ft. 8½ in. at heel of switch, widening to 4 ft. 9½ in. in 15 ft. and continuing through the turnout. At turnouts and crossovers in curved track, the gage will be in accordance with the rule for widened track on curves, and there must be a corresponding variation in placing the frog-guard rail; the distance from flangeway side of frog-wing rail to the flangeway side of guard rail must never exceed 4 ft. 5 in. Superelevation of outer rail to counteract the centrifugal force in the train cannot be introduced in turnouts from tangent track, since the main-track rails are in the same horizontal plane. In turnouts from curves, the rails must conform to the levels of the main track. In exceptional cases the inner rail of the turnout curve is depressed slightly.

Turnouts on Curves.—Turnouts should not be located on main-line curves, and turnouts from the outside of curves are sometimes prohibited, since there would be a reverse curve with sudden change in the direction of movement of the train and also reversal of superelevation. If placed on the inside of the main curve, the curvature is suddenly sharpened, and this arrangement cannot well be used on main curves sharper than 12 or 15 deg. Where a turnout runs from the inside of a curve, its degree will be the sum of the degrees of the turnout curve and main-track curve; when the turnout is from the outside of a curve, its degree will be the difference between the degrees of the two curves. Thus, with a No. 10 frog, the turnout from a tangent will be 6 deg.; from the inside of a 3-deg. curve it will be 6 plus 3 or 9 deg.; from the outside of a 3-deg. curve it will be 6 minus 3 or 3 deg. For a turnout on the inside, the frog number should be as high as possible in order to keep the degree of curve as low as possible.

The sharpest curves on which turnouts are placed ordinarily are those in which the degree of curvature on the turnout does not exceed the maximum curvature permissible for the class of track concerned. A striking case is a turnout of 36 deg. running from the inside of a 16-deg. curve; this installation has 12-ft. switch rails, a No. 6 frog, and a uniform gage of 4 ft. 8½ in. A crossover on a double-track main line has No. 18 curved frogs, and the gage is standard throughout. In industry tracks, turnouts on sharp curves may be required, but these are built usually with the same material as for turnouts from straight track. In exceptional cases, the switch rails may be bent to the curve. For such cases the equipment might be built specially, the frog having wide flangeways to reduce the wheel wear, and the switch perhaps having a guard rail in advance to guide "nosing" wheels into position, while a guard rail might also be placed behind the frog. The gage at the switch must be that of the curved track from which the turnout runs, and this is continued usually through the turnout.

Crossovers.—A crossover is a diagonal track connecting two adjacent and usually parallel tracks. It consists of two turnouts placed in opposite directions and connected by a short tangent, but where space is limited and all movements may be slow, the two turnouts may be so close together as to form reversed curves. Crossovers should not be located on curved track, especially on main lines, as the change of direction and of curve elevation is likely to result in derailment. There should be, furthermore, a long stretch of straight track beyond each end of the crossover.

On double track, the crossover should be laid with trailing switches, being used only by backward movements, but at certain places they are laid necessarily with facing switches, as in changing trains from high-speed to low-speed tracks. For such work, it is desirable to use long and easy crossovers and also to impose speed restrictions, as noted below.

The length of the crossover depends upon the frog number and the distance between centers of tracks. To find the distance between the frog points, measured parallel to the rails, on straight and parallel tracks, subtract twice the gage from the distance between centers of tracks and multiply the remainder by the frog number. Thus for No. 8 frogs and tracks 13 ft. c. to c., the distance will be 13 ft. minus 9 ft. 5 in. multiplied by eight, or 28 ft. 8 in. Another method is to take the distance between gage lines of the two tracks, subtract the gage, and multiply by the frog number. For a crossover with frogs of different numbers, find the distance for each frog; then half the sum of these distances will be the answer. For the diagonal distance between frog points, add the squares of distance between gage lines of the two tracks and distance between frog points measured along the track, and take the square root of the sum. The crossover is staked out as a tangent between the frog points.

For high-speed crossovers where the switches are operated from a tower, and also for ends of double track and at important junctions used at high speed, the Rock Island Lines use a No. 15 frog, with lead of 117 ft. $5\frac{3}{4}$ in. to point of frog and 52 ft. $3\frac{3}{4}$ in. between frog points in crossovers. Tracks are spaced 13 ft. c.-to c., and the switch angle is 1 deg. 9 min. For the turnout, the curve is 3 deg. 15 min. Beyond the frog, the tangent continues about 60 ft. and is followed by a 3-deg. curve running into the second main track. Formulas and tables for determining the distance between crossover frogs and locating the point of reverse curve are given in *Engineering News-Record* of Feb. 28, 1924, enabling such work to be set out with a tape.

Speed at Turnouts.—As to permissible speed through turnouts, that will depend largely upon the frog number and turnout curve, but in some cases a speed limit of 15 m. p. h. is prescribed for crossovers and turnouts. The rule on the Baltimore & Ohio R. R. is that No. 10 frogs are to be used at all turnouts and crossovers in main track (except as noted below), and also in sidetracks, yards, and industry tracks where conditions do not necessitate the use of frogs of smaller numbers; No. 16 frogs are specified for passing sidings and crossovers which are connected with interlocking, and No. 20 frogs at end of double track in high-speed main tracks and also at crossovers in such tracks. The Pennsylvania R. R. provides that in main tracks without superelevation, turnouts with No. 20 frogs are adapted for speeds up to 35 m. p. h.; No. 15, up to 25 m. p. h., and No. 10, up to 15 m. p. h. Where the turnout or crossover is from the inside of a curve, so that the superelevation is the same for both tracks, the speed should not exceed that required by the degree of the curve and the amount of superelevation on the turnout lead. For level turnouts, the A. R. E. A. recommends limits of 20 m. p. h. with frogs of Nos. 8 to 10, 27 m. p. h. with No. 14, 37 m. p. h. with No. 15, and 40 m. p. h. for higher frog numbers.

At all main-line turnouts, junctions, crossovers, and ends of double track, speed reduction rules are desirable but are not to be relied upon entirely, as serious accidents have happened through trains taking No. 8 and No. 10 interlocked crossovers at 40 to 60 m. p. h., although the speed limit was 15 m. p. h. In high-speed tracks, it is safer to use frogs over No. 20, and it has been suggested that where frogs of lower number are used in such connections, the switches should be kept set for the straight track and "Stop" signals displayed, the switches not being thrown or signals cleared for the diverging track until the train has come to a stop. This is specially the case at crossovers, with their reversed curves and necessary absence of superelevation.

Switches.—A method employed by the Pennsylvania R. R. for lessening the wear of switch rails is to cut a recess in the side of head of stock rail, $\frac{1}{4}$ in. deep at $1\frac{1}{2}$ in. ahead of the switch point, tapering to zero at 3 in. ahead of the point and 24 to 30 in. behind it. This recess must be cut by a milling machine and not by a chisel. In yards, the points of the switch rails may be staggered about 26 in. for the same purpose. Unusual wear of switch points should be investigated in order to discover and remedy the cause. Since $\frac{5}{16}$ in. wear of wheel treads removes the coning of the wheel so that the outside edge of the tire is practically on the level of the rail head, the switch rail must be removed when its head has become worn down level with the stock rail, unless the latter is worn correspondingly and is even with it for the full width of the widest wheel treads. Length of switch rails varies on different roads, but the following represents general practice:

Frog number	Switch rail	Frog number	Switch rail
4 & 5	10 or 11 ft.	10	$16\frac{1}{2}$ or 18 ft.
6	10, 13, 15 ft.	12	$16\frac{1}{2}$ or 24 ft.
7	13 or 15 ft.	12 (sidetrack)	16 ft.
8	13, $16\frac{1}{2}$, or 18 ft.	15	13 or 30 ft.
8 (sidetrack)		16	24 ft.
	13 or 15 ft.	20	30 or 33 ft.

Frogs and Crossings.—The numbers and varieties of frogs should be as few as possible, since these will usually suffice to meet all conditions and there will be an economy in the stock required to supply renewals and new work. Frogs and crossings of hardened steel construction are suitable at: (1) interlocking plants and ends of double track where trains are diverted at high speed; (2) at interlocking plants and connections with busy main lines where the traffic on the diverting side of the frog is more than 50 per cent of that on the main line; (3) where the traffic is slow and heavy in both directions through the frog, as at yard ladders and similar places. Spring-rail frogs are placed usually where the traffic is moderate and, in general, where traffic is much greater on one track than on the other; also for turnouts and crossovers located at high-speed points but used at comparatively slow speed. Built-up rail frogs are suitable for branch lines with moderate traffic and speeds, and also for yard tracks and sidetracks. Worn frog points may be welded or built up in the track, such repair being done before the point is badly damaged. Worn main-line frogs may be transferred to yards or low-speed tracks. Bolts should be kept tight and replaced promptly if broken.

Crossing Frogs.—In installing crossing frogs, especially those of hardened steel, the Pennsylvania R. R. requires that each crossing, consisting of four frogs, shall be handled as one piece with a crane. If two or more tracks cross one track, the adjacent crossings are bolted together when in position. Sliding a crossing into place and lowering with jacks is not permitted if a crane can be obtained, but if jacks are used they must be placed directly under the frogs and never under the arms of the crossing.

Different types of crossings for different track conditions may be used as follows: (1) Both tracks carrying heavy traffic, for angles less than 45 deg.; built-up rail crossings with hard centers or cast manganese inserts at intersections; for angles over 45 deg., solid cast manganese if roadbed is flexible, or solid manganese crossings with T-rail exterior arms if traffic is not exceptionally heavy. (2) One track carrying heavy and the other light traffic; built-up hard-center crossings for angles under 45 deg., and similar design with solid manganese running rails on the main track for angles over 45 deg. (3) Both tracks carrying light traffic, such as crossings of sidetracks; built-up rail crossings according to standard plans. (4) Heavy traffic main line crossing an electric railway; similar to type (3), with manganese running rails for the steam-road track; except that where the electric railway has heavy traffic, solid cast manganese construction is preferable. (5) Light-traffic line or sidetrack crossing an electric railway; built-up rail crossings are most satisfactory, except where the angle is greater than 45 deg. and the electric railway demands a built-up crossing with cast manganese inserts at the intersections.

Track Crossings.—The calculations for crossing frogs at track intersections are apt to be somewhat complicated. The crossings should always be put in under the direction of an engineer, stakes being set to mark the center lines of the two tracks, and lining of tracks at crossings should always be done with reference to the engineer's stakes. In the renewal of crossings, the angles should first be carefully measured, and the lines set out by instruments. It will very generally be found that the alinement has been affected by creeping of the rails, unless these have been anchored or secured by check plates. The gage of tracks should also be measured. If the alinement is bad, the proper location should be made before the frogs are ordered. Where the two tracks belong to different roads, this matter must be arranged before any work is done. If the two tracks have rails of different weights, the crossing should be built of the heavier rails, and a length or two of this rail laid beyond the frogs on the lighter track, so as to avoid excessive shock at the crossing.

Curved Crossings.—Special difficulties are involved at crossings of curved tracks. The degree of curve should be verified by measuring the tangent offset, and by taking supplementary angles, checked by measurements with a chain or steel tape. A satisfactory plan of a curved crossing cannot be made by taking measurements of the diagonal and of ordinates to the curve, but the better way is to run out the curve and get the angle of intersection, calculate the angles and distances, and check these from a large-scale drawing. The angle of crossing will be equal to the angle between the radii to the common point.

The practice on the Michigan Central Ry. is to take the angle of a crossing frog in the track with a transit, and then confirm the observed angle by tape measurements. The point is found at which the gage lines of the frog meet, and then a mark is made on the gage side of each rail forming the frog at a distance of 3 ft. from the intersection of the gage lines, measured toward the heel of the frog. The distance between these marks is measured and divided by six, which gives the sine of half the angle of the frog. This angle is, of course, the angle of the crossing if both tracks are straight lines. If the crossing is formed by one straight and one curved track or by two curved tracks, the intersection of gage lines at each of the four corners of the diamond is found, and also the angle of each frog in the manner stated above. All the connecting distances between the intersections of gage lines are measured to confirm the calculations, which are anything but simple.

Double Turnouts.—At double turnouts, with three-throw switches, which are undesirable features, the distance from main frog to crotch frog is the theoretical lead multiplied by 0.3, or approximately three times the frog number. In placing the crotch frog when both of the side frogs are of the same number, one turnout is lined first, and a track gage is moved along the track until its center is over the curve or curved lead, which point will be the position for the crotch frog.

Setting Out Turnouts.—In an ordinary turnout, the principal work of the engineer is to locate the frog point and set stakes for the frog point and perhaps switch point also, since a competent section foreman can then put in the turnout with the aid of his standard diagrams and plans, which often show the middle and quarter ordinates from main rail to curved lead for each different turnout. In some cases, the ordinates are given from a chord line drawn from the turnout wing of the frog to the heel of the switchrail.

To avoid cutting a rail, the actual frog point is marked at a distance from the rail joint equal to the length of the frog from the heel to the point. In giving the center line for the turnout, a stake is driven at half-gage from the gage side of the rail on the turnout side, the transit being set over the hub and a foresight taken parallel with the main track rail. An angle equal to the frog angle is then turned away from the track and a stake driven at about 10 ft. from the transit. From this second hub, the line is run as required. The degree of curve beyond the frog has no relation to that of the turnout curve, since they are separated by a tangent of at least the length of frog and of such greater length as may be desired, a minimum length of 10 ft. beyond the frog being desirable if it can be obtained.

Setting Out Turnouts by Offsets.—Single turnouts may be put in by offsets and measurements, but for several turnouts in series, the transit should be used, setting stakes for the points of switches and frogs and for the reverse and tangent points. For a number of parallel tracks, it is usual to stake out the first track and put in the others by measurement, but it is advisable to set out main-line turnouts and turnouts from curves with the transit.

To set the turnout curve, stretch a cord from the theoretical point of the frog, as marked on one rail, to the point of curve on the opposite rail, allowing for the spread at the heel of the switch rail. Divide this distance into four parts and (for standard-gage track) measure a middle ordinate of 1.177 ft. and two quarter ordinates of 0.883 ft. on the concave side of the turnout. From each of these points and from the frog point and point of curve, measure half the width of the gage, which will give five points on the turnout curve. Each side ordinate is three-fourths of the middle ordinate. At the middle point of the lead, the offset from main rail to turnout rail (gage to gage) is one-fourth the width of the gage, and at the quarter points, it is one-sixteenth and nine-sixteenth the gage, whatever may be the frog number or length of lead, or whether the turnout is from tangent or curve. The degree of the turnout curve is approximately 600 divided by the frog number when the turnout is from a tangent; if on a curve, adding or subtracting the degree of the main curve gives the turnout-curve degree when the turnout is from the inside or outside of the main curve, respectively.

Another method is to measure offsets from the gage side of the unbroken track rail, the track being carefully lined before the frog is placed. In plans of turnouts, the distance on lead rail between heel of switch and theoretical point of frog is sometimes divided into four parts, and the length of offsets is given to the curved lead rail, measured gage to gage. At the heel of the switch, there is a spread of about 6 in., and the ordinate at the frog point is 4 ft. 9 in. The intermediate offsets range from $12\frac{3}{8}$, 23, and $37\frac{7}{8}$ in. for a turnout with No. 7 frog, to $13\frac{1}{2}$, $24\frac{1}{2}$, and 39 in. with a No. 14 frog. In the former case, the lead is 61 ft. 8 in. and length of switch rail is 15 ft.; in the latter case, these figures are 108 and 20 ft. respectively.

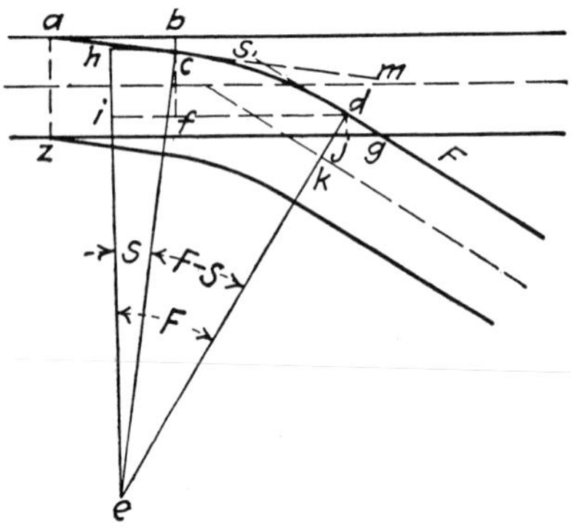

Fig. 99.—Turnout diagram.

Complicated Turnout Work.—In more complicated cases, where it is required to use lap leads, or where a track crosses a turnout between the heel of the switch and toe of the frog, which is a common occurrence in terminal layouts, it is necessary for the engineer to be able to determine the exact degree of curve of the turnout in order that he can figure the angles of the special frogs. The method employed for the special work of the Indianapolis Union Station is given below. Since the curved part of the turnout is tangent to the heel of the switch rail and the toe of the frog, it will be seen from Fig. 99 that the exact degree of curve is determined from the following dimensions, a change in any one of which will change the degree of curve: (1) thickness of switch rail at the point, usually $\frac{1}{4}$ in.; (2) length of switch rail, ac; (3) distance between gage lines at heel of switch, bc; (4) gage of track, cg; (5) length of frog from theoretical point to toe, dg; (6) frog angle, f. Other details in the diagram and formulas are the switch angle, S; and radius of center line of turnout, ek; lead from point of switch to point of frog, zg.

To find ek and zg: with ed as a radius perpendicular to dg, draw the arc dc tangent to ac at c and extend it to h, so that eh is perpendicular to the main track. Draw ce. Then deh equals the frog angle F, and hec equals the switch angle S.

G = gage
F = frog angle
S = bac
= switch angle
ac = length of switch point
bc = distance between gage lines at heel of switch
$\frac{1}{4}$ in. = width of point of switch rail
dg = length toe to point of frog
ek = radius of center line of turnout

To find:

ek and zg.

Solution:

$$\frac{bc - \frac{1}{4}\text{ in.}}{ac} = \sin bac = \sin hec = \sin \text{switch angle.}$$

$$ag\ (\sin F) = dj$$

$$F - S = ced$$

$$G - bc - dj = cf$$

$$ed \cos S - ed \cos F = cf$$

$$ed = \frac{cf}{\cos S - \cos F} = \frac{cf}{2 \sin \frac{1}{2}(F + S)\sin(F - S)}$$

$$ed \sin F - ed \sin S + dg \cos F + ac \cos S = zg = \text{length of lead.}$$

$$\text{Or, } zg = ab + fd + jg$$

$$ab = ac \cos S. \qquad jg = dg \cos \text{F}$$

$$fd = \frac{cf}{\tan cdf} = \frac{G - (bc + dj)}{\tan \frac{1}{2}(F + S)} = \frac{G - bc - dg \sin F}{\tan \frac{1}{2}(F + S)}.$$

$$\text{Then } zg = ac + dg \cos F + \frac{G + bc - dg \sin F}{\tan \frac{1}{2}(F + S)}$$

Turnout Work Simplified.—Turnout work is often regarded as a very complicated and difficult operation, due largely to the number and variety of formulas and mathematical calculations which have been devised. In ordinary practice, however, a good turnout may be laid down by almost any of the formulas and tables, since the work is done with rails which occupy considerable space instead of with thin lines on a diagram. In this connection is quoted the following concise statement written by A. M. Wellington on "The Art and Mystery of Laying out Turnout Curves.

The theoretical lead for standard gage = No. of frog × twice gage, = 9.42 × No. of frog. This assumes the curve to be a simple circular arc, which is not essential for a good curve, nor does it give the best, and considers the curve to extend back to the heel of the switch rail of a stub switch. The lead is always the same, whether the turnout be from a tangent or a curve. A difference of not exceeding 10 per cent in length, especially if the lead be made longer than above, has no appreciable injurious effect on the character of the turnout curve or on its radius. This is best seen by calling the turnout curve a parabola, and remembering that whether the tangents of a parabola be equal, or one 20 per cent longer than the other, will not affect the excellence of the curve, nor, materially, the sharpest radius. Whether the curve be called a circle or a parabola will not alter its position on the ground by more than a hair's-breadth.

To lay out the turnout curve, the frog being in place, and length of lead given, not differing more than 10 per cent from the theoretical lead. Practically, the best transit for running in the curve, and the only one much used for fixing points on it, is an experienced eye. On all kinds of turnout curves, whether from straight or curved main track: offset from gage side of main rail to gage side of lead rail at middle point of lead = $\frac{1}{4}$ gage, or 14 in.; offset at $\frac{1}{4}$ point of lead = $\frac{1}{16}$ gage, or $3\frac{1}{2}$ in.; offset at $\frac{3}{4}$ point of lead = $\frac{9}{16}$ gage, or $31\frac{3}{4}$ in.

Split Switches.—The gage side of the split rail is straight; therefore, it can only be considered, when in place, as a tangent to the true turnout curve. The point at which the theoretical turnout curve attains an offset of the width of a rail head (say $2\frac{1}{4}$ in.) from the main line is 0.8 of the theoretical lead from the frog. Hence to obtain the same turnout curve with a split switch as with a stub switch, the lead should be $(0.8 \times 9.42 \times N)$ or $7.54 \times N +$ the length. of the plain portion of the head of the point rail. Split-switch leads, in other words, other things being equal, should be a little shorter than stub-switch leads. But as all turnout curves are improved by being a little longer than a simple circular arc requires, a lead fixed by the rule of $9.4 \times N$ is good practice for split switches.

Radius.—By the principle of proportional triangles, the radius = lead × No. of frog = (No. of frog)^2 × gage. This holds only on turnouts from tangents.

Degree of Turnout Curve.—The frog angle = 57 deg. (or 3,420 min.) ÷ frog number. The degree of turnout curve =

$$\frac{\text{Frog angle}}{\text{Lead in stations of 100 ft.}};\ \frac{57\text{ deg.}}{N} \div \frac{2Ng}{100} = \frac{606}{N^2},$$

or, in round numbers, 600 ÷ square of the number of the frog. If the turnout be from a curved main track, add to the degree thus obtained the degree of the main curve, if the turnout is to the inside of the curve; and subtract it, if to the outside.

Terminal Plans.—For the complicated track layouts of large terminals, stations, junctions, and yards, careful surveys and close calculation and measurement are essential, as any slight variation or inaccuracy may cause serious trouble in construction or assemblage of the track work. For such cases, it is advisable to lay out the plan on a large scale and then take leads and other measurements from this plan or the tracing. Such complicated track work, like the special work of street railways, may be largely fitted up and assembled at the shops, so that it can be fitted to the lines staked out on the ground and put together with the same accuracy as bridge or structural steel work. At the Indianapolis Union Station, all frogs were located by placing a stake in the center of both the main track and the turnout at the heel of the frog. The frog was then centered between these stakes, and the rest of the turnout or slip crossing was built to conform to that location of the frog. It was found more satisfactory to place the stake at the heel of the frog, instead of at the point, as it gave a very definite location for the frog on the ground.

For the Southern Terminal Station at Boston the following methods were used in planning the layout, ordering material, and setting out the work for the track men. A plan was drawn to a scale of $\frac{1}{8}$ in. to 1 ft. upon which was a central base line with 100-ft. stations, so that points were designated at distances from these stations. On this plan was shown each tie (with its length), switch, frog, guard rail, and other details, with the station distance for each switch point and frog point, and the point of curve and radius for each piece of curved track; also all supports for interlocking apparatus, signal posts, and signal-bridge supports. The entire plan was about 24 by 7 ft., divided into several sheets for convenience. From this plan, all ties, special timbers, and materials were ordered and rails cut to lengths required. All calculations were made and checked in the office, so that the work in the yard was a careful reproduction of the plan, set out by transit and tape, with stakes set for line. After the track layout had been assembled, grade stakes were set, ballast was placed, and the tracks were raised to grade.

St. Paul Union Station.—To provide for laying out this intricate track system on the ground, the entire plan was based on cartesian coordinates. Before any field work had been done, the axes were established in the office, and from these a precise survey was made of the right-of-way lines. Every angle was measured at least five times. A standard tape was used and all measurements were made between tacks driven into wooden stakes. Temperature readings were taken every hour, and the elevation of the top of every stake was determined. From these data, calculations were made with corrections for temperature and grade, and the coordinate of every point was calculated.

In laying out the track work, the equation was calculated for every tangent line and the coordinates of every point of curve, point of intersection, and point of tangent. For switches, there were calculated the coordinates of intersection of center line of main track with center line of track at frog produced. No attempt was made to write the equations of curves, as these equations would be of second power and very cumbersome, with large figures. Seven-place logarithms were used in all calculations in order to avoid the effect of cumulative errors. The coordinates of all points were carried out to six significant figures. Most of the work lies in the fourth quadrant; none is in the second quadrant.

It is considered that no other method could compare with the coordinate system for such a complicated layout. Every feature was definitely and permanently located. This system has been of considerable value in designing intricate layouts, because in some cases it has been found in calculating that a layout which is feasible on the drawing table is not practicable in the field.

Field Work.—In the St. Paul work, no difficulty was experienced in laying out and checking the location of switches, slip switches, crossovers, and other special features as laid down on the plan. For staking turnouts, stakes were set for the heel and toe of the frog on both tracks, working usually from the point of intersection of the two center-line tangents through the frog. For the standard turnout with No. 8 frog, having a ½-in. frog point, the distance from this point of intersection to the actual toe of the frog measured along either tangent is 38 ft. If the two track tangents intersected at a standard frog angle, no stakes were set in the turnout side of the switch, except the heel and toe stakes mentioned, the proper curve being given to the turnout side of the switch by measurements from the gage line of the main-track rail. Where the switch was introduced on a curved main line, or in case it was necessary to introduce curvature outside the switch on the turnout track, the curves were worked out by making both tracks tangent through the frog.

For slip switches, in addition to heel and toe stakes at the frogs, a stake was set at the intersection of both center lines and also at the end of the knuckle rail or heel of end points to facilitate lining up the knuckle rails, which are the first parts of the slip laid in place. Since all the joints of the slip switch were numbered and stamped in the shop, as was also the case with crossings, it was an easy matter to assemble the various parts of the switch and connect them properly. Prints showing details of the slip switches on a scale of ¾ in. to the foot were given to the foreman, but after having installed one or two slips, the print, as a rule, was used only in checking up the tie spacing. There are several crossings in combination with double-slip switches and turnouts, but the spacing of these gave no special trouble as they are largely standardized.

The greatest difficulty was at the Y-connection with the north and south main line at the east end of the yard, in which are numerous turnouts and several crossovers. These latter occurred in curved tracks and tended to complicate matters, since it was necessary to locate switches so as to give required clearance as well as to keep the curvature as low as possible in the tracks connected and also through the crossover itself. Several base lines were established in connection with the track work, some parallel with the X-axis; and others parallel with the Y-axis. As the equation of the center line of the track was known, and as an added check on the location of the track, its intercepts could be readily computed and set off on the desired base lines. This furnished a very certain check on locations made. The whole system is dependent, of course, upon the accuracy of the original surveys.

Ladder Tracks.—These are diagonal tracks from which a series of parallel body tracks diverge for a yard or station layout. The angle between the ladder and body tracks conforms to that ot the first frog, and all the frogs should be of the same number. Putting in a ladder track at an angle with the body tracks greater than that of the frog angle necessitates varying the distance between centers of tracks, and requires special formulas for calculating this distance. For ordinary work, the same frogs should be used throughout and the body tracks should be lined in with a transit. Another method is to obtain from a table the perpendicular distance from the gage side of the main-track rail to the ladder rail at certain distances along the former, varying with the frog number. The distance between the frog points is calculated by dividing the distance between centers of tracks by the sine of the frog angle. This distance should be measured along the ladder rail, making allowance for the actual or blunt point in the first frog, and then measuring the distances, which will give the positions of the actual points. Stakes may then be set for the points of frogs and switches, using the theoretical leads. The frog point is placed opposite its stake and the point of switch allowed to fall where the short lead brings it. A string stretched across the ladder rail, parallel with the main rail at a distance equal to the specified distance between centers of tracks, will also mark the position of the frog. Stakes should be set to give the line of one of the body tracks, the others being put in by measurements made by the foreman.

The ladder track should not open from a main running track (especially where interlocking plants are used), but preferably from a siding or lead track whose main-track connection is at least 500 ft. in advance of the ladder-track connection. While No. 9 frogs are recommended for yard ladders, many railways use No. 8 or even No. 7, as these put the switches closer together and thus give a shorter distance for the cars to travel in switching. This is of importance in utilizing the space and in operating the yard; No. 7, however, should be the minimum for yards of modern design. Ladder track plans are shown in Fig. 100.

Ladder tracks should be laid out to the greatest possible angle for the frog used, and the following method has been used on the Pennsylvania R. R.: Take the minimum length of lead for the frog to be used, which is the distance from point of frog to point of switch. Add to this the clearance from the point of frog back to the point of the next switch. With a No. 7 frog, the lead is 60 ft.; the clearance distance will be 10 ft., making 70 ft. in all from switch to switch. The distance between centers of tracks divided by 70 will give the natural sine of the ladder angle, *A*, Fig. 100. An advantage of this method is that it puts a little curve behind the frog. This corrects all inaccuracies in laying out the track and makes the ladder look straight and perfect, and all the body tracks swing off from it similarly, while if the exact frog angle is used for the ladder, the slightest

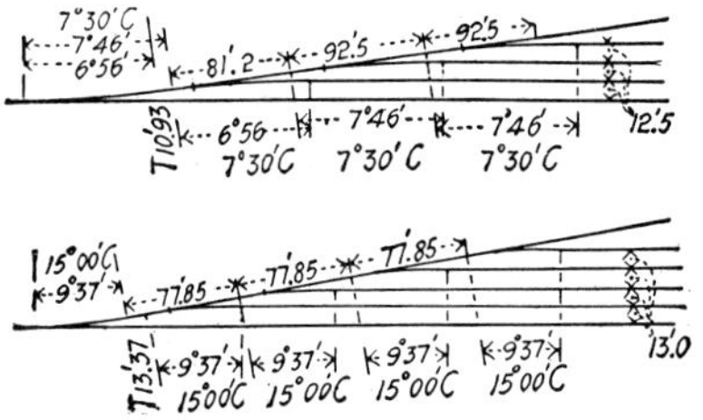

Diagram of Yard Entrances.

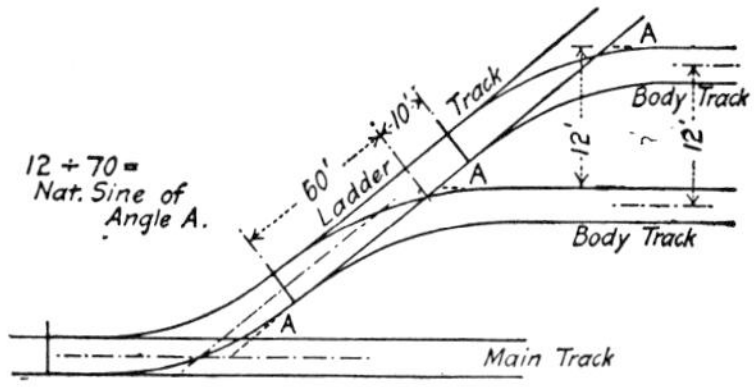

Ladder-Track Connections.

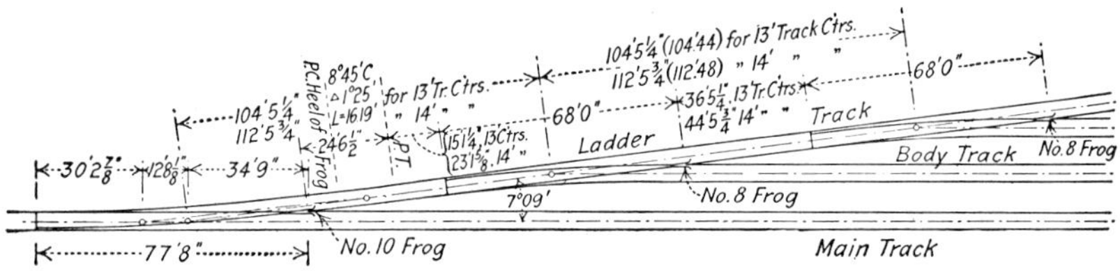

FIG. 100.—Ladder track layouts.

error in location is evident. In staking out ladder tracks, determine the ladder angle, and then turn it carefully with the transit from the track from which the ladder opens, and put in a line of stakes. On this line, the switches should then be located the exact distance apart, as called for by the original assumption. The switches will then be the least possible distance apart, which is an advantage. The angle of the ladder will be the greatest possible, which gives the maximum car capacity of the yard for the amount of space occupied.

Michigan Central R. R. Practice.—On this road, it is the practice to use a No. 11 frog in the main track for all yard leads where practicable. The angle of ladder track is 9 deg. 26 min.; the frogs are No. 9 and the body tracks are spaced alternately 13 and 15 ft. between centers, except that the first yard track is 20 ft. from the main track. The distance between switches on the ladder track is 85.43 ft., including 72.43 ft. for turnout lead and 13 ft. from point of frog to next head block. The triangle is figured, showing the location of the intersection of the yard track with the ladder track by using a 7-deg. curve beginning 10 ft. from the point of the frog. The curve of the last track in the yard will exceed 7 deg., because it must be equivalent to the average curve of a turnout.

The difference between the frog angle and the ladder angle, where the ladder intersects a track parallel with the yard, is taken up in the first one or two frogs, which are laid out with part of the curve in each direction. All the layout is figured to center lines, and the frog angle is shown at the intersection of the center lines of ladder and turnout. This angle can be divided in any proportion desired in the first and second frog to make the required curvature. Sufficient stakes must be set to locate the point of frog and also to define its direction. All tracks are laid out by running center lines, turning angles, and running curves, as shown on the plans. Offsets may be used for putting in the turnout only after the head block and point of frog are located. These offsets are furnished in standard form for each number of frog, but are exactly correct only for a turnout from a tangent track.

Rebuilding Ladder Tracks.—For rebuilding ladder tracks in large classification yards, which is a troublesome operation that must be done in limited time, all rails should be carefully measured, cut, drilled, and marked for location, and ties marked with their lengths. It is most convenient to distribute the material along the ground, opposite the ladder, but if this cannot be done, the material should be assembled on cars so that it can be unloaded in proper order. If the track is cut out for about 4 hr., a gang of about 40 men can renew three switches complete, and this arrangement is usually practicable by the cooperation of the yardmaster. Otherwise, the new rails must be laid on the old ties as opportunity permits, the ties being then renewed singly; but this method is expensive and is not likely to result in good work.

Yard Entrances.—At a freight yard, the ladder track serving a group of body tracks should not open from a running track, but from a lead track whose connection with the main track is at least a maximum train length from the ladder connection. The main-track connection should be made with a frog of No. 10 or No. 12, whether this is for the lead track or a ladder track. In Fig. 100 are shown typical yard entrances and a layout with ladder having a No. 10 frog for main-track connection and No. 8 frogs for the body tracks.

Placing a Turnout.—For placing a turnout in existing track the following description will serve as a guide: After locating the point of frog and point of switch, if the rail joints in the main track will interfere with the position of the switch rails and guard rails, new rails may be placed of such length as to bring the joints clear of the new work. The ballast is also removed from between the track ties for a sufficient distance to allow of placing the switch ties. Rails are cut for the leads, and if the switch points are to be directly opposite each other, the curved lead for the turnout will be slightly longer than the straight lead for the main track.

In spring-rail frogs, however, this distance is already provided for so that the rails may be cut to the same length; the Pittsburgh & Lake Erie R. R. also designs its rigid frogs in the same way. The sidetrack lead rail may be curved in a rail bender if necessary. Where the above difference in length occurs, the excess length of the curved rail is given by the Pennsylvania R. R. as follows (other roads give slightly shorter figures):

Frog number	Excess, inches	Frog number	Excess, inches
4	5¼	8	2½
5	4 3/16	10	1 15/16
6	3 7/16	15	1¼
7	2 13/16	20	⅞

The headblock is first placed at the point of the switch, together with its long gage plate or two tie-plates; one end of the head block is in line with the track ties, and a spike is partly driven to keep it from shifting endways. Then all the ties for the switch and frog are placed, with their slide plates, braces, heel plates, and guard-rail plates. The other turnout ties may also be placed, and all ties adjusted so as to be clear of the switch rods and to conform to the positions of present and future rail joints, including those at the heel of the switch and the ends of the frog. The plates and braces for the unbroken main rail are lined and full spiked, while those on the opposite side are secured temporarily or tacked by partly driven spikes.

The stock rail for the turnout, or knuckle rail, is then bent to give it a bend ahead of the switch point, with a deflection measured by the offset in a length of 10 ft. from the bend. The distance from switch point to bend varies with the length of the switch rail, and on one road it is as follows: 30-ft. switch rail, 15¼ in; 24-ft. rail, 13 in.; 16½-ft. rail, 9 in.; 15-ft. rail, 8⅛ in.; 13-ft. rail, 8⅜ in.; 11-ft. rail, 7⅛ in. On the Pennsylvania R. R., the practice is as follows: 30-ft. switch rail, bend 8 in. ahead of point, with deflection offset of 2 1/16 in.; 18-ft. rail, bend 4 13/16 in. ahead and offset 3½ in.; 10-ft. switch rail, bend 2 11/16 in. ahead of switch point and 6¼ in. offset for deflection angle.

The stock rail, main-track switch rail, lead rails, and frogs are then placed along the ends of the ties on the turnout side and bolted together, such cutting and drilling being done as may be necessary to complete the main track to proper length from point of switch to heel of frog. The old main rail is then disconnected and removed to allow of setting the switch rail, stock rail, lead, and frog; when the joints are bolted at the end of the stock rail and the heel of the frog, the parts are adjusted to gage, and spiked from frog to switch. Before cutting or opening the track, care must be taken that the unbroken main rail is in good line, especially where the switch point will be located. A flagman is then sent out in each direction to stop trains until he is notified that the track is closed and trains may proceed.

With a track gage set at the point of switch, the stock rail and the switch rail are thrown together, the former being then braced and spiked and the latter tacked in place by a partly driven spike so as to hold it to gage. The track gage is then shifted to the heel of the switch and the switch rail spiked to the gage, after which the stock rail is lined over with a bar so that the opening from gage to gage at heel of switch will be the prescribed distance, usually $5\frac{1}{2}$ or $6\frac{1}{2}$ in.

In fitting the switch, a slide plate is placed on the tie nearest to that part of the middle of the switch rail where the planing begins, and the stock rail is adjusted so that it does not bind against the switch rail and thus cause the latter to open at the point. This adjustment is tested by operating the switch rail so that it will be seen to touch the stock rail first at this part. When the adjustment is made, the first or middle slide plate is spiked, and then the remaining slide plates and braces are placed and spiked, working each away from the middle one. The slide plates must be put in exact position with a bar (not a pick), and no attempt must be made to adjust them to position or "draw the gage" by the driving of the spike. Before a train is allowed to pass, the turnout and of the stock rail must be spiked to hold it down, so that it will not move and perhaps strike some part of the train.

With this work completed, the remaining switch ties are placed, and the main track is lined and surfaced. Then the turnout switch rail and turnout lead are placed, and the switch rods are connected and adjusted to give the proper throw. After the switch rods are bolted, the switch is thrown for the turnout, and the heel of the turnout switch rail is spiked in position through the heel plate. Then the turnout lead rails are bolted up, lined to the curve, and secured temporarily at every third or fourth tie by spikes partly driven. This rail is then raised with a bar to allow the tie-plates to be slipped under it and spiked, holding the rail in its final position.

The switch is then set for the main track and spiked in that position for safety while the turnout rails are laid, connections made beyond the frog and stock rail, and the rails gaged and spiked. In this placing of the turnout rails beyond the frog, the curvature should run towards the main track and not away from it. Then the guard rail on the turnout side is placed, with its tie-plates, and is bolted or clamped in position.

As the final step, the switchstand is installed, and its connecting rod is coupled to the head rod of the switch. Then the switch is half thrown so that its rails are at exactly equal distances from the gage sides of the two stock rails, the switchstand lever being held vertical while making the measurements. After seeing that the switchstand is parallel with the main track, it is spiked securely to the head blocks. If the switch·

rails do not fit properly, the necessary final adjustments are made in the switchstand.

Moving Switches.—In renewing complicated special work, it is often desirable to renew the entire unit rather than to replace certain parts, on account of the irregularity due to worn and new parts. As such tracks usually carry intensive traffic, the renewal must be done quickly and this is often a difficult problem. Renewal of a 114-ft. length of a double slip switch 160 ft. long in a four-track main line by lifting it out bodily, with ties attached, by means of two 125-ton wrecking cranes which then placed a new slip already built up at the side of the track, has been accomplished in 1 hr. 18 min. on the Pennsylvania R. R. The weight was about 22 tons for the old slip with 100-lb. rails, and 27 tons for the new slip with 130-lb. rails. The ballast was shoveled out from between the ties, and a supply of fresh ballast was placed alongside the track ready for use. No planks or other supports were placed under the ties to stiffen the slip while being handled, but it was necessary to put blocks and wedges across the ends of the switch points to prevent them from lifting out of the seats. To each crane hook was hung a beam or spreader with chain at each end. The chains did not pass under the switch timbers, but those on the forward hold were applied round the rails at the heel blocks and those at the end next to the point were wrapped around the rails and switch points. With this arrangement, no distortion of the slip resulted.

After the wrecking cranes had been spotted in position, with two flat cars between them, they raised the old slip and laid it on the cars while the new slip was swung over it and lowered, after which they placed the old one clear of the track, to be dismantled. The load and the difficulties of landing the slip longitudinally were increased by the fact that the cranes were pulling diagonally against each other, so that heavy anchorage was required. As soon as the old slip had been removed, the roadbed was trimmed and the ballast dressed to an even bed. The end frogs were renewed while the slip was being handled. After the insulated joints had been connected up, some bar lining and surfacing was done and new stone ballast placed by hand before the track was put in service. For this work, there were 52 trackmen in addition to the wrecking-crane crews.

In yard improvement or enlargement, it is sometimes necessary to move a series of switches or a ladder, and if the distance is not too great each turnout may be moved by a locomotive or a derrick car with a cable. Tracks or other obstructions in the path are removed, ballast cleared from ends of switch timbers, and planks laid for timbers to slide on. All spikes are driven, and the ladder track is cut between turnouts. A chain bridle is attached to the turnout rails, and the hauling cable is attached to the middle of the chain, thus distributing the pull. As each turnout

TABLE XX.—TURNOUTS AND CROSSOVERS: BALTIMORE & OHIO R. R.

1	2	3	4	5	6	7	8	9	10	11	12	13	14	15	16	17	18	19	20
Frog number	Length of frog	Toe to ½-in. point	½-in. point to heel	Length of switch	Opening at heel of switch	Frog angle			Distance between switch points	Correction for E, H, and X, per foot	Degree of curve, outer rail	Lead on center line	Adjusted lead	Angle of switch rail	$\frac{9}{32}$-in. Pt. of Switch to Theo. Pt. of Switch	Distance center to center of ladder	Correction for D per foot	Frog number	Exact calculated lead
	B + C	B	C	S			E	H	X			A	L			D			
	Ft. In.	Ft. In.	Ft. In.	Ft. In.	In.	Deg. Min. Sec.	Ft.	Ft.	Ft.	Ft	Deg. Min. Sec.	Ft.	Ft.	Deg. Min. Sec.	In.	Ft.	Ft.		Ft.
4	8 0	3 0	5 0	11 0	5½	14 15 00	13.19	51.19	93.19	3.94	49 55 33	21.00	40 0	2 15 51	7⅛	52.81	4.06	4	38.39
4	15 0	6 0	9 0	11 0	5½	14 15 00	13.19	51.19	86.19	3.94	67 43 24	17.50	36 6	2 15 51	7⅛	52.81	4.06	4	35.85
5	10 0	4 0	6 0	11 0	5½	11 25 16	16.85	64.35	106.85	4.95	31 32 32	21.25	45 0	2 15 51	7⅛	65.65	5.05	5	44.09
5	15 0	6 0	9 0	11 0	5½	11 25 16	16.85	64.35	103.85	4.95	36 54 46	19.75	43 6	2 15 51	7⅛	65.65	5.05	5	42.41
6	10 0	4 0	6 0	13 0	5½	9 31 39	20.46	77.46	128.46	5.96	20 00 34	25.50	54 0	1 54 58	8⅜	78.54	6.04	6	53.17
6	15 0	6 0	9 0	13 0	5½	9 31 39	20.46	77.46	129.46	5.96	23 45 04	26.00	54 6	1 54 58	8⅜	78.54	6.04	6	51.50
7	12 0	4 6	7 6	13 0	5½	8 10 16	24.04	90.54	145.04	6.96	15 05 46	27.25	60 6	1 54 58	8⅜	91.47	7.04	7	58.89
7	15 0	6 0	9 0	13 0	5½	8 10 16	24.04	90.54	145.04	6.96	16 22 32	27.25	60 6	1 54 58	8⅜	91.47	7.04	7	57.62
8	12 0	4 6	7 6	13 0	5½	7 09 10	27.59	103.59	154.59	7.97	11 06 31	25.50	63 6	1 54 58	8⅜	104.41	8.03	8	64.52
8	15 0	6 0	9 0	13 0	5½	7 09 10	27.59	103.59	154.59	7.97	11 54 18	25.50	63 6	1 54 58	8⅜	104.41	8.03	8	63.36
8	12 0	4 6	7 6	16 6	6½	7 09 10	27.59	103.59	161.59	7.97	11 28 08	29.00	67 0	1 47 56	9	104.41	8.03	8	67.59
8	15 0	6 0	9 0	16 6	6½	7 09 10	27.59	103.59	161.59	7.97	12 18 39	29.00	67 0	1 47 56	9	104.41	8.03	8	66.37
10	15 0	6 0	9 0	16 6	6½	5 43 29	34.68	129.68	189.68	9.98	7 08 20	30.00	77 6	1 47 56	9	130.33	10.03	10	77.37
12	18 0	7 0	11 0	16 6	6½	4 46 19	41.73	155.73	214.73	11.98	4 42 00	29.50	86 6	1 47 56	9	156.27	12.02	12	86.64
12	18 0	7 0	11 0	24 0	6½	4 46 19	41.73	155.73	241.73	11.98	5 06 37	43.00	100 0	1 14 15	13	156.27	12.02	12	100.06
16	23 0	8 6	14 6	24 0	6½	3 34 47	55.80	207.80	296.80	15.99	2 40 34	44.50	120 6	1 14 15	13	208.21	16.02	16	119.92
20	28 0	10 0	18 0	30 0	6½	2 51 51	69.84	259.84	369.84	19.99	1 41 57	55.00	150 0	0 59 23	16¼	260.17	20.01	20	150.22

NOTE: All distances are calculated to the ½-in. frog point and to the $\frac{9}{32}$-in. switch point, and for tracks spaced 13 ft. between centers. For other track spacing, add to, or subtract from, E, H, and X the quantity in column 11; and add to, or subtract from D, the quantity in column 18 for each foot of variation.

is pulled, the connecting rails are laid and spliced. A ladder has been shifted in this way in a day, while to tear it up and relay it would have taken the same force about 10 days.

Frog and Switch Repair Work.—A notable development in modern track practice is the repair or building up of worn frogs in the track by autogenous welding. This requires careful work and should be undertaken before the parts are badly damaged. Much repair and rebuilding of switches and frogs is done at the railway blacksmith shop, which may be that of the mechanical department, the maintenance-of-way department, or the stores and reclamation department. In some cases, a traveling blacksmith shop has been found economical. This consisted of a box car for stores and with accommodation for the blacksmith and helper, and a flat car with forge, rail bender, and other tools. When a frog was removed for repairs, a piece of rail was spiked in place temporarily for the main track. This outfit, which was moved in local freight trains, saved much delay and expense in shipping heavy frogs to and from division shops. Such work as sharpening track tools and repairing hand cars was also handled.

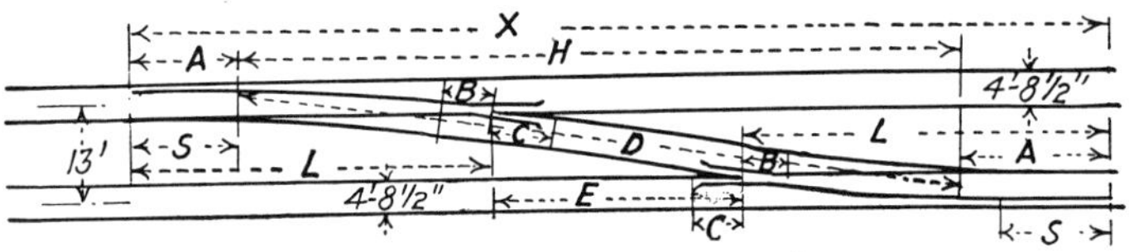

Fig. 101.—Diagram for turnouts and crossovers: Baltimore & Ohio R. R.

Turnout Tables.—A typical list of standard turnouts and crossovers is that of the Baltimore & Ohio R. R., shown in Table XX with the accompanying diagram, Fig. 101. The range of frog numbers is from 4 to 20, but the small frogs are little used, while frogs of oven higher numbers are used to a limited extent for high-speed work, as already explained.

CHAPTER XXII

BRIDGE WORK AND TELEGRAPH WORK

Bridge Work.—Maintenance, repair, and renewal of bridges, and trestles, and buildings on large railways is usually in charge of a separate department as noted in Chap. XV, but the maintenance-of-way department is closely connected with the work. Many railways consider it best and most economical to do all bridge renewal by company forces, on account of the difficulties and extra cost of work done under traffic and also on account of objection to having outside contractors' forces engaged on work under such conditions on an operated railroad. In some cases, foundation, masonry, and steel erection work are handled in the same way, except for the erection of large bridges and for difficult foundation work requiring special experience and machinery. Such a system calls for careful organization and complete equipment. The bridge gangs should have permanent work trains with good boarding and sleeping accommodations in order to attract and keep a good class of men.

For each division, there is usually a master carpenter or supervisor of bridges and buildings, who reports to the division engineer or the superintendent (or engineer) of bridges and buildings, and under whom are the foremen of the bridge and building gangs. These foremen, in charge of 100 to 200 mi. of line, must have intimate knowledge of the character and condition of all structures, carry on the routine and special repair work, and keep track in line and surface on bridge floors. In this last item, they must cooperate with the section foremen to insure that at the ends of the structure the track is in uniform line and surface on the bridge and the approaches. Surfacing is done mainly by placing shims under the ties, which requires careful work, all shims being carefully secured and being bored for the drift or screw bolts, so that they will not be split by driving bolts through them.

The bridge gang has usually an outfit train consisting of a foreman's car, dining and kitchen car, two or three bunk cars, and a combination tool and material car, additional material cars being added when required for large jobs. For heavy work, such as bridge renewal, there will be derrick cars and air-compressor outfits to serve riveting hammers and other pneumatic tools. This equipment is used also in emergency work, as in repairing washouts. On the Lehigh Valley R. R., the bridge gangs consist usually of 6 to 10 carpenters and iron workers; this road makes extensive use of locomotive cranes for distributing and handling material, and the cranes are equipped with suspended ways or leads for pile driving with a steam hammer. On the Illinois Central R. R., the gangs range from 6 to 20 men. Each division has a combination derrick car and pile driver, with drop hammer, steam hammer, and leads. This is generally handled by an individual crew, but on some divisions by the bridge gang of the district on which it is working, except that the engineman stays with the car.

Bridge Repair and Renewal.—Timber trestles constitute a considerable proportion of the bridge structures on many large railroads, and their repair is a continual activity. The number of trestles is being steadily reduced, however, by the policy of replacing them with solid fills or embankments having concrete or pipe culverts. The increasing use of treated timber has greatly increased the life of such structures, piles and decks having often a life of 20 years or more as compared with 7 to 10 years for those of untreated timber. For stream crossings in country subject to heavy storms, the foundations are the weak points in trestles, numerous accidents having occurred from the stream bed being scoured out so that piling gives way when a train gets on the trestle. If the structure remains in position, it is almost impossible to determine its safety until the flood has subsided. For timber repair work, it is usually best to have one main storage yard for timber and piles and to keep only an emergency stock on each division. Timber removed from old structures should be piled for examination as to its use or reclamation.

An air-compressor plant for boring tools and a portable wood-working outfit with gasoline engine are useful where extensive work is being done. For handling large timbers, a portable derrick may be rigged up on a push car, while a useful device is a dolly with a 6-in. grooved wheel journaled between two timbers about 18 in. long, or two tandem wheels with timbers 3 ft. long. This truck rides on the track rail, being held up by a handle or pole on one side. In making renewals with creosoted timber, all parts bored or cut for framing should be given several coats of hot creosote oil as soon as the cutting is done, all exposed surfaces being finally mopped with hot pitch. A repair gang on work of this kind should have a supply of oil and pitch and a 10-gal. heating pot.

In strengthening steel bridges, it is usually preferable to add additional material rather than to remove light members and replace them with heavier members, as the former plan avoids temporary weakening of the structure. Pile-driver cars or pile-driver leads hung to the boom of a locomotive crane or derrick car are used in placing falsework and for repairs to trestles. In replacing trestles and old steel spans with new steel girder spans, extensive use is made of large derrick cars and wrecking cranes of 100 to 125 tons hoisting capacity, as well as of lighter locomotive cranes. Girders or short girder spans run out on cars on falsework may be lifted off and lowered in place by a crane at each end, or a single derrick car can pick up a shorter girder, carry it forward, and set it in place without the aid of falsework.

mended for the next season as necessary or desirable. A report for each individual structure is made annually to the superintendent, the engineer of maintenance, the bridge engineer, or the chief engineer, according to the organization system of the railroad. Frequent inspection tends to increase the life and safety of the structures.

Bridge inspectors examine each structure at intervals of 1 to 6 months, according to the age and character of structures, in order to discover any defects or deterioration due to traffic, depreciation, derailments, misplaced car loads, or other causes. Report is made usually of each individual inspection, by a card or blank form, besides notes in the inspectors book. In addition, there is the almost daily inspection by the section gangs, who look for indications of defects or damage due to derailment, fire, or flood, but they make report only if such defects are discovered. Special inspections of individual structures are made where there is any sign of distress or serious condition, when the bridge has been subjected to flood and there is possibility of damage to foundations, or when the inspector's reports indicate the necessity for replacement or heavy repairs. If serious defects are discovered, the bridge engineer may be requested to send a representative.

Bridge inspectors should be familiar with design and have practical experience in shop and field work. Their qualifications depend somewhat upon the character of structures, and they may be selected from the bridge gangs and given some special inspection. Daily reports of each structure inspected are made either on ruled cards or printed forms bound in a book of pocket size. The forms must be comprehensive but simple, as it is not easy to fill out complicated forms in the field. Each report should be sufficiently complete to inform the bridge foreman what work needs to be done. Some reports have lists of items, each to be marked by the inspector, but in general a narrative form of report is preferred as being more easily made and understood.

Typical Inspection Systems.—An annual inspection of bridges is made on the Denver & Rio Grande Western R. R. by the division engineer in company with the master carpenter, the local bridge and building foreman, and one or two assistants to dig around sills and piles and remove ballast. The roadmaster sometimes participates in this inspection, a report of which is filed. The bridge engineer makes inspections of structures marked for replacement and of others which he wishes to give personal attention. All repair work is authorized by the division officers, except where the opinion of a superior officer is considered desirable. On the Lehigh Valley R. R., an annual inspection is made by the general bridge inspector, representing the bridge engineer, the latter making inspection in special cases and at times accompanying the inspection party over certain lines. On the Chicago & Northwestern Ry., an inspection on each division is made during the summer by the division engineer, the supervisor of bridges and buildings, and the foreman of the bridge crew, the division engineer then making a report to the engineer of bridges.

On the Illinois Central R. R., monthly inspection and report is made by each district foreman, and quarterly inspection and report by each division supervisor of bridges and buildings. An inspection made by the general bridge inspector accompanied by the division supervisor of bridges and buildings and the district bridge foreman, with sufficient bridge men to handle the motor car, averages intervals of about 8 months for each structure. He does not make a daily report, but his inspection notes are entered in a pocket book from which they are transcribed to blank forms giving description of structure, its condition, the necessary repairs within 1 year, and stating if renewal will be required within 2 years. These reports are bound by districts and are preserved only until the next inspection, as each report is complete as to the condition of the bridge at that time. Repair work is based upon the reports of regular inspections or of special examinations that may be required. Bridge inspectors on the Chicago & Northwestern Ry. report their movements daily by telegraph to the division engineer.

A daily report made by bridge inspectors on the Lehigh Valley R. R. for each bridge or structure is a sheet $8\frac{1}{2}$ in. high and $5\frac{1}{2}$ in. wide. At the top is the statement: "I have today inspected . . . at . . . and find its condition as follows: . . . " After a blank space is the heading: "The following work is required to maintain structure in good condition"; then another space and the heading: "The following work must be done to keep the structure safe." From these daily reports the division engineer's office compiles a monthly report on a sheet 13 in. high and 8 in. wide, with five columns headed as follows: "Bridge No. or Structure"; "Location"; "Day of Month Inspected"; "Condition"; Action Taken or Recommendation." Copies of these reports are sent to the bridge engineer, the engineer of maintenance-of-way, and the division superintendent. They are examined by the general bridge inspector who follows up any cases of serious defects. Plans for work required are then submitted for attention and approval of either the engineer of maintenance-of-way or the chief engineer, according to the importance or magnitude of the work.

Bridge Inspection Work.—The bridge inspector on his daily work may travel alone on a motor car or light hand car, but more generally he is accompanied on a larger motor car by one or two bridge men, and sometimes by the division supervisor of bridges and buildings and the foreman of a local bridge gang. His equipment will include a light hammer for testing rivets, a monkey wrench, and one or two $\frac{1}{2}$-in. steel bars for sounding timbers and piles, one end of the bar being pointed and the other end chisel edged. One end of the hammer may be pointed, for testing timbers. He will have also a plumb bob and line, a rule, a 50-ft. tape, and perhaps a mirror on special extension handle. The men will have cold chisels and scrapers for removing dirt and scale; a brace and bits, crank augers, and paint.

In boring timber to ascertain its condition, the holes (not more than $\frac{1}{4}$ in. in diameter) should be bored from the bottom or in upward inclination so as not to hold water. They should be plugged promptly. In testing rivets, a smart light side blow is struck on one head while the inspector holds his fingers on the opposite head and plate. Loose rivets should be marked at once with paint. The inspector should watch structures during the passage of trains, noting any undue deflection, swaying, vibration, or rattling of loose parts. The track should be in good line and surface on the deck and approaches, and well bedded on the latter so as to avoid heavy shocks from trains running into the bridge. If emergency repairs are required, the engineer, the superintendent, and the supervisor of bridges and buildings should be notified, and repairs made as soon as possible, and trains may be stopped if defects are found which impair the safety of the bridge.

Timber Structures.—For piles and trestle posts, it must be noted if there is splitting or decay, especially at the ground or water line; piles that have 7 in. in diameter of sound timber are marked for renewal with posts during the next season, but those with 6 in. or less diameter of sound timber are marked in red on the inspection notes as emergency piles to be replaced at once with posts. Trestle caps, and sills, corbels, stringers, girts, and sway bracing are noted as to width and bearing, the inspection notes showing number and size of members to be renewed. It must be noted if the bents are plumb and properly braced. The end bulkheads, track ties, guard rails, foot walks (required near switches), fire protection barrels, refuge platforms, and other parts are also examined. Besides the structure itself, the waterway is inspected to note if it requires clearing or enlarging.

Timber-box culverts are noted as to condition of timber and necessary clearing out or ditching at ends. For trusses it should be seen that they are vertical and have proper camber, with chord bolts snug and lateral rods in adjustment. Truss rods may be adjusted until the counterbraces have a firm bearing on the angle blocks and all the rods have uniform tension, but no adjustment is made while a load is on the bridge. Timbers, seats, and joints are examined for cracks, splits, and rot, and splices are noted for signs of pulling apart. Braces and counterbraces should have a square and even bearing on the angle blocks; any shifting from position indicates that the bridge needs adjustment, which must be made only under the supervision of the inspector.

Steel Structures.—Chords and trusses must be in line and have the proper camber; bed plates must be clean and level, and the rollers clean and free to move, with their axes at right angles to the line of bridge.

End shoes of girders and trusses must be sound and must have a uniform bearing on the rollers or bed plate. Tension members must be examined closely, and also the rods and bottom chords, especially where they are composed of more than one member. All members in any one panel should have an equal strain, and report must be made when one member is slack and the other tight. Compression members, such as end posts and top chords, must be straight, with tight joints, and no signs of buckling. Pins and nuts are examined for signs of wear and movement; chord pins are usually marked with a line of paint to detect such movement. Pockets in interior parts at panel points must have open drain holes or be filled. Rivets must be tested, particularly in end details, and in connections or stringers to floor beams and floor beams to trusses. Lateral systems, sway bracing, and portal bracing are examined. Note is made also of the condition of paint and of any serious corrosion, such as may be caused by brine dripping from refrigerator cars.

Masonry and Substructures.—Pipe, box, and arch culverts are examined for cracks, open joints, condition of paving or invert, and end approaches or waterways. Piers and abutments are noted for signs of settlement or displacement, disintegration, cracks, seepage, and shifting of the deck. Soundings may be taken to detect scour. Pedestals and bridge seats must be sound, without cracks or crushing. Deck construction and waterproofing is noted. It should be observed also if bridge watchmen and section gangs keep the bridge seats clean, the ballast back from the abutments, the approaches firmly bedded, and grass and brush cleared away from wooden structures.

Bridge Records.—It is important to have complete files of drawings of structures and also records giving type of construction, alterations, and present condition. These are filed usually in consecutive order, all bridges being numbered from some terminal point. Detailed descriptions and diagrams, as kept at headquarters, show for each structure its location, number, type, dimensions, waterway, height, foundations, substructure, and superstructure, with dates of construction, repair and renewal; also dates of inspections, with note of condition and references to inspection reports. Similar records may be kept at division offices and may be condensed in pocket-book size for the field use of engineers and inspectors. Every bridge, culvert, and opening, however small, should be numbered and included in the records.

In the card-index form of bridge record, (Fig. 103), there is a card 5 by 8 in. for each bridge, the back serving for additional information or a sketch. This card should show also the file numbers of drawings and correspondence, so that full particulars may be obtained readily when needed. Reports of inspections, filed behind the corresponding index cards, may be of a different color. To facilitate reference, the location of stations may be indicated by cards of another color or having a raised tab for the name. On some roads, different colors are used for different kinds of structures: buff cards for steel bridges, yellow for wood, pink for concrete arches, and green for culverts, all cards being numbered and arranged consecutively. The bridge record is sometimes a blueprint list, corrected periodically and furnished to certain officers. One form of tabular record of structures is a sheet 16 by 21 in., with main headings as follows: steel bridges, trestles, culverts, viaducts, tunnels. These are subdivided to show style of piers, trestles of treated timber, style of culvert, length of structure, and dimensions of tunnels.

No. Spans	Extr. Length	Design		BRIDGE No.	5860 13
Over	on		DISTRICT		DIVISION
Width	c. to c. Depth	No. Panels	Length	c. to c. =	
Built by	in	Chords = E	Beams = E	Stringers = E	

Weight of One Span	Sub-foundation
Cost per lb. cts.	Abutments of amt. cu. yds.
B. of R. to Top of Masonry	Piers of amt. cu. yds.
B. of R. to Underclearance	Ties , x x ft. lg. spaced c. to c.
Clearance, height, width	Alignment Super-elev. in.
Underclearance to High-water	Grade %. No. of Tracks
Drainage Area sq. ft.	Notes :--
Required Waterway sq. ft.	
Actual Waterway sq. ft.	
Correspondence file	
Profile Drawings	
Masonry Plans	
Sho[illegible]ils	Should be inspected times per year
Erection Plan	Span replaced 19

FIG. 103.—Bridge record card.

Typical Bridge Records.—On the Denver & Rio Grande Western R. R., bridge records are kept in large fixed-leaf books, two pages being given to each bridge. The left-hand page contains sketches made in chronological order showing all changes in the structure since its installation. The right-hand page gives a description of all repairs and replacements done at various times, with the cost. A single book contains the record of a main-line division, branch lines on a single division, or a combination of the two, the amount of territory covered being governed by the uniform size of the books. As an improvement on this system, new loose-leaf record books are being prepared, which will make better provision for a continuous record. All these records are arranged consecutively according to their mileage from Denver.

Original plans are filed in tough, bristol-board folders labeled with the bridge number or numbers of the structures, the plans of several bridges located in the same vicinity being grouped usually in a single folder. These folders are kept in drawers which in turn are part of a large filing case labeled to indicate the division or branch and extreme limits of mileage covered by the bridges. When renewals or repairs are of such a nature as to require a plan, this plan is filed in its proper folder. If no plan is required, the records kept in the bridge books indicate in sufficient detail the nature of the work done. A form is filled out by the foreman of the bridges and building gang which does work of any description on a structure; on this he shows sketches and gives full details, with cost, including a description of the physical condition of the structure. From this report, which is forwarded to the chief engineer, the entry is made in the bridge book.

On the Lehigh Valley R. R., bridge records are kept in the form of bridge lists for each division. At the end of each month, a blueprint copy is sent by the bridge engineer to the division engineers, who enter any changes made during the month. This corrected list, together with a typewritten note of the changes, is returned to the bridge engineer to enable him to bring his blueprint list and the tracings up to date. The original plans are filed under their numbers at the office of the bridge engineer, and are kept up to date by filing with them the plans made for any repairs or additions. For renewals, the new plans are placed in the file and the plans which have become obsolete are placed in a surplus file. If the material from the old structure is re-used, the old plans are filed with those of the new bridge in which it is used. If the old material is scrapped, the plans are kept for a time and then destroyed. Bridge record books are kept by the Chicago & Northwestern Ry; the structures are not classified, but are numbered in consecutive order, beginning at Chicago. Records of renewals are kept up to date by reports from division engineers as to all bridges where the new work makes a change from the old description. Plans for such new work, with dates, are filed when the work is completed. No record is kept of repairs, except those which require a bridge department plan. Repair work is based on the inspection reports of the division engineers and general bridge inspectors.

Loose-leaf books are used for the office record of bridges on the Illinois Central R. R., the bridges being shown in consecutive order according to bridge number, and the record being kept up to date by corrections based on the inspection reports. The original plans are filed by bridge numbers and by divisions and districts. The loose-leaf record sheets, 22 by 15 in., are ruled in 22 columns across the short width, with headings as follows; the last column being for "remarks."

General: (1) bridge number; (2) kind of structure; (3) length; (4), base of rail to ground; (5) year built; (6) high water to base of rail.

Timber trestles: (7) number of bents; (8) number of piles per bent; (9) kind of piles; (10) number of stringers in each panel.

Steel bridges: (11) description, with number and length of spans; (12) kind of piers and abutments.

Concrete trestles: (13) pile or piers and abutments; (14) length of span; (15) number of spans.

Culverts and arches: (16) span of arch or diameter of pipe; (17) length of barrel; (18) head walls.

Overhead bridges: (19) total length of bridge; (20) side clearance of track span; (21) overhead clearance.

Tunnels.—On railroads or divisions having many tunnels there may be tunnel repair gangs of 12 to 24 men, each with a train of camp and equipment cars. These gangs repair and strengthen old lining, put in new lining, enlarge the section to standard clearance limits, and do all kinds of repair work. Scaffolds for work on the roof may have hinged panels or platforms that can be dropped to clear trains. Trains approaching are flagged and notified to proceed with care. With portable telephone apparatus from each portal, the tunnel gang can be warned when to get clear of trains and can notify the foreman or flagman when all is clear. These forces are under an inspector of tunnels who supervises the work and also makes an annual inspection of each tunnel, noting and reporting its condition and making recommendations as to work that should be done during the next season.

TELEGRAPH WORK

Telegraph and Telephone Service.—Telegraph lines along the railways are built usually by the telegraph companies, while maintenance and repair work may be handled by the companies or the railways, or by both jointly. The telephone is now used almost as extensively as the telegraph for dispatching and other railway business, the same wires sometimes serving for both purposes. On the Pennsylvania R. R. there are telephone booths about a mile apart, so that a section foreman or a train conductor can get in communication with the block operator and, through him, with the dispatcher, if necessary. Another development on that road is the equipment of each section motor car with portable telephone apparatus and a 25-ft. extension pole for hooking over the wire, to enable the roadmaster or the foreman to get into communication with headquarters from any point on the road. It is very important that the roadmasters and section foremen should know exactly which are the train wires or dispatcher wires, used exclusively for communications relative to train movements, so that these can be given first attention and straightened out in case of the wires being down, owing to wind or snow storms, or other causes.

Telegraph Organization.—The Pennsylvania R. R. System owns and maintains the pole lines east of Pittsburgh and Erie, Pa., but west of these points, owing to the different contracts, there are jointly owned poles and wires used in the operation of the railway. There is a general superintendent of telegraph for the entire system, and the system is divided into three regions, each having a superintendent of telegraph and signals. Line maintenance, which covers clearing troubles and minor repairs, is in charge of local telegraph and signal maintainers having sections of $\frac{1}{4}$ to 15 mi. on the eastern region, 4 to 54 mi. on the central region, and 10 to 100 mi. (with an average of 70 mi.) on the western region. The maintainer looks after all wires, including signal wires; he clears all troubles on telegraph and telephone circuits and closes breaks in signal wires. Railway and commercial telegraph wires as well as signal wires are all carried on the same poles. Construction gangs for new work and extensive repairs range from 8 to 18 men, and have large motor cars for handling men and materials. Many of these gangs stop at hotels and boarding houses, but others have camp-car outfits. On the western region the standard 18-man gang comprises a foreman, an assistant foreman, a cook, an assistant cook, and 14 signalmen, assistant signalmen, and helpers.

Under the contract between the New York Central R. R. and the Western Union Telegraph Co., the latter furnishes the line-gang forces which attend to the construction and reconstruction of pole lines and are under the direction of a joint superintendent of telegraph, who has on his staff a telegraph and telephone engineer, a general foreman, and supervisors. For maintenance, the telegraph company also furnishes the section linemen, who, under the direction of the joint superintendent, perform the daily maintenance work and clearance of troubles on telegraph and telephone wires and equipment. On the Northern Pacific Ry. the telegraph company furnishes material for repairs, renewals, and reconstruction, while the railway company furnishes labor and freight. The total cost is then divided upon a wire-mileage, user-percentage basis, each company paying to the other according to the percentage of use. For maintenance, each company furnishes inspectors who work together in observing the condition of the plant; neither company is billed by the other for this service. On the Denver & Rio Grande Western R. R., all construction, repair, and maintenance is handled by the telegraph company.

Poles.—Telegraph poles are mainly of cedar and chestnut; but cypress redwood, spruce, and Oregon fir, or Norway pine are also used, the last being usually for very tall poles. Creosoted yellow pine poles treated by the pressure process are used extensively. Other kinds of poles frequently have preservative applied from the butt to a little above the ground line, either by painting or by dipping in open tanks, this treatment increasing the life by 50 to 100 per cent. The diameter should be at least 7 in. at top and 10 to 12 in. at the ground line for poles 20 ft. long. With not more than five cross-arms, these are spaced 24 in. apart; or 20 in. with six or more arms, the top cross-arms being 8 in. below the top of the pole, which is dressed to a ridge to shed water. Round poles (but not square poles) are grained or grooved to receive the cross-arms. Concrete poles are used to a limited extent; steel poles are rarely employed in railroad service.

Wood poles 20 to 25 ft. long are set 4 to $5\frac{1}{2}$ ft. in the ground, according to character of soil and number of wires carried; 40-ft. poles, $5\frac{1}{2}$ to $6\frac{1}{2}$ ft.; 50-ft. poles, 7 to $7\frac{1}{2}$ ft., and 75-ft. poles 8 to $8\frac{1}{2}$ ft. But in rock the holes are 3 to 6 ft. deep, with the poles securely wedged in place. The holes are dug with augers and long-handled tools. In backfilling, there should be one shoveler to three tampers to insure thorough tamping, and the earth should be rounded up around the post to about a foot above the ground.

Corner and terminal poles may have wood or stone footings. Poles are ordinarily vertical, but are inclined on curves to resist the pull of the wires. In soft ground, they may be held by wire guys attached to anchors, or where these cannot be used, a pole brace may be set against the pole, just below the lowest cross-arm, but the pole must not be cut or notched to fit the brace. In swampy ground, the pole may be set in a barrel filled with gravel, or may have a long sill bolted near the bottom, with braces extending up to the side of the pole. In loose ground, the hole may be backfilled with concrete. Where storms and sleet are frequent, a pair of light poles forming an inverted V with a spread of 5 to 6 ft. may be used instead of heavier single poles. Another special construction consists of pairs of vertical poles about 6 ft. apart, braced together, and having long cross-arms.

Pole Lines and Conduit Lines.—Poles may be set just far enough from the right-of-way line to permit of guying and to prevent the cross-arms extending over adjacent property; or they may be placed a pole length (above ground) from the track. The minimum distance from rail is ordinarily 12 ft., but where necessary, the distance may be 7 ft., provided the height to lowest cross-arm is at least 25 ft.; for an 8- to 12-ft. distance, this height may be 20 ft. On level ground, the lowest cross-arms should be 10 ft. from rail level. A clear head room of 27 ft. below the wire is required at railroad crossings, 18 ft. at street and highway crossings 15 ft. at private crossings, or 10 ft. at crossings used only by pedestrians or cattle. The wires must be kept 2 to 4 ft. from trees, buildings, or bridges, and 2 to 4 ft. above other wire lines at crossings, except that this vertical clearance must be 6 ft. over a trolley wire not protected by a trolley guard. At railroad crossings, the span should not exceed 100 or 150 ft. Lightning rods are placed at $\frac{1}{4}$- or $\frac{1}{2}$-mi. intervals for lines carrying up to 20 or more than 20 wires, respectively; but the number may be doubled where lightning occurs excessively, while the rods may be omitted in large towns and cities or where lightning occurs rarely. Wire guys are used on poles at corners and crossings, and also for those on open lines where heavy winds or sleet storms are prevalent. For this storm guying, side guys in both directions may be on every sixth to twelfth pole, and head guys in both directions on every fourth pole having side guys, according to local conditions.

In cities and towns where right-of-way conditions do not permit the use of pole lines, underground cables are laid in conduits. This construction is used along part of the Pennsylvania R. R. line between New York and Philadelphia, owing to the great number of wires required. It has been suggested, also, for use on open lines where sleet troubles are serious, but the high cost of such construction would rarely be warranted by these conditions. Telegraph, telephone, and signal wires are usually placed on the same poles, but where the poles are already heavily loaded, a separate pole line is sometimes erected for the signal wires. Iron wire of No. 8 B.w.g. or copper wire of No. 9 A.w.g. are commonly used, with No. 6 and No. 8, respectively, for long spans.

Spacing and Height of Poles.—Pole spacing depends upon the local wind and storm conditions, the number of wires, and the general character of construction of the line. The Western Union standard spans are 100, 130, 150, 175, and 200 ft., or 53, 40, 35, 30, or 25 poles per mile, but shorter spans are sometimes used on railroad lines. On the New York Central Lines west of Buffalo, the pole lines are constructed to carry ultimate loads of 30 or 50 wires. From Buffalo to Toledo, which is classed as 8-lb. wind-load territory, 70 poles per mile are required for 50 wires; while beyond Toledo, in 5-lb. wind-load territory, 60 poles are required. For 30 and 50 wires, the height of pole is 22 and 25 ft., respectively. Poles are spaced generally 44 to the mile in the eastern region of the Pennsylvania R. R. System, with an average height of 30 ft.; branch lines have 25-ft. poles, spaced 35 to 40 per mile. In the central region, the spacing is 100 to 182 ft., with poles generally 25 to 30 ft. long, but varying from 20 to 40 ft. In the western region, the usual spacing is 132 ft., but on heavy lines and in storm districts, it is reduced to 88 ft. The Northern Pacific R. R. uses 35 to 50 poles per mile, the closer spacing being for sleet-storm districts. Poles should be as short as possible in order to avoid undue strain on the poles, with a minimum headroom of 10 ft. under the lowest cross-arm; in irregular country, the height of cross-arm is increased as may be necessary to give safe clearance or head room under the wires.

Inspection and Repair.—Section foremen are required to note the general condition of telegraph and telephone lines, especially after high winds and sleet or snow storms, any defects being promptly reported. In case of poles broken or blown down, they should be propped up to prevent the wires from coming in contact with the ground or with each other. The division wire or train wire must be given first attention in clearing the lines. Poles should be inspected annually, the telegraph company doing this work for its own lines. With a small hatchet or a 48-in. bar of $\frac{5}{8}$-in. iron having one end pointed and the other chisel edged, the inspector sounds the pole, removing decayed wood; if necessary, a hole is bored with an auger; the character of the chips is noted and the hole is then plugged.

In stringing new wires, the reel of wire is usually carried by a motor car or push car on the track, although for a single wire on the outer or field side of the poles, the reel is mounted sometimes on a truck but more generally on a frame carried by men. Where a number of wires are to be strung, the reels may be stationary, while the wires are pulled out by a horse or team of horses. Work trains for stringing wires are rarely used except in case of storm breaks or other emergencies. Camp trains of the Pennsylvania R. R. have been noted already. Repair and construction outfits of the Western Union Telegraph Co. consists of three cars for a gang of 16 men with foreman, cook, and helper. One car has the foreman's room, eight double berths, a locker, and a sitting room; another has the kitchen, dining room, and cook's quarters; the third is a box car with about 10 ft. for refrigerator and food storage, the remainder being a storeroom for tools and materials.

RAILROAD AND POWER CROSSINGS

Where telegraph lines cross railroad tracks, specially heavy pole construction must be used, with a maximum span of 100 ft. for heavy construction or 125 to 150 ft. for lighter construction. The poles should be 12 ft. from the nearest rail on main tracks or 7 ft. on sidetracks. All wires or cables across the tracks must be at a minimum height of 27 ft., or 25 ft. if the wires are along a highway and are paralleled by the trolley wires of an electric railway. Trolley wires must be not less than 22 ft., and trolley feeders 25 ft. above the rails. The contact wire must have practically the same elevation throughout the crossing and for at least 100 ft. on each side; if this is not practicable, there must be trolley guards of such length that the trolley wheel of an approaching car will be within the guard when the front end is 50 ft. from the nearest rail, and will remain in the guard until the rear end of the car is 8 ft. beyond the last rail.

For electric light and power supply lines crossing over railroads, the A. R. E. A. specifies that where the distance from nearest support to farthest rail is not more than 75 ft., the clearance above the head of the rail must be 27 ft. for lines up to 750 volts and 28 ft. up to 7,500 volts. For higher voltages, the 28-ft. clearance must be increased 0.10 ft. for each 1,000 volts, with a minimum of 30 ft. When the distance exceeds 75 ft., the clearance height is to be increased 0.20 ft. for each additional 10 ft.

CHAPTER XXIII

RAILWAY IMPROVEMENTS AND BETTERMENTS

Intensive improvement and partial reconstruction of railroads, in order to secure increased efficiency and economy in operation as well as to meet the requirements of increasing traffic, have been conspicuous features in recent years and will continue for a long while. To illustrate its importance, this class of work involved expenditures of nearly $620,000-000 during the two years 1922 and 1923, and included the following items: new tracks, heavier rails, new ballast, and roadway machinery, $237,-440,000; bridges, viaducts, and tunnels, $70,458,000; stations, grain elevators, docks, and coal and ore piers, $51,870,000; grade and line revision and grade separation, $42,200,000; freight yards, $25,518,000; water and coaling stations, $21,247,000. Improvements of this sort are in a distinct class, so that the engineering work on existing railroads may be classed as follows: (1) construction of extensions and branches; (2) maintenance-of-way and structures; (3) improvements and betterments.

To a large extent, these improvements are undertaken owing to the enormous developments in traffic. Many railroads were built originally in thinly populated districts and with special regard to low cost and rapid construction to meet conditions of limited finances and light traffic, so that when heavy traffic developed, they were operated at a disadvantage. Even on lines of heavier construction, the growth of traffic and increase in train loads often necessitate extensive improvements to give greater facility and economy of operation. The economic aspect of such work on the Union Pacific Ry. is reviewed comprehensively by J. B. Berry in the 1904 "Proceedings" of the A. R. E. A., and committees have dealt with the subject in more recent issues.

Economy in cost of freight transportation is related closely to the concentration of traffic in heavy trains hauled by powerful locomotives, requiring in many cases the reduction of ruling or maximum grades and special means to avoid the increased liability to detentions in the movement of such trains. Heavier and longer trains involve lengthening of yard tracks and passing sidings and the improvement of other facilities. New regulations as to wages and working hours of train crews constitute another factor, as it is important that trains should make their runs within very definite time limits in order to avoid high charges for overtime.

Improvement works undertaken to expedite train movements and to promote the economic handling of traffic may include the following:

(1) double tracking to increase traffic capacity and reduce delays; (2) grade reduction and line revision to improve operating conditions; (3) extension and rearrangement of passing sidings to handle longer trains and to expedite movement; (4) construction and enlargement of freight yards and passenger terminals; (5) engine terminals, shops, and water and coaling stations of improved design and relocated to suit operating conditions; (6) large freight stations, warehouses, and grain elevators; (7) cutoff lines to enable through freight trains to pass around large cities and belt lines to facilitate interchange of through freight; (8) grade separation on city streets and on busy highways; (9) renewing bridges with heavier structures; (10) filling old trestles and viaducts with solid embankments; (11) widening cuts and fills; (12) installing block signals and interlocking plants to expedite train movements. Classes 1 and 2 are frequently combined in one project. Bridge work, passing tracks, yards, terminals, drainage, and signals have been discussed in previous chapters. Much of this improvement work is handled more or less by the maintenance-of-way department, which may also carry out such work as extending passing tracks, widening fills and cuts, making minor improvements in grade by filling sags (perhaps due to settlement), lowering summits, and also revising curvature.

Complications of Improvement Work.—Careful planning, systematic organization, and ready ingenuity are required in work of this kind to facilitate its progress, keep cost within limits, avoid interference with traffic, and provide for safety of trains and of the construction work. For example, a stretch of track elevation in Chicago consisted simply in building retaining walls and abutments, placing girder bridges, and making a sand fill. But this stretch included two busy double-track main lines approaching city terminals and in continual use by regular and light trains, engines, and switching movements. Main tracks had to be shifted from time to time to clear the construction work and then elevated gradually on temporary trestles across streets, while piles had to be driven in and between tracks. Every change in running tracks involved changes in turnouts, crossovers, and signals, with their mechanical and electrical connections. In addition, study had to be made of the distribution and handling of materials, mixing and placing concrete, placing bridge girders and slabs, building bridge floors, hauling away material excavated from foundations, and later bringing in the material for filling, all in the face of heavy railroad and street traffic. Similar conditions may obtain in varying degree on other improvement work, with added complications where contractors' trains are allowed to occupy main tracks.

Double Tracking and Additional Tracks.—On many main-line divisions, the traffic has reached the capacity of single track, resulting in congestion and delay in the operation of fast and slow trains in both directions, even with a liberal provision of passing tracks. Trains may spend considerable time on side tracks, waiting for opposing trains which may be late and thus increase the delay. An obvious method of relief is to provide a second track, in order to eliminate opposing movements and thus eliminate many facing switches, although passing sidings will still be required to permit fast trains to pass slow trains moving in the same direction. Furthermore, a smaller number of locomotives might then handle a greater amount of traffic with increased promptness and economy. In 1922, only 16 per cent of the total railway mileage was double track, and if branch lines are excluded, it may be estimated that double track does not exceed 22 per cent of main-line mileage. An alternative which is practicable and economical in many cases is the installation of automatic block signals to increase both the traffic capacity and the safety of operation on the single track at much less expense than the construction of a second track, with the necessary bridges and other works.

Double track and multiple tracks are needed particularly near large cities, where traffic concentrates and where suburban service may be included. At division points, also, double track should extend to the first passing track to avoid delays to freight trains arriving at, or getting away from, the yards. Some railroads have long stretches of triple track, with the middle track signaled for operation in both directions. Double tracking may be required on relatively short stretches where heavy grades or traffic conditions cause congestion. On the other hand, much work has been done in double tracking long stretches of main line, as on 750 mi. of the Santa Fe R. R. between Albuquerque, N. M., and Barstow, Cal., which serves two or three routes converging at each end.

Work of this kind might appear to be a fairly simple and inexpensive proposition of widening cuts and fills and widening or rebuilding bridges, although under the difficulties of construction work on operated lines. In reality, the proposition is much more complicated and costly, since possibilities of grade reduction and line revision usually enter into consideration, even in prairie or rolling country, with a view to providing a line more favorable for modern traffic conditions. Thus, the engineer may have to decide between different methods; (1) he may double track and slightly improve the present line; (2) if the present line is mainly favorable to traffic in one direction, he may build the second track in part as a separate single-track line favorable to movements in the opposite direction; (3) he may build new stretches of double track with improved location, the original line being abandoned or perhaps retained for local service.

Joint Double-track Lines.—In several cases, increased safety and facility of traffic have been secured by the operation of two separate but nearly parallel single-track railroads as a stretch of double track, one railroad operating all traffic in one direction and the other the traffic in the opposite direction. Interlocking plants are required at the ends of these sections. Such operation is limited usually to lengths of 10 to 20 mi., but it has been adopted for a 200-mi. stretch of the Central Pacific R. R. and Western Pacific R. R.

Traffic Capacity.—The economical limiting traffic capacity of single-track lines is placed in general at 50 trains daily, but this is an arbitrary approximation, the capacity being affected by ruling grades, character of profile, proportions of passenger and freight trains, and local operating conditions. In "Railroad Freight Transportation," L. F. Loree gives the capacity as 60 trains daily (of which 25 per cent are passenger trains), but considers that double tracking should begin when the traffic reaches 40 trains daily. On adjacent single- and double-track districts of the Rock Island Lines handling the same trains, the speed of tonnage freight trains was found to be 16.3 m. p. h. on double track as compared with 11.1 m. p. h. on single track. Elaborate studies of traffic capacity are given in the A. R. E. A. reports for 1922 to 1925.

In determining the operating and economic aspects of double tracking for the Meadville division of the Erie R. R., in 1915, estimates and comparisons were made on six locations, using varying portions and lengths of the existing line. Since such work is done almost invariably in connection with grade and curve reduction, the estimates of increased traffic capacity and economies in operation were based on the reduced grades, without attempting to capitalize the benefit derived from the additional tracks, so that the result of the comparison was on the safe side. The length, curvature, grade, rise and fall, and other physical characteristics of the old and new lines were first determined. Penalization values based on the operating cost per train mile for the division were computed either by Wellington's formulas or by inspection and determination of the effect on each individual item which goes to make up the total cost per train mile. Capitalization was taken at the rate at which money could be borrowed. The estimated cost of the proposed lines was then penalized according to present and estimated future round-trip trains, to obtain values for comparison. Taxes and additional line and structures to be maintained were also given a value. In this way, the most favorable line at the time, and at a future date of 10 to 15 years, was indicated. The annual saving effected by the new line was computed on the reduction in train miles and the increase or decrease in maintenance and taxes. This result represented a certain percentage of the investment; and this percentage, less the rate for money, indicated the percentage on the investment which would be saved by the construction of the new line.

Grade and Curve Reduction.—Under modern operating conditions and with the necessity of giving greater attention to operating economics, there are many cases where extensive expenditures are warranted in reducing grades and curvature to secure a more advantageous line. With lowei grades, a given traffic can be handled in fewer trains, thus increasing the capacity of the line as well as reducing transportation expenses. But the fewer trains will be heavier and longer, while if traffic increases, the number of trains may still be increased. The engineer has then a problem for improving the original location or relocating certain stretches, as noted above in regard to double tracking. Minor changes in grade may sometimes be introduced in the ballasting, at small expense.

Advantages of reducing and eliminating curves have been noted already, and this class of improvement is often practicable on lines in rough country where the original conditions warranted extensive curvature to avoid heavy grading, but where present traffic conditions warrant large expenditures in straightening the line.

It is sometimes assumed that 0.3 per cent is the economic limit for grade reduction, but special conditions may warrant a lower limit. Thus careful study and analysis of conditions on the Hocking Valley Ry. showed that it would be economical to incur increased expenditure in reducing the grades to a maximum of 0.2 per cent rather than 0.3 per cent against heavy coal movements. In many cases, the advantage of reduced grades lies in the ability to handle trains of standard tonnage rating without dividing or "doubling" or the assistance of pusher engines on account of heavy grades. While pusher grades cannot always be eliminated by relocation, the profile may be so modified as to reduce materially the trouble and expense of their operation. Grade reduction is sometimes for the purpose of eliminating steep grades, but it is more generally undertaken where the grades are only relatively steep, and thus limit the handling of trains of standard tonnage rating, or where the new profile will permit a material increase in this rating for an operating division. For example, the reduction of maximum grades from 0.8 to 0.3 per cent on the Cleveland-Indianapolis line of the Cleveland, Cincinnati, Chicago & St. Louis Ry., enabled the train load to be increased from 2,200 to 4,000 tons; and on the Cumberland Valley division of the Louisville & Nashville Ry., a reduction of the ruling grade from 1.25 to 0.65 per cent enabled the train load in the direction of heavy coal movements to be increased from 3,150 to 5,300 tons, with material lowering of operating expenses.

Raising Grade.—On lines in the lowlands of the Southwest, the grade is sometimes raised to be above flood level, the height of lift being from 1 to 5 ft. In such work done under traffic on the Illinois Central R. R., a lift up to 2 ft. on trestles was made by placing additional caps, well driftbolted. For greater lift, the general plan was the same, but an additional pile was driven at each end of the bent and the new caps were long enough to extend over the eight piles. In fills, material was placed by 20-car trains of air-operated dump cars controlled from the locomotive, 20-yd. cars being used on the smaller and 30-yd. cars in the higher fills.

When the fill came up to roadbed level, it was spread clear of the track by a spreader car and powerful locomotive. Raising the track was done in lengths of 1,000 ft., with 1-ft. lift in each length by a gang of 40 men. Thus for a 4-ft. raise, there would be four 1,000-ft. lengths in progress, the first at the original grade and the others 1, 2, and 3 ft. above it, all on a narrow fill. Behind this was the final dumping and filling for permanent grade and width, with a gang of 40 men adjusting the running tracks between this and the preceding section, which was still 1 ft. below final grade. The track was raised by jacks while earth was shoveled under the ties, and the dump-car train was then run over the track thus raised to deposit material for the next lift. In this way, the raising was kept in advance of the dumping, which was an advantage in wet weather.

Passing Sidings.—By lengthening passing tracks to accommodate maximum trains and by relocating them to suit changes in operating conditions, train movements in general may be expedited, besides particularly avoiding delays in handling long trains on old sidings of inadequate length. Automatic block signals and a well-arranged system of passing tracks will increase the capacity and effective operation of single-track lines very greatly. In the provision of passing tracks, uniform time for the successive runs is the main consideration, so that trains may proceed evenly. Thus, where a stretch of line between passing tracks has heavy grades or other features tending to reduce speed, these tracks should be spaced closer than for adjacent stretches having easy grades on which trains may make better speed.

In double tracking, the passing tracks are often linked in with the new second tracks. Passing tracks are still needed, however, on double-track lines. In such work on the Big Four Ry., the old passing tracks 6 mi. apart were linked in, and new ones for 100-car trains were provided about 10 mi. apart, their locations being determined by yards, terminals, and road crossings. Another division, with manual-block sections 15 mi. long, had double passing sidings 4,000 ft. long at the ends of the blocks; some of these were left as relief or passing tracks for the new double-track line.

Another improvement is the introduction of remote control of siding switches. Ordinarily, a train which is to take the siding must stop in advance, while a brakeman goes ahead to open the switch. All stopping and starting of heavy trains is costly, both in time and in wear or damage to track and equipment. With remote control, the operator at a station or interlocking tower can operate a switch at a distance of a mile or more, under orders from the dispatcher. An automatic signal indicates that the switch is open and that the train can proceed into the siding.

Filling Trestles.—The extensive use of timber trestling has been an important factor in the economy and rapidity of construction which characterizes the extension of the American railway system, but when trestles approach their life limit on account of decay, or become inadequate for modern heavy loading, it will frequently be found more economical to replace them with solid embankments which require little maintenance. In several cases, steel viaducts, too light for modern loads, have been filled in this way. Trestles are safe and substantial structures in themselves, especially where a ballasted floor is used, and those built of treated timber have long life, but in general there is an increasing cost of maintenance and repair after a comparatively few years. They are also liable to damage by fire and flood, and trestles over streams subject to freshets or floods should be replaced gradually by more permanent structures, although banks as well as trestles may be damaged by washouts.

Soft or wet clay is to be avoided, as it may slip and settle for several years, requiring frequent attention. Furnace slag and coal-mine refuse are available in certain districts. As a rule the material is shoveled, plowed, or dumped from cars on the trestle, forming ridges which unite as the filling rises. If the material is kept spread and leveled by scrapers and teams so as to build it up in horizontal layers, the strains on the trestle members will be equalized, and the bank will be so compact as to require little future maintenance. Banks built by simple dumping have more slips and settlement, requiring additional material to bring them to grade. For high trestles, inclined aprons may be built along the sides to throw the material clear of the posts and widen the fill, and a central ridge may be formed by dumping through the deck of the trestle upon a ridge-shaped apron or shield beneath it, the middle portions of alternate ties being cut away. With an additional rail outside of each main rail, a narrow-gage dumping track may be formed along each side of the trestle, or for very wide and high fills temporary low trestles may be built along each side of the main structure within the slope line of the fill. If the material contains many large boulders, an open apron like a picket fence along the edge of the trestle will serve to throw these clear, so that they will not damage the structure.

The trestle should be kept well braced longitudinally during the work, damage done by falling material being repaired; it must be watched for signs of distress or serious movement of the bents, the filling being distributed so as to avoid excessive local pressure. It is generally cheaper to put the trestle in good repair first than to be continually repairing it, strengthening weak spots, and adjusting distorted track during the work. Plowing material from cars should be done slowly, avoiding lateral strains due to side plowing or longitudinal strains caused by the plow encountering heavy boulders. For dumping, cars operated by compressed-air under control from the locomotive avoid the difficulties and dangers of operating dump cars by hand on trestles, but with either hand or air dumping it is sometimes necessary to anchor the car to the track by some simple device to prevent the shock from throwing the car off the structure. To fill trestles on soft ground, soundings and investigations should be made or the work may involve excessive cost and perhaps serious damage to the trestle or neighboring structures due to displacement of the soil. The dumping of material in swampy ground is likely to develop upheavals at a considerable distance, which may lead to suits for property damage.

With any method of filling, the trestle bents must remain in place, but the bracing should be removed as far as possible as the filling proceeds, so as to avoid holes or pockets and to leave the embedded timbers free to move with any settlement of the fill. Ties and stringers may be left in place for 1 to 3 years, while settlement takes place, the timbers being then removed and the bank filled to grade. The caps are sometimes removed at the same time.

Old steel viaducts on the Reading R. R. gave trouble when being filled, owing to the fill not being spread, the concentrated pressure causing distress of some of the columns and difficulty in keeping the track in line. Timber bents 60 to 70 ft. high were then erected, but as they were founded on the new fill, they were not stable and shifted considerably. After the girder connections to the column caps were cut, there was no more trouble with holding the track. As a result of this experience, the columns of other viaducts were reinforced with a concrete casing, the top of fill was kept level, and bracing was cut away as the filling rose.

Hydraulic Filling.—By this method, water under about 200-ft. head is discharged in 3- or 4-in. jets to break up loose material, the flow of water and earth being carried by flumes to the desired point. There a pool is formed by lines of planks, old ties, or marsh grass along the sides of the fill, the material settling and the water flowing quietly away; additional lines are placed to conform to the slopes as the fill rises. This method makes a solid bank, but the progress is slow. The top is usually finished to grade by dumping. In filling trestles 80 ft. high on the Soo Line, sand plowed from cars was leveled by $1\frac{1}{2}$-in. jets from $2\frac{1}{2}$-in. hose connected to steam pumps. This fill was carried up to a height of 20 ft., wide enough for two temporary trestles within the prism of the fill; dumping from the main trestle then filled the central space and brought the bank to grade.

Widening Cuts and Fills.—This work is included in various improvements, sometimes to give room for double track and sometimes to give better drainage or stability. In maintenance work, cuts are widened for enlarged ditches or to flatten slopes where material falls or slips, while banks are widened which were built too narrow or have been reduced by erosion. For heavy work in cuts, a steam shovel may widen one side, loading trains on the main track or a siding, and leaving a roadbed wide enough for another track, on which traffic may be handled while the steam shovel widens the other side or lowers the old roadbed to the new grade. Lighter work in maintenance may be done with scrapers and teams, with a ditching machine, or by hand, the excavated material in the last case being removed by wheelbarrows or by small cars on a narrow-gage track at the side of roadbed.

In widening old banks, the base should be cleared and the old slope stripped of grass and weeds, and then loosened or benched so as to bond with the new work. If the material for widening is simply dumped over an old and compacted slope, much of it will be lost; after heavy rains, long cracks will appear, and masses of the new fill will break away. A better plan, when the height does not exceed 5 to 10 ft., is to dump in place from a temporary track which is raised with the fill. For greater heights a temporary trestle may be built for dump-car trains. On the Cincinnati Southern Ry., a peg-leg trestle has been used for this purpose, each bent consisting of one batter post and a cap, the inner end of the latter resting on the edge of the old roadbed. For heavy work in widening comparatively low fills, it may be convenient to divert traffic to a temporary runaround track at the ground level.

For all widening of banks, it is good practice to spread and level the dumped material so that the fill will be built up in horizontal layers. This will greatly reduce settlement and consequent future attention or maintenance. In filling, allowance must be made for shrinkage due to: (1) settlement or compacting of the fill; and (2) subsidence of the original surface; this last will exist everywhere, except on solid rock, and will be at a maximum in swampy ground.

Grade Separation.—In many cities where tracks were built originally at street level, the danger and inconvenience due to heavy traffic over numerous grade crossings has necessitated separation of grades. This elimination of grade crossings of streets is mainly for the benefit of the community and the non-traveling public. It benefits the railroads in giving greater freedom of movement and greater safety, reducing delays, eliminating the cost of crossing protection, and also reducing the liability of accidents and damage suits. But a heavy proportion of the cost is paid by the railways, and, as a rule, these benefits are not commensurate with the cost. Much of this work is in the larger cities but the demand for grade separation in smaller cities is often reasonable from the fact that the railroad passes through the business section, interfering with traffic movements and property development and constituting an element of danger. Since the financial conditions of the railways will limit very strictly the amount of this costly work which can be undertaken at any one time, it is to the interest of the cities to assume a fair share of the cost in view of the importance of the work to the community. In most cases, the tracks are raised, the streets being depressed to some extent through subways in order to reduce the height of fill. More rarely the tracks are depressed, as on some of the work at Minneapolis, Detroit, and Milwaukee. This requires greater headroom at bridges; and also involves greater interference with sewers, water and gas mains, and other public utilities, for all of which the railway must usually share the cost. For residential districts, track depression has an advantage in avoiding obstruction to light and view, but extra maintenance will be required to keep the roadbed well drained and clear of rubbish.

As a rule, the elevated work consists of a solid fill between concrete retaining walls and bridge abutments, the streets being crossed on solid-floor bridges of steel girders or concrete slabs. In special cases, however, a concrete viaduct has been used, the spaces between streets being either left open or utilized for storage of goods or automobiles. In construction, the A. R. E. A. recommends that the work should be in entire charge of a superintendent of construction, with engineers to see to grading, masonry, bridges, and contract work; a roadmaster for the track work, a yardmaster for the switching movements, and a trainmaster and a dispatcher to handle traffic over the territory. It recommends also that the railway should handle with its own forces all work which may interfere with the movement of trains.

Grade separation is work of a special character and calls for careful planning and organization, in order to avoid excessive cost, to insure rapid but safe progress, to utilize the restricted working space to best advantage, and to minimize interference with both railway and street traffic.

Construction Work.—Methods and details of track elevation vary widely, but the following is more or less typical. Foundation piles for longitudinal retaining walls and transverse abutment walls at streets are driven by railway or other pile drivers, and the concrete footings poured by mixer plants on the street or on tracks, trenches for the abutment footings being spanned temporarily by heavy stringers to carry the running tracks. The track next to one of the walls, track No. 1, is given up to a concreting plant for building the wall. Pile trestles are built across the streets, for one or two tracks at a time. As each street is spanned, the approaches for track No. 1 are raised on sand deposited by work trains, and when the track is clear of the street, these trains are operated on it to continue the filling. The approach grade may be as much as 5 per cent for the work trains, but not more than 2 per cent for the running tracks.

With the wall completed and track No. 1 raised on the fill, the piles for bridging on track No. 3 are driven; track No. 2 is then abandoned, the track being cut and tilted up on the ends of the ties while the bank of No. 1 is widened by dumping from trains on that track. Then the old No. 2 track is pulled up the slope by derrick cars and relaid as elevated track No. 2, which is then utilized by work trains while No. 1 becomes a main line or running track. Remaining tracks are handled successively in the same way. Abutment walls are then built between the pile bents of the trestles, and the girders and bridge floors are set in place for one track at a time, the tracks being opened to traffic successively.

In some cases, the trestles for track No. 1 are only half the full height, blocking one or two streets temporarily. Filling for its approaches is placed first from trains on No. 2 (at street level) and track No. 1 is relaid on this fill at half-height. Trestles for Nos. 2 and 3 are built to full height. Track No. 1 is then raised gradually with material dumped from trains on this track, additional caps being put on the trestle as the work progresses. This fill is then widened by side dumping, but to avoid covering the low-level running tracks on narrow right-of-way, the toe of the fill may be retained temporarily by a cribbing of old ties built beside these tracks. One or two running tracks must be kept in service through the work all the time, according to traffic requirements, connected at one end with the two or four tracks already elevated, and at the other end with the old tracks at street grade. Cribbing of concrete units is sometimes used instead of a solid retaining wall or as a toe wall for earth fills.

Highway Grade Crossings.—Grade crossings of country highways are so numerous that when multiplied by an average cost of grade separation the total expense becomes prohibitive, even for the main highways. In spite of the expense being shared by federal, state, and local governments this work can proceed but slowly. As a rule, it presents little advantage to the railway, especially as it often involves an unfavorable change in profile to eliminate an isolated crossing. In many cases a highway can be relocated conveniently so as to eliminate a crossing.

Summit Tunnels and Tunnel Elimination.—On several lines carried over mountain ranges at high elevation by a steep and winding location to insure low cost and short time of construction, increased traffic has increased operating cost to such an extent as to warrant the driving of long tunnels at lower elevation. The Northern Pacific Ry. first crossed the Cascade range at an elevation of 3,675 ft. by a 7-mi. switchback line having grades of 5.6 per cent, but the Stampede tunnel at an elevation of 2,827 ft. reduced the distance to 3 mi. and the grade to 2.2 per cent. The Great Northern Ry. crossed the same range at an elevation of 4,055 ft. by a similar line 12 mi. long with grades of $3\frac{1}{2}$ and 4 per cent and 12-deg. curves, but its $2\frac{1}{2}$-mi. Cascade tunnel reduced the elevation to 3,350 ft., the distance to $3\frac{1}{2}$ mi. and the grades to 1.75 and 2.2 per cent. A recent case is the Denver & Salt Lake R. R., crossing the Continental Divide at an elevation of 11,660 ft. by a circuitous 30-mi. line having long 4 per cent grades and numerous 16-deg. curves. The 6-mi. Moffat tunnel line, started in 1924, will give a direct line of about 7 mi. with a maximum grade of 2 per cent and a summit at an elevation of 9,242 ft.

In line and grade revision, old tunnels are sometimes eliminated, as on parts of the Southern Pacific Ry., where originally short tunnels through mountain spurs were cheaper than deep sidehill cuts and long fills. Such work calls for careful study of the alternatives of line revision to eliminate tunnels or enlargement of tunnels which are too small for modern equipment, built originally to avoid the time and cost of deep cuts and high fills. As these tunnels were of comparatively small section, it was a problem in each case whether to enlarge and line the tunnel or to avoid it by relocation. The safety of traffic during operations was a factor, as well as the comparative cost, and in any case the handling of traffic added to the difficulty of the work. In some cases, a temporary "shoofly" or runaround line was built outside the spur while the tunnel was being caved in, and the new cut and approach fills were being built.

Resurveys.—A special class of work is the resurvey of a railway or division to check its alinement, profile, and property lines. For many old lines, perhaps built independently and later absorbed by other roads, the records are incomplete or missing as to original location and construction and as to subsequent variations and additions. For such a resurvey of parts of the Chicago, Milwaukee & St. Paul Ry., the party consisted of an assistant engineer in charge, an instrument man, a levelman, two draftsmen, three rodmen, and two chainmen, with a camp train and a gasoline motor car (see *Engineering News-Record*, Mar. 18, 1920). The field work included the survey, alinement, track profile, ground lines, and land lines. Alinement on a crooked division was checked by a party consisting of an instrument man, a rodman, and two chainmen, the work being facilitated by a diagram showing the degree of any curve when the intersection angle and external were known.

WORK TRAINS

Work Trains.—In maintenance and general improvements such as heavy ditching, ballasting, and filling, done by extra gangs and special forces, work trains or gravel trains are operated to transport the men and material. A work train is usually given orders authorizing it to work as an extra train on a specified track between certain points or stations and at certain times, being always protected in both directions by flagmen, except when running. It has to clear all regular trains, but no other extra or special train is allowed to pass over the specified stretch of track without being given orders to look out for the work train between certain stations.

In such operations, the dispatcher is informed as to the location of the work, the time probably required, and the movements to be made when the work is finished. The conductor is notified as to train movements, especially expected extras, and he may be required to report at a telegraph station for orders at a given time. When the train passes a telegraph station while running to a siding to clear regular trains, the conductor should stop and call up the dispatcher, who may be able to extend the working time.

The foreman of the gang and the conductor of the work train should work in harmony to insure effective work. They both make itemized daily reports (or a joint report) to the roadmaster or engineer as to work done, delays, idle time, number of cars loaded, and number of men in the gang. These reports are examined by both the maintenance and operating officers, so that insufficient work or excessive delays may be noted and investigated. If the work is interrupted while the train leaves to clear regular trains, the foreman should keep the men employed at clearing ditches, picking up scrap, or clearing station grounds. With a large gang there will be an assistant foreman and a timekeeper. The roadmaster may accompany the train (and is sometimes required to do so), or he may arrange to visit it during the day.

To insure efficient service, the work train should have an engine of ample power, equipment in good condition, and a capable crew, while the work should be planned carefully in advance and discussed with the foreman and conductor. Delays due to engine failures, broken draft gear, damaged cars, or other defects, not only interfere with the program of work but may result in blocking traffic. Where slow orders are issued on account of renewals or track work, the dispatcher should be notified when the work is done, so that the orders may be withdrawn or canceled, and unnecessary delay to traffic avoided.

Work-train Service.—When a work train is required, the roadmaster (or superintendent of bridges and buildings) arranges with the division superintendent and dispatcher, in order to provide for the necessary equipment and train crew, and to plan the best time and method of operating, keeping it as far as possible for periods of light traffic. For important work, a local operator may be established temporarily (as at a gravel-pit connection for ballasting or filling), having telegraph or telephone communication with the dispatcher, who can thus arrange for a maximum working time and a minimum interference in movements of regular traffic. There should be work sufficient for at least 1 day, and preferably 2 or 3 days. The operations should be planned in advance, the engineer or the roadmaster issuing an outline of each day's work to the foreman, the conductor, and the engineman.

Economy in Work-train Service.—These trains are necessarily expensive, but their expense is likely to be increased considerably if the operating officials take the attitude that the work train is a nuisance and must be kept off the main track as much as possible. With a train, a crew, and a gang of perhaps 20 to 50 men for distributing material, ballasting, or filling, the expenses go on whether much or little work is done, and in the latter case the cost of the work may be excessive. If a steam shovel is involved, the machine may stand idle for long periods because loaded trains cannot be taken away or empty trains brought in promptly, but such a condition is likely to be checked if the dispatcher is held accountable for delays to work trains. Even with proper consideration for such a train, the actual working time of its force may be only 40 to 50 per cent of the payroll time, owing to time lost in keeping clear of other trains.

For these reasons, it is very desirable that the dispatcher should realize that the work train is an important item in railway service, to be handled in such a way as to give it the maximum time for effective work without undue interference with traffic. Slight delays to local trains may be much less serious than stopping the work of a large gang of men. For extensive work, the local dispatcher and the roadmaster, and perhaps the division engineer and the superintendent, should consult together and cooperate in planning the work-train service. On a double-track line, it may be found that both traffic and improvement work can be handled most satisfactorily by giving up a section of one track to the work-train force for a day or half a day.

Work trains are maintained permanently on some divisions, while on others they are ordered only as required for special work. The latter system tends to reduce the demands for such service, as other and perhaps more economical methods may be devised for distributing material and performing some of the operations for which a permanent work train might be requisitioned. Thus for distributing material in moderate amount, local freight trains can be used to some extent, with care to avoid delay to trains between stations or to trackmen waiting at the unloading points; or the train may set out a car on a specified siding, to be unloaded and then picked up by another train. Such service requires cooperation between the roadmaster, the trainmaster, and the dispatcher. Ties, fence posts and other materials may be unloaded at stations and then distributed by motor section cars with trailers. This method can be employed also in some cases for transporting large gangs of men, where the only service of a train would be to take the men to and from work.

Example of Work-train Service.—On the Chicago & Northwestern Ry., the work trains are operated by the transportation department, except that on new work (such as track elevation) they are operated by the engineering department. The conductor and the foreman make a joint report by telegraph daily to the division engineer, who is thus kept informed of the progress of the work. In special cases, the conductor may act as foreman in a gravel pit, where the force is small. The trains are governed by telegraph train orders, issued each day, and confined to as short a territory as the work will permit. When scheduled trains are late, work trains are given the benefit of the time as far as possible, in view of the expense involved in work-train service.

On double-track districts where there is extensive work to be done, the work train may be given exclusive use of one track between certain points, usually between points at which there are crossovers. In still more extensive work, as for additional tracks or track elevation, the territory may be put in charge of local operators having communication with the train dispatcher of the division. An operator may be established also at a gravel pit where work-train or gravel-train movements are numerous. No distinction is made between districts having manual or automatic block signals and districts having no signals.

Work-train crews on the Lehigh Valley R. R. are in charge of the maintenance-of-way department. Each crew consists of an engineman, fireman, conductor and two brakemen, with an additional brakeman if the crew has to pilot trains for single-track operation in multiple-track territory. A foreman is in charge of the labor gang, with an assistant foreman and a timekeeper for a large gang. The conductor is charged only with the operation of the train, in conjunction with the dispatcher's orders and the usual rules. The foreman is responsible for the work of the gang and issues instructions to the conductor as to where work is to be done. Work trains are classed as extras and are assigned working limits at the beginning of the day's work. They are required to clear first class trains by ten minutes and all other trains with the least delay. When scheduled trains are late, work trains are given time by train order or message, the same as extra freight trains. They are also informed as to the movements of fast-freight trains or other trains to which special attention is to be given.

Camp and Boarding Trains.—The maintenance-of-way equipment usually includes boarding trains for extra gangs, construction gangs, and bridge gangs. These are under the supervision of the roadmaster, who selects the location, and inspects the boarding and sleeping accommodation. If no regular sidetrack is available, a temporary track will be laid, preferably at some distance from the main track. In the latter case, the track should be disconnected and the switch taken out; if a sidetrack in service is used, the switch should be spiked, wheels blocked, and a derail placed to prevent the possibility of the cars being moved so as to foul the main track. When the cars are placed adjacent to a running track, they should have no open doors, ladders, or walks on the track side.

Old box cars are commonly used, and too often they are dirty and unsanitary. Some roads have special 50-ft. bunk cars with six double berths on each side, accommodating 24 men; at the middle of the car is a 10 ft. space with side doors, seats, stove, coal bunker, and water barrel or cooler. The foreman's car is similar but has one end partitioned off and provided with a stove, table, and berths for the foreman and the assistant foreman; the other end of the car may be for tools and supplies. The kitchen car has one end arranged as a dining room for the foreman or officers, while the regular dining car has two tables 21 by 3 ft., with side benches. Bunk cars on the Big Four R. R. (Fig. 104) have a middle space opposite the doors and four double-deck bunks along each side. With a complete outfit, there is also a kitchen car, dining car (with central table and side benches), supply car (with ice box), commissary and tool car, and foreman's car; this last has a table and benches in one end and a room with bunks and desk at the other end. Each car (except kitchen and supply cars) may have a stove, this being generally in the center, but placed in the corner of the dining car. Windows with sliding sash are formed in the sides, and the bunk car has them in two tiers, giving one to each bunk.

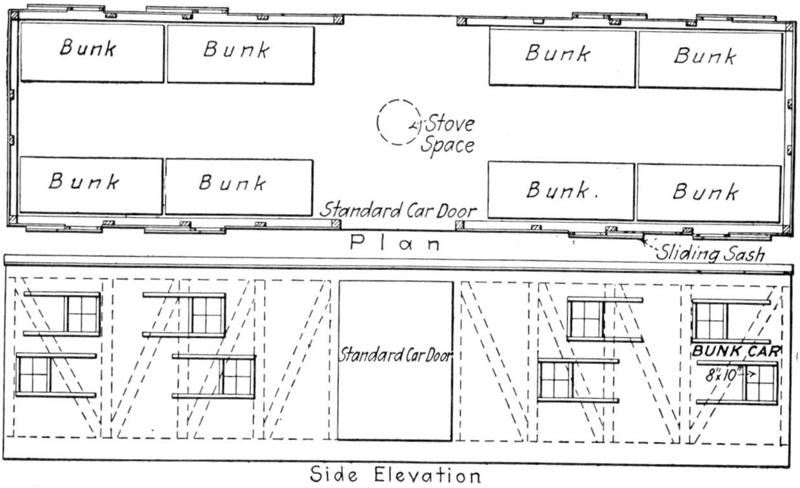

FIG. 104.—Bunk car; Cleveland, Cincinnati, Chicago & St. Louis Ry.

Cars should be kept clean and should be fumigated periodically. The surrounding ground also should be kept neat, toilets or privies provided and properly disinfected and cleaned, and all garbage and rubbish collected and burned or buried daily. Toilets and sanitary equipments at camps are noted under Buildings. Apart from sanitary conditions, the cars should be kept clean and well painted in order to present a good appearance to the community and the traveling public. In view of the importance of labor and the difficulty in keeping men on railway maintenance and improvement work, some railroads are giving more attention to providing adequate housing and boarding accommodation for the men. But it is still general to find old and unsanitary buildings and camp cars with unpleasant surroundings. Comfortable camps and good food help to solve the labor problem, especially in combination with assurance of permanent employment.

CONTRACTORS' CONSTRUCTION TRAINS

Construction trains operated by contractors on railroad improvements are in a different class from company work trains, since several trains operating in a short territory for a considerable time usually interfere with the operated tracks, and are under double control of the railroad and the contractor, while their operation must be handled by the former in such a way as to avoid undue cost or delay of the improvement. Where traffic is very heavy, the contractor may be required to build construction tracks over the entire work, but as a rule his trains operate to some extent on the railroad's main tracks.

In general practice, construction trains and also gravel trains for ballasting are under the direction of the engineer in performing their work, but are under the control of the operating department so far as their movements on operated lines are concerned, and they must conform to all train rules. Their movements are directed by train orders, usually in the form of work orders for the day, which permit them to operate between certain limits, protecting themselves against other trains. Where traffic is very heavy or other conditions warrant, an operator may be placed on the work, receiving instructions directly from the train dispatcher by telegraph or telephone. In certain cases, the contractor is under penalty for delay to traffic by his trains. It is more satisfactory, however, to have a clear understanding between the contractor and the railroad that traffic must not be delayed by the contractor but that he will have the active assistance of the railroad in the effective handling of his trains. These trains may be handled by the railroad company's engines and crews; but if the contractor uses his own engines and crews, each train while on the main line is in charge of a railroad conductor acting as pilot.

Construction Train Practice.—In the double tracking of 725 mi. of heavy-traffic line on the Santa Fe Ry., the construction work was mainly kept clear of the operated line. Where the contractor's trains had to operate on main tracks and had their own engines and engine crews, the railroad furnished train crews and charged their expense to the contractor. If engines and engine crews had been furnished, the railroad would have charged the expense of the crews and a daily rental for the engines, based on their tractive power. Where old fills were widened by side dumping from the operated track, the movement of both contractor's trains and regular trains was supervised closely by a construction trainmaster, but in most cases a separate parallel fill was made by dumping from a construction track on a temporary trestle. Where the work caused temporary interference with the main line, as when a construction track was benched so close as to foul the former for a short distance, "slow" order signs were placed, and the contractor was required to maintain flagmen.

Main-line trains were given orders to approach certain points under control and to proceed only by direction of the flagmen.

Widening and grade reduction on the Illinois Central R. R. near Chicago necessitated the use of busy main tracks by the contractor's trains in hauling excavated material to the site of extensive filling, the haul being from 6 to 12 mi. These trains, with 15 cars of 12- and 16-yd. capacity, had the contractor's engines and crews, but were in charge of railway conductors; they entered the main line at an interlocking plant and were operated as extras, under orders from the train dispatcher. On the other hand, in double tracking a portion of the Cleveland & Pittsburgh division of the Pennsylvania R. R. system, the entire construction work was handled on the contractor's tracks; this was practicable since the heaviest excavation was on a diversion from the old line, while no widening of old fills was required.

In depressing a busy piece of line at Milwaukee, Wis., the Chicago & Northwestern Ry. maintained one track for regular traffic in both directions and one for the construction trains, the traffic being so heavy that it was necessary to keep it clear of construction trains. These latter were handled by the railroad company's locomotives and crews. Movements of traffic and construction trains were controlled by two operators, one at each end of the work, in cabins having telephone communication with the dispatcher's office at the city station. Hand-flag signals were used to regulate train movements within the working limits. In handling construction trains on track elevation and other improvement works of the Rock Island Lines, the train crews report to the engineer in charge, but all movements on main-line tracks are controlled by the train dispatcher, and are made in accordance with regular train orders. Contractors are not allowed to operate trains on main-line tracks with their own crews, the railway furnishing crews in such cases, but if it is necessary to avoid interference with operated tracks, the contractor may be required to build special tracks for his own use.

CHAPTER XXIV

SIGNALING AND INTERLOCKING

Although signal equipment is not under the direction of the maintenance-of-way engineer, he must have a general knowledge of it in relation to railway operation. There are also certain phases in which he is directly concerned; for example, the ballast, track-maintenance work, insulated joints in track rails and switch rods, proper adjustment of switches as related to track circuits, electric wiring, electrical locking apparatus at switches, and the mechanical or electrical track appliances for train-control apparatus. Fixed signals governing train movements, independent of switchstand targets, may be classed as follows: (1) train-order signals, placed at stations to indicate whether a train is required to stop for orders; (2) block signals, indicating whether the block sections into which the line is divided may be used by trains, and if the blocks are clear or occupied by trains; (3) interlocking signals, at junctions, crossings, or other points, to indicate for which track or route the switches are set, whether this is a high-speed or low-speed route, and whether the route or block is clear.

Train Dispatching.—Under the written train-order system of operation, the train dispatcher at division headquarters issues written orders to the train crews to run between certain points at specified times, stopping or reporting at certain places or taking sidetrack to clear specified trains. These orders are given to outgoing crews at division points. Additional or supplementary orders to trains on the line are transmitted to way stations by telegraph or telephone, the station operator then displaying his train-order signal against the designated train, which stops or slows to enable him to deliver the order, according to the kind of train order and the practice of the road. Single-track operation by train orders is more difficult than for multiple track, since the former includes opposing movements while the latter has only following movements.

If all trains could be scheduled and stay on schedule time, there would be little occasion for issuing orders to trains during their runs, as all trains would meet and pass at the times and places designated, which would be an ideal condition. In practice, however, when one train gets late, the dispatcher must issue orders to change meeting points in order to avoid unnecessary delays to other trains, and these changes may involve several trains. If work trains, extra trains, or additional sections of scheduled trains are to be run, or if a scheduled train is canceled, additional orders are again necessary. Following trains are required to keep a certain number of minutes behind each other, according to rule, but this "time interval" is not maintained, as a leading train may break down or lose time, or the next train may run faster, thus closing up the interval. If the leading train should be stopped, a flagman should be sent back; or if delayed and losing time, a fusee should be dropped on the track to warn a following train to reduce speed. But these methods do not always furnish the protection desired.

Block System.—Under this system, the line is divided into blocks or sections. At the ends of the blocks, signals are provided by which indications are displayed to inform trains if they must stop or if they may enter the block. As a rule, the block system is superimposed upon the train-order system, and is supplementary to it, but to a limited (though increasing) extent, the train-order system is partly or entirely superseded when the block system is introduced. The use of the latter is increasing rapidly, as railways realize that for controlling traffic it has other benefits than that of safety, which is the one usually conceded.

Advantages of the block system lie in increased safety, facility, and economy of handling traffic. When used as a supplement to the train-order system, it provides a second line of protection, and when used in place of the latter, it serves to deliver the order directly to the train at the point of execution, the signal indication becoming the train order. For safety, it provides an actual "space interval" between trains, preventing collision. For facility (on both single and double track), it increases the number of trains that can be handled over a given length of time in a given time, partly by the closer spacing possible and partly by eliminating delays in stopping for orders. For economy, the expediting of train movements may reduce overtime payments for train crews, while the increased traffic capacity may postpone a more costly double tracking or construction of additional tracks.

Assuming plenty of train-order stations and ample dispatching capacity, the written train-order system alone may provide greater flexibility and traffic capacity than the block system alone or in combination with it (unless the blocks are very short), since trains may be run with less than braking distance between them, though this is at the expense of safety. But the average railroad cannot afford such an assumed number of train-order stations. This condition, together with the time required to deliver and execute the orders, produces the result that the block system will provide equal or greater capacity and increased safety at the same or lower annual expense.

Where a combination of train-order and manual-block system is in use, the same annual cost expended for a modified train-order system with automatic block-signal system superimposed may give a much greater road capacity; that is, many more trains can be run with equal facility in the same length of time. This is the usual practice at the present time. It is claimed, however, that by going a little farther and taking full advantage of the possibilities of signaling, the written train-order system may be superseded by signaling, even to the extent that train orders and train rules are eliminated, and still greater advantages and economies gained. This practice is already in use to some extent, the trains being handled exclusively by signals. Further developments are being given serious consideration. Such expediting of train movements, principally by reducing delays, may reduce overtime payments for train and engine crews, and also reduce the expense of fuel, supplies, and locomotive and car repairs, while the increased traffic capacity gained may postpone a more costly double- or multiple-tracking construction.

With the "absolute" block system, no train may be admitted to a block section until that section is clear or empty. Under the "permissive" block system, a second train may be permitted to enter the block, but must proceed with caution and be prepared to stop within range of vision. The average length of block section varies greatly, depending upon the amount of traffic, character of line, and location of stations and passing sidings; it ranges from about ½ to 2 mi. on busy main lines and 5 mi. or more for light traffic. Block-signal operation may be manual, controlled manual, or automatic.

Manual and Controlled-manual Signaling.—For manual block signaling, the signals are manually operated by signalmen, there being a signalman at each end of the block. A telegraph operator who also works the signals is classed under the rules as a signalman. There must be telegraph or telephone communication between adjacent stations and with the train dispatcher. On single track, before a train may be admitted to a block, permission must be obtained from the signalman at the farther end of the block before the signal at the entering end may be cleared to indicate that the train should proceed. On double track, the signalman may clear a signal for a train to enter a block immediately a train approaches, provided the block has been reported clear of the last train occupying it.

For illustration: On the approach of a train to signal station "A" (on double track), the block "A–B" having been reported clear, the signalman at "A" clears the signal for train to enter the block, and on the train passing the signal he reports that fact to the signalman at station "B." The latter records the fact on the block record sheet and, on the approach of the train to his station, clears the signal for train to enter the next block, "B–C," provided that block is clear. On the train passing signal "B" and entering block "B–C," the signalman at "B" reports to signalman at "C" that the train has entered that block and reports to the signalman at "A" that the train has cleared or passed out of block "A–B." The time of trains entering and leaving each block must be entered on the record by each signalman. The signals are kept normally in position indicating "stop," being cleared or moved to indicate "proceed" only on the approach of a train, although a few roads keep the signals normally indicating "proceed." The block sections may be 5 to 10 mi. long, or with the towers located at passing sidings.

With controlled-manual signaling, the block signal is under electric control from the station in advance and is automatically set at the "stop" position as the train passes, being then cleared only by the actions of the two signalmen in cooperation. Thus, when "B" is notified that a train is approaching from "A," he has to request the man at "C" to release the "B" signal, so that "B" can set it at the "proceed" position. But "C" will not do this if the last train admitted at "B" has not passed out of the "B–C" block at "C." After the train from "A" has passed "B," the signal "B" is automatically changed to the "stop" indication and the man at "B" cannot again set it at "proceed" until the train has passed out of the block "B–C."

Automatic Block Signaling.—In this system, signalmen are dispensed with and the signals are power operated. As the train passes each signal, it causes that signal to indicate "stop" and releases the signals in the rear to indicate "proceed." The system is economical for lines with both light and heavy traffic, the blocks in the latter case being so short that the expense for manual equipment and operators would be prohibitive, while communications between signal stations would consume excessive time. In any case, the blocks are usually much shorter than under the manual system.

The automatic system is not ordinarily absolute in its working. To avoid traffic being held indefinitely in case of failure of a signal, a train, after coming to a stop at a signal indicating "stop," may pass the signal and proceed cautiously to the next signal on double track. But on single track, it must send a flagman in advance and then proceed cautiously to the next clear signal. These signals are operated by power, usually electricity, through control by track circuits which cause the signal to display the proper indication according to conditions in the block which it governs. Each signal has mechanism to operate a semaphore arm or other moving part, or connections to vary the color or position of lights (or both) if the indications are given by lights.

Automatic Block Signaling on Single Track.—In order to facilitate traffic on single track without incurring the expense of operators or signalmen incident to the controlled-manual system, the absolute-permissive block system of automatic signals has been devised. Here, the signals restrict the movements of trains on the section of single track between passing sidings to one direction as long as the section is occupied, a "stop" signal being displayed at the farther passing siding to hold opposing trains at that siding, to which the train in the section is approaching. Trains may follow each other through the section of single track protected by the automatic signals, as is the case on tracks that are used in one direction. When the section of single track between passing sidings is clear of trains, it may then be used by a train running in either direction, the first train entering the single-track section establishing the direction in which trains may move until the section is again clear of trains.

Train-staff and Tablet System.—This system is used mainly to protect isolated single-track block sections, such as long bridges and tunnels or single-track links in double-track roads. The block station at each end of the block is provided with a case containing a series of staffs or metal tablets. No train may enter the block unless the engineman has the staff for that block. The cases are electrically controlled from the instrument at the other end of the block. To admit a train to block "A–B," the man at "B" releases the instrument at "A" so that one "A" staff can be taken out and given to the engineman. No other staff can be removed from the case at "A" or case at "B," as these instruments are automatically locked by the removal of this one staff. When the train reaches "B," the staff is delivered to the signalman, who puts it into his case and thus restores the "A–B" apparatus to its normal condition. Semaphore signals are used in connection with the staff to indicate to the engineman that the staff can be exchanged while the train is moving. Thus, if the signal at "B" indicates "proceed," the engineman would drop the staff for "A–B" and take that for "B–C." This can be done by means of apparatus similar to a mail crane and catcher, or by enclosing the staff in a pouch on a large brass ring which is hung on a post so that the engineman can thrust his arm into the ring.

Signal Equipment.—Since it might not be possible to stop a fast train at a "stop" signal after the signal came in sight, a distant signal is provided to indicate the position of the home signal, or each signal may be made to give an approach indication for the next signal in advance. When the home signal indicates "stop," the distant or approach signal indicates "prepare to stop at the next signal." Manual signals normally are held at the "stop" position, while automatic signals are generally arranged to indicate "proceed" as soon as a train has passed out of the block. The home or stop signal is placed at the entrance of the block and the distant or approach signal at braking distance in the rear.

Signals in general use are of two types: (1) a semaphore or pivoted arm giving the indications by position of the arm by day and having a spectacle casting with colored lenses moving in front of a lamp for indications by colors at night; (2) colored lights for both day and night indications, the lenses having hoods to obscure outside light; (3) two or more white or colored lamps in different positions giving indications by position and color of lights or only by positions of white lights. Along single and double tracks, the signals are mounted on posts on the right-hand side of the track, usually at a height of about 20 to 25 ft.; for three or more tracks, they may be mounted on steel bridges, each signal being directly over the track governed. Where there are three arms on a ground mast, the upper arm may be as much as 39 ft. above the top of the rail.

Where a track intervenes between a track to be governed and the post on which signals are to be placed, the signals are arranged on short masts placed on a bracketed platform on top of a 25-ft. main post (bracket post), the short masts being in the same relative position as the tracks which the signals govern. If the outer track is not signaled, the shorter (outer) post carries no operative arm, but is sometimes provided at night with a blue light to indicate that no signal is in use for the track adjacent to the bracket post. Dwarf or low signals about 2 ft. from the ground are used for sidings and reverse or back-up movements.

Semaphore signals are operated in either two or three positions and in the upper or lower quadrant. Two-position signals are being superseded by three-position signals on most roads, while semaphore signals are being superseded by light signals on a number of large railroads. In general, it may be said that for two-position signals the positions and indications are as follows: (1) horizontal, "stop;" (2) any position above or below the horizontal (usually 60 to 90 deg.), "proceed." With a three-position signal; (1) horizontal, "stop;" (2) inclined 45 deg., "approach next signal prepared to stop;" (3) vertical, "proceed." It is now general practice to use the upper quadrant system, the arm rising from the horizontal to the inclined or vertical position.

As to color, the faces of home-signal semaphore blades or arms are usually painted red, with a white square or stripe near the end, while two-position distant signals are yellow, with a black V-stripe, the end of the arm having a triangular notch parallel with the V. Manual-block or train-order signal blades are colored like home signals but have a convex round end and corresponding stripe. Automatic block-signal blades are colored similarly, but have a pointed end, with stripe of corresponding shape. The backs of all blades are painted black. Some roads prefer yellow to red for the blades, in which case the stripes are black instead of white. The colors by which the indications of three-position light signals are usually displayed are red for "stop," green for "proceed," and yellow for "prepare to stop at next signal." A purple light is generally used for "stop" on low or dwarf signals, except that red is used at terminals. There are two marked tendencies present in signal practice: (1) two-position signals are being superseded by three position signals, in which case the terms "home" and "distant" signals lose their significance; (2) light or lamp signals are superseding semaphore signals.

Automatic Train Stop.—To stop trains which are carelessly or accidentally allowed to pass a "stop" signal, mechanical track devices and ramp rails which engage brake-setting appliances on the train or engine have been used for several years on subway and elevated rapid-transit lines. Electrical contacts or inductive systems have been used to a limited extent on steam railways, but in 1922 the Interstate Commerce Commission ordered each of a number of railroads to install trial or experimental train-stop systems on one complete division as a means of preventing collisions. Some of these systems include a speed-control feature by which the brakes are applied when the speed of the train exceeds a prescribed limit or may be released when the speed has been reduced below that limit.

These train stops are of four general types: (1) mechanical trip—when the signal is at "stop," a trip on the track is raised to engage a lever on the engine or electric car and thus actuate the train brakes; (2) electrical contact—a ramp or inclined rail engages a contact arm on the engine, which, unless energized by the signal indicating "proceed," causes the brakes to be applied; (3) magnetic inductive—magnets in the track and on the engine are so arranged that when the signal is at "stop," an induced current between the magnets will actuate the control of brake-setting apparatus; (4) continuous inductive—this involves a continuous flow on the engine or car of a current induced from the current in the track rails; this current keeps the brakes off when the track is clear or causes their application when necessary; it also causes small signal lights in the cab to give indications continuously (instead of intermittently) as to conditions ahead. With such a system perfected, it would be practicable to dispense with the roadside signals, except where trains must stop on account of conditions or requirements other than that of the track being occupied. Train stops and speed-control devices are desirable mainly for lines of heavy traffic which are equipped with block signals, but in general the money for such equipment may be spent to better advantage in extending the use of block signals and interlocking plants.

Miscellaneous Signals.—Train-order signals are usually installed where there are no other signals by which trains may be held for the delivery of orders. Where a continuous system of automatic block signals is not in use and local protection as given by such signals is desired for switches, sidings, or station and terminal tracks, isolated automatic signals are generally installed, even in territory protected by manual block signals. Automatic signals so used are controlled by a track circuit and, if regular block protection is desired, the signals are arranged to give "stop" indications when track is occupied within the limits governed or when a switch is reversed. If no more protection is required than that given by a distant signal, the signals are arranged to give the indication "proceed with caution." With signals so used, a sign is placed to indicate the point at which protection provided by the signal ceases.

At points where there are no signalmen through whom trains may be instructed to take siding, several roads have installed indicators by which the instruction "take siding" may be displayed, thus facilitating train movements. Telephones are now being generally provided at main-line switches, and trainmen are required to get permission before entering upon, or fouling, any main line other than the one they occupy.

Interlocking.—This term is applied to the interlocking of the mechanical or electrical connections of switches and signals at terminals, crossovers, junctions, and grade crossings, in order to facilitate traffic and prevent clear signals being displayed for conflicting routes. Thus, for a single-track Y-junction at "B" at the end of a section "A–B," where two lines "B–C" and "B–D" diverge, a train from "C" cannot be signaled to proceed past "B" onto the joint line until the signals have been set to stop trains from "D" and "A" and the switches have been set for route "C–B–A." Both home and approach or distant signals are used, and the latter cannot be set to indicate "proceed" until the home signal is at "proceed."

It is desirable to arrange the apparatus so that after the train has passed the approach signal the signalman cannot change any switch of the route until the train has passed beyond the track circuit in which the switch is located, or until a predetermined time has elapsed. For this latter plan, a time release is installed. The time interval would allow the approaching train to come to a stop after the signal is placed at "stop;" or if the speed is such that the train cannot stop, the time interval would allow sufficient time for the train to pass over the switch. An interlocking home signal cannot be set at "proceed" until the switches for the required route are set and locked, and when this signal is set at "proceed," it locks all opposing and conflicting signals and releases its distant signal, so that the latter also may be set at the "proceed" position. In automatic block-signal territory the interlocking signals are usually made semi-automatic, so that they may be cleared under control of the signalman but will be restored automatically to "stop" as the train passes. In this way, continuous automatic block-signal protection is given, as well as interlocking protection.

An interlocking installation is, in effect, a block system for the section of tracks within the "stop" signal limits of the interlocking, and eliminates the delay and expense of stopping trains at crossings for safety and at junctions for orders to proceed. Such stops are estimated to cost an average of \$3 to \$5 each in fuel, wages, supplies, and wear and tear, depending upon such conditions as the weight of train, the grade of track, and the speed from which the stop is made. Should a train be separated by pulling out a drawbar in getting started again, the cost might reach \$50 to \$100 or more.

At yard entrances, an interlocking plant may save several minutes for each train. At drawbridges, the signals on the approaches are interlocked with the bridge mechanism. Thus, when the bridge is to be opened, the operator must set the signals at "stop" before he can release the bridge locks and start the machinery. When the bridge is again closed, the rail lifts and end locks must be returned to normal position before the signals can be changed to indicate "proceed."

With electric operation of switches and signals, these may be located and controlled at almost any distance from the signal station or operating point. For mechanical operation, a distance of about 2,000 ft. is the limit with wire connections, while with pipe connections it is about 1,000 ft. for signals and 800 ft. for switches, the distance being limited by the friction of the pipe on its carriers. On account of irregular working of wire connections, good practice usually calls for the use of pipe connections, or power operation. Pipe lines are usually of 1-in. pipe supported on roller carriers at intervals of about 7 ft.; wires are supported on small pulleys attached to stakes, and may be carried under streets and tracks in $\frac{1}{2}$-in. pipes filled with crude oil. Expansion and contraction of pipe lines is taken up by compensators of the double bell-crank pattern, one being sufficient for about 750 ft. and two for longer lines. Switches which are mechanically connected are locked in each position by a bolt or plunger. To prevent this lock from being released while a train is passing the switch, the lock may be mechanically connected to a detector bar on the lever operating the lock, or it may be electrically locked so that the lever is held and cannot be operated as long as a wheel of the train is on the circuit or detector bar. Derails at junctions and crossings to check trains which may pass a home "stop" signal are usually placed 300 to 500 ft. from the crossing, but the signal section of the American Railway Association is opposed to the use of derails on main lines.

Block Signals on Electric Railways.—Interurban railways are operated usually on the train-order system, the crews reporting to the dispatcher by telephone at stations and passing sidings. Connections on the poles for use with portable apparatus, or fixed telephones in cabins at sidings and crossovers, are sometimes provided for communication with the dispatcher at points between stations. Block signals, which are not used extensively, are of the automatic type, following steam railway practice in general, but the equipment is less elaborate and of lower cost to meet financial conditions, while the blocks are considerably shorter, usually $\frac{1}{2}$ to 1 mi. On account of electric propulsion, alternating-current track circuits are used. For single track, the block sections extend usually from one passing siding to the next, and in many cases have intermediate signals, to give permissive blocking for following trains and absolute blocking for opposing trains, as described already. In a typical system using three-position signals, each block includes a home signal at each end of the block, an intermediate signal for each direction near the middle of the block, and a preliminary track circuit to warn an opposing train just before entering the block.

With semaphore signals, the posts are about 20 ft. high, and where the overhead-wire conductor system is used, the blades are placed usually on the left-hand side of the post, as the trolley poles would otherwise interfere with the motorman's view of the signals. But a distinctive feature of electric railway signaling is the general tendency toward the use of color-light signals for both day and night indications. This system, which has advantages in simplicity and in low cost of installation and maintenance, is particularly adaptable to electric railways or electrified sections of steam railways on account of the electric-current supply being available for the entire length of the line. An objection as to the possibility of lights being out, with consequent absence of any signal, is provided against by the use of auxiliary lights or markers to denote the location of signals, together with the rule that an unlighted signal is to be regarded as a "stop" signal. Switch indicators may be installed to repeat the signal indications, and are sometimes equipped with a time lock which holds the "stop" indication for a definite period and then releases only if the block is clear. This arrangement is to provide against a train moving out of a siding simultaneously with a train entering the other end of the block section.

Simple block signals are often desirable on light-traffic, single-track, suburban lines and outlying extensions of street-car lines. A typical system divides the line into sections extending between the entering ends of adjacent passing sidings. At each siding, there is mounted on a post a box having two disks or targets and two lamps with colored lenses. An automatic switch on the trolley wire, 100 ft. in advance of the signal (that is, between the approaching car and the signal) is operated by the trolley wheel of the entering car, causing a green disk and light to be displayed in front of the car (if the block section is unoccupied), with a red disk and light at the other end of the section to stop an opposing car. The green signal remains while the car is in the block, warning a following car to proceed with caution. With three lights, green may indicate that the block is clear, yellow that it is occupied by a receding car, and red that it is occupied by an approaching or opposing car.

With the electric contacts placed on the trolley wires on the siding and the signal placed at the switch, the motorman leaving the siding sees the signal indication displayed before him, and can then stop his car before passing the switch if he gets a "stop" signal. To avoid cars meeting on a short stretch of single track, a simple automatic signal for absolute blocking may be installed. As a car enters the clear block, it sets a red signal behind it and at the opposite end of the block. At the same time a green pilot light displayed ahead of the car indicates to the motorman that the signals have been displayed, the corresponding pilot light at the opposite end being red.

CHAPTER XXV

RAILWAY EMERGENCY WORK

Emergencies.—In addition to the regular work of maintenance and improvement on the line, provision must be made for dealing with emergencies, which may be classed generally as follows: (1) train accidents, such as derailments and collisions; (2) slides and washouts, interrupting or endangering traffic; (3) snow storms threatening to block the road. Equipment must be held ready and general methods arranged for clearing tracks, closing gaps, reopening communication with the least possible delay, and rendering prompt and effective aid to persons injured or detained. In any case of accident, the nearest section foreman takes all his men to the scene, whether it is on his own or an adjacent section. With all such emergencies, the track and bridge and building forces are closely concerned, as they have to protect, clear, repair, and rebuild the line and structure, and to assist other forces in various ways. No set rules of working can be given, as the operations in each case will depend upon the character of accident, structure, location, traffic, and the equipment and material available, as well as upon the ideas and ability of the man in charge. The division engineer has usually a list of the equipment for such work, with its location, so that with the cooperation of the superintendent of bridges and buildings the necessary force, machinery and supplies can be forwarded promptly.

WASHOUTS AND BURNOUTS

Precautionary Measures.—In times of heavy rains and floods, precautions should be taken by keeping waterways clear of snow and drift, making frequent inspections, and putting on extra watchmen and trackwalkers to look out for any indications of the shifting of bridges, undermining of foundations, slipping of ground in cuts, and washouts at fills. As already explained, it is difficult to determine the safety of trestles during floods. Roadmasters and section foremen should keep the superintendent informed as to the conditions, thus insuring prompt attention to points of danger and avoiding the delay of running trains under slow orders where the road is safe. After a flood has subsided, very careful examination should be made of piers, abutments, and trestle bents to detect signs of scouring, shifting, or settlement. During protracted droughts, the fire barrels and buckets at wooden structures should be kept filled, and vegetation cleared away from around piles and bents, while bridge floors should be patrolled after the passage of trains.

On the discovery of a washout, slide, burned or damaged structure, or dangerous conditions, the foreman should send out flagmen to stop trains and then report at once to the roadmaster and the superintendent, giving the location, the number and name of structure, and the character and extent of damage, with particulars of train accident (if any). This information will enable the track and bridge forces to determine what equipment and material is needed; the wrecking train or pile-driver train and bridge gang can then be sent out promptly. Meanwhile, the foreman does what he can to check further damage and to prepare for the repair work. Methods of handling slides have been noted under "Drainage."

Drift and Ice Jams.—Damage to falsework and trestles by drift and logs may be prevented by booms extending from each shore to the main span or opening, men with poles being stationed, if necessary, to guide the floating material through the opening. Ice jams or gorges may be broken up by explosives, which should be done before they begin to move and exert dangerous pressures. One method is to put about 100 lb. of blasting powder in a 4-gal. can on a long rope, and to drop this through a hole, allowing it to drift under the ice, the rope being attached to a stake. The charge should be exploded by an electric blasting balley and not by a time fuse. If dynamite is used, a few sticks or cartridges may be tied together and one of them fitted with an electric blasting cap. The bundles of dynamite are dropped through holes about 30 ft. apart and the wires led to an electric blasting machine. If such a machine is not available, a fuse about 3 ft. long may be used, giving a little more than a minute before the explosion. Heavy charges must not be used near piers or other structures. Work of this kind should be handled only by careful and competent men.

Washouts.—Where a trestle has been carried away or a fill breached by flood, the widening of the breach may be checked by covering the slopes with stone, logs, brush, or trees interlaced to form a protective mat, while rough cribbing may be placed to cut off destructive currents and eddies. With a swift current flowing, it is generally useless to try to fill a gap with anything but loose rock, or such bulky and heavy articles as old cars. If the water is too high or too swift to allow of commencing repairs, the time may be spent in collecting material, building trestle bents or cribs, and filling sandbags. When the current flows parallel with the railroad and threatens to erode the fill or slope, piles may be driven along the toe or cribs lowered in place. Local conditions will govern the procedure, as a method proving successful in one case might tend to increase the erosion in others. If the track level at a washed fill is not more than 10 ft. above the ground left dry after the flood, inclines may be built at each end to a temporary track on the ground surface. This may take less time than the construction of a trestle.

Roadmasters and superintendents of bridges and buildings should have blueprint bills of material for trestling of different heights and panel lengths, as these will assist in ordering and preparing material for rapid reconstruction. Bridge timbers and lumber are usually on hand at division headquarters. Men in charge of such emergency work need judgment and self-reliance as well as skill, as they may often be thrown on their own resources. Great care should be taken to insure strength and safety of all temporary work, providing ample bracing to distribute pressures and thus relieve any unevenly supported bents. Trains should not be allowed on damaged bridges, especially truss bridges, until after inspection by competent authority, and trestles standing in floods should be regarded with suspicion, as the bents may be undermined by scour. At washouts, the liability of damage by heavy floating objects and by a second flood must be considered.

Temporary Trestles.—A track piledriver is usually available for building trestles across washouts, but if such equipment is not available, a hand derrick or gin pole may be used to handle piles, timbers, or framed bents. Men with rafts or boats guide the piles into the best position that can be found, and secure them with a horizontal plank at the water line; the top gang then puts on a brace or ledger board below the level of the caps. For 16-ft. panels, four- or six-pile bents may be used, with 32-ft. stringers or round timbers having the rear ends anchored or lashed to the caps; some of these stringers extend far enough for men to carry out posts or piles for the next bent. Sighting over the ledger boards gives the level for cutting off piles for the caps, which are secured by driftbolts or spiked battens, after which the ledger boards are removed and diagonal sway braces put on the bents.

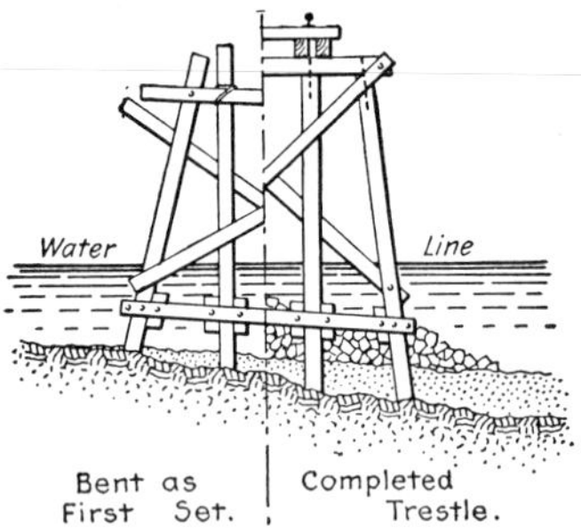

FIG. 105.—Temporary trestle bent.

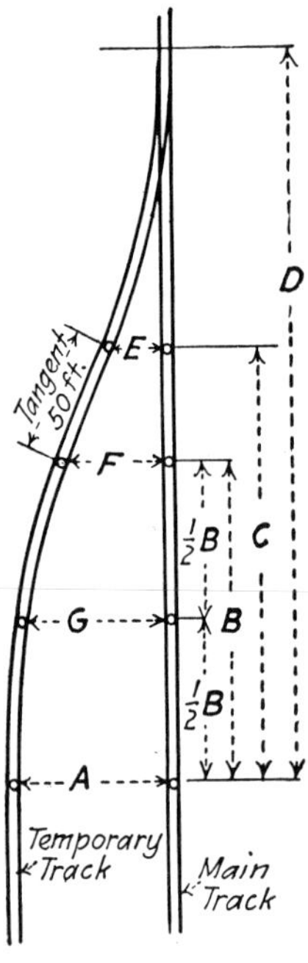

FIG. 106.—Temporary track; Southern Pacific Ry.

Longitudinal sash braces or girts are spiked or bolted along the outer piles, and diagonal bracing may be placed in the end panels.

If there is hard bottom for framed bents, soundings may be made for the legs of each bent; the posts are then cut to length, a cap is driftbolted to them, sway bracing is spiked on and also a horizontal plank or sill at about the water line. These bents may be floated out and set by a derrick car or gin pole, the lower end being weighted by old rails or splice bars spiked to the posts or sills, or by pig iron, stone, or scrap placed in a box formed by planking the sides of the bent at the bottom. Steel rods are sometimes used for diagonal bracing in high bents. Loose rock may be dumped into place to form a seat for bents with mudsills or to protect the piles and posts from scour. In the framed bent construction shown in Fig. 105, a double top brace is bolted to the batter posts and lashed to the plumb posts. The sill also is double, with spacing blocks to hold the piles in line, so that when the bent is placed, the posts can be driven down to a full bearing and then sawed off for a driftbolted cap. Where it is not desirable to use bolts, the timbers may be fastened together by dog irons or by plank battens spiked across the joint. The dog irons are rods about 12 in. long, with the ends bent inward to form 4-in. legs with chisel edges. Caps will be about 10 ft. long for 4-post bents with batter posts, or 14 to 16 ft. if all posts are vertical. Stringers, built up of two or three timbers breaking joint, may be held in place by triangular blocks spiked to the caps and stringers.

Trestle Construction.—To replace a washed-out fill 16 to 24 ft. high, a frame trestle 400 ft. long was built in $2\frac{1}{2}$ days by 37 men, using only carpenter tools and raising all bents by hand. To replace a steel viaduct wrecked by a derailed train, a frame trestle 210 ft. long, having a maximum height of 85 ft., was built in 3 days by 75 men, a wrecking car assisting to clear the site. An exceptional case is that of a pile trestle built in 20 ft. of water while the water was 15 in. above rail level. A pile driver put down the piles to a uniform level $4\frac{1}{2}$ ft. under water by means of a follower, and the cap was fastened with driftbolts driven by a follower in a 2-in. pipe. The track, which hung suspended in the gap, was raised on the bent by jacks and blocking to enable stringers to be placed under the ties, the stringers and ties being then secured by driftbolts driven as noted above. The pile driver then moved ahead. To enable the machine to reach this washout, track 12 to 15 in. under water was tested, lined, and surfaced for several miles.

Cribbing and Sheet Piling.—Cribs of ties, logs, or bridge timber are sometimes used instead of bents, being leveled at top to carry cap timbers supporting the track stringers. A single crib about 8 by 8 ft. will suffice for a height of 6 to 8 ft., while for greater heights there may be two cribs side by side or one larger crib. At wide openings, cribs with one end pointed to facilitate handling in the current may be towed or floated out and sunk on the natural bottom or on piles of dumped rock to form foundations for crib or frame piers. Cribs and piles may be placed along a fill narrowed by side wash, caps being laid with one end resting on the crib or pile and the other end on the standing part of the fill. Cofferdams for repairs to piers or abutments may be built with sheet piling driven by hand or a light (700-lb.) drop hammer in 12-ft. leads suspended from a derrick boom.

River-bank Protection.—In silt-bearing streams, bank erosion has been stopped in many cases by permeable jetties or retards 100 to 200 ft. long formed of trees with their butts anchored to piles and on shore so as to form a mass extending from the river bed to the water surface. The current being thus checked, the silt is deposited and builds up a protective beach or foreshore. For immediate protection when the Missouri River threatened the tracks of the Chicago & Alton R. R., trees were first placed along the bank, but as the current cut under them, old cars were set in the water by a wrecking crane and then placed about 20 ft. from the bank by a snagboat. Willow brush was piled between the cars and the bank and loaded with sandbags and riprap stone. Cribbing of old ties was also built up and filled with stone.

For permanent protection against erosion and undercutting of the river bank, the Wabash R. R. has made extensive use of mattress work. After dressing the bank to a slope of 1 on 2 or 1 on 3 by hydraulic grading with water jets, sheets of galvanized wire netting of 6-in. mesh are laid and sewed together. Two layers of willow-brush bundles are laid and secured by tie wires, after which the mattress is loaded with riprap stone of 20- to 125-lb. weight. Cables passing over and under the mattress are anchored to deadmen about 50 ft. from the top of the slope.

Shoofly Tracks.—For cases where it is desirable to build a temporary running track or "shoofly" around a wreck or washout, the Southern Pacific Ry. has a standard plan, as shown in Fig. 106, with dimensions for different curves and track spacing, as given in Table XXI.

Piledrivers and Derrick Cars.—Railway piledrivers, derrick cars, or locomotive cranes fitted with piledriver leads are part of the regular equipment on most divisions and are available for construction, improvement, and emergency work. The piledriver (Fig. 107) consists of a heavily built flat car, with a turntable carrying a revolving frame with leads at the outer end. The turntable may be at one end of the car, with engine and boiler at the opposite end; or it may be near the middle of the car, with the revolving frame extended beyond it to carry the engine and boiler as a counterweight. In addition to its rotation, the frame has a travel of 16 to 20 ft. over the turntable, to allow of setting piles 20 ft. ahead of the car or 10 to 25 ft. from either side of the center of the track. The machine should be able to work in a through bridge, so as to drive piles for falsework in renewal or repairs (see "Bridge Work"). The leads,

FIG. 107.—Railway pile driver.

TABLE XXI.—LAYING OUT TEMPORARY TRACKS—SOUTHERN PACIFIC RY.

10-degree Curves

A	B	C	D	E	F	G
10	53.6	103.3	156.9	2.5	7.5	9.4
20	84.0	133.5	217.5	6.3	13.7	18.5
30	107.3	156.5	263.8	10.3	19.7	27.5
40	127.3	176.0	303.3	14.4	25.6	36.5
50	144.8	193.1	337.9	18.7	31.3	45.4
60	160.3	208.3	368.6	23.0	37.0	54.4
70	174.5	222.1	396.6	27.4	42.6	63.4
80	187.8	235.0	422.8	31.8	48.2	72.3
90	200.2	247.0	447.2	36.2	53.8	81.2
100	211.9	258.3	470.2	40.7	59.3	90.2

15-degree Curves

A	B	C	D	E	F	G
10	42.2	92.0	134.2	2.3	7.7	9.5
20	66.2	115.4	181.6	5.7	14.3	18.6
30	85.0	133.7	218.7	9.5	20.5	27.7
40	100.8	149.2	250.0	13.5	26.5	36.7
50	114.7	162.6	277.3	17.5	32.5	45.7
60	127.2	174.4	301.6	21.5	38.3	54.7
70	138.7	185.6	324.3	26.0	44.0	63.7
80	149.4	195.6	345.0	30.3	49.7	72.7
90	159.2	204.8	364.0	34.6	55.4	81.7
100	168.5	213.5	382.0	39.0	61.0	90.7

40 to 50 ft. long, may be arranged to drive batter piles and are made to fold up or roll back so as to be within clearance limits for transportation. A steam pile hammer, a 3,000-lb. drop hammer, and a pump for jetting piles are usually provided, and there may be a boom for placing caps and stringers. The machine is usually self-propelling, at 5 to 10 m. p. h., so as not to require the service of a locomotive while at work, but this gear is disconnected during transportation in a work train or freight train.

The derrick car adapted for piledriving, shown in Fig. 108, has a short, heavy A-frame, but in many heavier designs a tall A-frame is pivoted so as to swing back to an inclined position for transportation. Such machines have booms 40 to 65 ft. long, handling loads of 25 to 45 tons on a reach of 30 to 35 ft., besides carrying the 25-ft. suspended leads about 20 to 25 ft. ahead of the car.

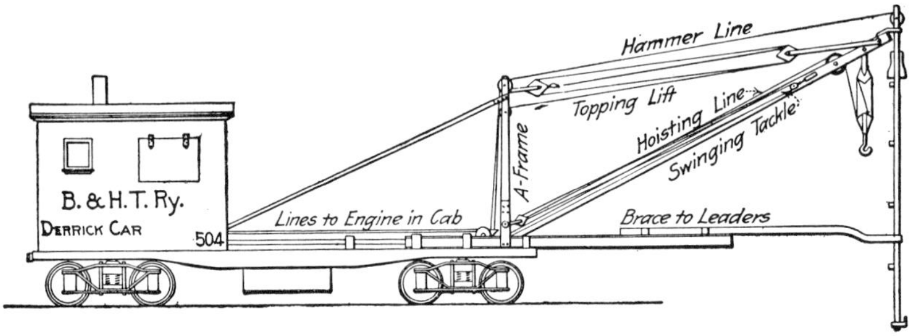

FIG. 108.—Railway derrick car with pile-driver.

Wrecking Train.—For clearing the line after accidents, each division has at headquarters a wrecking train held ready for immediate use, stationed on a yard track which is reserved exclusively for this train and is usually connected at both ends. No engine or car is allowed to be set on this track. The wrecking or relief train consists of a steam crane or derrick car of 60 to 175 tons capacity, a box car for tools, another box car for blocks and tackle and ropes, a flat car with blocking and track material, and a flat car with wrecking trucks or spare car trucks and a water tank to supply the boiler on the crane. There may be also a bunk car and a caboose, but usually the caboose is attached for the train crew assigned when the train is sent out.

Portable telephone or telegraph apparatus is carried in the tool car, for connection to the line wires at any point, and there may be an electric or acetylene-gas lighting plant, with portable lamp equipment, to give better lighting than the usual oil-flame torches. As a rule, the wrecking crane has headlights on the boom and the rear end of the cab, and sometimes carries additional lights. The train has a complete outfit of tools, appliances, and supplies, including material for engine and car repairs, and usually one man is responsible for the care of this outfit, of which a complete list is posted. On the return of the train to headquarters, all equipment is thoroughly inspected, cleaned, and made ready for use, and any necessary replacements are made at once. A report of the condition of the crane or derrick is made semi-monthly by the engineman and any repairs necessary are ordered promptly by the master mechanic.

The wrecking crew consists of a foreman or a wreckmaster with 8 to 10 car-repair men, and sometimes machinists to repair or dismantle a damaged locomotive. These men are assigned to the wrecking train and each has his special position or duty, but ordinarily they work in the car-repair yard or the shop. If an emergency arises during working hours, the dispatcher notifies the car foreman and he in turn notifies the crew. At night, the dispatcher notifies the telephone operator, who then calls the men, having their numbers and addresses on file. For a serious accident, two trains may be sent so as to work from opposite ends and thus clear the track as quickly as possible. When the wrecking train is ordered out, the first available engine, engine crew, and train crew are assigned to it, and it is sent out as a special train. At the work, the foreman is in charge of operations, but may consult with the roadmaster or the superintendent when they arrive.

In many cases, the wrecking crane is employed in the yard for handling and transferring loads, or it may be utilized on construction and improvement work or bridge renewal within easy reach of the terminal, the operating department being usually willing to assign it for active duty rather than have it laid up all the time. On the Chicago track-elevation work of the Chicago & Northwestern Ry., the wrecking crane is used regularly in handling girders and setting the 10- to 60-ton precast concrete floor slabs. This equipment, with its locomotive and crew, is subject to immediate recall by the operating department. On the St. Louis-San Francisco Ry., the crane is kept under about 20 lb. of steam during the winter, while in summer it has the fire laid so that steam pressure can be raised in about 30 min.

Pennsylvania R. R. Wrecking Trains.—Wrecking-train equipment on the Pennsylvania R. R. system includes wrecking cranes of 50 to 150 tons capacity, to suit traffic requirements, the most powerful machines being used on the main lines where the heaviest locomotives are in service. For the electrified New York terminal there is a special type of double-end electric crane of 50 tons capacity, on account of the tube tunnels and the limited clearance. A wreck train consists ordinarily of a steam crane; a car for spare trucks and the heavy lifting attachments required with the crane; a tool car for blocking, small lift attachments, jacks, gas-torch cutting apparatus, and other supplies; and a coach or bunk car for the accommodation of the wrecking crew. The tool cars and bunk cars are usually wooden cars retired from passenger service.

The wrecking cranes are used for general purposes, in which case arrangements are made by the division superintendent to keep in touch with the crew so that the outfit may be transferred promptly in the event of an emergency. For lighting at wrecks, electric flood lights on the crane are operated by a 500-watt turbo-generator and are so arranged as to give full illumination around the machine. For field lights on work in advance of the crane, oil torch lights operated by air pressure are used, but in view of the possible danger when using such lights around inflammable material, as at oil-tank cars, the adaptability of portable battery-operated electric lights is being investigated. For hand lights in such situations, dry-battery flash lights are used. Some of the wrecking trains are equipped with portable telephone apparatus for communication with the dispatcher or with headquarters.

The wreck-train crew is composed of the wreckmaster, the crane engineman, the fireman, two men in charge of the tools and commissary, and 35 shop and car-repair men who are called from their work or their homes when needed. The motive-power department keeps the train in condition, but its movements are under the direction of the operating department, while the wreckmaster is in charge of the work subject to the direction of the division superintendent, the trainmaster, or the track supervisor.

Wrecking Cranes.—A typical 100-ton wrecking crane of about 85 tons weight is shown in Fig. 109 with its boom lowered for transportation; in this position the boom extends over one of the flat cars of the train. On

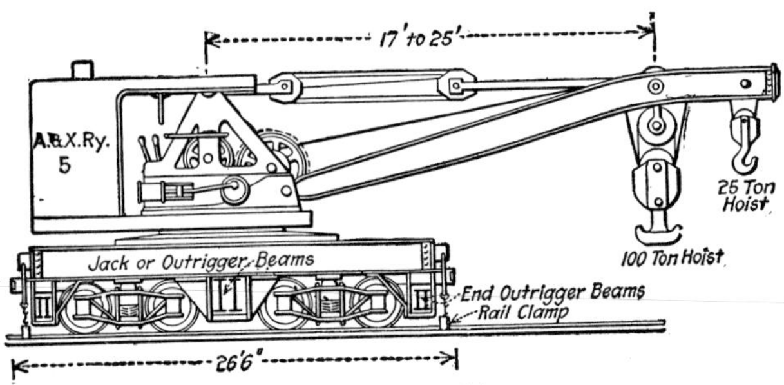

FIG. 109.—Railway wrecking crane.

the work, it is self-propelling, with a speed of 3 to 5 m. p. h., but the gear is disconnected for handling the machine over the road. The car or under-frame is of heavy steel construction and carries the turntable on which the revolving crane is mounted. At each end are clamps to anchor the frame to the rails and also outrigger beams on either side, supported by jacks or blocking to give an extended base of support. Heavier outriggers at the middle provide for blocking up the bedplate of the crane when making a heavy side lift. The main hoist can handle 100 tons on a radius of about 20 ft. and 80 tons at 25 ft., while the auxiliary hoist can handle 25 tons at a radius of 27 to 33 ft. Similar machines have 100-, 40-, and 15- to 20-ton hoists. Separate gears and brakes enable the loads to be raised or lowered while the boom is also raised or lowered to give the desired reach; spools outside of the engine frame provide for handling side lines. For lifting cars, a swing beam or spreader may be hung from the crane hook, having hooks at each end for the attachment of chains or slings.

Wrecking Operations.—While it is of importance to remove the wreckage and clear the track as quickly as possible, this should not lead to unnecessary rough handling or destruction of cars or freight. Note should be made of the type, name or initial, number, carding, and kind of freight of all cars damaged in the wreck or broken up and burned as wreckage. On double track, attention may be centered on clearing one track first. If cars of explosives are included, broken packages should be removed carefully and any spilled powder saturated with water. Where oil or gasoline is spilled from tank cars, it should be smothered with dry earth or cinders before work is done around it and before a locomotive or steam derrick is allowed to approach closely. Torches and open lights are dangerous in such cases.

In any case of accident, the nearest section foreman takes all his force to the scene, whether it is on his own or an adjacent section. He must assign men to watch property, prevent theft, and keep people back from dangerous places; he must also look out for broken rails, switches, wheels, or other parts that may have caused the accident, and he must be ready to repair or relay track promptly as the wreckage is cleared. In addition, the sectionmen help in the general work. A small supply of track material is carried on the wrecking train, while rails and ties are usually on hand at the roadmaster's headquarters. The Southern Pacific Ry. requires that at such points there shall be at least 1,000 ft. of rails available for temporary tracks and so placed as to be readily loaded on cars. Where the job is a long one and when weather is bad, prompt arrangement should be made to supply food and coffee; the bunk cars may have the necessary equipment and supplies, with one man assigned as cook.

For dealing with personal injury, the wreck train may carry first-aid supplies and stretchers, and the men may be instructed as to the temporary care of injured persons until medical or surgical aid can be obtained. In serious accidents, a special train is sent out as soon as possible with doctors, nurses, and supplies. For the care and transportation of the injured, an ordinary car is better than a parlor or sleeping car owing to the winding entrances of the latter. Stretchers may be placed across the reversed seat backs, giving about the right height for convenient medical attention. A few railways have hospital cars.

HANDLING AND CLEARING SNOW

Snow and ice are factors in maintenance-of-way work on a large proportion of the railway system and call for more or less serious attention as track problems. Roadmasters must see that their men have proper equipment for handling snow, and will assign men to accompany the snowplows and flanger cars, while during storms they must keep in communication with the division engineer and the train dispatcher. Section foremen attend to the removal of snow and ice from switches, frogs, interlocking and signal rods, stations, track scales, overhead bridges, and the entrances to tunnels and rock cuts. At water stations having no permanent attendants, the foreman may place day and night watchmen to see that ice formed on the track due to leakage, overflow, and spilling is kept cut away below the top of rails and clear of the flangeway space.

Difficulties in dealing with deep snow occur annually on many northern railways, which prepare for this situation by special organization and equipment. Where such difficulties occur only occasionally, there is often trouble due to makeshift methods and lack of preparation. Snow fences may be of little assistance in protracted storms, since changing winds vary the direction of the drifts, while successive portable fences erected on the drifts may soon be buried. A few roads, including the Soo Line, have planted trees as permanent snow fences and windbreaks. In deep cuts, there will be difficulty in getting rid of the snow, but this may be reduced by widening the cuts and flattening the slopes, or opening up through cuts into sidehill cuts. Drifts and slides containing earth and sand which have partly thawed and frozen repeatedly are very heavy and may cause derailment if "bucked" at high speed with a snowplow. This is specially the case in sidehill work, where the pressure is mainly on one side of the plow. Freshly fallen snow weighs about 12 to 15 lb. per cubic yard, while snow packed by weight or wind may weigh as much as 30 to 40 lb.

Snow 15 to 30 ft. deep is a serious proposition in railway operation, and if a train is once stopped, it may soon be stalled by the snow drifting across the tracks and blocking the wheels. In a great storm in Wyoming in January, 1917, about 45 through passenger trains were stalled in the snow, many of them at a distance from towns, and a 50-mi. stretch was buried for nearly 2 weeks, the snow filling the cuts within an hour after the rotary plows had got through. Snow is a trouble also in the northeastern states. With snow filled or packed along the rail heads, the engine wheels are likely to slip and the train pulls heavily, while derailment may be caused by the wheels riding up on the snow unless the rails are quickly cleared by flangers.

Snow Fences and Hedges.—Different types of fencing to protect the track from drifting snow have been noted in Chap. VII. On parts of the western prairie lines of the Canadian Pacific Ry., willow and poplar trees are planted to form snow fences. To check both snow and drifting sand on the eastern lines, particularly to prevent snow from drifting into cuts, this road plants pine, cedar, spruce, and balsam in two rows 6 ft. apart with the trees staggered and 3 ft. apart in each row. This plan is practicable only with the 200-ft. right-of-way, as the trees should not be less than 75 ft. from the track. On the Pennsylvania R. R., Norway spruce and Douglas fir have been found satisfactory, but to hold sand the fence must be closer or tighter near the ground than if it is to hold snow. This road has experimented also with hedges of various kinds, but barberry and osage orange are undesirable on account of being infested by scale. Different forms of privets have been tried, with varying success, while shifting sand has been held by planting cottonwood and willow. A study of drifting snow in relation to fences is given in *Engineering News-Record* of Nov. 5, 1925.

Snowsheds.—Protection against deep snow may be afforded by covering stretches of the line with snowsheds, which are used to some extent in open country but more generally on sidehill lines in mountain districts where deep drifts and great snowslides add to the trouble from heavy falls of snow. In the former case, the structures are rectangular, barn-like sheds, with sides strongly braced. On mountain lines the roof of the shed is usually sloped about 1 on 5 and the space on the uphill side is backfilled or cribbed to give a continuous slope from the top of the cut to the outer edge of the roof. In this way, slides containing rocks and trees will pass over the shed, with its deep cushion of snow on the roof. On hillsides where snowslides occur, split and glance fences are sometimes placed above the railroad to protect gaps or smoke openings in the roofs of snowsheds by breaking the slide and guiding the snow into ravines. The split fences are V-shaped, with a strongly braced and anchored crib at the point.

For the snowsheds, timber frame construction is generally used, with bents 6 to 8 ft. apart; plank sheathing is placed on the posts, while joists on the caps support the roof sheathing. In modern work, concrete footings are used and the feet of the posts are painted with preservative to prevent decay at the ground line. In a 30-mi. stretch of line on the Southern Pacific Ry., there are 25 mi. of snowshed. To divide such long stretches in case of fire, this road introduces portable or telescopic sections about 100 ft. long, at intervals of 5,000 ft. The movable portion is of standard dimensions, 16 ft. wide and 22 ft. high inside, mounted on wheels and rails, so that a locomotive with cable can haul it back into an enlarged stretch of the stationary structure, which is made 20 by $25\frac{1}{2}$ ft. to receive it. Double-track sheds with tracks 17 ft. c. to c. have a middle line of posts, but where the spacing is 13 ft., a single-span trussed roof is used.

These flat-roof sheds are designed for a load of 300 lb. per square foot; the snow is sometimes 20 to 25 ft. deep, but as a rule it is shoveled off when it gets about 10 ft. deep. Telephone boxes are placed about a mile apart, and fire trains are held on sidings, each consisting of a locomotive with pump and hose and some tank cars. Observation towers are maintained also, to note any fire along the sheds. Some large snowsheds on the Canadian Pacific Ry. have pipe lines with hydrants and coils of hose 400 ft. apart. In Fig. 110 are shown some different forms of snowsheds.

Concrete is being used extensively, and a design introduced on the Great Northern Ry., as shown in Fig. 110, has the footings and back wall of concrete, with timber posts, roof, and sheathing. Snow-sheds of precast concrete units used by the Union Pacific Ry. in open country

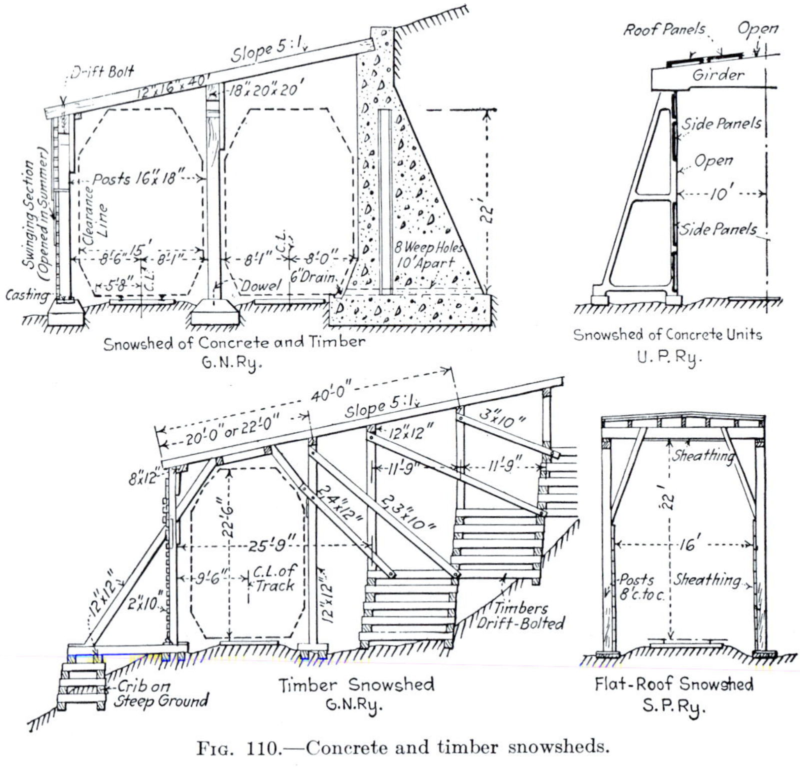

FIG. 110.—Concrete and timber snowsheds.

have bents composed of two A-frames (having the inner legs vertical) and a roof girder; these bents are 15 ft. apart, the spaces between them being closed by precast slabs. The units are erected by locomotive cranes. The inside width is 34 ft. for double track; for three tracks this is increased to 48 ft., the roof being then of steel-plate girders cased in concrete. The vertical clearance is 24 ft.

Fighting Snow.—Light snowfall and shallow drifts may be handled with pilot plows and flangers; but breaking and rotary plows and armies of shovelers may be necessary to clear deep drifts and heavy falls which threaten to blockade or which have blockaded the line. In heavy storms, the main thing is to keep on breaking and opening up the drifts with plows, their work being supplemented by wing plows, spreaders, and flangers. A deep cut with vertical sides, as left by the plow, is likely to fill quickly; it is, therefore, wise to break down the sides, throwing the loose snow into the cut, and then running a wing plow at high speed to throw the snow to a distance. In some cases, the snow cut is sloped back 30 or 50 ft.

In spite of snowsheds and snowplows, a great amount of hand shoveling is required, for which large gangs are employed. This may be to widen and clear the cuts, to clear turnouts and sidings, or to cut trenches across deep snow and thus loosen it in advance of the plows. If the trains are stalled during a storm, the wheels will soon be buried so that the engines cannot move the trains until they have been dug out by shoveling. In widening cuts by successive casting, the benches must be of such height that the men can throw up the snow readily. The transverse trenches may be about 10 ft. wide and 30 ft. apart, or smaller cuts 15 ft. apart with two men to each trench. In this trenching and in widening narrow cuts, watchmen should be stationed to give warning of the approach of trains and plows. On long cuts, it may be necessary to haul the snow away in cars and dump it over trestles. As shoveling snow is laborious work, especially with severe cold and high winds, the officers in charge must see that the men are physically fit for the work and are warmly clad, and must provide for supplying food and plenty of hot coffee.

Snowplows.—Pilot plows, composed of curved steel plates fitted to the engine pilots or attached in place of the pilot, can sometimes handle snow 2 to 3 ft. deep. Larger engine plows are of steel plate on frames 8 to 10 ft. high which replace the pilot and are braced to the engine frame and smokebox. These plows have been used effectively in drifts 15 ft. deep that had been loosened by cross-trenches. Similar plows may be attached to gondola cars or spreader cars.

The ordinary breaking or "bucking" plow resembles a heavily framed box car with wedge-shaped front end, the plow being driven by 2 to 6 or even as many as 12 locomotives. After each blow, when the plow has been driven until it is stalled, it is hauled back for another drive, but it may have to be dug clear of snow before it can be moved. In striking a heavy drift, the shock may derail the plow or engines; to reduce this danger some plows are made with a longitudinal sill or beam of heavy timbers having the coupler attached to the rear end while the front end is framed into the timbers of the plow so that the push from the engines is applied directly at the head of the plow. This timber is not a fixed part of the underframe, but its rear end has a slight lateral play to allow for movement in passing curves.

On both engine plows and breaking plows the plow itself is in the form of two wedges. A horizontal wedge, with its nose a little above the rails, drives into the snow and lifts it. A few feet back of this nose begins the vertical wedge, with a vertical or inclined nose and curving sides. For single track, this nose is in the middle of the horizontal wedge, and the snow is thrown to both sides; but for double track or sidehill work the vertical nose is at one side of the plow, so as to throw all the snow to the opposite side. This nose may be adjustable, to discharge to either side, as desired. As the horizontal edge of the plow is about 3 in. clear of the rails, it may be fitted with a drop plate that can be lowered to remove the snow to a depth of 2 or 3 in. below the rail heads; shoes beneath this plate ride on the rails.

Wings with concave faces sloping upward at about 30 deg. may be hinged to the side of the bucking plow for use in widening the cuts, the wings being normally housed within the clearance lines and being operated by hand gearing or compressed-air mechanism. After making the first cut through a drift, the plow may be run back and then driven through at high speed with wings extended, their spread being regulated by the operator on the plow. On multiple tracks, a push or bucking plow may not be of much service unless supplemented by others to clear the snow thrown from the first track. For such work, a bucking plow may be placed ahead of the engine and a wing plow behind; following this on the next track will be an engine with side-delivery plow in front and a wing plow behind with wings extended only on the outer side. Spreader cars may also be used to clear the softer snow thrown out by the plows.

Machine Snowplows.—Where deep snow is an annual trouble, the work of keeping lines open or opening them when blockaded is simplified by the use of a rotary plow. This machine is a large car containing boiler and engine geared to an $11\frac{1}{2}$-ft. wheel revolving in a vertical plane transverse to the track and mounted in a circular shell or drum at the head end of the machine. In front of the shell is a rectangular housing about 13 ft. wide and extending to within $3\frac{1}{2}$ in. of the rail head to gather in the snow and trim the sides of the cut. Radial knives on the wheel shave off the snow, which falls through into a fan wheel behind the cutter wheel and is discharged by centrifugal force through a top opening; the stream of snow falls at a distance of 50 to 200 ft. according to the speed of the wheel. A modified design dispenses with the fan wheel, each knife forming one edge of a slotted conical scoop or pocket which receives the snow and throws it out at the discharge opening. Flanging cars and spreader cars may follow to clear and widen the cut. The discharge hood, ice cutters, and flangers on the plow are adjusted by air cylinders controlled from an operating cab.

This machine, weighing 60 to 100 tons, is not driven violently against the snow, but is pushed steadily forward by one or more locomotives at such speed as not to clog and stall the wheel; this speed is usually 6 or 8 mi. an hour in snow up to 7 ft. deep, or 4 to 5 mi. in deeper snow.

The "cyclone" rotary plow, now little used, had a revolving cone $7\frac{1}{2}$ ft. long and a $7\frac{1}{2}$-ft. base diameter mounted diagonally in a housing at the head end of the machine, the 12-in. apex being at a bottom front corner and the center of the base at the opposite top corner. Special blades on the cone cut the snow and passed it to the rear end of the cone, from which it was discharged by centrifugal action through an opening at the top of the housing. The rotary plow may be followed by a wing plow run at high speed with wings extended to widen the cut or to throw out the loose snow shoveled into it by shovel gangs widening the cuts.

Flangers and Ice Cutters.—To clear snow and ice from the rail heads and to form flangeways for the wheels, flangers are fitted to the snowplows and sometimes to the engine pilots or to special flanging cars. The flanger is a notched horizontal bar mounted usually about $1\frac{1}{2}$ to 2 in. clear of the inside of the rail head; or it may be a vertical plate or blade placed diagonally across the rail and slotted to straddle it. This blade will cut a path 6 to 12 in. wide on each side, the path being about $\frac{1}{2}$ in. deep on the outside of the rail and $1\frac{3}{4}$ in. deep on the inside, for a flangeway. The blades or cutters are attached to a mechanism, operated by hand or by air, for raising them clear of the rails at switches and turnouts, but they are so attached as to be easily replaced if broken. For cutting ice or hard snow packed between the rails, a transverse beam under the snowplow or flanger car may be fitted with vertical steel teeth or cutters, the bar being raised and lowered in guides, like the flangers. Wire brushes are sometimes mounted behind the flangers or on the engine pilot, as they can be set almost in contact with the rail and will clear the head of ice and snow.

Clearing Yards and Terminals.—At city yards, section gangs and yard gangs must be supplemented by large gangs of shovelers in heavy storms, especially as standing cars are soon blocked and must be dug out before they can be moved. The work is slow, as each shovelful has to be carried some distance; better progress may be made by shoveling into boxes of about 7 cu. ft. capacity, having long, straight handles and being carried between two men. Yards subject to snow trouble should have plenty of clean ballast and be kept well drained. Trenches for switch connecting rods and interlocking mechanism must be kept open to prevent accumulation of water. Salt is used in clearing snow and ice from switches and frogs, but slide plates and moving parts must be kept well oiled or the salt water will cause rust and make the switches hard to move. Switches that are frozen up may be thawed out with steam jets from hose attached to yard engines or by portable oil torches. Track sweepers and double-end snowplows are sometimes used in clearing yards, but cars on the tracks often prevent any work except hand shoveling.

Clearing ice from flangeways and hard, packed snow between the rails with picks is slow work, and flanger cars and ice cutters, as described above, may be adapted to this work. Spreader cars may clear soft snow and may be fitted with the flangers and ice cutters mentioned, the cutters being mounted on the pilot plow at the head end of the car. In northern sections, turntables are sometimes enclosed, but another plan is to install heater coils and a vertical overflow pipe in the pit which is then filled with water that is kept warm. Snow cleared from the yard is thrown into the pit and thus disposed of. Snow may be removed from yards, driveways, and platforms by excavators or by large scrapers hauled by teams or light tractors.

Large terminals are sometimes partly or completely blocked by snow, gangs of shovelers being unable to keep the tracks, switches, and interlocking mechanism clear, while men able and willing to stand the severe work are not always available in sufficient numbers. This trouble is specially serious if suburban traffic is involved. Steam, gas, electric, and oil-burning heating apparatus is used to some extent to prevent the freezing up of switches and interlocking, the first being served either by a locomotive on a sidetrack or by a stationary boiler. The apparatus is disconnected and removed at the end of the season. Such methods may keep the tracks open and save the cost of a more or less efficient army of shovelers. At a point where, ordinarily, about 20 men were required to keep switches clear, the installation of heaters reduced the force to a foreman with three trackmen. For disposing of the snow removed from tracks, steam-heated concrete pits with sewer connections have been provided at some terminals.

There are possibilities for the mechanical handling of snow. Locomotive cranes with grab buckets have been used to some extent, but there are difficulties in handling such equipment in terminal yards. Track sweepers which deliver the material to conveyors loading cars coupled to the machine might be adapted to this work (see "Tools"); also endless-chain bucket excavators delivering to conveyors, as discussed in *Engineering News-Record*, Jan. 30, 1919.

To prepare for winter difficulties at the Chicago terminal of the Illinois Central R. R., a meeting is held early in the fall by the terminal roadmaster and the heads of the engineering, track maintenance, bridge and building, water service, and signal departments, together with representatives of the transportation department. Instructions are issued to section foremen and yard foremen, and after an inventory of tool equipment in stock, requisitions are made in order to have an ample supply of shovels, scoops, picks, brooms, and snow scrapers, as well as a store of rock salt. Steam-heater coils are placed at switches and in snow-melting pits, and five snowplows are placed at certain points in the terminal district. Several switch engines are equipped with dome attachments for $1\frac{1}{4}$-in. pipe, so that when necessary they can be installed at sidings and connected to the heater coils or to steam hose with nozzles;

steam fitters are assigned to accompany the engines when ordered out for this purpose. When large extra forces are required, their distribution is directed by one man at headquarters, according to requests and information telephoned to him, foremen and other offices on the work calling up at frequent intervals to report conditions and requirements.

Electric Railways.—On interurban lines, the passenger cars may be fitted with pilot plows, flangers, and rail scrapers, while larger plows and spreader wings may be attached to work cars and baggage cars. Large bucking plows with wings are used to some extent, and a few lines have rotary plows similar in principle to those already described, but having fewer knives on the 8-ft. cutter wheel. Flangers are sometimes held in contact with the rails by springs. Scrapers also are held down on the rail head by springs, and have a series of diagonal blades to cut ice and sleet. To clear trolley wires from sleet, the trolley pole may be fitted with a scraper or with a special contact wheel having ribs or lugs in the groove to cut the ice and jar it loose. Where the third-rail conductor is used, the cars are equipped usually with insulated wire brushes or scraper blades to clean the rail head, while in some cases a small stream of calcium chloride solution is fed upon the rail in advance of the contact shoe.

City and suburban cars very generally have scrapers or flangers attached to the truck frames. Track sweepers and spreader cars are also used in streets; while deep snow cleared from tracks is sometimes loaded into dump-car trains by small steam or electric shovels, or power loaders. Salt for clearing frogs, switches, special work, and grooved rails is sometimes spread from cars. On elevated railways, scrapers and wire brushes on the cars supplement the work of the trackmen in clearing the running rails and conductor rail; cars may be fitted with special brooms to clear snow from between the guard timbers and guard rails.

On the trolley-line electrified section of the Chicago, Milwaukee & St. Paul Ry. in mountain country, steam rotary plows are used, equipped with flangers. To a limited extent bucking plows and flanging cars are handled by electric locomotives. There is usually enough heat from current in the overhead wire to keep it clear of sleet and ice. Steam-driven plows are used also on the third-rail electrified line of the New York Central R. R., and to clear the under-running rail a special form of plow is attached to the front end of the multiple-unit suburban electric cars. Special plows and apparatus are used for clearing snow in terminal yards operated by electricity. Tests with different arrangements of the third rail have shown that there is less trouble with sleet when the shoe rides against the bottom of the rail and the top is covered by a shield.

CHAPTER XXVI

TRACK INSPECTION AND PRIZES

Routine Inspection.—Track is inspected daily, as a rule, either by a trackwalker or by the foreman or a member of the section gang going over the section on a motor car. Section foremen and roadmasters make frequent examination of the general condition of track and roadway, besides giving special attention to certain parts needing improvement in detail. Roadmasters and engineers also make general inspections of the divisions under their charge, and these inspections should be made frequently, in order to maintain close touch with the work and the working forces.

Official Inspections.—As a means of improving the morale of the maintenance-of-way forces and arousing in them an interest in their work, and thus aiding directly in securing good track, many railways have a system of prizes for the best track sections, these prizes being awarded on the basis of an annual (or semi-annual) inspection by superior officers in company with the engineers and roadmasters, and sometimes the section foremen also. This system of official inspection helps to give the employees a feeling of personal interest in the railroad as well as in their work, and in this way it produces highly satisfactory results.

As a rule, the inspection is made from the rear of a business car or special car, but in some cases this is supplemented by the use of a dynagraph car which charts the conditions of line, surface, and gage, and may also indicate bad spots by a jet of paint. Marking or rating of the track condition may be done by the superior officers or by the roadmasters, but in the latter case no man does any marking on his own territory. In some cases, the section foremen of a division do the inspection, or are taken over the division afterwards so as to see and note each other's work.

Prizes.—A prize of $25 to $100 on each division is enough to arouse interest and rivalry, but it is well to supplement the cash prize by a "Premium Section" sign, so that the laborers may feel that they are getting some recognition, as well as the foremen. Medals and motor cars form the prizes in some cases, as noted below. In making the awards, it is necessary to take into account the money expended upon each section for material, tool equipment, and extra gangs, since the condition of track may not be commensurate with the expenditure, while good track may be due to the work of the extra gang rather than to that of the section gang. Thus, the awards should be so based that the credit goes to the men who secure the best results per dollar of expenditure.

Marking for Track Condition.—The system of marking varies widely on different railroads, some having a few broad items while others prefer to detail a number of specific items. All the items may be marked by each man of the inspecting party, or they may be divided among committees, the latter being preferable when the items are numerous. In one plan there are four committees marking as follows: (1) line and surface; (2) joints, ties, and tie spacing; (3) ballast, spiking, switches, and ditching; (4) road crossings, station grounds, and policing. In the second method given below, there are three committees for the three markings. Five classifications for rating or marking are as follows:

1. Surface, 50; line and gage, 25; drainage, 25.
2. Surface and line, 40; roadbed and ditches, 35; policing, 25.
3. Surface, 25; line, 15; drainage, 15; neatness, 15; switches and frogs, 10; ballast, 10; yards and sidings, 10.
4. Surface, line, and gage, 25; spiking, tie lining, and spacing at switches, 25; ballast and drainage, 20; grass, weeds, and right-of-way,10; neatness at section houses and station grounds, 10; sidings, 5; fences and road crossings, 5.
5. Surface, 35; line, 35; ties, 6; rail anchors and joints, 6; ballast, 6; drainage, 6; general appearance, 6.

Track Inspection Cars.—Mechanical inspection and charting of track conditions by special apparatus is done on a few railroads, including the New York Central, the Erie, the Pennsylvania, and the Baltimore & Ohio. All of these roads use cars equipped with apparatus as designed originally by P. H. Dudley for a dynagraph car which he operated for several years, making inspection trips for the different railroads and submitting reports based upon the autographic charts. On these charts are recorded automatically the condition of surface of each rail, the gage of track, and the rolling motion of the car. Low places and joints, and places where gage is extremely wide are not only indicated on the chart but are marked on the track by paint automatically ejected, thus calling the attention of the foremen to points which may have been neglected or at which the defects may not be readily noted by casual inspection. This early work by Mr. Dudley aided materially in the development of his stiff rail sections and in bringing the tracks of the New York Central R. R. and Boston & Albany R. R. to a high degree of perfection.

In the car, a continuous sheet of paper moves across a frame fitted with stylus pens, the rolls on which the paper winds and unwinds being driven by gearing from one of the axles of the car, so that the speed of the paper varies with that of the train. For each indication there are two pens, one drawing a straight line or datum, and the other being shifted laterally by mechanism operated electrically in connection with the movement of the car as affected by track conditions. A specimen chart is shown in Fig. 111. Variations of $\frac{1}{4}$ in. or more in surface are indicated by the vertical movements of the ends of the middle axle of a six-wheel truck in relation to the other axles. For variations in gage, a two-wheel trailer has its wheels sliding on the axle but held against the rails by springs. As each wheel moves in and out due to variations of gage, these movements are recorded on the chart. A heavy pendulum in an oil bath records variations in transverse level of the track and a more sensitive pendulum indicates sudden lurching of the car due to defective track. Speed is indicated by a time-contact device marking the chart at 10-sec. intervals, while another pen makes a mark as an observer presses an electric button

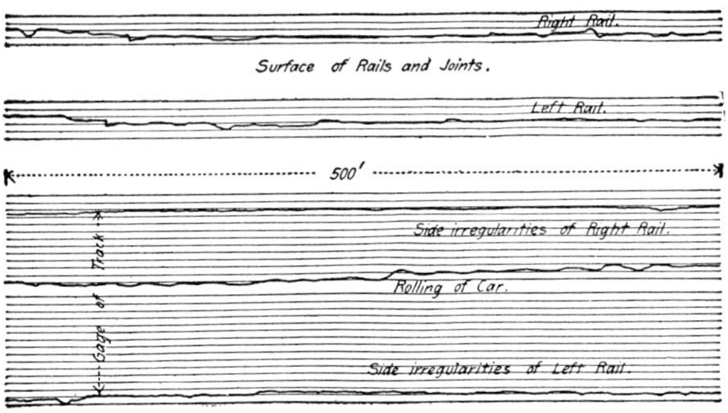

FIG. 111.—Track diagram from dynagraph car.

when the car passes mile posts and stations, so that the location of any point on the chart can be readily determined. The dynagraph car is handled usually on a special train at 30 to 35 mi. an hour.

Examples of Track Inspection Systems—Erie R. R.—For the annual official inspection in October, a special train is used, with a track inspection car having mechanism as above described to trace lines on a chart, showing the surface of each rail, the gage of track, and the cross-level; it also indicates on a recording dial the low joints and bad lurches. At one time, this car marked the low joints with white liquid, but the number of low joints has become so small that it is not practical to operate this device, as the liquid hardens and stops up the nozzle. A committee of three track supervisors rides the rear end of the car and judges the line, policing, and ditching on each section, the ratings of the three supervisors being averaged. The officers on the train are a representative of the executive vice-president, the regional vice-president, or his representative, the regional engineer, the division superintendent, the division engineer, and track supervisors. The results of the marking committee are worked up by the regional engineer, who also awards the prizes, subject to the approval of the two vice-presidents.

The prizes are awarded as follows: 50 points for perfect surface, 10 for each region, the difference being in judging the winners. On the Chicago region, they are awarded as follows: 50 points for perfect surface, 10 for perfect line, 10 for perfect policing, 10 for perfect ditching, and 20 for perfect on expenditure. This last means that the foreman used no more nor less than his allowance based on the equated mileage on the section, being penalized for having either more or less men than on his allowance. The prizes are as follows: first prize for best section on each operating division; $150; first prize for best section on each supervisor's territory $100. If the division prize goes to this subdivision, the second best section gets the $100 prize. Second prize on each supervisor's territory, $75. First prize for best supervisor's territory on region, $200; second prize for supervisor, $100. In addition, special prizes are given to track supervisors as thought best by the vice-president of the system, based on record. Generally if a supervisor has no low joints and no lurches, if his expenditures have been within his appropriation and if his policing and other conditions are good, he is given $100.

Pere Marquette R. R.—The annual inspection is made about the middle of October with a train consisting of office cars, a sleeping car, a dining car, and an inspection car which is open at the rear and has seats rising in tiers and accommodating about 32 persons. The committee includes the president, assistant to the president, chief engineer, assistant chief engineer, bridge engineer, signal engineer, superintendent of motive power, division superintendents, division engineers, and four or five track supervisors. The grading for signals and interlocking is based upon a field inspection by the signal engineer and members of his staff, just before or after the general inspection trip. A prize of $100 is awarded to the track supervisor having the highest average, and $100 to the supervisor who shows the greatest improvement over the previous year. On each supervisor's subdivision, a prize of $25 is awarded to the section foreman having the best general condition, as shown by the average of committee grades.

Marking is done by seven committees of three to six members, as follows: (1) line and surface; (2) roadbed and drainage; (3) ballast dressing and shoulder; (4) policing, right-of-way, and station grounds; (5) fences, crossings, and signs; (6) signals and interlocking; (7) shops, shop grounds, repair tracks, and enginehouses. Each member has a book of blueprint pages with four columns for section number, length, percentage, and remarks. The first two are filled out as information, the third is left blank for the rating, and the fourth contains special notes as to certain sections, such as "double track," "freight yards," "branch." For each item, the highest rating is 100. A sample of the marking is given below, exclusive of signals and shops.

Section number	Line and surface	Roadbed and drainage	Ballast dressing and shoulder	Policing, right-of-way and station grounds	Fences, crossings and signs	Average per cent
1	92.78	86.00	91.67	88.67	82.67	89.88
2	87.25	80.00	83.67	79.67	75.33	83.07
3	87.50	81.50	87.00	77.00	81.33	84.54
4	77.50	82.50	77.75	79.00	80.33	78.73

Delaware, Lackawanna & Western R. R.—Inspection for prizes and awards is made about the first of November, the markings being made by maintenance officers from other roads, who ride at the rear end of an observation car. This inspection is for surface (40), line (35), and general condition (25), and the results are entered on sheets 8 by $10\frac{1}{2}$ in. with 13 columns headed as follows: (1) cost per equated mile allowed; (2) section number; (3–6) field inspection ratings for surface, line, general condition, total rating; (7–9) cost of section maintenance per mile of main tracks, of third and fourth tracks, and of all other tracks; (10) total cost of section maintenance per equated mile; (11) total section equated mileage; (12) name of section foreman; (13) field inspection standing. At the head of the sheet are lines for the date and location and the names of the roadmaster, the supervisor, and three inspectors.

First and second prizes awarded to the section foremen are $100 and $50, with gold and silver medals 2 in. in diameter bearing the names of the foremen. Ornamental signs, "First Prize, 1924," are erected on the prize sections, and a section which has taken the first prize for three consecutive years gets an "E" sign for efficiency. The rules provide that foremen who have taken first prize on their sections for 3 consecutive years will be paid extra compensation to the amount of $10 per month in addition to their regular monthly salaries; the first and second prizes will then be awarded to the foremen standing second and third. This extra compensation will be paid the following year and each year thereafter so long as the foreman maintains his section with the same efficiency, and he will not be entitled to a prize. If one of these foremen does not maintain his section with the same high efficiency and drops to a lower rating, the extra compensation will be discontinued, and the foreman will again enter the prize contest as before. Any section exceeding the maintenance cost allowed by equated mileage will be eliminated from the prize competition for the year.

Pennsylvania R. R.—For the annual inspection of the main line, usually about the middle of October, the inspection parties are of two classes: (1) the "limited," which comprises officials above the rank of supervisor; (2) the "general," in which all officials down to and including assistant supervisors participate. The 36-ft. inspection car is open at the end, and the floor slopes up by steps to about the middle of the car, so that all the occupants may have a good view of the track. This part of the car has seats for 28 persons. The "limited" inspection party, traveling west from Jersey City to Pittsburgh, is made up of the general manager, chief engineer of maintenance-of-way, general superintendents, division superintendents, principal assistant engineers, and assistant engineers. The "general" inspection party, on the return trip, is made up of all the participants in the inspection. One train usually carries the general manager and his party; a second train is for the principal assistant engineer and assistant engineers of all the divisions; a third train for the supervisors; and a fourth train for the assistant supervisors. These are followed by the track-indicator car, in charge of the motive-power department, but the use of this car is not considered as part of the inspection and has no connection with the prize awards. Branch-line inspections are arranged by the division superintendents and usually follow the main-line inspection.

Four committees mark the ratings; (1) line and surface; (2) joints and tie spacing; (3) ballast, switches, and sidings; (4) ditches, road crossings, station grounds, and policing. The marking is from 1 to 10, and zero is used where there are no sidings or crossings. In awarding prizes, a committee is appointed on each main-line division, no officer of the division being included. The prizes awarded to supervisors are as follows, all being for the entire line between Pittsburgh and Jersey City: first prize of $1,200 is for main-line division (exclusive of yards) having best line and surface; $800 goes to the supervisor and $400 to the assistant supervisor. Second prize of $1,000 is for the division having the greatest improvement in line and surface during the year. Three prizes of $800 are awarded for main-line supervisors having best line and surface on each division, exclusive of the first prize section. Fourth prize of $100 is for the main-line yard supervisor having the best line and surface. These prizes are divided between the supervisor and assistant supervisor.

CHAPTER XXVII

MAINTENANCE RECORDS, REPORTS AND ACCOUNTS

It is important that there should be complete records of the physical characteristics of the railroad, with its structures and property, and of its maintenance work, improvements, contracts, purchases, and expenditures. It is equally important to have these records kept on a comprehensive system that enables information to be obtained readily and surely and is coordinated with the railway accounting system. The engineering records will include real estate and right-of-way properties, rails, ballast, ties, fences, bridges, signals, stations and other buildings, coaling and water stations, tunnels, yards, terminals, and also valuation details, together with the force, equipment work, and expenses of the maintenance-of-way department. Some of these items have been covered in previous chapters.

Classified lists of bridges, buildings, and other physical equipment are often kept in tabular form on large sheets or in books, but while these serve as office records, they lack availability for use or reference and are not conveniently kept up to date. Loose-leaf records are in many ways preferable, but with some means of checking against the loss or replacement of sheets removed for any purpose. Filing cases of various kinds are valuable for keeping together in some systematic arrangement the correspondence, drawings, accounts, reports, and other records for any one subject or class of work. In any case, there should be a simple but comprehensive index to show in what different files, books, etc., are the several records for any piece or class of work. Whatever system is employed, some person in the office should be assigned to the duty of keeping the files in order and having the necessary changes and additions entered to keep the records up to date.

Card-index Records.—The card-index system affords special facilities for simplifying the keeping and filing of records and makes the information more available than under any other method. It applies to permanent records, such as those of rails, bridges, structures, land, and valuation; and also to the periodical reports of labor, material, and progress in construction, maintenance, and operation.

For improvement work, records may be kept by cards of different colors for different items, all filed in station order between guide cards which have raised tabs with the station names in consecutive order.

Thus if ballasting, renewing rails, and installing signals are in progress between Brownville and Greytown, the respective records may be on buff, green, and pink cards behind the Brownville guide card. If it is desired to know what signal work is in progress on the division, all the pink cards can be taken out and a summary prepared. Each card would show the location of the work, number, or class of structure, brief description, the index number of correspondence or other files, the number of order authorizing the work, and the dates of ordering, commencing, and completing the work. Ledger accounts for individual contracts or pieces of work can be kept in this way, each card being a complete record for one account. Records of bridges, bridge inspections, and various other classes of structures and work can be filed conveniently in card-index form. Cards 3 by 5 or 5 by 8 in. are generally used.

Track Charts.—The engineering records usually include some form of map or chart showing the general plan and profile, all tracks and track connections, and the important physical features of the property; in many cases, the chart includes also certain features relative to operation. On some roads, these charts are simple, while on others they are very elaborate and give a great amount of information. They may be large sheets, perhaps bound in book form, for office records, and are generally arranged to allow for blueprinting in long narrow strips which can be folded for pocket use. Whenever changes or additions are made, they should be reported, so that corrections may be made on the original or file chart. Corrected copies should be issued at least once a year. Two types of track charts are shown in Fig 112.

The principal feature of the chart is a complete plan of all tracks; main and side tracks, passing tracks, junctions, crossovers, track crossings, industry spurs, engine terminal tracks, team tracks, and yard tracks. This plan is supplemented by a center-line plan giving location, length, and degree of curves, and also by a profile giving grades and elevations. Among the features shown on the chart may be such items as the following: mile posts, stations, freighthouses and other buildings, section houses, water and coaling stations, turntables, enginehouses, shops, signals and interlocking plants, signal towers, telegraph and telephone lines, fences, street and highway crossings, farm crossings, steam and electric railway crossings (with their angles), waterways, culverts, sewers, water mains and electric wires or conduits crossing the line, bridges, trestles, overhead structures, and tunnels. Railway and political boundaries may be shown, such as track sections, division points, right-of-way lines, corner and boundary posts, property monuments, and survey stations; also state, county, and township lines and city limits.

Weight and other particulars of rails, kind of ties, and kind of ballast or fencing may be indicated by colors or symbols on straight lines along the chart. Varied information shown in writing or by symbols may include the following: car capacity of sidetracks, number of enginehouse stalls, capacity of water or fuel tanks, bridge numbers, dimensions and materials of buildings, type and spans of bridges, dates of construction, telegraph calls of stations or offices, and names of supervisors and foremen; also the ruling grades, the limiting weight of engine that may be

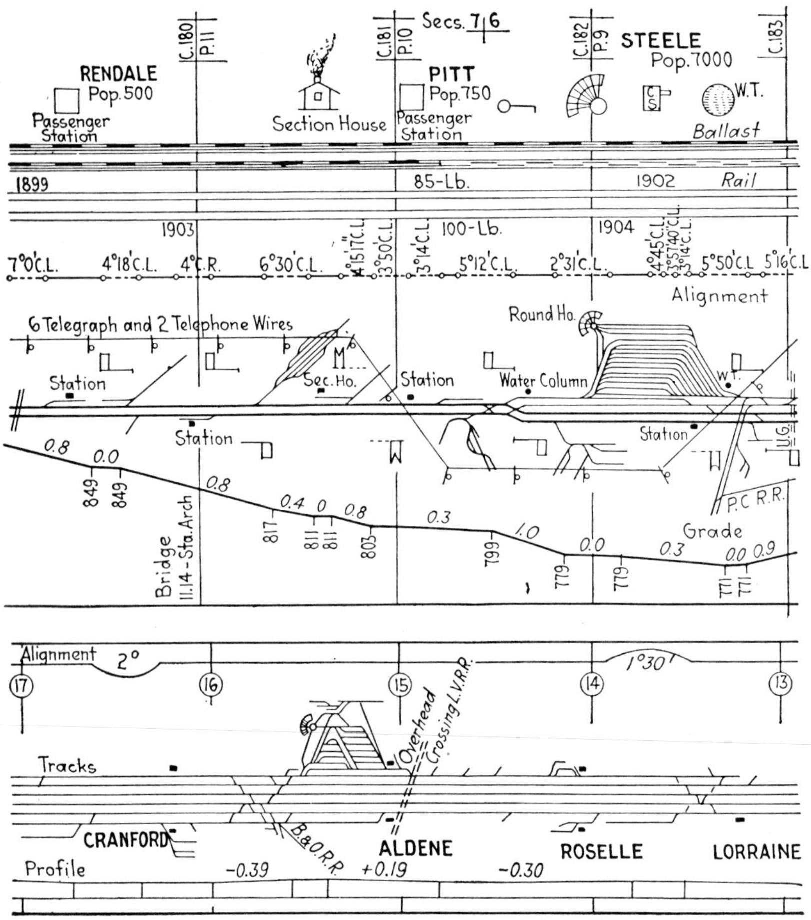

Fig. 112.—Typical track charts.

run (as governed by track construction or bridges), and the rated train loading for freight engines. Supplementary to the track chart are station and yard plans, right-of-way maps, and real-estate records.

Scales for track charts range from 500 ft. to 1 mi. to the inch. Scales recommended by the A. R. E. A. are ½ mi. per inch for track charts; 100, 200, or 400 ft. for right-of-way and track maps; 50 to 200 ft. for station maps and track layouts; and 50, 100, or 200 ft. for individual right-of-way or property maps. For profiles, it recommends a horizontal scale of 1 in. to 400 ft. with a vertical scale of 1 in. to 20 ft. ordinarily or 1 in. to 4 ft. for ballast profiles.

Examples of Track Charts.—A track chart used for each division of the Chicago & Northwestern Ry. is on sheets 34 by 52 in. inside border lines, with a horizontal scale of 1 mi. to the inch and a vertical scale of 100 ft. to the inch. Along the top are marked the track sections and mile posts, with a vertical line across the entire chart at each mile. Below these divisions is the alinement plan and then the track plan, with profile below it. Along the bottom of the chart are marked the weight and make of rails, with date of laying. These charts are corrected annually. Blueprints are framed for office use, and are cut into strips and folded for pocket use. A chart used by the Pennsylvania R. R. is divided vertically by sections instead of miles, and has at the top the names and addresses of track supervisors and section foremen. Both of these charts give a variety of information by lettering and symbols. A simpler form (Fig. 112) which is almost exclusively a track chart, as used by the Central R. R. of New Jersey, is on a scale of 2 in. to the mile, but with the track spacing exaggerated to permit of showing every siding, spur, crossover, track crossing, and yard track; also stations and roundhouses. This plan is in strips about 9 by 48 in., with light vertical lines indicating the miles. Above the track plan is an alinement plan, indicating length and degree of curves. Below is a profile, with grades and elevations.

Rail, Tie, and Ballast Charts.—In addition to the track chart, many railroads have separate chart records of rails, ties, ballast, and fences. In one form, sheets 8 by 13 in. are divided vertically by lines at mile intervals on a scale of 1 in. to the mile, with five lengths of track across each sheet. For each length there are three lines; the middle line indicating the rail and ballast and the two side lines the fences; colors, numbers, letters, and symbols indicate the weight of rail, kind of ballast, kind of ties and spacing, kind of fence, and other information. A rail chart used by the Central R. R. of New Jersey consists of blueprint sheets 8½ by 11 in., across each of which are four or more plans on a longitudinal scale of 2 in. to the mile. On a broad line for each track, different colors represent rails of different weights, while against each line is written the date of laying and the character of rail; O. H. for open-hearth, B for bessemer, N for new, and R for relaid. Stations and miles are also indicated (see "Improvement Records").

Bridge and Building Records.—Bridge records have been noted in Chap. XXII; building records may be on the same general scheme. If on the card-index system, they will be filed usually in consecutive number, but may be of different colors for stations, roadway structures, fuel and water stations, shops, and enginehouses. With these cards may be filed the periodical inspection cards, usually filled up in the office from the inspectors' record cards. If the latter are so designed that the inspector indicates the condition of the several features of the building, then only one inspection card need be filed for each structure, since it will show the condition throughout at the date of inspection. As each inspection report comes in, therefore, its notes are entered on a card which is placed in the file, the old card being then removed. The inspectors' reports are filed as records.

A suitable distribution of items for building inspection records may be as follows, with a separate line for each: foundations, walls, paving, floors and supports, partitions, roofing, gutters and downspouts, doors and glazing, windows and glazing, finishing hardware, heating appliances, plumbing, drains, lighting appliances, exterior painting, interior painting, platforms, miscellaneous. The card gives also the name and location of the structure, date when built, and its general condition as a whole (good, fair, or bad).

For water stations, the A. R. E. A. form of record gives extensive details as to water supply, wells, pipe lines, lift or head, pumps, tanks, water columns, water-treating plant, and buildings. A separate record gives an itemized statement of the cost of pumping water.

Valuation Records.—A specially important record is that of the valuation work carried on by the railways in conjunction with the Interstate Commerce Commission. Valuation records of the Chicago & Northwestern Ry. are kept on a card-index system with cards 5 by 8 in., on which are entered the notes from the field books of valuation parties. These cards serve also as permanent records of the company's property. Besides the cards, the valuation records include the field books, profiles (1 in. to 400 ft. horizontal and 1 in. to 20 ft. vertical), maps of alignment and right of way (1 in. to 400 ft), station terminal and yard maps (1 in. to 100 ft.), mass diagrams of grading quantities and overhaul, detail plans of structures, and copies of all valuation reports to federal, state, and other authorities (see *Engineering News*, Oct. 28 and Dec. 16, 1915).

Improvement Records.—Records of road additions, betterments, and retirements are kept by the Hocking Valley Ry. in a loose-leaf ledger, with sheets 12 in. high and 19 in. long. This ledger is kept in consecutive station order. For listing permanent improvements, the Chicago, Rock Island & Pacific Ry. uses small index cards 5 by 3 in., giving the location, division, A. F. E. (authority for expenditure), number and date, estimated expenditure (distributed to "A & B" and other accounts), salvage, actual expenditure, and date of completing the work. On the back is a brief record of the dates on which the form is sent to, and received from, the several approving officers.

This road has also an annual chart on which the authorized rail renewals, ballasting, and roadbed widening are shown in yellow. Monthly progress reports are plotted on the office chart in black, and at the end of the year the chart is corrected to show all work done during the year. The chart consists of a series of horizontal lines with a scale of 1 in. to the mile. Separate sets of lines show, for each track, the rails (kind, weight, year rolled, year laid, brand, and kind of joint), new rails (authorized and laid), present ballast (kind and depth), new ballasting (authorized and completed), present roadbed width, and roadbed widening (authorized and completed).

Progress Records.—For a graphic record of progress on construction work, the engineering department of the Chicago, Burlington & Quincy R. R. uses a chart having a column at the left for a list of items of work, 10 columns for percentage of progress and a right-hand column for totals. The list is given for the entire work and is repeated for each contract section. Horizontal lines plotted for the several items show the percentage of work completed, and are extended by different colors every month. Thus, for the work as a whole, and for each contract section, the percentage of progress is shown for each item.

With this chart there is a tabulated "monthly procedure statement," a special feature of which is that it shows for each contract section the yardage of the month, total yardage to date, yardage remaining, and the average monthly yardage necessary to complete the work within the time limit. Thus in the following example, the average monthly progress required for section 1 is 8,679 yd., while the January record was only 7,943 yd., indicating that the contractor must increase his progress materially. From the column of equipment, both the engineer and the contractor can judge whether an additional shovel or a larger shovel is desirable; it also helps to show whether the progress is in proper relation to the equipment in use.

Work section	Excavation					Equipment
	Total, cubic yards	Done this month, cubic yards	Total completed, cubic yards	Remaining, cubic yards	Required rate per month, cubic yards	
1	86,400	7,943	25,644	60,756	8,679	One $\frac{3}{4}$-yd. shovel
2	56,500	6,237	21,448	28,815	5,715	One $\frac{3}{4}$-yd. shovel
3	296,800	37,033	31,151	265,649	37,950	Two $\frac{3}{4}$-yd. shovels
4	406,200	38,419	133,060	273,140	39,020	One $1\frac{3}{4}$-yd. shovel
5	99,800	10,123	33,698	66,102	9,444	One 1-yd. shovel

Improvement work on the Kansas City Southern Ry. is covered by work orders, and the division engineers report each month the percentage completed during that period on each project, with its total progress at the end of the month. This percentage is based on the physical progress rather than the financial status of the work, material delivered on the ground not being reported until it has been used.

MAINTENANCE REPORTS

Forms of Reports.—Numerous periodical reports from the several maintenance-of-way officers to their superiors are necessary in order to show the work of the department and to enable its progress and cost to be determined. These reports relate largely to labor, work, and material, and from them the costs are distributed in the accounting systems. They are largely tabulated statements on ruled forms in books or separate sheets, but the style of report varies widely on different railways. It is desirable to have as few sizes as possible for the sake of uniformity and of convenience in filing. For this reason many report forms are of the standard letter size, $8\frac{1}{2}$ by 11 in., so that they can be placed in letter files, either separately or in conjunction with the correspondence. In many cases, the reports are progressive, those received by one officer being compiled or summarized in his own report to the next superior officer; thus, the foremens' reports for the several sections may be condensed into the roadmaster's report for his territory, and the several roadmasters' reports may be the basis of the report by the division engineer or engineer of maintenance. This compilation must be done with care, however, the originals being examined in the light of personal acquaintance with the work and checked as to inaccuracies or omissions.

Report forms for the use of foremen of section, bridge and building or extra gangs, and work trains should be simple and clear. These men are usually of limited education and have little understanding of an accounting system, but upon their reports is based the accounting of the maintenance-of-way department. Complicated analysis or distribution of work is not required as a rule, although some time books have a very detailed distribution or classification. It is important, however, to itemize the different kinds of work done and the time spent on each; also the amount of regular and overtime work, in view of wage conditions. This information the foremen alone can give, but it is the basis of distribution in the accounting for expenditures.

In general, the aim should be to have report forms on which the men can make simple and definite statements, from which the analysis and the distribution for statistical or accounting purposes can be made in the office of the engineer or auditor, where also the payrolls are prepared from the time books. If such compilation is required of the roadmaster, he must have the necessary office force. A summary of the reports in general use is given below, but the list will vary on different railways.

MAINTENANCE-OF-WAY REPORTS

Reports to Supervisor or Roadmaster

Time book	By section foreman	Monthly or half monthly
Work performed	By section foreman	Daily, weekly, or monthly
Fencing	By section foreman	During renewals
Rails replaced	By section foreman	Monthly
Ties replaced	By section foreman	Weekly
Switch work	By section foreman	Monthly
Material and scrap	By section foreman	Monthly
Tool list	By section foreman	Monthly
Broken rails	By section foreman	Promptly
Work-train service	By work-train foreman	Daily
Extra-gang work	By extra-gang foreman	Daily
Steam-shovel work	By steam-shovel operator	Daily

Supervisor to Engineer (Monthly)

Distribution of payroll
Work performed
Rails laid and removed
Ties laid and removed
Track ballasted
Sidetracks laid and removed
Material and scrap
Tools
Fencing
Broken rails and joints
Gravel and earth handled
Switches and frogs replaced
Extra-gang work
Work-train service
Bridge inspection
Train and other accidents

Work and Material Reports.—In ordinary practice, the section foreman makes daily note of time and labor, usually in a pocket book, and at the close of the day enters these figures in a book or form which is sent monthly to the roadmaster. The latter either compiles these into a statement for the division or sends the reports (when examined and checked) to the engineer's office, where the expense items are classified according to the accounting system.

Daily Reports.—Several railroads now have daily reports sent in by the foremen in order to have a more accurate distribution of time and material than is given by the monthly report or time book, and also to have the information compiled promptly for the monthly report to the auditor. On the Bessemer & Lake Erie R. R., the daily report is a form $8\frac{1}{2}$ by 11 in., having numerous printed items, with space for figures to be inserted. The foreman sends this report to the supervisor on the first passenger train after the close of the day's work; he sends a copy also to the engineer of track or engineer of bridges and buildings according to the class of work done. Some of the principal items of the report are as follows:

Number of men (including foreman) worked today..............................
Number of hours worked today:..
Number of men.......Number of hours.......Worked at (kind of work)..........
Number of men.......Number of hours.......Worked at (kind of work)..........
Number of teams.....Number of hours.......Worked at (kind of work)..........
Number of feet of rails used............. Which track.........................
Number of ties used..................... Which track.........................
Number of piles driven................. Number of bents capped..............
Trestle completed........... feet....... Number of bridge ties placed..........
Excavation completed....... cubic yards Concrete completed....... cubic yards
Work-train work (kind, hours, location)..
Steam-shovel work (cars loaded, material)..
Number of loaded cars on hand (filling, ballast, other material)..................
Number of cars unloaded (car number and initial, contents)......................
Locomotive crane: cars loaded.........; cars unloaded..........; material......

In the daily material report of the same road, which also is $8\frac{1}{2}$ by 11 in., there are four classifications: material used, removed, received, and shipped, together with the necessary description, quantity, and location; also whether it is new, relayer (or second-hand), or scrap material. The sheet is sent every day, but with the statement "Nothing to report," if necessary. This report is for stock material only: rails, ties, frogs, switches complete, switch rails, guard rails, splice bars, switchstands, switch lamps, and bridge steel and timber. The bridge item is reported by carpenter foremen. Section foremen report all other material, including that used by extra gangs on their sections.

A daily report of labor made by section gangs and extra gangs on the Hocking Valley R. R. has one side ruled for the name, occupation, hours, and rate of pay for each man, with a final column for "Deduction or Remarks." The other side gives the classification of work, with the number designating the account for each item, and a column for the "Total Time" put in for such items as have been covered in the day. Work on tracks is divided into: main tracks, tracks for freight only, and other tracks. A similar sheet (pink) provides for time distribution on construction work, the classified list of accounts being different from that of the maintenance report. The daily material report shows material used, recovered, received, and shipped, that of the first two classes being separated as to "main tracks," "yard tracks," "freight tracks," and "other tracks." On another division of the sheet, the foreman notes whether material chargeable to "increased weight" or "improved quality" was handled to point of use by hand car, push car, work train, or regular train. This material report is sent daily to the engineering accountant. The same form is used by foremen in the bridge and building, water supply, telegraph, and scale departments. All these reports are printed forms, $8\frac{1}{2}$ by 11 in.

Weekly and Monthly Reports.—A form of weekly report of rail and tie renewals ($8\frac{1}{2}$ by 11 in.) is shown in Fig. 113, the foreman including his own time. Another work report, $5\frac{3}{4}$ by $9\frac{1}{2}$ in., is not ruled in columns, but has at the left-hand side a list which includes the following items: number of men, total days of work (as shown by time book), frogs and switches examined, rails laid, ties laid, track surfaced (with its location), track ballasted, ditch made or cleared, condition of section cars.

SECTION No. .... SUB-DIV. No. .... DIVISION.

WEEK ENDING ....

RENEWING AND SPACING CROSS TIES.

	RENEWALS			SPACING TIES		
	No. Cross Ties	No. Hours Work	Kind of Ballast	No. Ties Moved	No. Hours Work	Kind of Ballast
Number Cross Ties—Main Tracks.						
" " " —Sidings.						
Total.						
Average Number Cross Ties Per Hour.						

REPORT OF MARKED TIES REMOVED FROM TRACK.

REMOVED FROM MAIN TRACKS				REMOVED FROM SIDINGS			
Number Removed	Year Put in as Marked on Tie	Kind of Timber	Cause of Removal	Number Removed	Year Put in as Marked on Tie	Kind of Timber	Cause of Removal

NEW RAILS LAID IN RENEWALS.

	NEW RAILS LAID			Kind of Angle Bars	CROSS TIES MOVED	
	Ft. 90 lb.	Ft. 80 lb.	Number Hours Work		Number Ties Moved	Number Hours Work
Number Feet New Rails in Main Tracks.						
Number Feet New Rails Laid in Sidings.						
Total.						
Average Number Per Hour.						

Unusual Work Done:

Business Sidings Built:

Progress of Work:

.... Foreman.

INSTRUCTIONS:—Report to be sent to Supervisor on 7th, 14th, 21st and last day of month. Foreman's time to be included in distribution.

FIG. 113.—Weekly report of rails and ties.

A monthly report of track material, recommended by the A. R. E. A., is a sheet 11 by 17 in., shown in Fig. 114. From his daily record of materials received, used, and shipped, the section foreman makes out this report, which is ruled in vertical columns. In the first columns, he enters the kind of material, and the unit (feet, number, etc.) with the amount on hand at date of previous report. Material received is classified as: by shipment; from track renewals; from tracks taken up (for both main and side tracks); total. Material used and shipped is classified as:

renewal or repair of main and side track; new work; shipped away; total. The last column shows the amount on hand at date of report.

Description of Material	Unit	On Hand at Date of Last Report	Received During the Month						Total to Account for	Used and Shipped During the Month				Total Disposed Of	On Hand at Date of this Report
			By Shipment	Out of Track by Renewals		From Track Torn Up				Renewals or Repairs		Construction or New Work	Shipped Away		
				Main Track	Side Track	Main Track	Side Track			Main Track	Side Track				

Fig. 114.—Monthly report of track material.

Combined Report.—While separate reports covering different items of work are largely used, the Bessemer & Lake Erie R. R. combines a number of these in the "Foreman's Daily Record of Material Handled and Work Performed." This is a tall book, $4\frac{3}{4}$ by 11 in., with stiff paper covers, with pages arranged as follows:

1. List of men, with date of hiring.
2. Rails received and shipped.
3. Rails laid and removed from track.

4 to 5. Ties received, placed, removed, and shipped.

6 to 9. Frogs, switches, angle bars, steel switch ties.

10 to 11. Wood switch ties placed and removed (classified by length).

12 to 17. Other materials, classified as for ties.

18 to 19. Ballasting, ditching, widening fills, raising track out of sags, track surfaced, lined, and trimmed.

20 to 25. Diary for records of work done, accidents, slides, personal injuries and other occurrences. Also, "safety, sanitation or welfare work done each day, with number of hours spent on this work."

26 to 29. Tools; on hand, received, shipped, used, on hand.

30 to 31. Cars handled; date, car initial and number, contents; loaded or unloaded, and at what point.

32. Freight transferred; kind and quantity; reason for transfer; loss or damage in transfer.

Tools.—Reports of tools and equipment on hand are of a similar character to the material reports. A monthly inventory prepared by the roadmaster has printed lists of the various items, with five columns for the number of each item; on hand at date of last report, received, shipped, used, and on hand at date of this report. Second-hand tools are not included in this list, except for certain large tools which are subdivided into "new" and "second-hand" in the printed list. These inventories are sent to the engineer or auditor.

Ballasting.—For report of extensive work the A. R. E. A. has a loose-leaf sheet, $8\frac{1}{2}$ by 11 in., to be made out in triplicate daily by the engineer in charge. The original is sent to the chief engineer and one carbon copy to the division engineer, the other being retained by the engineer in charge. This report gives in tabular form: the quarry or gravel pit; the number and kind of cars, with cubic yards of ballast; and then the mile-post location and length in feet for: (1) delivered and distributed: (2) roadbed prepared and skeletonized; (3) ballast placed, first lift; (4) second lift; (5) track finished (lined and dressed). The number of foremen and laborers, time and cause of delays, and weather report are entered at the bottom of the sheet, where also a sketch is made of the track ballasted, with cross-section showing depth under ties. A somewhat similar report is used for tracklaying on new lines. Monthly reports of main track ballasted, made by the roadmasters, may have columns headed as follows: mile-post location, station stakes, which track, length of track ballasted (feet), kind of ballast, cubic yards placed.

A monthly ballasting report made by superintendents to the engineer of maintenance-of-way on the Missouri Pacific Ry. shows the locations at which ballast has been applied for patching, with the number of yards applied and the depth under the ties at each stretch or patch.

Ties and Rails.—Reports of ties and rails, made weekly or monthly by the section foremen (Fig. 113) show the number received, laid, removed, shipped and (ties) burned, the numbers being separated for main and side tracks, and the kind of wood. For treated ties removed, a special report of 10 columns may be made, showing (for main and side

CHIEF ENGINEER'S DEPARTMENT.

REPORT OF RAIL LAID IN MAIN TRACK.

On the __________ Division. Month of __________ 190_

	RAIL LAID.							RAIL REMOVED.		
	SURVEY STATION. From.	SURVEY STATION. To.	Northbound or Southbound	Feet of Track.	BRAND AND DATE OF RAIL	Weight per Yard.	New or Old.	BRAND AND DATE OF RAIL.	Weight per Yard.	
1										1
2										2
3										3

FIG. 115.—Monthly report of rails.

track separately): the number of removals, the year marked on the ties, kind of wood, kind of treatment, and cause of removal. The first column gives the number of ties bearing each date noted in the second column. To avoid mistakes, if the ties removed are untreated, a dash is marked under "treatment" to show that this item is not omitted by oversight. For test sections of track, special forms of reports may be made, as noted in the chapter on "Ties." Reports of the Bessemer & Lake Erie R. R., which uses steel ties extensively, has separate columns for steel and wood ties and two columns for "damaged" and "scrap" steel ties. During rail renewals, monthly reports may be made by the roadmaster on such a form as Fig. 115. This is $8\frac{1}{2}$ by 11 in., with 15 numbered lines ruled in columns and then 15 numbered lines for corresponding remarks or explanations.

Switches, Frogs, and Sidetracks.—A detailed report for every switch and frog removed, as in Fig. 116, serves as the basis of an annual summary of the life of these appliances. The report, $8\frac{1}{2}$ by 11 in., is made by the roadmaster after personal inspection. Monthly reports of all main-line turnouts and their conditions, as sometimes required from roadmasters, are ruled in columns for: location; track; facing or trailing; kind and

.... DIVISION.

REPORT OF FROGS AND SWITCHES REMOVED FROM TRACK.

Date, ... 19....

Section No. .... Station .... Survey Station .... + ...

Name of Track .... Main or Side Track ....

FROGS	SWITCHES
No. .... Length .... Weight of Rail	Length .... Weight of Rail ....
Spring or Rigid .... Right or Left Hand	Split or Special ....
Bolted or Keyed .... New or Old When Laid	Main or Side Track Point .... Right or Left Hand ....
If Special Make, give name	Part Damaged, Broken or Worn Out ....
Name of Manufacturer	New or Old When Laid ....
Show Location of Break on Sketch. If Spring Frog, mark Spring Rail.	Distance of Break from Point ....
	Condition of Stock Rail ....
	Name of Manufacturer ....

CAUSE OF REMOVAL—BROKEN.—State Cause and Condition of Point Broken ....

....

DAMAGED OR WORN OUT.—State Exactly in What Way ....

....

DEFECTIVE.—Locate any Flaws and State Cause ....

....

CONDITION.—Fit for Use, Scrapped or can Repairs be Made? ....

By Whom can Repairs be Made? ....

State if any Accident or Delay was Caused ....

....

Kind of Ballast .... Condition of Ballast ....

Condition of Ties .... Line .... Surface ....

Date Laid—Year .... Month .... Date Taken Up—Year .... Month .... Life in Track—Years .... Months ....

INSTRUCTIONS.
Supervisors will make this report from personal inspection for every Frog and Switch of over 71-lb. rail removed from track. Date report day Frog or Switch is removed from Track.

Correct.

....
Supervisor.

FIG. 116.—Frog and switch report.

throw of switchstand; condition of switch point; kind and condition of frog; width of guard-rail throat; gage at frog; remarks.

In semiannual inspections by the engineers and superintendents on the Pennsylvania R. R., each switch is examined and operated, and report made on a printed form including such items as: condition of point, throw, switchstand, connections, guard-rail spacing, bolts, and cotters.

The foreman is with the party and is given a carbon copy of the report. Another plan provides for a weekly report by the section foreman and a monthly report by the roadmaster to the division engineer and the

..................................DIVISION.

MONTHLY SIDE TRACK REPORT FOR..............................190.....

Name of Siding..............................*connecting with*..............................

Located at mile-post..........*plus*..........*feet north, running in*..............................*direction*

..................................*side of main track.*

DATE			LENGTH IN FEET					
AUTHORIZED	BEGUN	COMPLETED	NEW	ORIGINAL	EXTENDED	TAKEN UP	PRESENT	CLEAR

KIND OF MATERIAL		NEW	OLD		PRICE	AMOUNT	
Steel rails pounds per yard	Feet						
Joints	No.						
Track Bolts	Pounds						
Spikes	"						
Rail Braces	No.						
Cross Ties, kind	No.						
Switch Ties	Feet						
Frogs, kind No.	No.						
Switches, kind	No.						
Switch Stands, kind	No.						
Switch Locks	No.						
Switch Lamps	No.						
TOTAL							

KIND OF LABOR	HOURS						
	Section Men	Extra Gang	Work Train				
Grading							
Putting in switch							
Laying rails and ties							
Hauling ballast							
Putting in ballast							
TOTAL							
GRAND TOTAL							

INSTRUCTIONS.

A sketch of the track, showing connections, must accompany this report.

The columns "Price" and "Amount" are for use in M. of W. office only.

All lengths except "Clear" to be figured from point of switch.

The above statement is correct.

..............................ROADMASTER.

FIG. 117.—Report of sidetrack work.

superintendent, showing the condition of each switch. In his inspection, the roadmaster notes any defects in a book and sends a carbon copy to the foreman, who signs and returns this copy when the repairs are made.

Reports of sidetrack work show each track put in or taken up, and give the location, length, weight of rail, and kind of switch. A roadmaster's monthly report of new sidetracks is shown in Fig. 117, on which is made also a sketch of the track and its connections. The last two columns are filled out in the engineer's office. On the Missouri Pacific Ry., a monthly report of sidetracks laid or removed is prepared by the assistant engineer and forwarded through the division engineer to the chief engineer. After record has been made the report is referred to the valuation engineer.

Fencing and Ditching.—A roadmaster's monthly report of fence work on his division is a letter-size sheet ruled as shown in Fig. 118. Reports of fence and ditch work made weekly by section foremen have a line of each day's work and the total. They are ruled in columns for date, location by station stakes, side of track, hours of work, and remarks. The fence report has columns for length built (rods), number of posts set, and rods of wire used. The other report shows length of ditch built, length of tile laid, and number of tile laid. A daily report for a fence gang shows the number of men, length and kind of fence built and repaired, post holes dug, posts set, kind of soil, and number and kind of cattleguards built and repaired.

FENCE CONSTRUCTION ON .... DIVISION DURING...... 19..

Location by station stakes		Which side of track	Length, feet	Kind of fence	Cost			New or old material	Remarks
From	To				Labor	Materials	Total		

FIG. 118.—Report of fence work.

Work-train and Extra-gang Reports.—Daily reports are made as a rule by the foremen of work-train gangs and extra gangs, the reports being sent either to the roadmaster or the engineer. One form is simply a letter-size sheet with ruled lines and a right-hand column for the hours of work on each kind of work entered. On this the foreman makes a complete diary, describing the work, force, progress, delays, and other details. Other roads use ruled forms. Work-train conductors send reports each night to the roadmaster or the engineer, giving the number of the locomotive, time of going on and off duty, summary of work done, and particulars as to cause and time of delays. Samples of daily reports are shown in Figs. 119 and 120. Similar reports by steam-shovel operators will include: time of stopping and starting; time for spotting; delays (length and cause); working time; number of men at pit; number of men with trains; type and capacity of cars; number of cars loaded; kind of material; average haul; supplies consumed.

Miscellaneous Reports.—Monthly reports of equipment and condition of highway crossings may be ruled in columns to show for each crossing its name, location, kind of gate, condition, and repairs made.

CHIEF ENGINEER'S DEPARTMENT.

DAILY REPORT OF EXTRA FOREMAN.

On what piece of work engaged 190

Work Train Engine No. .... *Reported for duty at* .... *M. Went off duty at* .... *M.*

CARS LOADED.					CARS UNLOADED.				
Initials.	Numbers.	Sec. No.	Contents.	At What Point.	Initials.	Numbers.	Sec. No.	Contents.	For What Purpose.

HANDLED FROM STEAM SHOVEL.

No. of Flat Cars Loaded .... *Unloaded* .... *Kind of Material* ....
" *Gondolas* " .... " .... " ....
" " .... " .... " ....
" " .... " .... " ....

DAILY DISTRIBUTION OF LABOR.

Extra Foremen should fill out this blank each night and forward to their superior.
The Distribution of Time should be taken from the book in which it is kept, the object being to obtain, daily, a record of how much time is being expended on each piece of work, or portion of it.

DISTRIBUTION OF LABOR. (EXCLUSIVE OF TEAMS AND DRIVERS.)	Section Number.	Number of Men Working.	Total Hours Worked.	AMOUNT. $	Cts.
Teams and Drivers—Give Number					
Total,					

.... *Extra Foreman.*

NOTE.—Time of Foreman should always be included.

FIG. 119.—Work-train report.

Daily or weekly reports are made by water-station attendants. A daily card report suggested by the A. R. E. A. is a card $5\frac{1}{2}$ by 6 in. having printed on one side a set of instructions and the name and address of official to whom the report is sent. The other side is ruled for entries as to coal or gasoline used, time of pumping (started, stopped, running time), and supplies received and on hand. Where a water-softening plant is in use, a special report is made, including routine tests of raw and treated water, quantities pumped and delivered, chemicals used. Periodical reports will be made also by the chemists and superintendents of water service. Reports of condition, inspection, and repair work on

DAILY REPORT OF EXTRA FORCE, ... DIVISION.

.... 189......

This report should be forwarded to Division Engineer daily. "Extra Men" includes all men paid on Road Master's rolls outside of those on Section and Fence Gang rolls. Each gravel pit and each piece of important work, such as a section of second track construction, change of grade, a new yard, or extensive ballasting of old main track, should be kept separate. Gangs or work trains engaged in ordinary miscellaneous work should be shown as doing such.

At What Point, or On What Piece of Work.	Days' Work Performed by Extra Men.					Excavators at Work. Numbers of	Number of Work Trains.	Locomotives, Numbers of	Cars in Service of Work Trains at End of Day.			Remarks.
	Conductors and Brakemen.	Men on Excavator Rolls.	Teamsters with Teams.	All Other Extra Men.	Total.				Flats.	Gondolas.	Dumps.	

DAILY REPORT OF GRAVEL AND EARTH HANDLED, ... DIVISION, ... 189...

This Report should be Forwarded to Division Engineer Daily. The material handled on each piece of Second Track, Change of Grade, or other important work, as well as in each Gravel Pit, should be Shown Separately. Gravel unloaded for ordinary main line ballasting on divisions where such ballast has been received from elsewhere, must be likewise treated.

Loaded								Unloaded or Shipped.							Remarks.
At What Point, or On What Piece of Work	By What Means, Excavator or Hand.	Gravel			Earth.			On What Piece of Work, or To What Point.	Gravel			Earth.			
		Flats.	Gondolas.	Dumps.	Flats.	Gondolas.	Dumps.		Flats.	Gondolas.	Dumps.	Flats.	Gondolas.	Dumps.	

FIG. 120.—Reports of extra gangs.

buildings are noted under "Buildings." Reports of serious train accidents, as made by roadmasters, specify the train time, location, and kind of accident, together with such information as the following:

Main, side, or private track	Condition of track
At switch or frog	Gage of track
Trailing or facing switch	Damage to train
On curve or tangent	Damage to track
Curve; degree, elevation, gage	Damage to structures
Track construction	Cause of accident

Time and Labor Distribution.—Two reports of fundamental importance in maintenance-of-way records and accounting are the daily records of labor time and distribution of the track forces, as made by the section foremen, since upon these figures are based the accounting and distribution of classification of expenses. The time report is comparatively simple, consisting usually of a book or sheet having a line for the name and occupation of each man, followed by columns for the number of hours worked each day, the rate of pay, and note of any deductions or special conditions. With present wage conditions and the high extra rate for overtime, it is necessary to separate the regular and overtime hours and pay, requiring two lines opposite each man's name.

These figures are recorded daily for each man, and are then posted on a time sheet or time book, which is usually sent to the roadmaster at the end of the month as a basis for the payroll. In states which have laws requiring semi-monthly wage payments, the time books must be prepared accordingly. An extension of this system on some roads requires the "hours worked" to be given for morning and afternoon separately, but this is not usual. The foreman must be instructed as to the necessity of making up his report every day, and not leaving it to be filled out from his notes and memory at a later time. In fact, some roads require a time report to be sent in daily. The time book should be carried by the foreman while on duty, so that it can be examined by the roadmaster in making his trips over the line. On the Chicago & Northwestern Ry., the time roll is a loose-leaf sheet 8½ by 14 in.

The distribution of labor as to the various kinds of work done is more complicated, and although the distribution or classification is governed to some extent by the Interstate Commerce Commission accounting system, there is great variation in detail on different railroads. As a rule, the foreman describes the work in his own way in his pocket book, and then enters it against the appropriate item on the printed list or sheet forming his monthly report. As with the time report, the figures are entered daily and totals are made at the end of the month. On the Chicago & Northwestern Ry., the monthly distribution of track labor is made on a sheet 8½ by 14 in., having at the left a list of items of work.

Then come narrow columns for the hours spent each day on these different items, with another column for the total of the month, and a final column giving the number of the account to which the item or work is chargeable. At the bottom of each daily column is entered the total hours worked. This list of items, with explanatory remarks is given below:

Distribution of Track Labor—Chicago & Northwestern Ry.

Roadway Maintenance:—Cleaning drift, cinders, dirt, and scrap from tracks, roadways, ditches, and right of way. Cutting weeds and brush. Removing slides. Sloping cuts. Restoring shoulders. Dressing ballast and sod lines. Oiling roadbed. Plowing fire guards. Putting out fires along right-of-way. Patrolling tracks. Sunday track inspection. Protecting banks, riprapping, and repairing washouts. Laying and removing temporary tracks at washouts.

Bridges:—Repairing and watching, cleaning water channels under bridges. Filling water barrels. Filling abandoned bridges and placing riprap at bridge abutments.

Ballast:—Unloading. (Does not include putting under.) See "Applying Ballast."

Tunnels:—Watching at tunnels, repairing or whitewashing tunnels. (Does not include repairs to roadway and track in tunnels.)

Other Track Material:—Repairing frogs, switch points, etc., by oxyacetylene.

Track Repairs—General:—Surfacing and alining track. Gaging, shimming, tightening bolts, etc. Repairs to tracks damaged by wrecks, washouts, or derailments. Taking up tracks.

Applying Ballast:—Putting ballast in track. (Does not include unloading.) See "Ballast."

Applying Track Ties:—Unloading and distributing; also gathering up and loading old ties and destroying same.

Applying Switch Ties:—Unloading and distributing; also gathering up and loading old ties and destroying same.

Applying Rails.—Unloading and distributing; also loading rail removed from track.

Right-of-Way Fences:—Repairing fences, farm gates, cattleguards, and wing fences. Repairing snow and sand fences when used as right-of-way fences.

Snow and Sand Fences and Snowsheds:—Repairing of permanent and portable snow fences except where they take place-of-right of way fence. (Setting up and taking down portable snow fence should be shown under "Removing Snow, Ice, and Sand.")

Crossings and Signs:—Repairing farm passes, highway crossings, track signs, crossing gates, crossing signal; also repairs to over- and under-grade crossings.

Station Buildings:—Cleaning stockyards, filling with cinders, etc.

Telegraph and Telephone Lines:—Repairing lines, resetting poles, loading or unloading telegraph or telephone material. (Repairs to telephone lines must be separated from repairs to telegraph lines and detailed report shown on back of this sheet.)

Signals and Interlockers:—Repairing interlocking plants, automatic or other signals. Unloading material to be used in repairs to signals. (Show name of interlocker or location of signals.)

Roadway Machines:—Repairs made to hand cars, motor cars, and push cars.

Small Tools and Supplies:—Repairs to section tools.

Removing Snow, Ice, and Sand:—Clearing roadway and track of snow, ice, and sand; distributing, setting up, and taking down portable fences.

Station Labor:
- Transferring, loading, and reloading freight. Disinfecting stock-yards or pens. Cleaning freight stations.
- Carrying mail, janitor work at passenger stations, miscellaneous labor at passenger stations.
- Cleaning stations other than exclusive passenger or freight stations or other labor at stations used for both passenger and freight business. Sprinkling station grounds.
- Tending switch lamps. (Show name of station.)
- Unloading coal at stations. (Show whether for station, pumphouse, flag shanty, or interlocker tower.)

Fuel for Locomotives:—Loading or unloading coal at chutes. (Show name of station.)

Fuel—Coal:—Labor loading or unloading coal into storage. (Show name of station.)

Enginehouse Expense:—Loading cinders at cinder pits. Loading or unloading coal for enginehouse use. (Show name of station.)

Other Supplies for Locomotives:—Unloading sand for locomotive use. Drying sand for locomotive use.

Train Supplies and Expenses:—Cleaning and disinfecting stock and freight cars. Icing refrigerators.

Signal and Interlocker Operation:—Cleaning interlocking plants. Working as emergency levermen.

Crossing Protection:—Working as crossing flagmen or gatemen.

Damage to Livestock:—Removing and burying cattle or livestock killed on right-of-way. (Show number of train striking stock.)

Clearing Wrecks:—Rerailing engines and cars. Building temporary tracks, transferring passengers or baggage at wrecks, loading or unloading or transferring contents of wrecked cars. (Show whether passenger or freight train wreck.) (Repairing track on account of wreck must be reported under "Tracklaying and Surfacing.")

Other Expenses:—Time attending safety committee meetings.

Gravel Pits:—Clearing, ditching, draining, and stripping. Laying and repairing tracks. Loading gravel. Laborers and watchmen around steam shovel.

Authorized Work:—Includes all new work and improvements, and should be reported in total here. Show A.F.E. number, location, nature of work, and hours engaged. Total hours worked each day.

Units of Cost and Work.—Itemized unit costs of track work are not available under the ordinary time-book distribution of track labor, which gives only the items of work and the time spent upon each, omitting entirely the amount of work done during that time. In view of the large proportion of operating expenses represented by maintenance-of-way, it is very desirable that unit costs should be determined. Such records would enable an approximate average or standard to be developed for each item of work, and comparisons made as to the relative efficiency and economy of the work or different sections of a division and on different parts of a railway system.

Such cost records must be based upon the daily reports of the foremen of section and extra gangs, as already noted. To meet this situation, a few railways have introduced forms of reports which are designed to give the necessary basic information and at the same time to be sufficiently simple to insure accurate figures being given. The distribution of labor under the new system varies on different railroads, but it is probable that as the use of the system extends, the reports used will gradually be reduced to some approximately standard form.

Unit Cost Report—St. Louis-San Francisco Ry.—In the system of track reports and accounting devised by F. G. Jonah, chief engineer of the St. Louis-San Francisco Ry., the foreman is provided each month with a pocket book 4 by $6\frac{3}{4}$ in., in which he enters the names of his men and enters daily for each man the hours of straight time and overtime. He enters also the work done each day and the amount or unit of work, one page being ruled in lines for a description in his own words and the opposite page having columns for: kind or name of track or structure; location by mile post or station; track letter; number of units; and number of hours. The lower half of one page is for "material used" and the other for "material released." Other pages are for rails, ties, and other materials received and disposed of, as in reports previously mentioned.

At the close of the day's work, the figures are transferred from the pocket book to the monthly time book, which has pages 10 by $9\frac{1}{2}$ in. In the labor distribution or "kind of work," there are 46 items for each of which is given the time and amount for each day, as in the following example:

Cleaning ditches....................hours;......................linear feet
Inserting ties.........................hours; ....number of ties
Clearing weeds......................hours;......................feet of track
Applying ballast....................hours;.....lift, inches........feet of track

For accounting purposes, these 46 items, which are listed in Table XXII, are arranged under the designation of the various accounts to which they are chargeable. They are divided also for "maintenance labor," "transportation labor," and "labor on new work." The second of these divisions is to show what proportion of the section gang's time is taken up by labor chargeable to transportation accounts, such as cleaning stations, unloading coal for stations, or caring for switch and signal lamps. The third division includes grading, unloading ballast, tracklaying, and removing tracks. In order to secure cost data on the essential items of track-maintenance work, the headings for distribution of time are given so plainly that the foremen should have no difficulty in transferring the figures correctly from his pocket book to the distribution sheet. (See also Chapter XXVIII.)

Another feature of this system is the dividing of tracks into 26 classes, beginning with: A, first and second main track; B, passing tracks, and so on down through storage, freighthouse, repair, shop, and enginehouse tracks to Z, company material tracks. The corresponding letters are entered in the "track letter" column of the pocket book noted above. Some other roads classify simply: (1) main tracks; (2) tracks used by freight only; (3) other tracks.

TABLE XXII.—CLASSIFICATION OF TRACK WORK—ST. LOUIS-SAN FRANCISCO RY.

Item		Cents
1. Cleaning weeds from track	A	1.00
2. Cleaning ditches in cuts	B	9.00
3. Cleaning surface ditches	B	2.00
4. Mowing and burning right-of-way	A	0.70
5. Unloading ballast, per car		$ 4.12
6. Applying chats and gravel ballast	A	7.00
7. Applying slag and rock ballast	A	10.00
8. Applying cinder, sand, or other ballast	A	10.00
9. Inserting ties in chats and gravel ballast	C	44.00
10. Inserting ties in slag and rock ballast	C	42.00
11. Inserting ties in sand, cinders, or earth	C	35.00
12. Surfacing track on chats or gravel	A	7.00
13. Surfacing track on slag or rock	A	11.00
14. Surfacing track on sand, cinders, or earth	A	9.00
15. Unloading rail	B	7.00
16. Loading rail	B	9.00
17. Laying rail	A	9.00
18. Lining and gaging track	A	3.00
19. Applying rail anchors, each		3.00
20. Applying tie-plates, each		6.00
21. Repairing signs		

22. Repairing right-of-way fences
23. Repairing cattleguards and wing fences
24. Repairing grade crossings
25. Unloading coal for stations
26. Unloading coal for water stations
27. Patrolling track
28. Repairing roadbed damaged by washouts
29. Loading and unloading ties
30. Applying frogs and switches
31. Applying other track material
32. Repairing track damaged by washouts
33. Repairing track damaged by wrecks
34. Cleaning and scrubbing stations
35. Transferring and handling freight
36. Care of switch lamps outside switch yards
37. Care of switch lamps in switch yards
38. Care of signal lamps
39. Watching crossings
40. Clearing wrecks, transferring freight and baggage
41. Removing and burying stock killed
42. Removing snow, ice, and sand from platforms and tracks
43. Signals and interlocking
44. Care of station grounds
45. New work
46. All other work not classified above

These semi-monthly reports of the foremen go to the roadmasters and then to the bureau of accounts, where the payrolls are made up and where the statistics relating to unit costs are compiled. The results are entered on large sheets having the 46 items at the left, with two lines for each item. On one line is the unit (number of ties, feet of track surfaced, etc.) and on the next line is the unit cost. For the first 20 items, however, only the cost of time, in hours, is shown. On one sheet, the columns show the figures for each section on a roadmaster's territory, with an average for the whole territory. Another sheet shows the average for the several roadmasters' territories on each division, and finally there is a sheet showing the average for each division in a district, together with an average for the entire system. Thus for each item of work, there is the unit cost on each section, division, and district and on the railway as a whole. These figures are reviewed by the district engineers, who call attention to work that appears to be of excessive cost on any division, territory, or section.

To illustrate: One divisional report shows that on 11 divisions the cost of applying gravel ballast ranged from 4 to 12 cts. per foot of track, with an average of 7 cts. for the entire railroad. Inserting ties in gravel ballast cost 38 to 55 cts. per tie, and laying rails cost from 6 to 35 cts. per foot of track. In the accompanying list the average cost is given for some of the items in a certain month: *A* indicates cost per foot of track; *B* indicates cost per lineal foot; *C* indicates cost for each tie.

Cost-record System: Illinois Central R. R.—A somewhat similar system on the Illinois Central R. R. has only 24 items of work and gives the units in man-hours instead of cents. Thus, one roadmaster's report shows that for putting in ties, one section placed 248 ties in 82 hr.; another placed 114 in 57 hr., and so on, with a total of 5,059 ties in 1,636 hr., or an average of 3.08 ties per man-hour. A report for the entire division gives units and time for each roadmaster's territory, with an average of 2.03 ties per man-hour for the division. The forms made out by the section foremen are consolidated at the end of the month, and a typewritten tabular summary is prepared, showing the average man-hours consumed for each item and also a standard average as issued by the engineer of maintenance-of-way. Where there is great disparity between the actual and standard averages, the roadmaster is notified, and he in turn takes up the matter with the section foreman and tries to have the average improved for the next month. As a matter of interest, some averages for one division (for a certain month) are given below:

	Average per man-hour
Spotting in ties; rock and cementing gravel	1.27 ties
Spotting in ties; bank-run, washed gravel, and cinders	1.62 ties
Spotting in ties; while surfacing	2.03 ties
Surfacing track; rock and cementing gravel	4.9 ft.
Surfacing track; bank-run gravel	5.3 ft.
Dressing track; rock and cementing gravel	12.2 ft.
Dressing track; bank-run gravel	18.0 ft.
Dressing track; washed gravel and cinders	21.4 ft.
Unloading rail and fastenings	93.00 ft.
Laying new rail	3.76 ft.
Laying second-hand rail	9.70 ft.
Applying rail anchors, number	25.00
Applying tie-plates, number	13.60

Requisitions for Materials.—In the handling of requisitions for supplies for the track department, there has been frequent trouble owing to the requisitions being disapproved or cut down by superior officers without knowledge of the actual requirements. An erroneous idea sometimes entertained is that it is economy to cut down a requisition, but officers with authority of this kind now more generally realize the economic relations between expenditures and the results obtained therefrom. If the superior officer considers that the requisition is too large, he should at least consult the engineer who has approved it and thus ascertain the actual conditions before making any change. Arbitrary cutting of requisitions is discouraging to men who have prepared them carefully with a view to actual needs, and if the practice is general, it is likely to result in careless estimates or in requisitions purposely enlarged in order to meet requirements even if reduced. The roadmaster or the engineer who is responsible for the work should keep himself informed as to business and financial conditions, preparing his estimates and requisitions with these conditions in view, and being prepared to show that they are reasonable and necessary in case they are called in question.

Systematic handling of requisitions is necessary for record of materials and their cost. Printed forms are generally filled out by the section foreman or the roadmaster and forwarded to the engineer. Separate forms of requisitions for standard stock material and special material are made out by section foremen on the Bessemer & Lake Erie R. R. The former, after approval by the supervisor and engineer of track, are sent to the stores department for shipment, the requisitions being sent ordinarily twice a week, except when it is necessary to have material promptly in order to avoid delaying the work. For special material, a requisition is sent usually by the engineer of track to the chief engineer for signature; after this it is sent to the stores department, which makes the necessary purchase. On the Chicago & Alton Ry., requisitions for materials are sent by the division roadmasters to the superintendent and on approval they go to the general storekeeper. A copy is sent also by the roadmaster to the general supervisor of maintenance, and if he considers any modification desirable he notifies the storekeeper to hold the requisition until an investigation is made. Small quantities or materials required immediately are furnished directly by the storekeeper on request of the roadmaster, who then conforms it by a requisition in the usual way.

On the Missouri Pacific Ry., requisitions made by roadmasters, supervisors and foremen are sent on a form which covers materials carried in stock by the storekeeper. The material is furnished by the supply car or in some other way on receipt of this form approved by the division engineer and superintendent. In some cases the storekeeper makes a purchase requisition on another form to cover material requested in excess of the usual stock, this form being first approved by the chief engineer or engineer of maintenance-of-way and the general manager. The supply agent also approves the requisition, to indicate that the material cannot be supplied from store stock. In special cases such requisitions may be referred to the president for approval. At the beginning of the season, the engineer of maintenance-of-way advises the purchasing agent of the material which will be required for the year's program for laying rails, tie-plates, rail anchors and splice bars, the purchasing agent then arranging for the purchase of the materials and for their delivery on request. Each month the engineer informs the purchasing agent by letter of the amount of each kind of material which will be required during the following month at each location where work is in progress, and the supply agent arranges for delivery as specified. This system is followed for the purpose of reducing to a minimum the amount of matetial actually carried in stock.

Authority for Expenditure.—For any proposed change in railway property involving an extension, addition, or betterment, it is necessary to request authority to make the necessary expenditure, for which a careful estimate of cost must be made in advance. This request is made on a special form giving the location and description of the project, reasons and necessity for the work, and an estimate of cost, with credit for salvage. The form is signed by successive officers recommending its approval, and when signed by the highest approving officer, it becomes a work order or "authority for expenditure," known as an A.F.E.

This A.F.E. authorizes the expenditure of a specific amount on a specific piece of work, such as a length of second track, a length of rail renewal, a new coaling station, or the renewal of a bridge. As a rule, the order is good only for the year of the issue, a new A.F.E. being required for the succeeding year, so that the annual expenditures may be classified properly. The expenditure must be kept within the amount authorized, and if this amount should prove insufficient, a separate and supplemental A.F.E. must be obtained. As the progress reports are received, the amounts are entered on the A.F.E. form. Such expenditures are classified as: (1) maintenance; and (2) additions and betterments, known as "A & B." Thus, replacing a switch with a new one of the same weight is classed as maintenance, while installing an additional switch is classed as "A & B."

Maintenance Instruction Books.—It is usual for a railroad to have a book of instructions for the information of roadmasters and section foremen as to the methods of maintenance and track work, the standard plans and practice, the duties and responsibilities of officers and men, the company's regulations, and the safe handling of the work and the railway traffic. In introducing this plan, the books may be small and inexpensive pamphlets for different lines of work. After these have been in use for sufficient time to develop criticisms or desirable changes, the material may be improved and compiled in permanent book form, of convenient size for the pocket. An average size is 4 by 6 in., but the contents range from 130 to 300 pages. The matter should be clearly and concisely written and well printed in good-sized type, so as to be easily read and understood.

These books include the rules and instructions governing roadmasters, section foremen, work-train conductors, extra-gang foremen, water-department employees, and bridge and building foremen. Methods of performing ordinary work are given, such as surfacing, lining, and rail renewals. Train rules and train signals are usually included, together with such material as tables of superelevation on curves, table of ordinates for bending rails, and bills of timber for switches and turnouts. Illustrations may be given of roadbed cross-sections, turnout diagrams, methods of piling ties, and fences at road crossings. These illustrations should be clear, with lettering easily read, and are preferably printed on the pages, as folded sheets are likely to be torn and lost. The books are regarded as railroad property, the same as the tools and appliances furnished. They are numbered, and record is kept of their issue, the foreman being required to sign a receipt and to turn in the book with other property if he leaves the employ of the railroad.

Correspondence and Form Reports.—While the duties of roadmasters and supervisors keep them out on the line for the greater part of the time, prompt and careful attention must be given to communications from superior officers. Replies should be definite and concise. The letters should be filed for reference, with copies of the replies. Special care should be taken in the proper filling out of report forms and blanks. Each officer receiving reports of this kind should note them carefully and call attention to anything which indicates that they are incomplete or inaccurate, or that there has been excessive cost, so that such matters may be corrected or improved.

CHAPTER XXVIII

ECONOMIC ASPECTS OF TRACK MAINTENANCE

Budget Systems for Track Work.—A great obstacle to economical and effective management of track maintenance is the lack of a permanent force, owing to the general custom of seasonal employment and the unfortunate tendency of managements arbitrarily to order work stopped and forces reduced whenever there is a decline in traffic. That the seasonal system is a disadvantage and is largely unnecessary has been shown already. The practice of arbitrary changes during the working season disrupts careful plans, disorganizes forces, discourages men and officers, and causes a loss of money and waste of material that probably exceed any saving in wage payments; and such loss cannot be made good when orders are given to resume work. This situation can be overcome only when executive officers realize the economic aspects of the situation. For both maintenance and improvement work, there should be a budget or definite appropriation for a term of 3 to 5 years (or at least 1 year), and all work then carried out under the plans and appropriations of the budget. In this way, and with the funds authorized sufficiently in advance of the program, the simultaneous supply of money, men, and materials can be arranged and the work done to the best advantage and with the highest economy. On each district or division, the local officers would plan in advance the distribution of their part of the appropriation, assigning a certain amount for each month and giving the larger proportions to the months in which most work can be done. The roadmasters would be informed of the amounts allotted to them and would also be given monthly statements of expenditures to show how nearly they are working to the prescribed amounts.

This budget and distribution system should be used also in planning the work of extra gangs. It has been suggested by Mr. Morse, chief engineer of the Rock Island Lines, that a budget made for each division and subdivided for each track section should cover the expense of four classes of work: (1) minimum organization to maintain track, exclusive of tie renewals; (2) additional seasonal help for renewing ties and mowing right-of-way; (3) additional help for improvements that reduce future maintenance cost, such as widening cuts and fills and cutting ditches; (4) exceptional expense due to storms, wrecks, and snow.

Distribution of Allowance for Maintenance.—On the Chicago, Rock Island & Pacific Ry., each division engineer sends a request to the superintendent for an allowance for material and labor on all work to be done during the coming month. He sends copies also to the engineer of maintenance-of-way for his district and to the engineer of maintenance-of-way of the system. After considering these several requests, the district engineers of maintenance-of-way submit their recommendations to their general managers, who pass them on to the chief engineer. Then the engineer of maintenance-of-way for the system, after conferring with the chief engineer, prepares recommendations which are submitted to the vice-president in charge of operation. Finally, the vice-president, at a meeting with the chief engineer and general managers, decides the amount which is to be spent during the coming month.

When the district engineers of maintenance-of-way are notified of the amounts assigned to their districts, they proceed to allot the amount for each division and send a copy of this to the chief engineer. Each division engineer distributes his allotment to the roadmasters, the master carpenter, the signal supervisor, and other subordinate officers, who in turn instruct their foremen as to the amount which can be expended. Each division allowance is split into four parts, for four periods of the month. At the end of each period, a report of the amount spent during that period is sent by the division engineer to his district engineer of maintenance and the engineer of maintenance-of-way for the system. If the allowances are being overrun without apparent reason, action is taken to investigate the situation and control the expenditures.

For maintenance work on the Kansas City Southern Ry., each division engineer's office submits an estimate of its requirements during the coming month. These estimates are passed upon in the chief engineer's office, to be revised, reduced or increased as may be considered necessary. A consolidated program is then prepared and submitted to the management, and when this is approved it is distributed to the divisions as an allowance for the month. Using this program as a basis, estimates are prepared four times a month by the division offices, showing the approximate expenditures under each heading for the period reported. These reports enable the chief engineer's office to maintain a close check on expenditures during the month and to prevent the allowances from being exceeded. Improvement work is budgeted each year, so that monthly programs are not prepared, but a statement in letter form is sent weekly from each division office to the chief engineer, showing the estimated charges during the week for betterment or improvement work.

For railroads of the smaller class, a less elaborate system is suitable. In some cases, there are no monthly estimates for maintenance expenses, but at the end of each year a budget is prepared covering cost of maintenance of track and structures and another budget for construction work. When the budgets are authorized, the work progresses accordingly, and if it appears that expenses will exceed the estimates and budget allowances, additional authority is secured to cover these expenditures before they are incurred.

Program or Schedule of Work.—To supplement the budget and provide for the best use of the funds thus made available, a program or schedule of work must be prepared so that materials may be ordered and forces organized accordingly, either for general maintenance or for some large improvement job. It is desirable to have the season's work scheduled in advance for each division and for each track section, to insure economy and systematic progress and to avoid having too much time spent on the smaller routine work. With the amount of appropriation known, the division or maintenance engineer and the roadmaster should go over each section with its foreman to consider the condition and needs as to rails, track material, ballast, drainage of soft spots, widening fills, ditching, and sidetrack and yard requirements. The engineer and the roadmaster should then draft a schedule of the work for each month and the date for completion. They should also prepare a similar schedule or program for the extra-gang work. Such schedules can be shown on charts for duplication by blueprinting.

In a system for scheduling and recording the season's work on a division, as proposed by the A. R. E. A., the program is decided early in the year and then laid out on a chart to show the amount of work of various kinds on each section for each month, and for each track separately on double track. This program is marked in yellow, while on monthly progress charts it is marked in red and the actual performance in green. Near the end of the month, the division engineer calls a meeting of the roadmasters to plan the work for the next month, with its location and the relative order in which items are to be undertaken. This program is plotted on the chart, with dates for commencement and completion of the items. The quantity of work to be accomplished is computed from standard schedules and the force is allotted accordingly.

Each roadmaster has a copy of the chart for his own territory and issues to the foreman work orders covering all or a part of the month's work. A carbon copy of each order is posted on a board in the roadmaster's office, so as to show the location of the gangs and the work that is being done. On completion of the work specified, the foreman dates and signs the order and returns it to the roadmaster. The latter must see that the schedule is followed, and before the end of each month he will note and record the progress on the several specific works, so that at monthly meetings with the engineer the schedule and actual progress can be compared and steps taken to speed up any backward work or check any undue expense. This plan does not relieve the foremen from the responsibility of carrying on repairs and routine work. In continuation of this system, the foremen should be instructed and encouraged to plan each week's work in advance, for unless a man knows each day what he has to do, there will be much waste of time in going to and from small jobs, with neglect of the more important work.

Unit Costs and Performance Standards.—Attempts to increase the efficiency and economy of railway track maintenance and general improvement work face a serious difficulty in the absence of means for determining and showing the output per man or per gang or the unit cost of the items of work performed. In the first place, the men usually receive a regular and uniform wage. In the second place, the only basis for cost data (with few exceptions) is in the entries of working time made by the section foreman or gang foreman in his time book. It has been stated by Mr. Ford, assistant chief engineer of the Rock Island Lines, that: "In general the productivity of railway maintenance labor has not been commensurate with results secured in other lines of industry where measures have been used for gaging output." Some special forms of work reports for determining units of cost and performance have been noted in the previous chapter. Various cost-record systems have been tried, generally in an experimental and inconclusive manner, but it should be practicable to devise a system that would be complete, simple, fair, and efficient and would provide for determining unit costs and the relative efficiency of the different gangs. As the section gangs work as units, it is their efficiency that is desired, rather than that of the individual men.

It has been objected that the varying local conditions of track and traffic make general comparative data of questionable value, but this objection does not apply to comparisons of track sections on one division or to divisions having practically uniform conditions. Cost keeping performs two useful purposes: (1) to show actual unit costs under known conditions; and (2) to enable comparisons to be made of costs at different periods or different locations, assuming that a uniform system is used.

Time Studies.—As a basis for performance records and costs, standard schedules must be established for a given amount of work in a given time. This is done by time studies of the individual items of work. In a system proposed by the A. R. E. A. and largely following that of the Baltimore & Ohio R. R., described later, a daily record would be made of the performance of each gang, with its percentage relation to the standard units. The standard, or 100 per cent efficiency, would be the performance of a first-class gang, consisting of the most effective number of men, following an approved method of doing the work and working at a speed that can be maintained without difficulty or strain. The standards would be determined by careful studies of work done by such gangs. Since loss of time in going to and from work and in clearing the track for passing trains is beyond the control of the foreman, this time is classed separately as "detention" and is expressed in percentage of total time. In grading each foreman's record for comparison with the standard schedule, he is credited with the percentage, so that detentions do not affect the rating of his work.

As an example, assume that a foreman with a gang of men has to renew ties without raising track and that the standard schedule for such work, with this force, is 8 ties in 10 hr. The foreman's report of actual work shows 64 ties removed in 100 hr., while the standard allowance would be 80 hr. However, 10 per cent of his time is charged to detentions leaving 90 hr. for actual work, which represents an efficiency of 89 per cent. Monthly records compiled from the totals of the daily records would show the efficiency percentage for every gang on each class of work, together with the average of every gang for all work combined and the average for the roadmaster's territory. It is proposed that each supervisor would have an assistant devoting his entire time to the study of the schedules and actual performance and to the instruction of the foreman as to the standard practice. A specimen time study, omitting the detailed explanations is given in Table XXIII.

TABLE XXIII.—TIME STUDY: RENEWING TIES

Operation	Minutes per tie	Operation	Minutes per tie
1. Cribbing ballast	9	7. Tie-plating	1
2. Drawing spikes	3	8. Spiking	4
3. Removing ties	1	9. Tamping	10
4. Preparing bed	2	10. Dressing ballast	8
5. Carrying new tie	1	11. Piling old tie	1
6. Placing tie	2	Foreman's time	6
Hours per tie	0.8	Minutes per tie	48

Seven men, 8 hr	56 man-hours
Actual time renewing ties	49 man-hours
Average performance 49 × 0.8	61 ties per day

Time studies of tie renewal work on the St. Louis-San Francisco Ry. revealed excessive time spent due to faulty methods, the correction of which resulted in reducing the cost per tie from 44 cents to 21½ cents in one case, and from 34 to 20 cents in another case. Careful observations were made of several gangs, noting the conditions of work and the number of men on the several processes, and taking the time for each process with a stop-watch. An analysis of numerous records of this kind enabled methods to be improved, time and cost reduced, and a general average of unit time and cost for each process to be determined. A summary of the records for four gangs is as follows, the figures given being the average time in minutes for each tie renewed:

	1	2	3	4
	Mins.	Mins.	Mins.	Mins.
A. Pulling spikes	.....	1.8	2.7	1.0
B. Digging at side of tie	13.7	6.0	5.5	6.0
C. Removing old tie and inserting new tie	8.2	5.5	6.0	5.8
D. Tamping	13.5*	4.8	6.8	3.9
E. Spiking	6.5	3.8	2.9	4.3
F. Dressing ballast	.....	3.3	3.8	3.9
G. Tie-plating	.....	0.6	None	0.8
Total time per tie	41.9	25.8	27.7	25.7
Total cost, cents per tie	32.0	20.3	22.6	19.4

* Includes dressing ballast.

Equating Track Conditions.—In order to reduce the track conditions on different sections to a uniform basis for the allotment of the number of men and the amount for maintenance expense, the A. R. E. A. gives the following equivalents for 1 mi. of single main track: 1.15 mi. of second track; 1.33 mi. of third or fourth track; 2 mi. of branch-line track, passing track or thoroughfare track in yards; 3.33 mi. of yard tracks; ½ mi. of track tank; 4 mi. of ditch; 12 main-track switches; 20 sidetrack switches; 10 track crossings; 12 street crossings; and 25 to 50 country highway crossings. Thus, if a 5 mi. section on a single-track road has a mile of passing track and six main-track switches, it would have the same force and allowance as if it was a 6-mi. section of single track without any such features.

Economics of Track Labor.—Although labor is the largest item of expense in the maintenance of way and structures, comparatively little has been done, except in very recent years, to provide for the most effective and economical use of track labor. It has been shown already that the section foreman's daily distribution of time is still the usual basis of payment. In a system which was employed by the Baltimore & Ohio R. R. and the Pennsylvania R. R. for some years, but discontinued under war conditions, the mileage of each track section was equated according to the preceding paragraph. A monthly track-condition sheet was prepared also, based on inspection, with values given in relation to the following standard values: roadbed, 5; ties, 14; ballast, 15; rail and other track material, 12; line, surface, and gage, 50; right-of-way and grounds, 4. The equivalent mileage (determined as above) divided into the total percentage of condition gave the equated mileage for each section. The ratio of this mileage to the total equated mileage of the division determined of men to be allotted to that section out of the total number of men allotted to the division.

In this way, the number of men on each section was proportioned to the equivalent length of track and the relative condition of track on that section. Supplementing this distribution of force, daily records of track-work performance were kept in accordance with the system noted under time studies. At the end of the month, a monthly record was compiled from the totals of the daily record, this monthly record showing the percentage of efficiency made by every gang on each class of work, the average of every gang on all work, and the average for the road-master's subdivision. The standard performance divided by the actual performance gave the percentage of efficiency.

Bonus System.—Improvements effected by the performance-record system described above, were due mainly to better management and supervision of the work. For further improvement by personal effort on the part of the men, a bonus system of payment was introduced, beginning with work of 67 per cent of standard efficiency in order to stimulate efforts to reach the standard. From a small increase of pay for section gangs having 67 per cent efficiency, the scale rose to a bonus of 20 per cent of the wages for a gang having 100 per cent efficiency and 50 per cent bonus when a high efficiency of 130 per cent was attained. In addition to the bonus to foremen and laborers, the supervisor also received a bonus in order to enlist his interest in the general scheme of improvement in the track work and condition of track. In four years, the efficiency of typical subdivisions rose from 72 and 74 per cent to 91 to 101 per cent, or an increase of 17 to 30 per cent, while the efficiency of the division increased about 21 per cent.

The Contract System.—A special development in maintenance-of-way practice is the introduction of the contract system as a substitute for the railway force-account system. Actual experience has been too limited to permit of definite conclusions as to practical or economical advantages, although experience on a large scale in a few instances indicates that it can be operated successfully under proper management. In an extended discussion of the subject in *Engineering News-Record* of Oct. 26, 1922, it was pointed out that much depends upon the way in which the specifications and contract are drafted and the way in which the work is handled and supervised. For success, it is essential that on both sides there shall be not only competence and experience, but also integrity and an attitude of fairness and cooperation.

The contract system has been applied mainly to such work as is done usually by extra or floating gangs. This includes heavy ditching, widening cuts, tile drainage, renewing ties and rails, ballasting, surfacing, filling trestles, relaying switches or turnouts, building fences, and also renewing spikes, rail joints, tie-plates, and rail anchors on long stretches of line. To a much smaller extent, it has been applied to the routine work handled usually by section gangs, such as daily inspection and policing, respiking, tightening bolts, clearing ditches, cutting weeds, maintaining track in line and surface, and repairing fences and crossings.

If such work should be let by contract, it probably would be done by large floating gangs rather than by small local gangs.

Examples of Contract Work.—Track maintenance by contract on 400 mi. of the Canadian Pacific Ry. included practically every item of maintenance work, and in this case the contractor used the station-gang system to a large extent, letting small subcontracts for tie renewals and ditching. On the Missouri Pacific Ry., ballasting by contract on a large scale, which included the excavating, loading, handling, distributing, and placing of the ballast, showed a marked saving in cost over similar work done by railroad forces. There was no difficulty in handling the contractor's trains. The extent to which the contract system has been employed on individual railroads varies from small jobs to 80 per cent of the ballasting and 75 per cent of the rail renewals in a given season.

Supervision and Equipment.—Contract work is usually under the control of the contractor's foreman or superintendent, subject to supervision by railway inspectors or timekeepers who are on the work continually, while frequent visits should be made by the roadmaster and division engineer. It is desirable also to have occasional inspection by superior officers, such as the chief engineer or maintenance engineer. Camp cars and work-train service for distributing material are furnished, as a rule, by the railway; it may also supply ditching machines, spreader cars, locomotive cranes, and even motor cars and push cars, depending largely upon the contractor's equipment. Distribution of material and collection of old material after the work (as in rail and tie renewals) is done sometimes by the railway and sometimes by the contractor. Objections to the contract system on account of difficulty in handling the contractor's trains without interference with traffic appear to have little practical basis, since the conditions would differ very little from those of improvement works carried out by contract on all railroads (see "Permanent Improvements").

Specifications.—In the A. R. E. A. specifications for tracklaying, maintenance, and renewing rails by contract, the railway company is to furnish materials and stake out the work, while the contractor furnishes foremen, labor, and tools. Speed and other restrictions for the contractor's trains must be arranged with the operating department, and the contractor is required to provide flagmen for the protection of the work and the regular traffic. The Association's specification for ballasting by contract (1925) provide that the contractor will furnish all labor, tools, supplies, and camp equipment, while the railway will furnish hand cars, signal lanterns, flags, torpedoes, bunk cars, heating stoves, and fuel. Payments are to be made on unit prices per mile of track skeletonized in gravel; per mile ballasted with 8- , 10-, or 12-in. lift; and per tie installed. The unit prices are to be based on a specified rate for common labor (cents per hour), the prices being increased or decreased from time to time in the same ratio as the rate actually paid by the contractor varies from the basic rate, but the rate of pay is not to be changed without approval of the chief engineer.

Payment for Contract Work.—The cost-plus-percentage system has been used in spacing and renewing ties and for gathering up old material after rail renewals. Relaying rails has been paid for by cost-plus-fixed fee, the fee ranging from a minimum when the total cost was high to maximum when the cost was low. Unit prices are sometimes per mile and sometimes per piece. The former method, in which the price covers all work of the specified character within a specific territory, has been applied for rail renewal, for spacing ties in advance of rail renewal, for ballasting, and for surfacing track. The price for surfacing varies with different amounts of lift. In piece-work payment, the unit price may be per tie renewed, per switch or turnout relaid, and per tie-plate or rail anchor placed. Unit prices are likely to be low when work is let by competition based on the railway's detail specifications, stating explicitly what part of the work the contractor is to do and what part the railway will do. As contractors become more experienced in work of this kind, the contracts and prices will tend to become stabilized. Payments are made usually on the division engineer's monthly or semi-monthly estimates of work done, but small jobs may be paid for on completion. In either case, a percentage may be retained for a time to protect the railway company against liens that may be filed for labor or materials.

Company or Contract Labor.—Ultimate economy and relief from the labor problem are the main purposes of employing the contract system, but the degree of economy realized has been very variable. The contractor can obtain men when there is a shortage of railway labor. This fact, together with his greater freedom in regard to wages and working hours and in varying his forces according to the work, is considered one of the main factors resulting in economy. But the change in labor conditions due to the increasing use of machinery may offset this advantage to some extent, and in one case where a railway introduced a ditching machine, it abandoned the former contract system for team and hand work in ditching and widening cuts. It has been objected that contract laborers would be less skilled in track work and have less loyalty to the railroad than permanent employees. An obvious answer to this objection is that under present conditions much railway labor is of a transient character, and it is often difficult to organize and hold a force of railway laborers. For a large gang to handle extensive rail renewal or track surfacing, it is doubtful if there would be much difference between a railway force and a contractor's force. The further objection that the contract system may put work on an operated line into the hands of outsiders, with consequent difficulty in placing responsibility for train delays or safety of operation, is offset by the fact that under very similar conditions this has caused little or no trouble and has given satisfactory results. In general, however, there is little tendency to increase the small amount of track work now done by contract.

Standardization of Track.—In the preparation of standard or recommended designs and methods of practice relative to railway material and work, great advances have been made during recent years, especially by the work of the A. R. E. A. In general, standardization reduces the time and cost of both production and distribution, it eliminates confusion due to numerous designs, and it results in higher efficiency and economy. It does not or need not retard progress, since standards are always subject to revision to meet changing conditions. For materials manufactured or produced under specifications, it has particular advantages; as for rails, frogs, switches, ballast, bolts, spikes, concrete, and the preservative treatment of ties. It may be applied, also, in many lines of work, such as width of roadbed, and depth and cross-section of ballast.

In the actual use of the adopted standards and methods, progress has been much slower than in their formulation and adoption. The variations from standards are in some cases quite distinctive, but in the majority of cases they are too small to produce any appreciable benefit over the standard. They serve rather to mark an individuality that usually involves higher cost without corresponding advantage. Three reasons for this situation may be suggested: (1) inertia, or indifference to new developments; (2) prejudice or definite preference for present individual practice; (3) failure to realize the economic advantages. Objections due to immediate personal trouble and temporary complications that may arise from a change to improved or standard practice are often seen more clearly than the ultimate economy and benefit of the new practice.

When it is assumed that certain special designs are necessary or are best suited to the conditions of a particular railroad, investigation may show that they conform so nearly to standard designs that the difference serves only to increase the cost of manufacture and the time necessary for production. As a matter of fact, the adopted A. R. E. A. standards are adapted to meet all ordinary conditions of railways of the class for which they are designed, and exceptional conditions are so rare as to be almost negligible. An important development, therefore, for track design and maintenance, is in securing a more general use of the present standard designs, types, and methods.

Railway Maintenance: Work and Accounting.—All the numerous methods and improvements which have been described have for their purpose the better organization and procedure of work in order to secure both efficiency and economy, not only in the engineering and maintenance departments but also in the general operation of the railway. In the same way, all the various records, reports, estimates, appropriations, schedules, and budget systems which are here described or outlined are part of the general system of accounting for railway expenditures, but are for the purpose of insuring and showing that the maintenance work is handed economically, and that a proper output or performance is given for each item of expenditure.

It is coming to be more generally recognized that accounting requires much more attention from engineers than it has received from them in the past. Railway engineers have often failed to realize the importance of an accounting system for their own work and especially its important relation to the general business and accounts of the railway. In other words, the engineer has not infrequently regarded his work as an independent and complete unit, overlooking the fact that his department is only one of several departments of the railway company, all interlocking with the general purpose of the railroad, which is that of furnishing transportation while, at the same time, handling its business and operations in such a way as to show profit for the transaction.

It is evident that in future the engineer in railway maintenance service will have to deal more extensively with the financial and economic aspects of his work. He will have to consider not only the work itself, with its details and management, but also the methods of recording its progress and cost, and of accounting for the expenditures on the work done under his direction.

INDEX

A

Accidents, 12, 242, 408
  bridges, trestles, 145, 364
  care of injured, 417
  crossings, 126, 130
  curves, 325
  equipment for, 205
  reports, 449
  snow, 418
  track and, 289
  train, 6, 205
  washouts, 205, 409
Accounting, A. F. E. and A. B., 436, 456
  budget, 458
  maintenance, 289, 467
  railway, 12
  street railway, 234
  track, 9, 221, 449
  treated ties, 45
A. F. E. records, 436, 456
Alinement, improving, 307
Ashes, ashpits, 180

B

Backsight, lining, 307
Ballast, 20
  bridge deck, 22, 137
  chart records, 435
  cleaning, 210, 267
  clear of rails, 24
  concrete, 230
  cross-sections, 24
  curves, 25, 27
  distributing, 259, 266
  drainage, 18
  earth, 21, 25, 297
  grass on, 24
  hand-dressed, 25, 296
  oiling, 21, 288, 308
  poor, surfacing, 297
  quantities, 23, 27, 28, 29, 30
  shells, 21, 24, 31
  shoulders, 24
  shrinkage, 21
Ballast, size, 22
  sub-ballast, 17, 21, 23, 26, 283
  tamping (see also *Tamping*), 7, 8
  toe lining, 25, 296
Ballasting, cost, 265
  earth, handling, 297
  from trains, 266
  lining track, 306
  new lines, 259, 263
  reballasting, 264
  reports, 442
  stakes, 216
Betterments (see *Improvements*).
Blast plates, bridges, 142, 366
Blasting, ice jams, 409
Block system, 399
Bolting, 305
Bolts, bridge track, 139
  oiling, 305
  rail joint, 81, 84, 89, 90, 92
  removing, 270
Boundaries, signs, 149
Brakes, track, 159, 160, 194
Bridge and building dept., 249
    bridge tracks, 134
    emergencies, 408
Bridge, floors, 32, 140
    ballasted, 22, 134
    curves, 144, 355
    fireproofing, 137, 142
    rails free, 70, 304
    solid, 137, 138
    spacing blocks, 145
    thin, 136
    waterproofing, 142
  gangs, 363
  inspection, 368
  plans, 367, 372
  records, 372, 435
  warnings, 156, 168
  work, maintenance, 363, 365
    material for, 250
    tools, 208, 218
    traffic at, 365
    wrecking cranes, 415

Bridges, ballast for, 22, 137
  blast plates, 142, 366
  Boone viaduct, 144, 145
  brine corrosion, 47, 139, 307, 372
  classification, 368
  clearance, 367
  concrete slab, 138
  curves on, 135, 144, 335
  design and construction, 367
  draw, rail joints, 87
    signals, 405
  expansion joints, 88
  floors and track, 134
  gantletted track, 143
  maintenance, 2, 363
  marks, monuments, 366
  masonry, foundations, 363, 372
  Memphis, exp. jt., 88
  N. Y., N. H. & H., track, 143
  numbering, 149, 372
  Ohio, track, 143
  painting, 366
  Quebec, 88, 143, 144, 146
  rails free, 70, 304
  refuges, 291
  renewing, 365
  rerailers, 147
  safety girders, 145
  signs, 149
  strengthening, 364
  Thebes, track, 32, 88, 143
  timber, fire protection, 143
    testing, 218
    treated, 48
  track elevation, 138
  track on, 134, 143, 291
  train loadings, 368
  watchmen, 246
  water barrels, 143
Brine, corrosion, 47, 139, 307, 372
Budget system, 458
Buildings, 162
  bunk houses, 164, 165
  cabins, 166
  concrete, 162, 165
  engine houses, 182, 196
  floors, 164
  freight houses, 198
  maintenance, 2, 162, 249
  numbering, 162
  paint, whitewash, 162, 367
  railway, 162
  section houses, 162, 164
Buildings, shed at slides, 285
  stucco, 162, 165
Bumpers, bridge protection, 146
  hydraulic, 158
Bumping posts, 146, 158

C

Cable railways, grades, 230
Cables, electric, conduits, 378
Camp cars, 363, 376, 379, 394
Camps, sanitation, 167, 395
Capstans, electric, 184
Car dumpers, 199
  tracks, curves, 325, 330
  transfers, 199
Card index, 372, 432
Cars, ballast, 266
  brine from (see *Brine*).
  camp, 334, 363, 376, 379, 394
  derrick, 205, 364, 412
  dump, 386
  flanger, 423
  inspection, 218
  motor, section, 290
  moving by capstan, 184
  passenger, water service, 170
  rails on, 277
  section, 217, 290, 292
  sharp curves, 338
  spreader, 422
  track capacity, 189
    inspecting, 427
  transfer to boats, 199
  unloading at fill, 386
  weighing, 185
  weight, 3
  work train, 394
  wrecked, record, 417
Cattle, at crossings, 124, 126
Cattleguards, 124
Chairs, rail supports, 33, 52, 60, 86, 97
    133
  stub switch, 105
Charts, progress, 460
  track, ballast, 427, 435
Clearance limits, 168, 183, 367
  tracks, 150
Closets, camp, 167
Coal mines, tracks, 199
Coal storage, 178
Coaling stations, 177
Cofferdams, 411

Concrete, bridge decks, 137, 138
buildings, 162
bumping posts, 158
drinking troughs, 170
fence posts, 121
footings, 210
paving, 128, 231
poles, 224
posts, signs, slabs, 148
snowsheds, 419
ties, 57, 192
track, 7, 9, 10, 19, 32, 33
base, 127, 132, 133
water tower, 171
white, 148, 152, 153
Conduits, electric, 34, 140, 378
Construction, forces, 389
Contract work, bridge painting, 366
bridges, 363
fencing, 309
handling trains, 396
track work, 464
Correspondence, handling, 457
Costs, ballasting, 265
curves, traffic, 325
ditching, 218, 219
stopping trains, 139, 269, 405
track, 12, 218
units, 451, 461, 466
Cranes, bridge work, 364, 415, 416
lifting track, 309
locomotive, 181, 205, 260, 273
moving switches, 360
Creeping, bridges, 304
rails and track, 43, 53, 70, 75, 98, 132
Creosoting, 47, 170
Cribbing, 45, 280, 411
Crossings, highway and street:
accidents, 126
approaches, 126, 127
concrete base, 127
eliminating, 126, 131
gates, cabins, 166
near stations, 131
paving, 127
protection, 127, 129
rails and joints, 87
reports on, 447
signs, signals, 129, 150
spikes for, 95
track, concrete base, 132
continuous rail, 113
Crossings, track, creeping, 98, 132
curved, 348
derails, 117
electric, 379
eliminating, 132
frogs, 111, 348
protection, 131, 379
setting out, 348
train delays, 131
wires over, 377, 379
Crossovers, 114, 202, 203, 344, 345
for track work, 269
high-speed, 345, 346
spring-rail frogs, 110
table of, 361
Culverts, 137, 280, 364
inspection, 371
open-deck, 136
pipe, 148, 281
Curvature, reducing, 325, 383
Curve elevation, 139, 327, 331, 335
bridges, 135
electric railway, 225
elevated railway, 333
measuring, 215
runout, 334
signs, 340
turnouts, 344
Curves, railway, 321, 338
ballast section, 29
bending rails, 212, 303
bridges, 144, 335
cleaning, oiling, 233
compound, 331
cost of maintenance, 325
cubic parabola, 328
degree, 321
divergence, 324
economics of, 324
electric railway, 224, 233
flangeway, 108
forces on, 326, 331
frogs, switches, 101, 343
gage widened, 108, 215, 336
grade compensation, 318
guard rails, 339
lemniscate, 328
lining, 299, 306
locomotives on, 332
loop, 197, 338
marking, 152, 340
narrow-gage, 317
oiling wheels, 338

Curves, railway, parabolic, 330
  portable railways, 11
  rail braces, 98
    joints, 86, 271
    stresses, 73
    wear, 75, 324
  rails, 68, 71, 271, 322, 338
  setting out, 322, 340
  sharp, 225, 325, 338
  shifting rails, 306
  sidetracks, 338
  signs, 152, 154, 340
  spiking, 304
  spiral, 327
  station layouts, 200
  study needed, 319, 324, 333
  superelevation (see *Curve elevation*).
  tie bars, 97
  track, 337, 339
    laying, 260, 264
    spacing, 18, 30
  train resistance, 324
  transition, 327, 335, 336
  turnouts on, 101, 344
  vertical, 264, 319
  wheel action, 73, 325, 331, 334
  yard, 192
Cuts, slopes, 280
    grass on, 288
  sliding, 408
  wet, 283
  widening, 124, 387

D

Derails, 116, 293, 406
Ditches, cleaning, 282
  irrigation, bridge, 136
  outlets, 282
  surface, 280
  track, 17, 18, 28, 280, 281
Ditching, branch lines, 210, 290
  machine, 205, 285
  reports, 446
Divisions, length, 252
Double tracking, 11, 381
Drainage, roadbed, 17, 26, 280
  slides, 284
  street track, 227
  subdrainage, 17, 284
  track, 17, 26, 282
    crossings, 132
  tunnels, 32
Drift catcher, 281
  removing, 409
Dynagraph car, 427

E

Earthwork, shrinkage, 388
Economics, bridge floors, 137
  curves, grades, 317, 319, 324, 383
  double tracking, 383
  labor and tools, 218, 464
  maintenance-of-way, 1, 2, 3, 11, 12, 458
  rails, 61
  railway improvements, 380, 382
  scrap reclamation, 315
  tie-plates, 50
  ties, 35, 40, 45, 54, 57
    treatment, 45
  track labor, 218, 463
    maintenance, 458
    work, 1, 6, 289, 458
  water supply, 169, 175
Electric lighting, 192, 291, 414, 417
Electric railways, 223
    bridge ties, 135, 144
    compound rails, 71
    conduit, 224
    cost of rebuilding, 229
    crossings, 132, 279
    interurban, 223, 224
    maintenance, 223, 231
    organization, 224, 226
    passing sidings, 224
    pole spacing, 223
    rail corrugation, 75, 228
      expansion, 70
    reconstruction, 229, 231
    signals, 406
    snow handling, 425
    special work, 229
    switches, 103, 116, 229
    third-rail, 139, 141, 224
    tie-rods, 97
    track, 58, 97, 224, 230
      work, 227
    wires at crossings, 133
Electrolysis, 223
Elevated railways, curves, 333, 335, 338
  floor and track, 139, 140
  maintenance, 291
  snow removal, 425
  third-rail, 141
Emergency work, 245, 364, 408
  telegraph lines, 378

Rail renewals, 269, 272
  cost, 218, 219
  electrification, 270
  elevated railways, 276
  practice, L. V. R. R., 273
  machinery, 272
  reports, 440, 443
  winter, 269
Rails, alloy steel, 67, 68, 101
  bending, 212, 303
  bolt holes, 84
  breakage, 66, 76, 292
  bridge section, 60
  bull-head, 7, 71, 86
  canting, 51, 72
  chairs (see *Chairs*).
  chart records, 435
  compound, 71, 228
  contact area, 74
  continuous, 64, 111
  corrugation, 75, 228
  creeping, 43, 53, 70
  cropping, machine, 77
  curve wear, 75, 324, 326, 339
  curves, 68, 71, 271, 338
  cutting, 211, 212, 304
  damage by locomotives, 74
  deflection, 86
  design, manufacture, 61, 66
  double-head, 7, 71, 86
  Dudley, 6, 61
  easer, 111, 132
  elevated railway, 139
  expansion, 69, 271
  flat-top, 73
  foreign, 7, 60, 62, 63, 71
  handling, 211, 276
  heat treatment, 68
  heavy, 9, 61, 63, 64
  history, 60
  inspection, 61
  laying, 69, 258, 269
  length, 78
  life, 74
  lift, at bridges, 87
  light, 7, 10
  loads, traffic, 66, 279
  long, 8, 64, 87, 225, 228, 271
  manganese steel, 68, 101
  Manning, 71
  old, classified, 66, 77
  on yielding base, 17
  punch and drill, 211, 212
Rails, reconditioning, 77
  relation to operating, 61
    to track, 8, 64
  relayer, 75, 77, 289
  rerolled, 77
  scarf and miter ends, 86
  second-class, 66
  shifting, 306
  short, 64, 271
  spare, 160, 417
  specifications, 69, 101
  stock, 99
  street railway, 70, 71, 133, 227
  stresses, 5, 7, 72, 74
  switch, 99, 100, 106, 343
  tests, 68, 69
  third-rail, 224, 225
  tie-plates and, 62
  traffic and, 63, 75
  transverse fissures, 66, 76
  tunnels, 69
  wear, 62, 73, 74, 75, 317, 324, 326, 339
    templets, 76
  weight, 7, 16, 63
  welding, 77, 89
  wheel load, 61, 62, 72, 79
  wheels and, 1, 3, 73
  worn, renewal, 294
Rail-way, 1
Railways, cut-offs, 11
  electric, track, 35, 70, 71, 75
  English, rails, 71
  extension, 11, 380
  finance and track, 289
  foreign, 71, 316
  gages, 316
  improvements, 380, 459
  industrial, 10, 58, 201
  logging, 267, 317
  military, 10
  portable, 10, 58
  regulation, 14
  underground, 32
  U. S., statistics, 1, 15, 35
Rapid unloader, 267, 386
Rates and revenue, 14
Realinement, 383
Reconstruction, 2, 381
Records, bridge and building, 372, 435
  card index, 432
  charts, 437
  railway, 432
  section foremen, 164, 241

Records, time books, 241, 442
  tools, 205
Relocation, double-track, 382
Reports, bridge inspection, 369
  cost records, 452, 454
  forms of, 438
  progress, 460
  roadmasters', 457
  section foremen, 164, 241, 439
  tie renewals, 56
  time books, 241, 442
  track condition, 463
Requisitions, handling, 454
Rerailing devices, 146
Resurveys, 391
River-bank protection, 412
Rivets, testing, 371
Roadbed, 17, 280
  concrete, 19, 20, 32
  crowning, 17, 30
  grading, 317
  rolling, 17, 283
  soft spots, 280, 283
  tunnels, 32
  water pockets, 283
  width, 18
Roadwasters, 238, 240, 257
  electric railways, 225, 233
Rock gang, 256
Rocks, removing, 205, 246, 256
Rules, book of, 456

S

Sand, drifting, 287, 418
    drying, for engines, 180
  track, 160
Sanitation, camp, stations, 167
Scales, track, 185
  wagon and truck, 185, 192
Scrap, 205, 290, 314
Section foremen, 240, 244
  at wrecks, 417
  duties, 174, 244
  reports, 164, 241, 439
Sections, track:
  French system, 246
  length, 252, 254
  marking, 154, 426
Septic tanks, 167
Sheet piling, 411
Shimming, 305, 363
Shops, maintenance dept., 205
Sidings, 187, 199
  ballasting, 30
  concrete ties, 59
  electric railway, 224
  passing, 199, 201, 224, 385
  reports, 446
  signals, 404
  spring switches, 103
  street railway, 230
Signal department, 251
Signals, 11, 398
  automatic and manual, 400, 401
  bridge work, 365
  car repair, 293
  curve, 340
  double-track, 382
  electric railways, 225, 406
  lamps, colors, 117, 403
  remote-control, 406
  sidings, 203, 385
  switch, 104, 115, 117
  track crossings, 113
    work, 291
  train, 293
  train-staff, 365, 402
  train-stop, 403
Signs, 148
  colors, 149
  crossing, 129
  slow, stop, 148, 152, 292
Slides, snow, 418
Sliding ground, 284
Slopes, turfed, 19, 28, 30, 282, 284
Snow, electric line, 425
  fences, 122, 295, 418
  fighting, 419
  melting pits, 184, 424
  plows, 205, 421
  removing, 184
  track work, 421
  widening cuts, 124
Sodding, ditches, slopes, 19, 28, 30, 31, 282, 284
Special work, track, 229
Specifications, rails, 69, 101
  ties, 36, 39, 42
  track contracts, 465
Speed control, 404
  limits, signs, 154
  reduction, 290, 346, 365
Speeds, at crossovers, 345, 346
  maintenance and, 4
  momentum grades, 318

Spike holes, plugging, 45
Spikes, auxiliary devices, 96, 97
  bored holes, 93, 96
  clear of rails, 51, 52, 93, 96
  clips, 96, 97
  corrosion, 47
  cut and screw, 8, 92, 94, 229
  damage to ties, 8, 37, 40
  experimental, 96
  for tie-plates, 96
  Goldie, 96
  hammering, 93
  invention, 60
  lateral thrust, 93, 95
  long, planks, 128
  placing, 304
  pulling, 208
  resistance, 92, 94
  screw, 7, 34, 92, 94, 139, 141, 142, 229
  tests, 94
  wood, 97
Spiking, 304
Spirit level, 216
Splice bars, 80–90
Spreader, 205
  snow, 422
Stakes, ballast, 264
  curve, 340
  ditching, 283
  lining, 298, 307
  track crossings, 348
  turnout, 354, 355
Standards, track, 5, 61, 467
  track work, 461
  turnout plans, 342
Standpipes, 171
Station floors, 141
  grounds, 313
  platforms, 167, 197
    elevated railway, 140
    shelters, 168
Stations, Boston, tracks, 196, 353
  Chicago, 104, 142
  facilities at, 197
  Indianapolis, 141, 243, 350, 353
  New York, loop, 338
  Pitts. & W. Va. Ry., 142
  St. Paul, 105, 142, 197, 353, 354
  sanitation, 167, 314
  signs, 154
  slip switches, 104
  track drains, 280
    floor, 33, 141
Stations, track forces, 243
    plans, 197, 200, 204, 353
    stops, 158
  trainsheds, 168
Steel, copper content, 51
Stockyards, 162, 170
Stop blocks, 159
Stores department, 221
  scrap handling, 314
  track materials, 221, 454
Street railways (see *Electric railways*).
Stremmatograph, rail tests, 73
Structures (see also *Buildings*).
  maintenance, 12
  snowsheds, 419
Subdrainage, 18, 280
Subgrade, load on, 7
Subways, maintenance, 291
Supervisor (see *Roadmaster*).
Surfacing, 296, 309
  on bridges, 363
Switch locks, 102
  ties and timbers, 113
Switchbacks, 340
Switches, angle, 100, 342
  continuous rail, 104
  curves, 101
  derailing, 116
  electric railway, 229
  facing, trailing, 100, 201, 342
  footguards, 109
  gage, adjusting, 102
  head blocks, 114
  indicators, 407
  junctions, 342
  leads, 99, 106
  locking, 102
  manganese steel, 101
  McPherson, 104
  melting snow, 424
  moving, 360
  point rails, 99, 100, 347
  power-operated, 192
  protection, 101
  remote-control, 104, 201, 385, 406
  renewing, 309, 360
  repairs, 362
  reports, 443
  right- and left-hand, 100
  sand track, 160
  slip, 104, 354, 360
  split, 99
  spring, 103, 201

Switches, stub, 99, 105
switchbacks, 341
targets, lamps, 117
three-throw, 105, 349
throw, 99
tie-plates, 97, 102
tie-renewal, 301
tie-rods, 102
trailing through, 103
trap, 341
wear, 346
Wharton, 104
wheels at, 101, 110
yard, 189, 193, 346
Switching, yard, 193
Switchstands, 102, 103, 115, 192
numbered, 192
placing, 359
targets, 115
yard, 115, 192

T

Tamping, 7, 279
joint ties, 80
machine, 219, 300
shovel, 265
space for, 43
Tanks, concrete, steel, wood, 170
oil storage, 180
painting, 366
track, 174
winter protection, 171, 174
Team yards, 192
Telegraph, companies, 375
and telephone lines, 375
wreck trains, 414
Telephone, track warning, 293, 375
train orders, 406
Tenders, capacity, 180
filling tanks, 172, 174
Terminals, 11, 187
clearing snow, 423
freight, passenger, 196, 198
loop, 197
rail and water, 198, 199
rapid-transit, 197
slip switches, 104
track plans, 350, 352
Third rail, 139, 224, 225
clearing snow, 425
Tie-dressing plants, 44
Tie-plates, applying, 52, 302
Tie-plates, as rail braces, 51
copper-steel, 51
curves, 337
economics, 40, 50
elevated railways, 139
felt, 52
rail renewals, 270
rails and, 62
screw spikes, 7
set screw, 53, 98
shimming, 305
shoulder, 97
street railway, 229
switches, 97, 102
wood, 52
Tie plugs, 45, 304
Tie renewals, 4, 55, 300
economics, 40, 45
estimates, 55
numbers, 35, 54
reports, 56, 440, 443
steel, 443
switches, 115, 301
time study, 462
Tie-rods, curves, 337
elevated railway, 227
switch, insulated, 102
track, 97
Ties, adzing, boring, cutting, 37, 44, 48, 93, 96, 220, 302, 305
ballast bearing, 7
beam strength, 42
blocks, bowls, 7, 59
bridge, 35, 135, 145
burning old, 45, 301
checks, S-irons, 39
concrete, 7, 57, 59, 192
consumption, 16, 35
cost, 45
crossings, 131
crossovers, 114
cutting, trees, 38
damage by spikes, 8, 37, 40, 93
dapping, bridge, 136
date marks, 56
decay, wear, 40, 115
distributing, 300
economics, 45, 54, 57
elevated railways, 139
European, 7, 37, 43
foreign timbers, 39
handling, 211
hewed, sawed, 38, 136

Ties, indenting, 44
inspection, 39, 40, 43, 54
joint, 40, 80, 86
life, 40, 45, 55, 57
loads on, 35
metal, 7, 57
bridges, 139
cast-steel twin, 86
corrosion, damage, 57
electric railway, 229, 230
fastenings, 97
foreign, 59
life, 57
portable track, 10, 58
reports, 443
numbers, 8, 35
old, disposal, 45, 301
plugging, 45, 304
protection, 37, 40
records, 40, 55, 435
relation to track, 8
respacing, 43, 84
rigid, flexible, 7
seasoning, 43, 48
spacing, 42
specifications, 36, 39, 42
substitutes, 7, 57, 300
switch, 35, 113, 115
taper for curves, 336
test sections, 55
treated, 36, 44
cutting, boring, 48, 302
economics, life, 45
electric railway, 229
marking, 56
steaming, 49
woods for, 37, 39
Timbers, boring test holes, 371
bridge, treated, 48, 364
crossings, 127, 131
defects, 39
old, second growth, 37
poles, posts, 119, 120, 376
resources, U. S., 35
switch, 35, 113, 115
Time books, 245, 442, 449
Time studies, 461
Tools, bridge and building, 218, 363
inspection, 370
distribution, 221
ditching, 282, 285
dolly car, 304
electric, pneumatic, 219
Tools, from scrap, 205
heavy, 205, 218
labor-saving, 218, 260
motor cars, 290
pole inspection, 378
post hole, 377
rail saw, 304
records, reports, 221, 442
section gang, 205, 226
sighting boards, 216, 297, 298
tie-plate gage, 53, 216
track thrower, 220, 309
sweeper, 220, 313
trackwalkers', 246
wasted, 314
wrecking train, 414
Torpedoes, track signals, 292
Towers, water tank, 171
Track, brakes, 160, 194
charts, 433
circuits, 24, 242, 404
depression, 389, 397
elevation, 381, 388
forces, combining, 292
electric railway, 225, 232
emergencies, 408
extra gangs, 252
renewing rails, 274
responsibilities, 247
stations, 350, 353
training 244, 247
inspection, 241, 244, 426, 444
layouts, complicated, 352
material, distributing, 393
sections, length, 242
tanks, 174
Track work, bolting joints, 89
budgeting, 458
clearing, mowing, 211, 220, 295, 312
closing track for, 269, 274, 315
combining gangs, 248, 290
contract, 464
costs, 218, 451, 461, 466
ditching, 218, 219, 282
emergency, 408
extra gangs, 242, 246, 248
fencing, 310
gaging, 50, 299
labor distribution, 449
lighting, 291
lining, 207, 265
long rails, 225
lowering grade, 297

Track work, machinery, 218, 221, 272, 387
planning, 294, 460
policing, 310
protecting, 2, 3, 6, 270, 283, 291, 292, 301, 375
raising track, 216, 296
records, reports, 438, 467
respacing ties, 84
routine, 295
safety, 270
seasons, 243, 293
section gang, 242, 295
shifting tracks, 220
sighting line, 298
single track, 269, 274, 365
spiking, 50, 209
standards, 461
steel ties, 57
street railway, 229
Sunday, 247
surfacing, 216, 275
tamping, 207, 219
telephones on, 375
time studies, 461
track tanks, 175
tunnels, 291, 375
turnouts, 357
unit cost, 451, 461, 466
units, 461
weed killing, cost, 218
winter, 245, 247, 269, 273, 295
Tracklayers, 259
Tracklaying, broken joints, 86
curves, 260, 324
force, methods, 69, 258, 259
records, speed, 258
Tracks, abandoned, salvage, 315
accounting, 9, 45, 458, 467
bridges, 32, 60, 134
car capacity, 189
center bound, 35, 300
classification, 6
clearing, policing, 9
concrete base, 7, 9, 10, 19, 32, 132, 133, 141, 227
cost and maintenance, 10, 11
creeping (see *Creeping*).
curves, 337
defects, 289
deflection, 8, 35, 73
design, 4, 7
electric railway, 224
Tracks, engine house, 180, 182
parking, 196
engineers for, 1, 239
English, 52, 97, 264
gantletted, 143, 160, 185, 365
ice removal, 174, 175
improvements, 9, 11
industry, 200, 333
labor, materials, 289
ladder, 18, 104, 109, 189, 355
load distribution, 7, 8
loop, 197
mines, steel ties, 58
moving on cars, 309
narrow-gage, 388
oil unloading, 179
operating conditions, 463
passing (see *Sidings*).
paved, 133, 223, 226
piers, 198
policing, 313
portable, 10, 58
raising, 385
relation to traffic, 4, 6
repair, 2, 189
shifting, 308
shoofly, 391, 412
spacing, 17, 18, 30, 189, 192, 197, 201, 224, 225, 229
special types, 2, 9, 10, 32
staking out, 258
standards, 5, 61, 467
station, layout, 197
street railway, 58, 225
stresses, 5, 7, 327
studies, 6
telephones on, 375
terminal, 104, 196
test sections, 7, 56
traffic capacity, 4, 6, 383
tunnels, subways, 9, 32, 60
typical 7, 9
unit structure, 64
unsafe, 289, 292
wave motion, 73
wear, 3
wires over, 133
without ballast, 9, 19, 20, 33
without ties, 227
Y-, 184, 196
yard, 189, 192
Trackwalkers, 118, 209, 215, 245, 255

Traffic, at bridge work, 365
capacity, single track, 383
cost on curves, 325
on construction work, 397
operating, 380, 399
relation to track, 4, 6
Train, dispatching, 398
loads, grades, 318
resistance, 6, 7, 61, 319, 324
runs, overtime, 380
Train-staff and tablet, 402
Train-stops, automatic, 403
Trains, at water stations, 173
camp and boarding, 394
construction work, 396, 465
cost of delay, 131, 187, 269, 405
drinking water, 170
long and heavy, 380
mail service, 157
movements, 11
protection, 270, 398, 410
runaway, checking, 117, 341
sidings and, 201
signals, 293
stopped at switches, 201, 346, 385
stores, 221, 241
through freight, 187
time study, 197
wrecking, 410
yard delays, 187
operations, 190
Transfer platform, 187
tables, 184
Transportation, economy, 380
Trees, snow breaks, 418
timber supply, 35
Trenails, 97
Trenches, rails over, 281
Trestles, construction, 411
filling, 136, 364, 385
fire protection, 137
flood danger, 281, 364, 408
floors, 134, 136
inspection, 371
peg-leg, 388
raising grade, 384
repair, 364
temporary, 388, 390, 411
treated timber, 364
Tunnels, Cascade, Moffatt, 390
clearance, 168
drain, 285
eliminating, 391
Tunnels, grade compensation, 318
rail wear, 75
repair gangs, 375
rerailers, 147
spiral, 341
summit, 390
telltales, 168
track, 32
work, 290, 296
watchmen, 246
Turnouts, 99, 342
complicated, 349
electric railway, 224
gage at, 344
industry tracks, 109, 201, 344
putting in track, 351, 357
section car, 161
setting out, 349, 353, 357
sharp, 109
sidings, 201, 202
simplified, 351
table of, 361
Turntables, 183, 196
clearance, 367
end-pivoted, 184

U

Unit costs, 461

V

Valuation, railway, 11, 436
Viaducts, filling, 385

W

Walls, toe, 30, 280
Warehouse, 198
Washouts, 205, 409
Watchmen, bridge and tunnel, 246
crossings, 130
floods, 408
Water columns, 172, 195, 200
Water service, organization, 169, 250
sanitary code, 170
stations, clearing ice, 418
locating, 170
records, reports, 436, 447
supply, 171, 250
boiler, treatment, 169, 175, 250
drinking, 170
pipe line, 176

Waterproofing, bridge floors, 142
Waterways, 280, 364
Weeds, killing, 294, 295, 310, 312
Welding, rails and frogs, 77, 87, 127, 228, 231, 233
Wheels, coning, 51, 72, 73
  contact area, 74
  curves, 324, 325, 334
  cylindrical, 73
  electric railway, 227
  elevated railway, 139
  flanges, 1, 99, 337
  frogs and, 99, 105, 108, 111
  gage, 74
  guard rails, 145
  hammer-blow, 223
  loads, 61, 62, 72, 79
  oiling flanges, 338
  switches, 101, 104, 346
  wear, 73
  worn, effects, 74, 101, 104, 111
Whistle signs, 155
Winter work, 245, 247, 269, 273, 295, 367
Wires, over tracks, 133
  sleet removal, 363
  telegraph, 378
Work trains, bridge, 363
  contractors, 396
  ditching, 286
  electric railway, 225
  material, 294
Work trains, operation, 391
  reports, 439, 446
  telegraph, 376, 379
Wreck train and work, 414, 417
Wrecking cranes, 415, 416
  bridge work, 364

Y

Yards, 11, 185, 187
  caboose tracks, 190
  car movements, 11, 188
  coach, 198
  concrete ties, 59
  curves, 97, 339
  drainage, 280
  entrance, 356, 357
  guard rails, 108
  hump, 185, 193
  interlocking, 405
  lighting, 192
  main tracks, 188
  maintenance, 290
  section force, 243, 290
  signs, 155
  snow removal, 184, 423
  spring switches, 100, 103, 116
  team, 192
  track brakes, 194
  track spacing, 187
  train delay, 187

INDEX OF RAILROADS

A

Africa, sand drifts, 287
American, 1, 15, 35
A., T. & S. F., bunk house, 165
  cost, maintenance, 12
  double tracking, 382, 396
  rail joints, 82
    renewals, 276
  switchbacks, 341
  tie-plates, 51
  ties, 37, 46, 47, 54, 56
  track work, 258, 306
  turntable, transfer, 183, 184
  vertical curves, 319
Australia, sand drifts, 287
  ties, 39, 45

B

B. & O., bridge floors, 136, 144
  canting rails, 74
  cost of track, 11
  organization, 251, 254
  rail renewal, 274
  reports, 461, 463
  speed at turnouts. 346
  ties, 46, 55
  time reports, 461
  tools, track, 205, 220
  track forces, 243
    tanks, 174
  turnouts, 362
  wheel loads, 72
Baltimore Belt, tunnel, 32
Bess. & L. E., reports, 439, 442, 447
Boston & Alb., track, 427
Boston Elev., curves, 333, 338
  rails, 68, 338
Buff., R. & P., tie-plate gage, 53, 216

C

Can. Pacific, bridge guards, 146
  contract work, 465
  curves, 332, 335
  grades, 318
Can. Pacific, rails, 72, 276
  snow, 418, 420
  spiral tunnels, 341
  switches, 104
  tracklaying, 258
  water tank, 172
  wheels on curves, 326
Cen. N. J., heavy rails, 63
  track chart, 435
Cen. Pacific, double track, 383
Chi. & Al., bank protection, 412
  requisitions, 455
  winter work, 246
Chi. & E. Ill., bridge floor, 138, 144
  signs, 152
  telltale, 156
Chi. & N. W., Boone viaduct, 144, 285
  bridges, 135, 144, 369, 370, 374
  checking slides, 285
  engineers, 238
  guard rails, 108
  labor classification, 450
  rail joints, 85, 91, 94
  signal department, 251
  station track, 142
  tie plates, 51
  time book, 449
  track chart, 435
    depression, 397
  valuation records, 436
  water tank, 171
  welded rails, 87
  work trains, 393
  wrecking train, 415
Chi., B. & Q., ash handling, 182
  bridge floors, 144
  concrete plank, 129
  cost, track, 12
  grass on sand, 288
  machines, 262, 272
  organization, 252
  rail anchors, 98
    joints, 89
  records, 46, 437
  separating yard, 191
  track forces, 243

Chi. Elev., curves, 333, 335
loops, 198
switches, 101
track, 68, 86, 139, 276
Chi. Jn., concrete track, 19
Chi., M. & St. P., bridge deck, 139
bridge classification, 368
derrick car, 365
rail laying, 269
rail-wear gage, 76
resurveys, 391
snow handling, 425
tie records, 46
vertical curves, 320
water treatment, 177
Chi., R. I. & P., bridge deck, 137
budget, 458
construction trains, 397
crossovers, 345
rail renewal, 275
records, 436
sidings, 202
station plans, 200
telltale, 156
tie treatment, 49
track capacity, 383
Chi., St. P., M. & O., fences, 119
Chi. Union Sta., platforms, 167
track, 20, 104
Cin. So., bridge renewal, 365
construction trestle, 388
C., C., C. & St. L., bridge deck, 139
bunk cars, 394
engineers, 238
grade reduction, 384
section houses, 163, 164
slides, 285
tools, 205
track work, 292
Cuba, ties, 39

D

Del. & H., ballasting, 265
Del., L. & W., ballast, 31
oiling track, 308
rail joints, 84, 86, 92
spikes, 94, 96
track inspection, 430
tunnel track, 34
D. & R. G. W.; bridge work, 369, 373
engine house, 184
telegraph work, 376
Den. & S. L., Moffat tunnel, 390
Dul., M. & N., bridge dept., 250

E

El Paso & S. W., pipe line, 176
Electric, Baltimore, 231, 232
Chi, A. & E., 225
Chi., N. S. & M., 226
Detroit, 230
Ill. Traction, 224
Indianapolis, 230
Kan. City, 230, 234
Milwaukee, 230, 234
New Orleans, 229
New York, 32
Philadelphia, 233
English, track, 33, 43, 52, 175, 264, 285
Erie, double tracking, 383
organization, 254
track forces, 243
inspection, 428

F

Fla. W. & N., tracklaying, 262

G

German, track, 7, 86
Gt. Nor., cost, track, 12
section houses, 163, 164, 165
snowsheds, 420
switchback, tunnel, 341, 390
tools, 206
water dept., 250

H

Hock. Vall., grade reduction, 384
records, 436, 440

I

Ill. Cen., ballast, 27
bridge work, 364, 370, 374
cost records, 454
grade revision, 384, 397
guard rails, 108
snow handling, 424
switchstands, 115
tracklaying, 259
transition curves, 329
weed killing, 311
Ind. Harbor, brake, 160
Ind. Union, tracks, 141, 350, 353
trainshed, 167

K

K. C. So.; ballast, 25, 31
budget, 459
records, 438
roadbed, 283
Ky. & I. T., bridge floor, 143

L

L. E. & W., curves, 338
Lake Sh. El., bridge floor, 144
Le. Vall., bridge work, 363, 369, 370, 374
rails, 63, 71, 273
joint, 86, 92, 94
screw spikes, 94
tie renewals, 54
winter work, 246, 269
work trains, 394
Long Id., concrete track, 20
Los A. & S. L., oiling sand, 288
L. & N., grade reduction, 318, 384
guard rails, 339
section house, 164
tools, 206
transfer table, 184

M

Mich. Cen., bunk house, 165
bridge floors, 144
cattleguard, 125
crossings, 348
fencing, 119
hump grades, 194
ladder tracks, 356
organization, 251, 256
track forces, 243
Mich. El., bridge floors, 144
M., St. P. & S. S. M., bridge dept., 250
filling trestles, 387
M.-K.-Tex., hump, 194
Mo. Pacific, bridge floors, 144
contract work, 465
requisitions, 455
track reports, 443, 446

N

Nash., C. & St. L., organization, 257
rail joint, 87
wear gage, 76
section house, 165
septic tank, 167
slides, 285
N. Y. Cen., ballast, 310
concrete track, 20
loops, 198, 338
rail joints, 81, 82, 83, 84, 85, 90, 92
rails, 61, 64, 65, 66, 69, 75
snow handling, 425
stores train, 222
telegraph work, 376, 378
ties, tie plates, 43, 51
track signs, 152, 155
tanks, 175
N. Y., N. H. & H., bridge floors, 143
grass slopes, 288
rail renewal, 275
shifting track, 308
tools, 206
N. & West., guard rails, 108
rail gage, 76
renewal, 276
switchstands, 115
winter work, 246
Nor. Pacific, concrete track, 9, 19
creosoting, 48
rail wear, 76
switchback, tunnel, 341, 390
telegraph, 376, 378
ties, 40
tracklaying, 262

P

Paris, L. & M., track studies, 6
Paris-Orleans, maintenance, 246
Penna., ballast, 21, 29, 268
curves, 320, 327, 333, 335, 336, 338
ditching, 287
double tracking, 397
engineers, 238
hump grades, 194
ladder tracks, 355
longitudinals, 9, 44
organization, 256
plantations, 36
rail joints, 82, 92, 94
laying, 271, 277
rails, 52, 62, 64, 68, 70
reports, 444, 463
screw spikes, 94
snow hedges 419
speed, turnouts, 346
switches, 105, 346, 347, 360, 444
telegraph, 375, 378
tie-plates, 52

Penna., track chart, 435
forces, 243
inspection, 430
spacing, 18
sweeper, 220, 313
training men, 244
wheel loads, 72
wrecking train, 415
Peo. & Pekin, Y-track, 184
Pere Mar., concrete track, 9, 19
engine terminal, 195
track inspection, 429
Phila. & Read., bridge joint, 366
concrete bumper, 158
filling viaducts, 387
flower beds, 314
long rails, 64
tie treatment, 47
Phila. Elev., deck, 140
Phila. Underg., track, 33
Philip. Ids., ties, 39
P. & L. E., ash handling, 182
guard rails, 108
switches, 101
Pitts & W. Va., station track, 142

R

Rich., F. & P., coal storage, 179
Rock Id. (see *Chi., R. I. & P.*).

S

St. L.-S. F., time studies, 462
unit costs, 452
wreck train, 415
St. P. Union, switches, 105
track layout, 142, 197, 353, 354
Seaboard A. L., cost, track, 12
South American, ties, 39
Southern, fence gangs, 310
So. Pacific, ballast, 25
bridge floor, 134, 135, 136, 137, 144
concrete ties, 59
organization, 253
So. Pacific, painting, 366
rail braces, 98
joints, 87
rails, 64, 76, 338, 417
rerailer, 147
shoofly, 412
slides, 285
snowsheds, 419
street crossings, 127
tie treating, 47
track forces, 243
tracklaying, 258
tunnels, 168, 391
turntables, 183

T

Tex. & Pacific, weed killing, 311
Transandine (Chile), switchbacks, 341

U

Un. Pacific, ballast, 28
bridges, 365, 367
curves, 333, 336
draining slides, 285
oiling sand, 288
reconstruction, 380
sidings, 200, 202
snowsheds, 420
ties, tie-plates, 49, 54
track work, 161, 246, 259, 292
training men, 244
U. S. (see *American*).

V

Virginian, transition curve, 329

W

Wabash, bank protection, 412
bridge floors, 144
West. Pacific, double track, 383
W. U. Tel. Co., railway work, 376, 378
W. & L. Erie, hump grades, 194

INDEX OF ENGINEERS, INVENTORS AND AUTHORITIES

A

Abbott, rail joint, 83, 85
Adams, rail joint, 60
Alfred, track design, 9, 19
Am. Ry. Assn., 52, 61, 73, 117, 130
A. R. E. A., 4, 5, 6, 7, 22, 26, 36, 39, 42, 51, 52, 56, 57, 61, 70, 72, 73, 82, 91, 95, 96, 108, 119, 121, 127, 133, 137, 149, 155, 181, 194, 197, 201, 205, 258, 265, 324, 328, 333, 336, 367, 368, 389, 434, 436, 441, 442, 447, 460, 461, 463, 465, 466
Am. Soc. C. E., 7, 60, 61, 62, 72, 73
A. S. T. M., 67, 69
Am. Wood P. Assn., 50

B

Balkwill, crossings, 112
Berry, U. P. Ry. work, 380
Bessemer, rails, 65
Bierd, track thrower, 309
Birkinshaw, rail, 60
Bonzano, rail joint, 81
Brunel, rail, 60

C

Clyde, tracklayer, 261
Crandall, transition curve, 328
Creese, track thrower, 309
Cuenot, track studies, 6

D

Dudley, rails, 6, 61, 73, 81, 427

E

Eymon, track crossing, 108

F

Ford, track labor, 461
Fowler, curve tests, 324

G

Goldie, spike, 96

H

Haarmann, steel ties, 59
Hale, ballast shrinkage, 21
Harris, tracklayer, 262
Harvey, bolt, 90
Haskell, track crossing, 113
Holbrook, transition curve, 327, 328
Holman, tracklayer, 262
Hurley, tracklayer, 263

I

Int. Comm. Comm., 6, 11, 14, 45, 405

K

Kimball, concrete tie, 58
Kruttschnitt, train resistance, 6

L

Lakhovsky, spike, 96
Leighton-Haskell, crossing, 113
Leslie, rail joint, 87
Lindenthal, track, 9
Loree, traffic, 383
Loweth, bridges, 368

M

Manning, rail, 71
M. C. B. Assn., wheels, 52
McPherson, switch, 104
Molitor, transition curve, 329
Morse, budget, 458

N

Neafie, rail joint, 85, 308
Nordberg, track thrower, 220

P

Post, steel ties, 59

R

Rail Mfrs. Assn., 67
Railway Mail Service, 157
Ramsbottom, track tanks, 174
Rendel, steel ties, 59
Ripley, locomotives and track, 5
Roach, frog substitute, 111
Roberts, track layer, 261
Rutledge, track lining, 306

S

Sandberg, rails, 68
Schaub, track design, 9
Searles, spiral, 328
Shearer, concrete tie, 58
Stevens, rail, spike, 60, 80
Strickland, rail, 60
Sullivan, curves, 326, 329

T

Talbot, transition curves, 328
Thiollier, spike, 96
Tratman, metal ties, 57

U

U. S. Forest Laboratory, 35, 55, 56
U. S. Mail Service, 157

V

Vautherin, steel ties, 59
Vignoles, rail, 60

W

Ware, tie-plate gage, 53, 216
Webb, grade compensation, 319
Wellington, 75, 319, 328, 330, 352, 383
Westcott, tracklayer, 263
Wharton, switch, 104